Duxbury titles of related interest

ADA, *DPL 4.0 Decision Programming Language for Windows*

Bell & Schleifer, *Decision Making under Uncertainty*

Bell & Schleifer, *Risk Management*

Berk & Carey, *Data Analysis with Microsoft Excel*

Clauss, *Applied Management Science and Spreadsheet Modeling*

Clemen, *Making Hard Decisions: An Introduction to Decision Analysis, 2nd Edition*

Devore, *Probability & Statistics for Engineering and the Sciences, 4th Edition*

Farnum, *Modern Statistical Quality Control and Improvement*

Fourer, Gay & Kernighan, *AMPL: A Modeling Language for Mathematical Programming*

Gass, *Linear Programming, 5th Edition*

Higgins & Keller-McNulty, *Concepts in Probability and Stochastic Modeling*

Kao, *Introduction to Stochastic Processes*

Kirkwood, *Strategic Decision Making: Multiobjective Decision Analysis with Spreadsheets*

Kuehl, *Statistical Principles of Research Design*

Lapin, *Modern Engineering Statistics*

Lapin & Whisler, *Cases in Management Science*

Larsen, Marx & Cooil, *Statistics for Applied Problem Solving and Decision Making*

Middleton, *Data Analysis Using Microsoft Excel: Updated for Windows 95*

Newton & Harvill, *StatConcepts: A Visual Tour of Statistical Ideas*

Ostle, Turner, McElrath & Hicks, *Engineering Statistics: The Industrial Perspective*

SAS Institute Inc., *JMP-IN 3.2: Statistical Discovery Software*

Savage, *INSIGHT: Business Analysis Software for Microsoft Excel*

Schleifer & Bell, *Data Analysis, Regression and Forecasting*

Schleifer & Bell, *Decision Making under Certainty*

Schrage, *Optimization Modeling Using Lindo, 5th Edition*

Schruben, *SIGMA: Graphical Simulation Modeling & Analysis for Windows*

Siegrist, *Interactive Probability*

Stata Corporation, *StataQuest 4 for Windows 95*

Vining, *Statistical Methods for Engineers*

Winston, *Operations Research: Applications & Algorithms, 3rd Edition*

Winston, *Introduction to Mathematical Programming, 2nd Edition*

Winston, *Simulation Modeling Using @Risk*

Winston & Albright, *Practical Management Science: Spreadsheet Modeling & Applications*

To order copies contact your local bookstore or call 1-800-354-9706. For more information contact **Duxbury Press** at 10 Davis Drive, Belmont CA 94002 or go to: **www.duxbury.com**

Modern Industrial Statistics

Modern Industrial Statistics

Design and Control of Quality and Reliability

Ron Kenett

KPA Ltd and *Tel Aviv University*

Shelemyahu Zacks

Binghamton University

Duxbury Press
An Imprint of Brooks/Cole Publishing Company
I(T)P® An International Thomson Publishing Company

Pacific Grove • Albany • Belmont • Bonn • Boston • Cincinnati • Detroit • Johannesburg • London
Madrid • Melbourne • Mexico City • New York • Paris • Singapore • Tokyo • Toronto • Washington

Sponsoring Editor: *Curt Hinrichs*
Marketing Team: *Marcy Perman, Michele Mootz*
Editorial Assistant: *Rita Jaramillo*
Production Editor: *Timothy Wardell*
Manuscript Editor: *Carol Dondrea*

Interior Design: *William Baxter*
Cover Design: *Terri Wright*
Cover Art: *Zsiba-Smolover*
Cover Photo: *George Erml*
Typesetting: *SuperScript*
Printing and Binding: *Courier Westford*

For more information, contact Duxbury Press at Brooks/Cole Publishing Company at:

BROOKS/COLE PUBLISHING COMPANY
511 Forest Lodge Road
Pacific Grove, CA 93950
USA

International Thomson Editores
Seneca 53
Col. Polanco
11560 México, D.F., México

International Thomson Publishing Europe
Berkshire House 168-173
High Holborn
London WC1V 7AA
England

International Thomson Publishing GmbH
Königswinterer Strasse 418
53227 Bonn
Germany

Thomas Nelson Australia
102 Dodds Street
South Melbourne, 3205
Victoria, Australia

International Thomson Publishing Asia
221 Henderson Road
#05-10 Henderson Building
Singapore 0315

Nelson Canada
1120 Birchmount Road
Scarborough, Ontario
Canada M1K 5G4

International Thomson Publishing Japan
Hirakawacho Kyowa Building, 3F
2-2-1 Hirakawacho
Chiyoda-ku, Tokyo 102
Japan

10 9 8 7 6 5 4 3 2 1

Library of Congress Cataloging-in-Publication Data

Kenett, Ron.
    Modern industrial statistics : design and control of quality and
reliability / Ron Kenett. Shelemyahu Zacks.
        p.   cm.
    Includes index.
    ISBN 0–534–35370–3
    1. Industrial statistics.   2. Quality control—Statistical
methods.   3. Reliability (Engineering)—Statistical methods.
Zacks, Shelemyahu   II. Title.
TS156.K42 1998
658.5′62—DC21   97–43141

Brief Contents

Foreword

A decade ago, Motorola began a journey that would shape the company's success for decades to come. The issue was survival; the challenge was quality; the solution would come from our people. We didn't have a detailed map to guide us but we set out to achieve one very large but simple goal: to improve the way we serve our customers.

Motorola's goal is virtual perfection in everything we do. Perfect quality. Perfect delivery. Perfect reliability. Perfect service. We believe that no defect is tolerable and that perfection is operationally possible. We call it Six Sigma quality.

Where quality was once measured by the number of defective parts per thousand, our Six Sigma goal translates into a defect rate of only 3.4 parts per million in each step of our processes. The goal is a visionary one but our initiative of continuous improvement reaches out for change, refinement, and even revolution in our pursuit of the highest possible quality.

Today it is not a secret that we are all awash in an overload of information, but in reality we cannot have too much relevant data. Too often companies operate in a vacuum—not knowing how their products and services truly stack up against the competition.

At Motorola we use statistical methods daily throughout all of our disciplines to synthesize an abundance of data to drive concrete actions. Only with constant monitoring and continual follow-up can you get a true picture of where you stand and a clear view of where you need to go. We use statistical data to measure both our successes and our failures. More importantly it helps us to manage our future.

How has the use of statistical methods within Motorola Six Sigma initiative, across disciplines, contributed to our growth? Over the past decade we have reduced in-process defects by over 300 fold, which has resulted in a cumulative manufacturing cost savings of over 11 billion dollars. Employee productivity (worldwide) measured in sales dollars per employee has increased three fold or an average 12.2 percent per year over the same 10 year period. Our product reliability as seen by our customer has increased between 5 and 10 times.

Each year over 6,000 Total Customer Satisfaction empowered teams are formed, representing more than 60,000 of our 138,000 employees to solve problems and drive continuous improvement. Over half of these teams are from non-manufacturing disciplines, such as finance, legal, personnel, engineering, and service, all using statistical tools to drive perfection in everything we do.

If you don't set out to achieve perfection, you probably aren't going to achieve it.

Robert W. Galvin
Chairman of the Executive Committee
Motorola, Inc.

Preface

Modern Industrial Statistics: Design and Control of Quality and Reliability provides the tools for those who drive to achieve perfection in industrial processes. Learn the concepts and methods contained in this book and you will understand what it takes to measure and improve world-class products and services.

The need for constant improvement of industrial processes, in order to achieve high quality, reliability, productivity and profitability, is well recognized. Further, management techniques, such as total quality management and business process reengineering, are insufficient in themselves to achieve the goal without the strong backing of specially tailored statistical procedures, as stated by Robert Galvin in the Foreword.

Statistical procedures designed for solving industrial problems are called Industrial Statistics. Our objective in writing this book was to provide statistics and engineering students, as well as practitioners, the concepts, applications, and practice of basic and advanced industrial statistical methods, which are designed for the control and improvement of quality and reliability.

The idea of writing a text on industrial statistics developed after several years of collaboration in industrial consulting, teaching workshops and seminars, and courses at our universities. We felt that no existing text served our needs in both content and approach so we decided to develop our notes into a text. Our aim was to make the text modern and comprehensive in terms of the techniques covered, lucid in its presentation, and practical with regard to implementation.

Features

Modern Industrial Statistics employs several unique characteristics. Among the most significant is its integration and use of the computer in solving real industrial problems.

- **Resampling techniques**, or *bootstrapping*, which occupy a central position in the practice of statistics, have freed the researcher and the practitioner from imposing on a problem theoretical assumptions that are difficult to verify. They can also substitute complex analytical methods for assessing the properties of statistical decision and control procedures. Many regard bootstrapping methods as the inferential and decision-making tools of the 21st century.

- **Simulators**, which perform simulations, have been specially developed for the book to assess the operating characteristics of various process-control procedures such as the cycle speed of a piston in a gas turbine and the output current of an electrical circuit. These simulators are used in various parts of the book to generate observations under specific conditions. Our book is unique in its use of the computer as a tool for analysis, design, and control.

- MINITAB is used throughout the book. This package was chosen because of its wide availability, in student or professional versions for Windows, mainframes, or UNIX platforms. Moreover, MINITAB is user friendly, interactive, and easy to teach. An appendix covering the basic features and commands of MINITAB is included.

- In addition, S-PLUS, which has become a package of choice for the more advanced user, is included as well. We furnish all the S-PLUS functions and routines needed for the book in a special appendix.

- Included in the book is a CD-ROM containing all the data sets and the programs that were specifically developed for the book. The programs are either executable files (.EXE) or MINITAB macros (.MTB) stored in subdirectories corresponding to the chapters in which they're first introduced. The S-Plus functions are provided on the CD-ROM as well. The data sets, the simulators, the executable programs, the macros, and the S-PLUS functions are all described in the appendices.

- The book is easily adaptable to other software packages. Faculty interested in using MATLAB, SAS, JMP-IN, STATGRAPHICS, EXCEL, or SPSS can read the data files into other formats by following the data library on the Duxbury Resource Center at **www.duxbury.com**

Organization and Coverage

The book is comprised of fourteen chapters, organized in two parts. Part I consists of eight chapters, emphasizing statistical thinking in industrial settings, employing data analysis, graphical techniques, and data-based simulations. Most examples in the book are from actual industrial problems, illustrating the versatility and applicability of the statistical methods. Part I includes chapters on probability models, statistical distributions, and estimation and testing of hypotheses. It contains a chapter on sampling from finite populations (needed for bootstrapping and for sampling inspection and auditing of quality), a chapter on bootstrap and nonparametric methods, and a chapter on multiple regression and categorical data analysis. Part II is devoted to six chapters covering basic and advanced methods of industrial control and improvement of quality and reliability. The Solutions Manual contains the complete solutions to all of the problems in the text.

Students taking courses in industrial statistics and using this book will need knowledge of calculus and some matrix theory (linear algebra). We developed the material for upper division undergraduates and beginning graduate students in the fields of Industrial Engineering and Statistics. Students of electrical and mechanical engineering can also benefit from the book. The text can also provide a reference to practitioners working in the areas of quality management, quality engineering, and reliability.

A two-semester course can cover a good part of the book. However, the book was designed to be flexible enough for one-semester courses and several differing course needs. Examples of how the book may be used in one-semester courses are:

Introduction to Industrial Statistics: Chapters 1, 2, 3, 4, 6, 7

Sampling Plans and Reliability: Chapters 4, 5, 6, 7, 9, 14

Statistical Process Control: Chapters 1, 4, 6, 7, 10, 11

Statistics for Quality Design and Improvement: Chapters 4, 6, 8, 10, 12, 13

Advanced Industrial Statistics: Chapters 7, 8, 9, 11, 12, 13, 14.

Acknowledgments

The authors wish to thank their families for their encouragement, the administration of Binghamton University for its support, and the various industries at which they worked or consulted for the opportunity to solve industrial problems. The examples presented in the text have many people behind them and it is impossible to acknowledge all contributors. Among those contributors we should, however, mention several key individuals: Mr. Richard Buetow from Motorola, Mr. Haim Rosen, Mr. Meir Sperling and Mr. Yakov Bantay from Tadiran Telecommunications and Professor Tom Kelley from Binghamton University. We thank the following reviewers for their thoughtful feedback to earlier drafts of the manuscript: James A. Alloway, E.M.S.Q. Associates; Robert L. Armacost, University of Central Florida; Michael Martin, Australian National University, Canberra; P. Simon Pulat, University of Oklahoma; and Derek Rollins, Iowa State University. We wish to thank the editor, Mr. Curt Hinrichs; the editorial staff and accuracy checkers for their excellent suggestions, corrections, and editorial work; and for production editing, Timothy Wardell at Brooks/Cole. Special thanks are due to Mrs. Marge Pratt for her excellent typing of the text.

S. Zacks   R.S. Kenett
Binghamton, NY   Raanana, Israel

To Sima, Dolav, Ariel, Dror and Yoed with whom I share everything
Ron S. Kenett

To my wife Hanna, our sons Yuval and David and their families with love
Shelemyahu Zacks

Contents

Principles of Statistical Thinking and Analysis

Part I is an introduction to the role of industrial statistics in modern industry and to statistical thinking in general. Typical industrial problems are described, and basic statistical concepts and tools are presented through case studies and computer simulations.

Chapter 1 is an overview of the role of statistical methods in industry. It presents a classification of statistical methods in the context of various management approaches. We call this classification the **quality ladder** and use it to organize the methods of industrial statistics covered in Part II.

Chapter 2 presents basic concepts and tools for describing variability. Both Chapters 2 and 3 emphasize graphical techniques for exploring and summarizing variability in observations. Chapter 3 is on the multivariate techniques required to handle observations taken on several variables simultaneously. In particular, the chapter covers computerized multivariate graphical techniques, including dynamic graphics. These techniques have recently become widely available to practitioners through inexpensive and powerful software. Regression methods are described in detail using several case studies.

Chapter 4 is a basic introduction to probability, with a comprehensive treatment of statistical distributions that are commonly used in industrial statistics.

Chapter 5 is a special chapter dedicated to sampling for estimation of finite population quantities with a natural expansion to bootstrap methodology.

Theoretical properties of estimators and inferential methods that are based on parametric models are presented in Chapter 6. Distribution-free, nonparametric inference is discussed in Chapter 7. This chapter covers basic methods of statistical inference from a modern approach that relies on computer-intensive bootstrap methods. With this approach, most statistical procedures used in making inferences from a sample to a population are handled by software programs, thus eliminating the need for traditional mathematical assumptions and models. The chapter also presents several special software procedures we developed to analyze data using bootstrapping methods.

Chapter 8 deals with linear models in one and several dimensions. Relationships for both continuous variables and discrete categorical data are analyzed and regression diagnostics and prediction intervals are also covered.

1

The Role of Statistical Methods in Modern Industry

The Different Functional Areas in Industry

The basic theme of this book is that the tools and concepts of industrial statistics have to be viewed in the context of their applications. These applications are greatly affected by management style and organizational culture. Therefore, to lay the foundations for the book, we begin by describing key aspects of the industrial setting.

Industrial organizations typically include departments for product development, manufacturing, marketing, finance, human resources, purchasing, sales, quality assurance, and after-sale support. Marketing personnel typically determine customer requirements and levels of satisfaction using surveys and focus groups. Sales personnel are usually responsible for providing forecasts to purchasing and manufacturing departments. Purchasing specialists analyze world trends in the quality and prices of raw materials so they can provide critical input to the product developers. Budgets are prepared by the finance department using forecasts that are validated periodically. Accounting experts rely on auditing and sampling methods to ascertain inventory levels and the integrity of databases. Human resources personnel track data on absenteeism, turnover, overtime, and training needs. The quality departments commonly perform audits and special tests to determine the quality and reliability of the product. Research and development engineers perform experiments to solve problems and improve products and processes. Finally, process controls of production operations are typically developed and maintained by manufacturing personnel using control charts. These are only a few general examples of problem areas in modern industrial organizations. **Industrial statistics** is used to resolve problems in each of these functional units.

To provide more specific examples, we first take a closer look at a variety of manufacturing industries. Then we discuss examples from these types of industries.

There are basically three types of production systems:

1 **Continuous flow production**
2 **Job-shop operation**
3 **Discrete mass production**

Examples of continuous flow production include papermaking and chemical transformations. Such processes typically involve expensive equipment that is very large in size, operates around the clock, and requires very rigid manufacturing steps. Continuous flow industries are both capital intensive and highly dependent on the quality of the purchased raw materials. Rapid customizing of products in a continuous flow process is virtually impossible, and new products are difficult to introduce.

Job-shop operations are, in many respects, exactly the opposite. Examples of job-shop operations include metalworking of parts and injection molding of plastic components. Such operations allow production of custom-made products and are very labor intensive. Job shops usually use general purpose machinery, which can frequently be idle.

Discrete mass production systems can be similar to continuous flow production if a standard product is produced in large quantities. Flexible manufacturing is achieved when the system can switch from mass production to job-shop operation.

Machine tool automation began in the 1950s with the development of numerical control operations. In the automatic or semiautomatic machines used in these operations, computer commands position tools for a desired cutting effect. Today's hardware and software capabilities can make a job-shop manufacturing facility as automated as a continuous-flow enterprise. *Computer-integrated manufacturing (CIM)* is the integration of *computer-aided design (CAD)* with *computer-aided manufacturing (CAM).* The development of CAD has its origins in the evolution of computer graphics and *computer-aided drawing and drafting,* often called *CADD.*

As an example of how these systems are used, we will follow the creation of an automobile suspension system designed on a computer using CAD. The new system must meet testing requirements under a battery of specific road conditions. After coming up with an initial design concept, design engineers use computer animation to show the damping effects of the new suspension design on various road conditions. The design is then iteratively improved on the basis of simulation results and established customer requirements. In parallel to the suspension system design, purchasing specialists and industrial engineers proceed with specifying and ordering the necessary raw materials, setting up the manufacturing processes, and scheduling production quantities. Throughout the manufacturing of the suspension system, several tests provide the necessary production controls. Ultimately the objective is to minimize the costly impact of failures in a product after delivery to the customer. Statistical methods are employed throughout the design, manufacturing, and servicing stages of the product. Methods for the statistical design of experiments are used to optimize the design of the suspension system (see Chapters 12 and 13). Control of the manufacturing process is required to ensure that the product is manufactured according to specifications. Methods of statistical process control are employed at various stages of manufacturing to identify and correct deviations from process capabilities (see Chapters 10 and 11). The incoming raw materials have to be inspected often for adherence to quality level (see Chapter 9). Finally, to assess the reliability of the suspension system and provide early warnings of product deterioration, the product is tracked and field failures are analyzed (see Chapter 14).

CAD systems provide an inexpensive environment in which to test and improve design concepts. CIM systems typically capture data necessary for process control. Computerized field-failure tracking systems and sales forecasting are very common. The application of industrial statistics within such computerized environments allows the practitioner to concentrate on statistical analysis as opposed to repetitive numerical computations.

Service organization can be either independent or complementary to manufacturing-type operations. For example, a manufacturer of electronic telephone switching systems can also provide installation and after-sale service to its customers. The service takes the form of actually installing the telephone switch, programming the system's database with an appropriate numbering system, and responding to service calls. The delivery of services differs from manufacturing in many ways. The output of a service system is generally intangible. In many cases, the service is delivered directly to the customer, and there is no opportunity to correct "defective" transactions before delivery. Some services involve very large numbers of transactions. Federal Express, for example, handles 1.5 million shipments per day, to 127 countries, at 1650 sites. The opportunities for error are many, and process error levels must be maintained at levels of only a few defective parts per million. Such low defect levels might appear at first highly expensive to maintain and therefore economically unsound. In the next section, we deal with the apparent contradiction between maintaining low error levels and reducing costs and operating expenses.

1.2
The Quality–Productivity Dilemma

In order to reach World War II production goals for ammunition, airplanes, tanks, ships, and other military materiel, American industry had to restructure and raise its productivity while adhering to strict quality standards. This was partially achieved through large-scale applications of statistical methods, following the pioneering work of a group of industrial scientists at Bell Laboratories. Two prominent members of this group were Walter A. Shewhart, who developed the tools and concepts of statistical process control, and Harold F. Dodge, who laid the foundations for statistical sampling techniques. Their ideas and methods were instrumental in the transformation of American industry in the 1940s, which had to deliver high quality and high productivity. During those years, many engineers were trained in industrial statistics throughout the United States.

After the war, a number of Americans were asked to help Japan rebuild its devastated industrial infrastructure. Two of these consultants, W. Edwards Deming and Joseph M. Juran, distinguished themselves as effective and influential teachers. Both Drs. Deming and Juran witnessed the impact of Walter Shewhart's new concepts. In the 1950s, they taught the Japanese the ideas of process control and process improvements, emphasizing the role of management and employee involvement. The Japanese were quick to learn the basic quantitative tools for identifying and realizing improvement opportunities and for controlling processes. By improving blue collar and white collar processes throughout their organizations, the Japanese were able to reduce waste and rework, thus producing better products at a lower price. These changes occurred over a period of several years, leading eventually to significant increases in market share for these products.

In contrast, American industry had no need for improvements in quality after World War II. There was an infinite market for American goods, and the emphasis shifted to high productivity, without assuring high quality. This was reinforced by the Taylor approach, which split the responsibility for quality and productivity between the quality and production departments. Many managers in U.S. industry did not believe that high quality and high productivity could be achieved simultaneously. The **quality–productivity dilemma** was born, and managers apparently had to make a choice. By focusing attention on productivity, they sacrificed quality, which in turn had a negative effect on productivity. Increasing emphasis on meeting schedules and quotas made the situation even worse. In the meantime, Japanese industrialists proved to themselves that, by implementing industrial statistics tools, managers could improve processes in every industrial organization and thus resolve the quality–productivity dilemma. In the 1970s, several American companies began applying the methods taught by Deming and Juran, and by the mid-1980s, several of these U.S. companies were reporting outstanding successes. Quality improvements generate higher productivity because they allow higher-quality products to be shipped more quickly. The result is better products at lower costs—an unbeatable formula for success.

The automobile industry provides many examples of this dilemma resolution. Consider, for instance, the success of cars manufactured by Toyota in Japan, and in Lexington, Kentucky. Table 1.1 shows how Toyota compares to General Motors on quality (assembly defects per 100 cars) and productivity (assembly hours per car). Toyota cars are manufactured more efficiently by a factor of 2 and have fewer defects by a factor of 3. A comparison of Japanese and U.S. companies in

TABLE 1.1
General Motors' Framingham assembly plant versus Toyota's Takaoka assembly plant, 1986

	GM	TOYOTA
Gross assembly hours per car (total effort/total # of cars)	41	18
Adjusted assembly hours per car (standardized assembly job)	31	16
Assembly defects per 100 cars[a]	130	45

Source: From Womack, Jones, and Roos (1990).
[a] Powers (1987).

the software industry leads to similar conclusions (see Table 1.2). The key to this achievement is **total quality management** and the application of industrial statistics, which includes analyzing data, understanding variability, controlling processes, designing experiments, and making forecasts.

TABLE 1.2

Comparison of Japanese and U.S. software development companies

		U.S.	Japan
Fortran equivalent	Mean	7300	12,500
SLOC^a/work-year	Median	3000	4700
Failures/1000 SLOC during first 12 months	Mean	4.44	1.96
in the field	Median	0.83	0.20
Number of companies surveyed		20	11

Source: From M. Cusumano (1991).
^{a} SLOC = source lines of code

The effective implementation of industrial statistics depends on the management approach practiced in the organization. We characterize different styles of management by a **quality ladder**, illustrated in Figure 1.1. Management's response to the question: "How do you handle the inconvenience of customer complaints?" determines the position of an organization on the ladder. Managers might respond by describing an approach based on waiting for complaints to be filed before initiating any corrective actions. They might implement strict supervision of every activity in the organization, requiring several signatures of approval on every document, or they might take a more proactive approach. The four management styles we identify are: (I) fire fighting, (II) inspection and traditional supervision, (III) processes control and improvement, and (IV) quality by design. Industrial statistics tools are used in the context of the top three steps in the quality ladder. Levels III and IV are more proactive than II. When management's style consists exclusively of fire fighting, there is typically no need for industrial statistics methods. The foundation for the quality ladder is the use of data and statistical thinking, on which we elaborate in the next section.

1.3
Fire Fighting

Fire fighters specialize in putting out fires. Their main goal is to get to the scene of a fire as quickly as possible. To meet this goal, they activate sirens and flashing lights and have their equipment organized for immediate use at a moment's notice. Fighting fires is also characteristic of a particular management approach that

FIGURE 1.1
The quality ladder

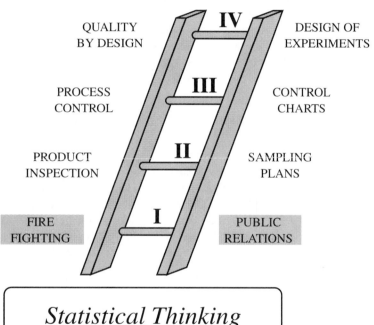

focuses on heroic efforts to resolve problems and crises. The seeds of these problems are often planted by the same managers who work the extra hours required to fix them. **Fire fighting** has been characterized as an approach where there is never enough time to do things right the first time, but always sufficient time to rework and fix problems once customers complain. This reactive approach to management is rarely conducive to serious improvement, which relies on prior thought, data, and teamwork. Industrial statistics tools are rarely used under fire-fighting management. As a consequence, decisions are often made without the causes for the failures being known.

In Chapter 2, we study the structure of random phenomena and present the basic statistical tools used to describe and analyze such structures. The basic philosophy of **statistical thinking** is the realization that variation occurs in all work processes and the recognition that reducing variation is essential to quality improvement. Failure to recognize the impact of randomness leads to unnecessary and harmful decisions. One example of the failure to understand randomness is the common practice of adjusting production quotas for the next month based on the current month's sales. Without appropriate tools, managers have no way of knowing whether the current month's sales are within the common variation range or not. Common variation implies that nothing significant has happened since last month and therefore no quota adjustments should be made. Under such circumstances, changes in production quotas create unnecessary, self-inflicted problems. Fire-fighting management, in many cases, is responsible for avoidable costs and quick temporary fixes that have negative effects on the future of the organization. The true challenge for the manager is to prevent "fires" from occurring. One approach to preventing fires is relying on thorough and frequent inspection and traditional supervision. The next section provides some historical background on methods of inspection. [Note that the American Society for Quality Control has recently issued a *Special Publication on Statistical Thinking* (see the ASQC web site at www.asqc.org).]

1.4
Inspection of Products

In medieval Europe, most families and social groups made their own goods—cloth, utensils, and other household items. The only saleable cloth was woven by peasants who paid their taxes in kind to their feudal lords. The ownership of barons or monasteries was identified through marks put on the fabric; these marks were also an indication of quality. Because no feudal lord would accept payment in shoddy goods, the products were carefully inspected before the mark was inscribed. Surviving documents from the late medieval period indicate that bales of cloth frequently changed hands repeatedly without being opened, simply because the marks they bore were regarded everywhere as guarantees of quality. In the new towns, fabrics were made by craftsmen who went in for specialization and division of labor. Chinese records of the same period indicate that silks made for export were also subjected to official quality inspections. In Ypres, the center of the Flemish wool cloth industry, already existing weavers' regulations were put into writing in 1281. These stipulated both the length and width, as well as the number of warp ends and the quality of the wool to be used, in each cloth. A fabric had to be of the same thickness throughout. All fabrics were inspected in the drapers' hall by municipal officials. Heavy fines were levied for defective workmanship, and the quality of the fabrics that passed inspection was guaranteed by affixing the town seal. Similar regulations existed elsewhere in Belgium, France, Italy, Germany, England, and Eastern Europe.

The trademarks used by the modern textile industry as a guarantee of quality originated in Britain. They first found general acceptance in the wholesale trade and then, from the end of the 19th century onward, among consumers. For a time, manufacturers still relied on in-plant inspections of their products by technicians and merchants, but eventually technological advances introduced machines and processes that ensured that certain standards would be maintained independently of human inspectors and their know-how. Industrial statistics plays an important role in the textile industry. In fact, it was the first large industry that analyzed its data statistically. Simple production figures, including percentages of defective products, were being compiled in British cotton mills early in the 19th century. The basic approach during the preindustrial and postindustrial period was to guarantee quality by proper inspection of the cloth (see Juran, 1995).

In the early 1900s, researchers at Bell Laboratories in New Jersey developed statistical sampling methods that provided an effective alternative to 100% **inspection** (see Dodge and Romig, 1929). Their techniques, called **sampling inspection**, eventually led to the famous MIL-STD-105 system of sampling procedures used throughout the defense industry and elsewhere. These techniques implement statistical tests of hypotheses in order to determine if a certain production lot or manufacturing batch is meeting acceptable quality levels. Such sampling techniques are focused on the product, as opposed to the process that makes the product. Details on the implementation and theory of sampling inspection are provided in Chapter 9. The next section introduces a quality assurance approach that focuses on the performance of the processes throughout the organization.

1.5
Process Control

In 1924, in a memorandum to his superior at Bell Laboratories, Walter Shewhart documented a new approach to statistical process control (see Godfrey, 1986). The document, dated May 16, includes a sketch of a technique designed to track process quality levels over time. Shewhart labeled this a **control chart**. The technique was developed further, and Shewhart published his methodology two years later (Shewhart, 1926). Shewhart realized that any manufacturing process can be controlled using basic engineering ideas. Control charts are a straightforward application of engineering feedback loops to the control of work processes. The successful implementation of control charts requires management to focus on processes, with emphasis on **process control and improvements**. When a process is able to produce products that meet customer requirements and a system of process controls is subsequently employed, product inspection is not necessary. The industry can deliver its products without time-consuming and costly inspection. These are the prerequisites for **just-in-time deliveries** and increased customer satisfaction. No off-line inspection implies quicker deliveries, less testing, and therefore lower costs.

Shewhart's ideas are essential for organizations seeking improvements in their competitive position. As mentioned earlier, W. Edwards Deming and Joseph M. Juran were instrumental in bringing this approach to Japan. Deming emphasized the use of statistical methods, and Juran created a comprehensive management system, including the concepts of breakthroughs, the quality trilogy, and the strategic planning of quality. Both were awarded a medal by the Japanese emperor for their contributions to the rebuilding of Japan's industrial infrastructure. Japan's national industrial award, called the Deming Prize, has been awarded since the early 1950s. In the United States and Europe, the U.S. National Quality Award and the European Quality Award, respectively, have been awarded since the early 1990s to recognize companies that can serve as role models to others. Notable winners include

Motorola, Xerox, Milliken, Globe Metallurgical, AT&T Universal Card, and the Ritz-Carlton. Similar awards exist in Australia (the Juran Prize), Israel, Mexico, and other countries. Chapters 10 and 11 deal with implementation and theoretical issues in process control techniques. The next section takes the ideas of process control one step further.

1.6

Quality by Design

Quality by design is an approach relying on a proactive management style, where problems are sought out and products and processes are designed with "built-in" quality. The design of a manufactured product or a service begins with an idea and continues through a variety of development and testing phases until production begins and the product is made available to the customer. Process design involves planning and designing the physical facilities, as well as the information and control systems, required to manufacture the good or deliver the service. The design of the product, and the associated manufacturing process, determines the product's ultimate performance and value. Design decisions influence the sensitivity of a product to variations in raw materials and work conditions, which, in turn, affects manufacturing costs. General Electric, for example, has found that 75% of failure costs are determined by a product's design. In a series of bold design decisions, IBM developed the Proprinter so that all parts and subassemblies would snap together during final assembly without the use of fasteners. Such initiatives yield major cost reductions and quality improvements. These examples demonstrate how design decisions affect manufacturing capabilities with an eventual positive impact on the cost and quality of the product.

Another design decision, reducing the number of parts, is also a statistical problem that involves clustering and grouping similar parts. For example, many organizations find themselves purchasing hundreds of different types of aluminum bolts for very similar applications. Multivariate statistical techniques can be used to group together similar bolts, thereby reducing the number of types of bolts required and thus eliminating potential mistakes and lowering costs.

In the design of manufactured products, technical specifications can be precisely defined. In the design of a service process, however, quantitative standards may be difficult to determine. In service processes, the physical facilities, procedures, people's behavior, and professional judgment all affect the quality of the service. Quantitative measures in the service industry typically consist of data from periodic customer surveys and information from internal feedback loops such as waiting time at hotel front desks or supermarket cash registers. The design of products and processes, both in service and manufacturing, involves quantitative performance measurements. We can now describe the basic concepts and tools of quality engineering aimed at achieving **quality by design**.

A major contributor to modern quality engineering has been Genichi Taguchi, formerly of the Japanese Electronic Communications Laboratories. Since the 1950s, Taguchi has advocated the use of statistically designed experiments in industry.

In 1959, the Japanese company NEC ran 402 planned experiments. In 1976, Nippon Denso, a 20,000-employee company producing electronic parts for automobiles, reportedly ran 2700 designed experiments. In the summer of 1980, Taguchi came to the United States to "repay the debt" of the Japanese to Shewhart, Deming, and Juran and delivered a series of workshops at Bell Laboratories in Holmdel, New Jersey. His methods slowly gained acceptance in the United States. Companies like ITT, Ford, and Xerox have been using Taguchi methods since the mid-1980s with impressive results. For example, an ITT electrical cable and wire plant reported reduced product variability by a factor of 10. ITT Avionic Division conducted more than 2000 designed experiments between 1984 and 1986. Over a period of 30 years, Taguchi developed a comprehensive approach to quality engineering, including an economic model for optimization of products and processes. Central to his approach is the statistical design and analysis of experiments, which he learned from Western statisticians and applied in the Japanese industry. Chapter 12 presents a comprehensive treatment of the principal methods of design and analysis of experiments. Chapter 13 presents the tools and concepts of quality engineering, as developed by Taguchi, with case studies and examples.

An essential complement to quality engineering is quality planning. Planning in general is a basic engineering and management activity. It involves deciding, in advance, what to do, how to do it, when to do it, and who is to do it. Quality planning is the process used in the design of any new product or process. Table 1.1 showed that, in 1986, General Motors cars averaged 130 assembly defects per 100 cars. A cause-and-effect analysis of car assembly defects would point out causes ranging from production facilities, suppliers of purchased material, manufacturing equipment, engineering tools, and so on. Other choices of suppliers, different manufacturing facilities, and alternative engineering tools would have produced a higher—or lower—number of assembly defects. Planning usually requires careful analysis, experience, imagination, foresight, and creativity. Planning for quality was formalized by Joseph Juran as a series of steps:

1   Identify the customers of the new product or process.
2   Determine the needs of those customers.
3   Translate the needs into technical terms.
4   Develop a product or process that can respond to those needs.
5   Optimize the product or process so that it meets the needs of the customers, including their economic and performance goals.
6   Develop the process required to produce the new product or to install the new process.
7   Optimize that process.
8   Begin production or implement the new process.

For further reading on this subject, see *Juran's Quality Control Handbook* (Juran and Gryna, 1988).

The effective use of industrial statistics tools parallels the rise of organizations up the quality ladder (see Figure 1.1). As the use of data becomes gradually integrated into the decision process, at both the short-term operational level and at the long-term strategic level, different tools are needed. The ability to plan and forecast successfully is a result of accumulated experience and proper techniques. Figure 1.2 presents the 14 chapters of this book in the context of the quality ladder.

FIGURE 1.2
Organization of the book in the context of the quality ladder

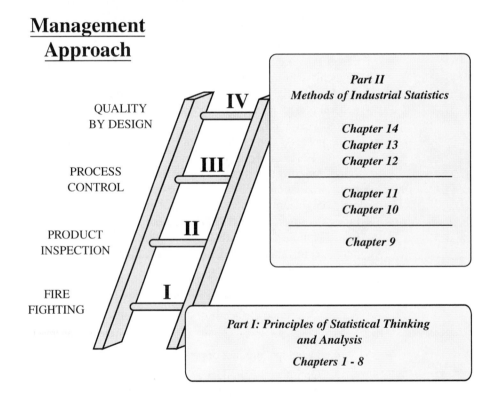

Organization of *Industrial Statistics:*
Design and Control of Quality and Reliability

Management
Approach

QUALITY
BY DESIGN

IV

Part II
Methods of Industrial Statistics

Chapter 14
Chapter 13
Chapter 12

PROCESS
CONTROL

III

Chapter 11
Chapter 10

PRODUCT
INSPECTION

II

Chapter 9

FIRE
FIGHTING

I

Part I: Principles of Statistical Thinking
and Analysis

Chapters 1 - 8

1.7
Chapter Highlights

Modern industrial organizations (manufacturing or service) are described and classified. Different functional areas of a typical business are pre-sented, as are problems typical for each area. The potential benefits of the methods of industrial statistics are then introduced in the context of these problems. The main theme here is that the apparent conflict between high productivity and high quality can be resolved through improvements in work processes by introducing statistical thinking. Shewhart's, Deming's, and Juran's contributions to industries seeking a more competitive position are outlined. Different approaches to the management of industrial organizations are summarized and classified using a quality ladder. Industrial statistics methods are then categorized according to the steps of the ladder. These consist of fire fighting, inspection, process control, and quality by design. Corresponding to each management approach is a particular set of statistical methods.

The important terms introduced in this chapter include:

- Industrial statistics

- Production systems:

    Continuous flow production

    Job-shop operation

    Discrete mass production

- Quality–productivity dilemma
- Total quality management
- Quality ladder
- Management approaches:

    Fire fighting

    Statistical thinking

    Inspection

    Process control and improvement

    Just-in-time delivery

- Quality by design

Exercises

1.1 Describe three work environments where quality is assured by 100% inspection.

1.2 Search periodicals such as *Business Week, Fortune, Time,* and *Newsweek* and newspapers such as the *New York Times* and *Wall Street Journal* for information on quality initiatives at Motorola, Xerox, Milliken, Globe Metallurgical, and the Ritz-Carlton.

1.3 Provide examples of the three types of production systems.

1.4 What management approach cannot work with continuous flow production?

1.5 What management approach characterizes each of the following:

a A school system

b A military unit

c A football team

1.6 Provide examples of how you, personally, apply the four management approaches

a As a student

b In your parents' house

c With your friends

Understanding Variability

Random Phenomena and the Structure of Observations

Many phenomena that we encounter are only partially predictable. It is difficult, for example, to predict the weather or the behavior of the stock market. In our discussions, we are more interested in industrial phenomena, such as performance measurements of a product, or the sales volume of a given model during a specified period. Measurements of such phenomena often reveal a degree of variability. The objectives of this chapter are to present variable phenomena and introduce methods for analyzing them in order to understand the structure of variability and enhance our ability to control, improve, and predict the future behavior of such phenomena. We start with a few simple examples.

EXAMPLE **2.1**   A piston is a mechanical device that is present in most types of engines. One measure of the performance of a piston is the time it takes to complete one cycle. We call this measure cycle time. In Table 2.1 we list 50 cycle times of a piston operating under fixed operating conditions (the data are stored in file CYCLT.DAT). The differences in these numbers is quite apparent, and we can make the statement "cycle times vary." Such a statement, in spite of being true, is not very useful. We have only established the existence of variability—we have not yet characterized it, and we are still unable to predict and control the future behavior of the piston.   ∎

EXAMPLE **2.2**   Consider an experiment in which a coin is flipped once. Suppose the coin is fair, in the sense that it is equally likely to fall on either one of its faces. Furthermore, assume that the two faces of the coin are labeled with the numbers 0 and 1. In general, we cannot predict with certainty on which face the coin will fall. If the coin falls on the face labeled 0, we assign to a variable X the value 0; if the coin falls on

TABLE **2.1**

Cycle times of piston (in sec) with control factors set at minimum levels

1.008	1.117	1.141	0.449	0.215
1.098	1.080	0.662	1.057	1.107
1.120	0.206	0.531	0.437	0.348
0.423	0.330	0.280	0.175	0.213
1.021	0.314	0.489	0.482	0.200
1.069	1.132	1.080	0.275	0.187
0.271	0.586	0.628	1.084	0.339
0.431	1.118	0.302	0.287	0.224
1.095	0.319	0.179	1.068	1.009
1.088	0.664	1.056	1.069	0.560

the face labeled 1, we assign to X the value 1. Because the values that X will obtain in a sequence of such trials cannot be predicted with certainty, we call X a **random variable**. A typical random sequence of 0, 1 values that can be generated in this manner might look like the following:

$$0, 1, 1, 0, 1, 0, 1, 0, 1, 1, 1, 1, 1, 0, 1, 0, 1, 1, 1, 1,$$
$$0, 0, 1, 1, 1, 1, 1, 0, 0, 1, 0, 1, 1, 1, 0, 0, 0, 1, 0, 1$$

In this sequence of 40 random numbers, there are 15 0s and 25 1s. In a large number of trials, because the coin is fair, we expect 50% of the random numbers to be 0s and 50% to be 1s. In any particular short sequence, the actual percentage of 0s will fluctuate around the expected number of 50%.

At this point, we can use the computer to simulate a coin tossing experiment. There are special routines on the computer for generating random numbers. We illustrate this by using the MINITAB$^{©}$ software package. The following commands generate a sequence of 50 random binary numbers (0 and 1):

```
MTB> Random 50 C1;
SUBC> Bernoulli .5.
MTB> print C1
```

The first command generates 50 random numbers and puts them in a column labelled C1. The subcommand is that these numbers should be either 0 or 1, with equal likelihood: 0.5. The computer actually generates a real number, U, in the (0, 1) interval. If $U \leq 0.5$, then the value 0 is stored; otherwise the value 1 is stored. Execute these commands to see another random sequence of 50 0s and 1s. Compare this sequence to the one given earlier. ∎

EXAMPLE **2.3**   Another example of a random phenomenon is illustrated in Figure 2.1: 50 measurements of the length of steel rods. These data are stored in the file STEELROD.DAT.

FIGURE 2.1

Length of 50 steel rods (in cm)

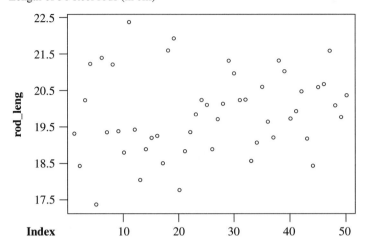

Steel rods are used in the car and truck industry to strengthen the structures of these vehicles. Assembly line automation has created stringent requirements as to the physical dimensions of parts. Steel rods supplied by Urdon Industries for Peugot car plants, for example, are produced by a process that creates rods that are 20 cm in length. However, due to natural fluctuations in the production process, the actual length of the rods varies around the nominal value of 20 cm. Examination of this sequence of 50 values does not reveal any systematic fluctuations. We conclude that deviations from the nominal values are random, and thus it is impossible to predict with certainty what the values of additional measurements will be. However, we shall learn later that we can analyze these data further and ascribe a higher likelihood to new observations falling close to 20 cm rather than strongly deviating from the nominal value.

It is possible for a situation to arise in which, at some time, the process will start to malfunction, causing a shift to occur in the average value of the process. The pattern of variability might then look like the one in Figure 2.2. An examination of this figure shows that a significant shift has occurred in the level of the process after the 25th observation but that the systematic deviation from the average value has persisted. These deviations from the nominal level of 20 cm are first just random and later systematic and random. The steel rods obviously became shorter. A quick investigation revealed that the process was accidentally maladjusted by a manager who played with machine knobs while showing the plant to important guests. ■

In formal notation, if X_i is the value of the ith observation, then

$$X_i = \begin{cases} O + E_i & i = 1, \ldots, 25 \\ N + E_i & i = 26, \ldots, 50 \end{cases}$$

FIGURE 2.2

Level shift after the first 25 observations

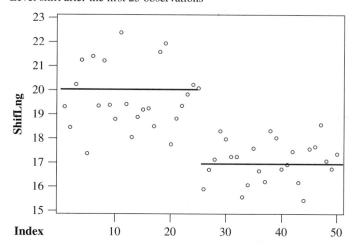

where $O = 20$, the original level of the process; $N = 17$, its new level after the shift; and E_i is a random component. Note that O and N are fixed and, in this case, constant nonrandom levels. Thus, a random sequence can consist of values that have two components: a **fixed component** and a **random component**. A fixed nonrandom pattern is called a **deterministic pattern**. As another example, in Figure 2.3 we present a sequence of 50 values of

$$X_i = D_i + E_i \qquad i = 1, \ldots, 50$$

FIGURE 2.3

Random variation around a deterministic trend

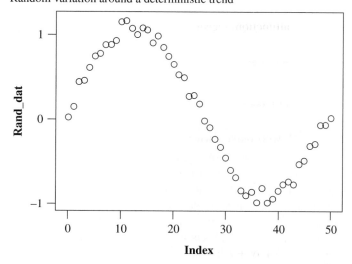

where the D_i's follow a sinusoidal pattern and E_i's are random deviations with the same characteristics as those in Figure 2.1. The sinusoidal pattern is $D_i = \sin(2\pi i/50)$, $i = 1, \ldots, 50$. Because this component can be determined exactly for each i, it is deterministic, whereas E_i is a random component. If the random component could be eliminated, we would be able to predict exactly the future values of X_i. For example, by following the pattern of the D_i's we could determine that X_{100} would be equal to 0. However, because of the random component, an exact prediction is impossible. Nevertheless, we expect that the actual values will fall around the deterministic pattern. In fact, using methods that will be discussed later, certain prediction limits can be assigned.

2.2
Accuracy and Precision of Measurements

Different measuring instruments and gauges (such as weight scales and voltmeters) may have different characteristics. For example, we say that an instrument is **accurate** if repetitive measurements of the same object yield an average equal to its true value. An instrument is **inaccurate** if it yields values whose average is different from the true value. **Precision**, on the other hand, is related to the dispersion of the measurements around their average. In particular, a small dispersion of the measurements reflects high precision; large dispersion reflects low precision. It is possible for an instrument to be inaccurate but precise, or accurate but imprecise. Other properties of measuring devices or gauges, such as reproducibility, stability, and so on, will not be discussed here. In this connection, note that instruments need to be calibrated periodically relative to an external standard. The National Institute of Standards and Technologies (NIST) is responsible for such calibrations.

EXAMPLE **2.4**   Figure 2.4 shows weight measurements of an object whose true weight is 5 kg. Ten measurements were taken on each of three instruments. Instrument A is accurate (the average is 5.0 kg.), but it shows considerable dispersion. Instrument B is not accurate (the average is 2.0 kg.), but it is more precise than A. Instrument C is as accurate as A, but it is more precise than A.   ∎

2.3
The Population and the Sample

A **statistical population** is a collection of units with a certain common attribute. For example, the set of all the citizens of the United States on January 1, 1996, is a statistical population. Such a population is composed of many subpopulations—for example, all males in the age group 19–25 living in Illinois. Another statistical population is the collection of all concrete cubes of specified dimensions that can be produced under well-defined conditions. The first example, all the citizens of the United States on January 1, 1996, is a *finite* and *real* population, whereas the

F I G U R E   2.4

Samples of 10 measurements on three instruments

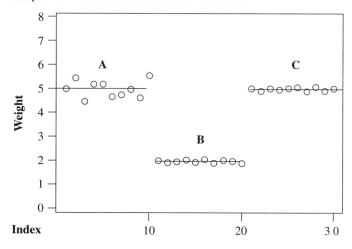

population of all units that can be produced by a specified manufacturing process is *infinite* and *hypothetical*.

A **sample** is a subset of the elements of a given population. A sample is usually drawn from a population for the purpose of observing its characteristics and making some statistical decisions concerning the corresponding characteristics of the whole population. For example, consider a lot of 25,000 special screws that were shipped by a vendor to factory A. Factory A must decide whether to accept and use this shipment or reject it (according to the provisions of the contract). Suppose it is agreed that, if the shipment contains no more than 4% defective items, it should be accepted and, if there are more than 4% defectives, the shipment should be rejected and returned to the supplier. Because it is impractical to test each item of this lot (although it is finite and real), the decision as to whether to accept the lot is based on the number of defective items found in a **random sample** drawn from the lot. Such procedures for making statistical decisions are called *acceptance sampling methods*. Chapter 9 presents these methods in greater detail.

In the next section, we discuss how a random sample is drawn in practice.

2.3.1   Drawing a Random Sample from a Finite Population

Consider a finite population consisting of N distinct elements. We first make a list of all the elements of the population, which are labeled for identification purposes. Suppose we wish to draw a random sample of size n from this population, where $1 \leq n \leq N$. We distinguish between two methods of random sampling: (a) *sampling with replacement* and (b) *sampling without replacement*. A sample drawn with replacement is obtained by returning the selected element, after each choice, to the population before the next item is selected. In this method of sampling, there are altogether N^n possible samples. A sample is called **random sample with replacement (RSWR)** if it is drawn by a method that gives every possible sample the same chance (probability) to be drawn. A sample is without replacement if an element drawn is not replaced and hence cannot be drawn again. From a population of size N, $N(N-1)\cdots(N-n+1)$ possible samples of size n can be drawn without replacement. If each of these has the same chance of being drawn, the sample is called **random sample without replacement (RSWOR)**.

Practically speaking, the choice of a particular random sample is accomplished with the aid of random numbers, which can be generated by various methods. For example, an integer has to be drawn at random from the set $\{0, 1, \ldots, 99\}$. If we had a 10-faced die, we could label its faces with the numbers $0, \ldots, 9$ and cast it twice. The results of these 2 rolls (say, 1 from the first roll and 3 from the second roll) would yield a two-digit integer—say, 13. Because, in general, we do not have such a die, we could instead use a coin and flip it seven times to generate a random number between 0 and 99. To do this, we let X_j ($j = 1, \ldots, 7$) be 0 or 1, depending on whether a head or tail appeared on the jth flip of the coin. We then compute the integer I, which can assume one of the values $0, 1, \ldots, 127$, according to the formula

$$I = X_1 + 2X_2 + 4X_3 + 8X_4 + 16X_5 + 32X_6 + 64X_7$$

If we obtain a value greater than 99, we disregard this number and flip the coin again seven times. In a similar manner, a roulette wheel could also be used to generate random numbers. However, all these methods are more theoretical than practical. In actual applications, we use ready-made tables of random numbers or computer routines.

EXAMPLE **2.5**   The following 10 numbers were drawn by using a random number generator on a computer: 76, 49, 95, 23, 31, 52, 65, 16, 61, 24. These numbers form a random sample of size 10 from the set $\{0, \ldots, 99\}$. If by chance two or more numbers are the same, the sample would be acceptable if the method were RSWR. If the method were RSWOR, any number that had already been selected would be discarded. The MINITAB command for drawing a RSWR of 10 integers from the set $\{0, \ldots, 99\}$ and storing them in column C1 is:

```
MTB> Random 10 C1;
SUBC> Integer 0 99.
```

■

2.3.2 Drawing a Random Sample from an Infinite Population

If the population is infinite (hypothetical)—for example, all the products that can be obtained from a given production process—then any sample can be considered random if we can assume that the structure of the observations is

$$X_i = C + E_i \qquad i = 1, \ldots, n$$

where $X_1, \ldots, X_n$ are the observed values and $E_1, \ldots, E_n$ are independent random error components generated by an identical randomness mechanism and C is a certain constant (average level of the process). The notions of independence and identical randomness mechanisms will be treated in more detail in Chapter 4.

2.4
Descriptive Analysis of Sample Data

In this section, we discuss the first step for analyzing the results collected in a sampling process. One way of describing a distribution of sample values, which is particularly useful in large samples, is to construct a **frequency distribution** of the sample values. We distinguish between two types of frequency distributions—namely, frequency distributions of: (1) **discrete variables** and (2) **continuous variables.**

A random variable X is called *discrete* if it can assume only a finite (or at most a countable) number of different values. For example, the number of defective computer cards in a production lot is a discrete random variable. A random variable is called *continuous* if, theoretically, it can assume all possible values in a given interval. For example, the output voltage of a power supply is a continuous random variable.

2.4.1   Frequency Distributions of Discrete Random Variables

Consider a random variable X that can assume only the values $x_1, x_2, \ldots, x_k$, where $x_1 < x_2 < \cdots < x_k$. Suppose that we have made n different observations on X. The frequency of x_i $(i = 1, \ldots, k)$ is defined as the number of observations having the value x_i. We denote the frequency of x_i by f_i. Notice that

$$\sum_{i=1}^{k} f_i = f_1 + f_2 + \cdots + f_k = n$$

The set of ordered pairs

$$\{(x_1, f_1), (x_2, f_2), \ldots, (x_k, f_k)\}$$

constitutes the frequency distribution of X. We can present a frequency distribution in tabular form:

Value	Frequency
x_1	f_1
x_2	f_2
$\vdots$	$\vdots$
x_k	f_k
Total	n

It is sometimes useful to present a frequency distribution in terms of proportional frequencies p_i, which are defined by

$$p_i = f_i/n \qquad i = 1, \ldots, k$$

A frequency distribution can be presented graphically in a form called a *bar-diagram*, as shown in Figure 2.5. The height of the bar at x_j is proportional to the frequency of this value.

In addition to the frequency distribution, it is often useful to present the **cumulative frequency distribution** of a given variable. The cumulative frequency of x_i is defined as the sum of frequencies of values less than or equal to x_i. We denote this by F_i and the proportional cumulative frequencies by

$$P_i = F_i/n$$

A table of proportional cumulative frequency distributions could be represented as follows:

Value	p	P
x_1	p_1	$P_1 = p_1$
x_2	p_2	$P_2 = p_1 + p_2$
$\vdots$	$\vdots$	$\vdots$
x_k	p_k	$P_k = p_1 + \cdots + p_k = 1$
Total	1	

The graph of a cumulative frequency distribution is a step function and looks typically like that in Figure 2.6.

FIGURE 2.5

Bar diagram of a frequency distribution

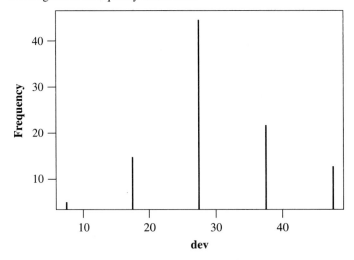

FIGURE 2.6

Step function of a cumulative frequency distribution

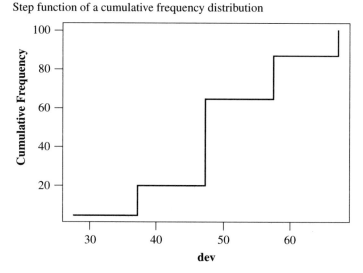

EXAMPLE **2.6**   A U.S. manufacturer of hybrid microelectronic components purchases ceramic plates from a large Japanese supplier. The plates are visually inspected before screen printing. Blemishes will affect the final product's electrical performance and its overall yield. In order to prepare a report for the Japanese supplier, the U.S. manufacturer decided to characterize the variability in the number of blemishes found on the ceramic plates. The following measurements represent the number of blemishes found on each of 30 ceramic plates:

$$0, 2, 0, 0, 1, 3, 0, 3, 1, 1, 0, 0, 1, 2, 0$$
$$0, 0, 1, 1, 3, 0, 1, 0, 0, 0, 5, 1, 0, 2, 0$$

Here the variable X assumes the values 0, 1, 2, 3, and 5. The frequency distribution of X is displayed in Table 2.2.

TABLE 2.2

Frequency distribution of blemishes on ceramic plates

x	f	p	P
0	15	.50	.50
1	8	.27	.77
2	3	.10	.87
3	3	.10	.97
4	0	.00	.97
5	1	.03	1.00
Total	30	1.00	

We did not observe the value $x = 4$, but because it seems likely to occur in future samples, we include it in the frequency distribution, with frequency $f = 0$. The bar diagram and cumulative frequency step function are shown in Figures 2.7 and 2.8, respectively. ■

FIGURE 2.7

Bar-diagram for number of blemishes on ceramic plates

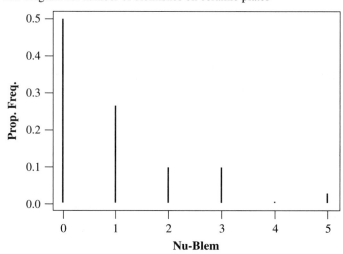

FIGURE 2.8

Cumulative frequency step-function for number of blemishes on ceramic plates

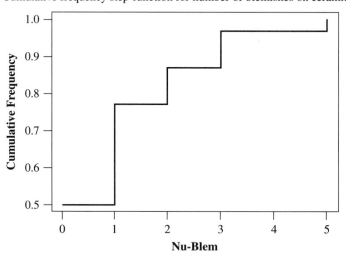

2.4.2 Frequency Distributions of Continuous Random Variables

For the case of a continuous random variable, we partition the possible range of variation of the observed variable into k subintervals. Generally speaking, if the possible range of X is between L and H, we specify numbers $b_0, b_1, b_2, \ldots, b_k$ such that $L = b_0 < b_1 < b_2 < \cdots < b_{k-1} < b_k = H$. The values $b_0, b_1, \ldots, b_k$ are called the limits of the k subintervals. We then classify the X values into the interval (b_{i-1}, b_i) if $b_{i-1} < X \le b_i$ $(i = 1, \ldots, k)$. (If $X = b_0$, we assign it to the first subinterval.) Subintervals are also called *bins*, *classes*, or *class intervals*.

In order to construct a frequency distribution, we must consider the following two questions:

1   How many subintervals should we choose?

2   How large should the width of the subintervals be?

In general, it is difficult to give to these important questions exact answers that would apply in all cases. However, the general recommendation is to use between 10 and 15 equal-width subintervals in large samples. The frequency distribution is given then for the subintervals, where the midpoint of each subinterval is a numerical representation of the interval. A typical frequency distribution table might look like the following:

Subintervals	Midpoint	Frequency	Cumulative frequency
(b_0, b_1)	$\overline{b}_1$	f_1	$F_1 = f_1$
(b_1, b_2)	$\overline{b}_2$	f_2	$F_2 = f_1 + f_2$
$\vdots$	$\vdots$	$\vdots$	$\vdots$
(b_{k-1}, b_k)	$\overline{b}_k$	f_k	$F_k = n$

EXAMPLE **2.7**   Nili, a large fiber supplier to U.S., South American, and European textile manufacturers, has tight control over its yarn strength. This critical dimension is typically analyzed on a logarithmic scale. This logarithmic transformation produces data that are more symmetrically distributed. Consider $n = 100$ values of $Y = \ln(X)$, where X is the yarn-strength (lb/22 yarns) of woolen fibers. The data are stored in file YARNSTRG.DAT and shown in Table 2.3.

The smallest value in Table 2.3 is $Y = 1.1514$ and the largest value is $Y = 5.7978$. This represents a range of $5.7978 - 1.1514 = 4.6464$. To obtain approximately 15 subintervals, we need the width of each interval to be about $4.6464/10 = 0.46$. A more convenient choice for this class width might be 0.50. The first subinterval could start at $b_0 = 0.95$ and the last subinterval could end with $b_k = 5.95$. The frequency distribution for these data is presented in Table 2.4.   ∎

A graphical representation of the distribution is given by a **histogram**, as shown in Figure 2.9. Each rectangle has a height equal to the frequency (f) frequency (p)

TABLE   **2.3**

A sample of 100 log (yarn-Strength)

2.4016	1.1514	4.0017	2.1381	2.5364
2.5813	3.6152	2.5800	2.7243	2.4064
2.1232	2.5654	1.3436	4.3215	2.5264
3.0164	3.7043	2.2671	1.1535	2.3483
4.4382	1.4328	3.4603	3.6162	2.4822
3.3077	2.0968	2.5724	3.4217	4.4563
3.0693	2.6537	2.5000	3.1860	3.5017
1.5219	2.6745	2.3459	4.3389	4.5234
5.0904	2.5326	2.4240	4.8444	1.7837
3.0027	3.7071	3.1412	1.7902	1.5305
2.9908	2.3018	3.4002	1.6787	2.1771
3.1166	1.4570	4.0022	1.5059	3.9821
3.7782	3.3770	2.6266	3.6398	2.2762
1.8952	2.9394	2.8243	2.9382	5.7978
2.5238	1.7261	1.6438	2.2872	4.6426
3.4866	3.4743	3.5272	2.7317	3.6561
4.6315	2.5453	2.2364	3.6394	3.5886
1.8926	3.1860	3.2217	2.8418	4.1251
3.8849	2.1306	2.2163	3.2108	3.2177
2.0813	3.0722	4.0126	2.8732	2.4190

TABLE   **2.4**

Frequency distribution for log yarn-Strength data

(b_{i-1}, b_i)	$\bar{b}_i$	f_i	p_i	F_i	P_i
(0.95, 1.45)	1.2	4	.04	4	.04
(1.45, 1.95)	1.7	11	.11	15	.15
(1.95, 2.45)	2.2	18	.18	33	.33
(2.45, 2.95)	2.7	21	.21	54	.54
(2.95, 3.45)	3.2	16	.16	70	.70
(3.45, 3.95)	3.7	15	.15	85	.85
(3.95, 4.45)	4.2	8	.08	93	.93
(4.45, 4.95)	4.7	5	.05	98	.98
(4.95, 5.45)	5.2	1	.01	99	.99
(5.45, 5.95)	5.7	1	.01	100	1.00

of the corresponding subinterval. The area of the rectangle is proportional to the frequency of the interval along the base.

A cumulative frequency distribution can be represented graphically by a **frequency polygon,** whose height begins at 0 and ends at n. Over the ith subinterval it is a line segment F_{i-1} with F_i $(i = 1, \ldots, k)$. The frequency polygon of the data in Table 2.3 is given in Figure 2.10.

F I G U R E   2.9

Histogram of log yarn-strength (from Table 2.4)

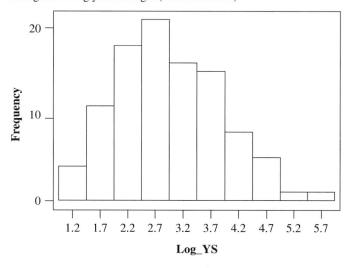

F I G U R E   2.10

Cumulative frequency polygon (from Table 2.3)

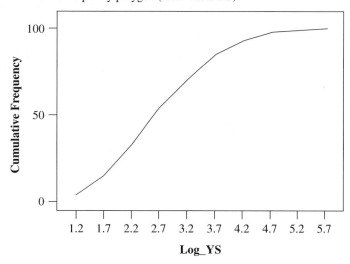

Typically, computer software selects a default midpoint and width of class interval, but allows users the option of changing these choices. The shape of the
histogram depends on the number of class intervals chosen. Students can experiment with the data set YARNSTRG.DAT, by choosing a different number of class
intervals, starting with the default value. The MTB command

```
MTB> Histogram C1;
SUBC> MidPoint;
SUBC> Bar.
```

yields six class intervals. The command

```
MTB> Histogram C1;
SUBC> MidPoint;
SUBC> N Interval 11;
SUBC> Bar.
```

yields the histogram shown in Figure 2.9.

2.4.3   Statistics of the Ordered Sample

In this section we identify characteristic values of a sample of observations that have been sorted from smallest to largest. Such sample characteristics are called **order statistics**. In general, **sample statistics** are computed from observations and are used to infer characteristics of the population from which the sample was drawn. Statistics not based on sorted sample values are discussed in Section 2.4.4.

Let $X_1, X_2, \ldots, X_n$ be the observed values of some random variable, as obtained by a random sampling process. For example, consider the following 10 values of the shear strength of stainless steel welds (lb/weld): 2385, 2400, 2285, 2765, 2410, 2360, 2750, 2200, 2500, 2550. What can we do to characterize the variability and location of these values?

The first step is to sort the sample values in increasing order. That is, we rewrite the list of sample values as: 2200, 2285, 2360, 2385, 2400, 2410, 2500, 2550, 2750, 2765. These ordered values are denoted by $X_{(1)}, X_{(2)}, \ldots, X_{(n)}$, where $X_{(1)} = 2200$ is the smallest value in the sample, $X_{(2)} = 2285$ is the second smallest, and so on. We call $X_{(i)}$ the *i-th order statistic* of the sample. For convenience, we can also denote the average of consecutive order statistics by

$$X_{(i.5)} = (X_{(i)} + X_{(i+1)})/2 = X_{(i)} + .5(X_{(i+1)} - X_{(i)}) \tag{2.4.1}$$

For example, $X_{(2.5)} = (X_{(2)} + X_{(3)})/2$. Some characteristic values that depend on these order statistics include the sample minimum, the sample maximum, the sample range, the sample median, and the sample quartiles. The **sample minimum** is $X_{(1)}$ and the **sample maximum** is $X_{(n)}$. In our example, $X_{(1)} = 2200$ and $X_{(n)} = X_{(10)} = 2765$. The **sample range** is the difference $R = X_{(n)} - X_{(1)} = 2765 - 2200 = 565$. The "middle" value in the ordered sample is called the **sample median**, denoted by M_e. The sample median is defined as $M_e = X_{(m)}$, where $m = (n+1)/2$. In our example, $n = 10$ so $m = (10 + 1)/2 = 5.5$. Thus

$$M_e = X_{(5.5)} = (X_{(5)} + X_{(6)})/2 = X_{(5)} + .5(X_{(6)} - X_{(5)})$$
$$= (2400 + 2410)/2 = 2405$$

The median characterizes the center of dispersion of the sample values and is therefore called a **statistic of central tendency**, or *location statistic*. Approximately 50% of the sample values are smaller than the median. Finally, we define the **sample quartiles** as $Q_1 = X_{(q_1)}$ and $Q_3 = X_{(q_3)}$, where

$$q_1 = \frac{(n+1)}{4}$$

and

$$q_3 = \frac{3(n+1)}{4}$$

(2.4.2)

Q_1 is called the *lower quartile* and Q_3 is called the *upper quartile*. These quartiles divide the sample so that approximately one-fourth of the values are smaller than Q_1, one-half are between Q_1 and Q_3, and one-fourth are greater than Q_3. In our example, $n = 10$ so

$$q_1 = \frac{11}{4} = 2.75$$

and

$$q_3 = \frac{33}{4} = 8.25$$

Thus, $Q_1 = X_{(2.75)} = X_{(2)} + .75 \times (X_{(3)} - X_{(2)}) = 2341.25$ and $Q_3 = X_{(8.25)} = X_{(8)} + .25 \times (X_{(9)} - X_{(8)}) = 2600$.

These sample statistics can be obtained from a frequency distribution using a cumulative frequency polygon, as shown in Figure 2.11, based on the log yarn-strength data of Table 2.3. Using linear interpolation within the subintervals, we

FIGURE 2.11

A cumulative frequency polygon with linear interpolation lines indicated

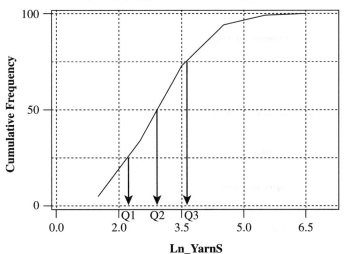

obtain $Q_1 = 2.3$, $Q_3 = 3.6$, and $M_e = 2.9$. These estimates are only slightly different from the exact values $Q_1 = X_{(25.25)} = 2.2789$, $Q_3 = X_{(75.75)} = 3.5732$, and $M_e = X_{(50.5)} = 2.8331$.

The sample median and quartiles are specific forms of a class of statistics known as **sample quantiles**. The *pth sample quantile* is a number that exceeds exactly $100p\%$ of the sample values. Hence, the median is the .50th sample quantile, Q_1 is the .25th sample quantile, and Q_3 is the .75th sample quantile. We may be interested, for example, in the .90th sample quantile. Using linear interpolation in Figure 2.11, we obtain the value 4.5, whereas the value of $X_{(90.9)} = 4.3019$. The *p*th sample quantile is also called the *100pth sample percentile*.

The MINITAB command

MTB> Describe C1

yields, among others, the following statistics of the data in column *C1*: Median, Min, Max, Q_1, and Q_3. Applying this command to the piston cycle time of file CYCLT.DAT, we find $n = 50$, $M_e = 0.5455$, $X_{(1)} = 0.1750$, $Q_1 = 0.2982$, $Q_3 = 1.0718$, and $X_{(50)} = 1.1410$.

2.4.4 Statistics of Location and Dispersion

Given a sample of n measurements, $X_1, \ldots, X_n$, we can compute various statistics to describe the distribution. The **sample mean** is determined by the formula

$$\overline{X} = \frac{1}{n} \sum_{i=1}^{n} X_i \tag{2.4.3}$$

Like the sample median, $\overline{X}$ is a measure of central tendency. In physics, the sample mean represents the center of gravity of a system consisting of n equal-mass particles located on the points X_i on the line.

As an example, consider the following measurements, representing component failure times in hours since initial operation:

45, 60, 21, 19, 4, 31

The sample mean is

$$\overline{X} = (45 + 60 + 21 + 19 + 4 + 31)/6 = 30$$

To measure the spread of data about the mean, we typically use the **sample variance**, defined by

$$S^2 = \frac{1}{n-1} \sum_{i=1}^{n} (X_i - \overline{X})^2 \tag{2.4.4}$$

or the **sample standard deviation**, given by

$$S = \sqrt{S^2}$$

The sample standard deviation is used more often because its units (cm, lb) are the same as those of the original measurements. In the next section, we will discuss some ways of interpreting the sample standard deviation. For now, simply note that data sets with greater dispersion about the mean have larger standard deviations. The computation of S^2 is illustrated in Table 2.5 using the failure time data.

The sample standard deviation and sample mean provide information on the variability and central tendency of observations. For the data set (number of blemishes on ceramic plates) in Table 2.2, $\overline{X} = 0.933$ and $S = 1.258$. Looking at Figure 2.7 we note a marked asymmetry in the data. In 50% of the ceramic plates there were no blemishes and in 3% there were five blemishes. In contrast, consider the histogram of log yarn-strength in Figure 2.9, which shows remarkable symmetry, with $\overline{X} = 2.9238$ and $S = 0.93776$. The difference in shape is obviously not reflected by $\overline{X}$ and S. Additional information pertaining to the shape of a distribution of observations is derived from the **sample skewness** and **sample kurtosis**. The sample skewness is defined as the index

$$\beta_3 = \frac{1}{n} \sum_{i=1}^{n} (X_i - \overline{X})^3 / S^3 \qquad \text{(2.4.5)}$$

The sample kurtosis (steepness) is defined as

$$\beta_4 = \frac{1}{n} \sum_{i=1}^{n} (X_i - \overline{X})^4 / S^4 \qquad \text{(2.4.6)}$$

These indices can be computed in MINITAB by the following commands:

```
MTB> Let k1 = std(C1)
MTB> Let k2 = mean((C1 − mean(C1)) ∗ ∗3)
MTB> Let k3 = k2/(k1 ∗ ∗3)
MTB> Let k4 = mean((C1 − mean(C1)) ∗ ∗4)
```

TABLE 2.5

Computing the sample variance

X	$(X - \overline{X})$	$(X - \overline{X})^2$
45	15	225
60	30	900
21	−9	81
19	−11	121
4	−26	676
31	1	1
Sum 180	0	2004

$$\overline{X} = 180/6 = 30$$
$$S^2 = 2004/5 = 400.8$$

```
MTB> Let k5 = k4/(k1 * *4)
MTB> Print k3 k5
```

Notice that the letters $k1, k2, \ldots$ represent MINITAB constants, whereas $C1, C2, \ldots$ represent columns (arrays). The **LET** command defines a constant k or a column C in terms of operations performed on other constants or columns. ** stands for an exponentiation of constants or of terms of a column. In the preceding commands $k3$ is the skewness and $k5$ the kurtosis. These two statistics are provided by most statistical computer packages. If a distribution is symmetric (around its mean), then skewness = 0. If skewness > 0, we say that the distribution is positively skewed or skewed to the right. If skewness < 0, then the distribution is negatively skewed or skewed to the left. In distributions that are positively skewed, $\overline{X} > M_e$; in those that are negatively skewed, $\overline{X} < M_e$. In symmetric distributions, $\overline{X} = M_e$.

The steepness of a distribution is determined relative to that of the normal (Gaussian) distribution (described in Section 2.5 and discussed in detail in Section 4.4.2.1). In a normal distribution, kurtosis = 3. Thus, if kurtosis > 3, the distribution is called *steep*. If kurtosis < 3, the distribution is called *flat*. Schematic representations of these shapes are given in Figures 2.12 and 2.13.

To illustrate these statistics, we computed $\overline{X}$, S^2, S, skewness, and kurtosis for the log yarn-strength data of Table 2.3. We obtained

$$\overline{X} = 2.9238$$
$$S^2 = 0.8790 \quad S = 0.93776$$
$$\text{Skewness} = 0.4040 \quad \text{Kurtosis} = 2.8747$$

The sample mean is $\overline{X} = 2.9238$ for values on a logarithmic scale. To return to the original scale (lb/22 yarns), we use the measure

$$G = \exp\{\overline{X}\}$$
$$= \left(\prod_{i=1}^{n} Y_i\right)^{1/n} = 18.6177 \qquad \textbf{(2.4.7)}$$

where $Y_i = \exp(X_i)$, $i = 1, \ldots, n$. The measure G is called the **geometric mean** of Y. The geometric mean, G, is defined only for positive variables. One can prove the following general result:

$$G \leq \overline{X}$$

Equality holds only if all values in the sample are the same.

Other statistics that measure dispersion are

1 **Interquartile range** (IQR)

$$\text{IQR} = Q_3 - Q_1 \qquad \textbf{(2.4.8)}$$

2 **Coefficient of variation**

$$\gamma = S/|\overline{X}| \qquad \textbf{(2.4.9)}$$

F I G U R E   2.12
Distributions skewness

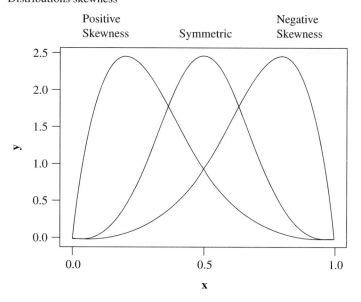

F I G U R E   2.13
Distribution steepness

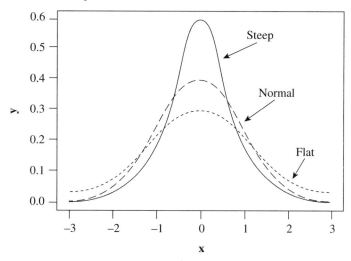

The interquartile range (IQR) is a useful measure of dispersion when there are extreme values (outliers) in the sample. It is easy to compute and can yield an estimate of S; for more details, see Section 2.6.4. The coefficient of variation is a dimensionless index used to compare the variability of different data sets when the standard deviation tends to grow with the mean. The coefficient of variation of the log yarn-strength data is $\gamma = 0.938/2.924 = 0.32$.

2.5

Prediction Intervals

When the data $X_1, \ldots, X_n$ represent a sample of observations from some population, we can use the sample statistics discussed in the previous sections to predict how future measurements will behave. Of course, our ability to predict accurately depends on the size of the sample.

Prediction intervals based on order statistics are very simple and are valid for any type of distribution. Because the ordered measurements partition the real line into $n + 1$ subintervals,

$$(-\infty, X_{(1)}), (X_{(1)}, X_{(2)}), \ldots, (X_{(n)}, \infty)$$

we can predict that $100/(n + 1)\%$ of all future observations will fall in any one of these subintervals; hence, $100i/(n + 1)\%$ of future sample values are expected to be less than the ith order statistic $X_{(i)}$. It is interesting to note that the sample minimum, $X_{(1)}$, is *not* the smallest possible value. Instead, we expect to see one out of every $n + 1$ future measurements to be less than $X_{(1)}$. Similarly, one out of every $n + 1$ future measurements is expected to be greater than $X_{(n)}$. For large values of n, one can determine values i and j, $1 \leq k < j \leq n$ so that $(X_{(i)}, X_{(j)})$ is a prediction interval expected to contain $100j\%$ of future observations.

Predicting future measurements using sample skewness and kurtosis is a bit more difficult because it depends on the type of distribution the data follow. If the distribution is symmetric (skewness $\cong 0$) and somewhat bell-shaped, or normal* (kurtosis $\cong 3$), as in Figure 2.9, for the log yarn-strength data, we can make the following statements:

1 Approximately 68% of all future measurements will lie within *1 standard deviation* of the mean

2 Approximately 95% of all future measurements will lie within *2 standard deviations* of the mean

3 Approximately 99.7% of all future measurements will lie within *3 standard deviations* of the mean.

The sample mean and standard deviation for the log yarn-strength measurements are $\overline{X} = 2.92$ and $S = 0.94$. Hence, we predict that 68% of all future measurements will lie between $\overline{X} - S = 1.98$ and $\overline{X} + S = 3.86$; 95% of all future observations will be between $\overline{X} - 2S = 1.04$ and $\overline{X} + 2S = 4.80$; and 99.7% of all future observations will be between $\overline{X} - 3S = 0.10$ and $\overline{X} + 3S = 5.74$. For the data in Table 2.3, exactly 69, 97, and 99 of the 100 values fall in these intervals, respectively.

When the data do not follow a normal distribution, we may use the following result:

---

*The normal, or Gaussian, distribution is defined in Chapter 4.

Chebyshev's inequality: For any number $k > 1$, the percentage of future measurements within k standard deviations of the mean will be at least $100(1 - 1/k^2)\%$.

This means that at least 75% of all future measurements will fall within 2 standard deviations ($k = 2$). Similarly, at least 89% will fall within 3 standard deviations ($k = 3$). These statements are true for any distribution; however, the actual percentages may be considerably larger. Notice that for data which are normally distributed, 95% of the values fall in the interval $[\overline{X} - 2S, \overline{X} + 2S]$. The Chebyshev inequality gives only the lower bound of 75%, and is therefore very conservative in some cases.

Any prediction statements using order statistics or the sample mean and standard deviation can be made only with the understanding that they are based on a sample of data. They are accurate only to the degree that the sample is representative of the entire population. When the sample size is small, we cannot be very confident in our prediction. For example, if a prediction is based on a sample of size $n = 10$, and we find that $\overline{X} = 20$ and $S = 0.1$, then we might make the statement that 95% of all future values will be between $19.8 = 20 - 2(0.1)$ and $20.2 = 20 + 2(0.1)$. However, a second sample may produce $\overline{X} = 20.1$ and $S = 0.15$. The new prediction interval would be wider than 19.8 to 20.2 and would represent a considerable change. Also, a sample of size 10 is not large enough to establish that the data have a normal distribution. With larger samples, say $n > 100$, we may be able to draw this conclusion with greater confidence.

In Chapters 6 and 7, we will discuss theoretical and computerized statistical inference whereby we assign a "confidence level" to such statements. This confidence level depends on the sample size. Prediction intervals that are correct with high confidence are called *tolerance intervals*.

2.6
Additional Techniques of Exploratory Data Analysis

In this section, we introduce modern graphical techniques that are quite common today in exploratory data analysis. These techniques are the *box and whiskers plot*, the *quantile plot*, and the *stem-and-leaf diagram*. We also discuss the problem of sensitivity of the sample mean and standard deviation to outlying observations, and introduce some robust statistics.

2.6.1 Box and Whiskers Plots

The **box and whiskers plot** is a graphical presentation of data that displays various features such as location, dispersion, and skewness. A box is plotted with its lower fence at the first quartile $Q_1 = X_{(q_1)}$ and its upper fence at the third quartile $Q_3 = X_{(q_3)}$. Inside the box, a line is drawn at the median, M_e, to indicate the statistics of central location. The interquartile range, $Q_3 - Q_1$, which is the length of the box, is a measure of dispersion. Two whiskers are extended from the box. The lower whisker is extended toward the minimum, $X_{(1)}$, but not lower than 1 1/2 the interquartile range—that is,

$$\text{Lower whisker starts} = \max\{X_{(1)}, Q_1 - 1.5(Q_3 - Q_1)\} \qquad \textbf{(2.6.1)}$$

Similarly,

$$\text{Upper whisker ends} = \min\{X_{(n)}, Q_3 + 1.5(Q_3 - Q_1)\} \qquad \textbf{(2.6.2)}$$

Data points beyond the lower or upper whiskers are considered outliers.

EXAMPLE 2.8   In Figure 2.14, we present the box-whiskers plot of the log yarn-strength data of Table 2.3. For these data, we find the following summarizing statistics:

$$X_{(1)} = 1.151$$
$$Q_1 = 2.279$$
$$M_e = 2.833$$
$$Q_3 = 3.573$$
$$X_{(100)} = 5.798$$
$$Q_3 - Q_1 = 1.294$$

In the box-whiskers plot, the endpoint of the lower whisker is at $\max\{1.151, 0.338\} = X_{(1)}$. The upper whisker ends at $\min\{5.798, 5.4514\} = 5.4514$. Thus, $X_{(100)}$ is an outlier. We conclude that this one measurement of yarn strength, which seems to be exceedingly large, is an outlier (could have been an error of measurement).   ∎

FIGURE 2.14
Box and whiskers plot of log yarn-strength data

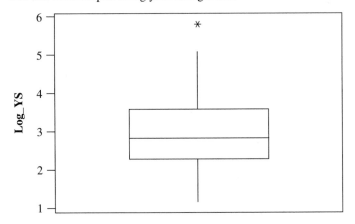

2.6.2   Quantile Plots

A **quantile plot** is a plot of the sample quantiles x_p against p, $0 < p < 1$, where $x_p = X_{(i_p)}$, where $i_p = \max(1, [p(n + 1)])$. In Figure 2.15 we see the quantile plot

FIGURE 2.15

Quantile plot of log yarn-strength data

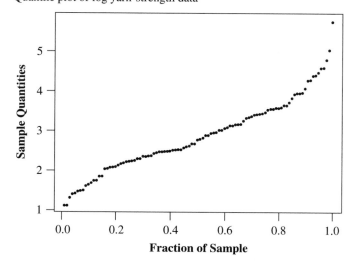

of the log yarn-strength. From such a plot, we can obtain graphical estimates of the quantiles of the distribution. For example, from Figure 2.15, we immediately obtain the estimate 2.8 for the median, 2.23 for the first quartile, and 3.58 for the third quartile. These are close to the values presented earlier. We also see in Figure 2.15 that the maximal point of this data set is an outlier. Portions of the graph which are linear indicate uniform distribution of data on corresponding intervals. Thus almost 50% of the data points ($x_{.4}$ to $x_{.9}$) are uniformly distributed. The data between $x_{.1}$ and $x_{.4}$ tend to be larger (closer to the M_e) than those of a uniform distribution, and the largest 10% of the data values ($> x_{.9}$) tend to be also larger (further from the M_e) than those of a uniform distribution. This explains the slight positive skewness of the data, as seen in Figure 2.14.

2.6.3 Stem-and-Leaf Diagrams

The following is a stem-and-leaf display of the log yarn-strength data:

Character Stem-and-Leaf Display

Stem-and-leaf of log y, $N = 100$, leaf unit $= 0.10$

```
   5  1  11344
  15  1  5556677788
  34  2  00111122222233344444
 (21) 2  555555555566677888999
  45  3  000011112223344444
  27  3  5556666677789
  14  4  00013344
   6  4  5668
   2  5  0
   1  5  7
```

In order to construct a **stem-and-leaf diagram**, the data are classified into class intervals of equal length, as in a histogram. The 100 values in Table 2.3 start with $X_{(1)} = 1.151$ and end with $X_{(100)} = 5.798$. The stem-and-leaf diagram uses only the first two digits at the left, without rounding. All values between 1.0 and 1.499 are represented in the first class as 1.1, 1.1, 1.3, 1.4, 1.4. There are five such values, and this frequency is written at the left. The second class consists of all values between 1.5 and 1.999. There are 10 such values: 1.5, 1.5, 1.5, 1.6, 1.6, 1.7, 1.7, 1.7, 1.8, 1.8. All other classes are represented similarly. The frequency of the class to which the median, M_e, belongs is enclosed in parentheses so that it stands out. The frequencies below or above the median class are cumulative. Because the cumulative frequency (from above) of the class right after that of the median is 45, we know that the median is located right after the fifth largest value from the top of that class—namely, $M_e = 2.8$—as we saw before. Similarly, to find Q_1, note that $X_{(q_1)}$ is located at the third class from the top. It is the tenth value in that class, from the left. Thus we find $Q_1 = 2.2$. Similarly, we find that $X_{(q_3)} = 3.5$. This information cannot be obtained directly from the histogram. Thus, the stem-and-leaf diagram is an important additional tool for data analysis.

Figure 2.16 presents the stem-and-leaf diagram of the electric output data (OELECT.DAT).

F I G U R E   2.16
Stem-and-Leaf Diagram of Electric Output Data

Character Stem-and-Leaf Display
Stem-and-leaf of Elec_Out, $N = 99$, leaf unit $= 1.0$

```
    5  21  01111
   10  21  22333
   19  21  444445555
   37  21  6666666677777777777
 (22)  21  8888888888889999999999
   40  22  00000000011111111111
   21  22  22233333
   13  22  44455555
    5  22  6777
    1  22  8
```

2.6.4 Robust Statistics for Location and Dispersion

The sample mean, $\overline{X}_n$, and the sample standard deviation are both sensitive to extreme observations. To illustrate, suppose we made three observations on the sheer weld strength of steel and obtained the values 2350, 2400, 2500. The sample mean is $\overline{X}_3 = 2416.67$. What happens if the technician mistakenly punches into the computer the value 25,000 instead of 2500? The sample mean will come out as 9916.67. If the result is checked on the spot, the mistake will likely be discovered and corrected. However, if it is not checked immediately, the absurd result will remain and cause difficulties later. Also, the standard deviation will be recorded wrongly as 13,063 rather than 76.376 (the correct value). This simple example shows how sensitive the mean and the standard deviation are to outliers in the data.

To avoid such problems, a more **robust statistic** than the sample mean, $\overline{X}_n$, can be used. This statistic is the **α-trimmed mean**. A proportion α of the data is trimmed from both the lower and the upper end of the ordered sample. The mean is then computed on the remaining $(1 - 2\alpha)$ part. Let us denote by $\overline{T}_\alpha$ the α-trimmed mean. The formula of this statistic is

$$\overline{T}_\alpha = \frac{1}{N_\alpha} \sum_{j=[n\alpha]+1}^{[n(1-\alpha)]} X_{(j)} \tag{2.6.3}$$

where $[\cdot]$ denotes the integer part of the number in brackets—for example $[7.3] = 7$—and $N_\alpha = [n(1 - \alpha)] - [n\alpha]$. For example, if $n = 100$ and $\alpha = .05$, we compute the mean of the 90 ordered values, $X_{(6)}, \ldots, X_{(95)}$.

MINITAB, under the command **DESCRIBE**, also yields the trimmed mean $\overline{T}_{0.05}$, which is labeled TRMEAN.

EXAMPLE 2.9 Let us now examine the robustness of the trimmed mean. We either read C:\ISTAT\MTW\OELECT.MTW or import data file OELECT.DAT to column $C1$ of MINITAB. The command **DESCRIBE** $C1$ yields

```
MTB> Read 'C:\ISTAT\DATA\OELECT.DAT' C1.
99 ROWS READ C1
    215.406 213.616 214.756 221.706 . . .
MTB> describe C1
          N     MEAN    MEDIAN   TRMEAN   STDEV   SEMEAN
C1       99    219.25   219.10   219.21   4.00    0.40

         MIN     MAX      Q1       Q3
C1     210.90  228.99   216.80   221.71
```

We see that $\overline{X}_{99} = 219.25$ and $\overline{T}_{.05} = 219.21$. We order the sample value by using the command **SORT** $C1$ $C2$. The largest value in $C2$ is $C2(99) = 228.986$. Let us now change this value to $C2(99) = 2289.86$ (an error in punching the data) by the command **LET** $C2(99) = 2289.86$. When we apply now the command **DESCRIBE** $C2$, we obtain

```
MTB> let C2(99) = 2289.86
MTB> describe C2
```

	N	MEAN	MEDIAN	TRMEAN	STDEV	SEMEAN
C2	99	240.1	219.1	219.2	208.2	20.9

	MIN	MAX	Q1	Q3	
C2	210.9	2289.9	216.8	221.7	

By comparing the two outputs, we see that $\overline{X}_{99}$ changed from 219.25 to 240.1, S_{99} (STDEV) changed dramatically from 4.00 to 208.2 (and, correspondingly, SEMEAN $= S/\sqrt{n}$ changed).

However, M_e, $\overline{T}_\alpha$, Q_1, and Q_3 did not change at all. These statistics are called **robust** (nonsensitive) against extreme deviations (outliers). ∎

We have seen that the standard deviation S is very sensitive to observations in the extremes. A robust statistic for dispersion is

$$\tilde{\sigma} = \frac{Q_3 - Q_1}{1.3490} \tag{2.6.4}$$

The denominator 1.3490 is the distance between Q_3 and Q_1 in the theoretical normal distribution (see Chapter 4). Indeed, Q_3 and Q_1 are robust against outliers. Hence $\tilde{\sigma}$, which is about three-fourths of the IQR, is often a good statistic to replace S.

Another statistic is the **α-trimmed standard deviation**:

$$S_\alpha = \left(\frac{1}{N_\alpha - 1} \sum_{j=[n\alpha]+1}^{[n(1-\alpha)]} (X_{(j)} - \overline{T}_\alpha)^2 \right)^{1/2} \tag{2.6.5}$$

The $\alpha = .05$-trimmed STD for the OELECT data is $S_\alpha = 3.5969$. Because S_α is the standard deviation of a trimmed sample, it is robust against outliers. The value $\tilde{\sigma}$ for the sample of OELECT is $\tilde{\sigma} = 3.6324$. These two robust statistics, $\tilde{\sigma}$ and S_α, yield close results. The sample standard deviation of OELECT is $S = 4.00399$.

2.7
Chapter Highlights

The chapter focuses on statistical variability and on various methods of analyzing random data. Random results of experiments are illustrated with distinction between deterministic and random components of variability. The difference between accuracy and precision is explained. Frequency distributions are defined to represent random phenomena. Various characteristics of location and dispersion of frequency distributions are defined as well, and the elements of exploratory data analysis are presented.

The main concepts and definitions introduced in this chapter include:

- Cycle time
- Random variable
- Fixed component
- Random component

- Deterministic pattern
- Accuracy
- Precision
- Statistical population
- Sample
- Random sample
- Random sampling with replacement
- Random sampling without replacement
- Frequency distributions
- Discrete random variables
- Continuous random variables
- Cumulative frequency distribution
- Sample statistics
- Sample range

- Sample median
- Measures of central tendency
- Sample quartiles
- Sample quantiles
- Sample mean
- Sample variance
- Sample standard deviation
- Sample skewness
- Sample kurtosis
- Prediction intervals
- Box and whiskers plot
- Quantile plot
- Stem-and-leaf diagram
- Robust statistics

Exercises

2.1.1 Generate at random 50 integers from the set $\{1, 2, 3, 4, 5, 6\}$. To do this, use these MINITAB commands:

```
MTB> RANDOM 50 C1;
SUBC> INTEGER 1 6.
```

Using this method of simulation, count the number of times the different integers are repeated. This counting can be done by using the MINITAB command

```
MTB> TABLE C1
```

How many times do you expect each integer to appear if the process generates the numbers at random?

2.1.2 Construct a sequence of 50 numbers having a linear trend for deterministic components with random deviations around it. Use these MINITAB commands:

```
MTB> Set C1
```

```
DATA> 1(1 : 50/1)1
DATA> End.
MTB> Let C2 = 5 + 2.5 * C1
MTB> Random 50 C3;
SUBC> Uniform −10 10.
MTB> Let C4 = C2 + C3
MTB> Plot C4 * C1
```

Plot $C4$ against $C1$ to see the random variability around the linear trend.

2.1.3 Generate a sequence of 50 random binary numbers $(0, 1)$ when the likelihood of 1 is p, by using the command

```
MTB> RANDOM 50 C1;
SUBC> Bernoulli p.
```

Do this for the values $p = 0.1$, 0.3, 0.7, 0.9. Count the number of 1s in these random sequences by the command

SUM(C1)

2.2.1   The following two sets of measurements of an object's weight come from two different weighing instruments. The object has a true weight of 10 kg.

Instrument 1:

[1]	9.490950	10.436813	9.681357	10.996083	10.226101	10.253741
[7]	10.458926	9.247097	8.287045	10.145414	11.373981	10.144389
[13]	11.265351	7.956107	10.166610	10.800805	9.372905	10.199018
[19]	9.742579	10.428091				

Instrument 2:

[1]	11.771486	10.697693	10.687212	11.097567	11.676099	10.583907
[7]	10.505690	9.958557	10.938350	11.718334	11.308556	10.957640
[13]	11.250546	10.195894	11.804038	11.825099	10.677206	10.249831
[19]	10.729174	11.027622				

Which instrument seems to be more accurate? Which instrument seems to be more precise?

2.2.2   The quality-control department of a candy factory uses a scale to verify compliance with the weight requirements of packages. What could be the consequences of problems with the scale's accuracy, precision, and stability.

2.3.1   Draw a random sample with replacement (RSWR) of size $n = 20$ from the set of integers $\{1, 2, \ldots, 100\}$.

2.3.2   Draw a random sample without replacement (RSWOR) of size $n = 10$ from the set of integers $\{11, 12, \ldots, 30\}$.

2.3.3

a   How many words of 5 letters can be formed if $N = 26$ and $n = 5$?

b   How many words of 5 letters can be formed if all letters are different?

c   How many words of 5 letters can be written if the first and the last letters are x?

d   An electronic signal is a binary sequence of 10 zeros or ones. How many different signals are available?

e   How many electronic signals in a binary sequence of size 10 are there in which the number 1 appears exactly 5 times?

2.4.1   For each of the following variables, state whether it is discrete or continuous:

i   The number of heads resulting from 10 coin flips

ii   The number of blemishes on a ceramic plate

iii   The thickness of ceramic plates

iv   The weight of an object

2.4.2   Data file FILMSP.DAT contains data gathered from 217 rolls of film. The data consist of the film speed as measured in a special lab. Prepare a histogram of the data.

2.4.3   Data file COAL.DAT contains data on the number of yearly disasters in coal mines in England. Prepare a table of frequency distributions of the number of coal mine disasters. (You can use the MINITAB command **TABLE** C1.)

2.4.4   Data file CAR.DAT contains information on 109 different car models. For each car, there are values on five variables:

i   Number of cylinders (4, 6, 8)

ii   Origin (1, 2, 3)

iii   Turn diameter (m)

iv   Horsepower (hp)

v   Number of miles/gallon in city driving (mpg)

Prepare frequency distributions for each variable.

2.4.5   Compute the following eight quantities for the data in file FILMSP.DAT:

i   Sample minimum, $X_{(1)}$

ii   Sample first quartile, Q_1

iii   Sample median, M_e

iv   Sample third quartile, Q_3

v Sample maximum, $X_{(217)}$

vi Sample quantiles x_p for $p = .8, .9,$ and $.99$

Show how you get these statistics by using the formulas. (The order statistics of the sample can be obtained by first ordering the values of the sample. For this use the MINITAB command

```
MTB> SORT C1 C2
```

Certain order statistics can be put into constants by the commands—for example

```
MTB> Let k1 = 1
MTB> Let k2 = C2(k1)
```

The sample miminum is $C2(1)$; the sample maximum is $C2(217)$; and so on.)

2.4.6 Compute with MINITAB the indices of skewness and kurtosis of the FILMSP.DAT. Interpret the skewness and kurtosis of this sample in terms of the shape of the distribution of film speed.

2.4.7 Compare the means and standard deviations of the number of miles per gallon/city for cars by origin (1 = United States; 2 = Europe; 3 = Asia) according to the data of file CAR.DAT.

2.4.8 Compute the coefficient of variation of the turn diameter of U.S.-made cars (origin = 1) in file CAR.DAT.

2.4.9 Compare the mean $\overline{X}$ and the geometric mean G of the turn diameter of U.S. made and Japanese cars in CAR.DAT.

2.5.1 Compare the prediction proportions to the actual frequencies of the intervals

$$\overline{X} \pm kS \qquad k = 1, 2, 3$$

for the film speed data given in FILMSP.DAT file.

2.6.1 Present side by side the box plots of miles per gallon/city for cars by origin. Use data file CAR.DAT.

2.6.2 Prepare a stem-and-leaf diagram of the piston cycle time in file OTURB.DAT. Compute the five summary statistics $(X_{(1)}, Q_1, M_e, Q_3, X_{(n)})$ from the diagram.

2.6.3 Compute the trimmed mean, $\overline{T}_{0.10}$, and trimmed standard deviation, $S_{0.10}$, of the piston cycle time of file OTURB.DAT.

2.6.4 The following data are the time (in sec) to get from 0 to 60 mph for a sample of 15 German-made cars and 20 Japanese-made cars:

German-made cars			Japanese-made cars			
10.0	10.9	4.8	9.4	9.5	7.1	8.0
6.4	7.9	8.9	8.9	7.7	10.5	6.5
8.5	6.9	7.1	6.7	9.3	5.7	12.5
5.5	6.4	8.7	7.2	9.1	8.3	8.2
5.1	6.0	7.5	8.5	6.8	9.5	9.7

Compare and contrast the acceleration times of German- and Japanese-made cars, in terms of their five summary statistics.

2.6.5 Summarize variables Res 3 and Res 7 in data set HADPAS.DAT by computing sample statistics, histograms, and stem-and-leaf diagrams.

2.6.6 Are there outliers in the Res 3 data of HADPAS.DAT? Show your calculations.

3

Variability in Several Dimensions

When surveys or experiments are performed, measurements are usually taken on several characteristics of the observed elements in the sample. In such cases, we have multivariate observations, and the statistical methods used to analyze the relationships among the observed values are called *multivariate methods*. In this chapter, we introduce some of these methods. In particular, we focus on graphical methods, linear regression methods, and the analysis of contingency tables. Linear regression methods explore the linear relationship between a variable of interest and a set of variables. By using these methods, we try to predict the values of the variable of interest. Contingency table analysis studies the association among qualitative (categorical) variables on which we cannot apply the usual regression methods. We start the chapter with multivariate graphical analysis, using methods available in computer packages. We then introduce the concepts of multivariate frequency distributions, and marginal and conditional frequency distributions. Following this we present the most common methods of correlation and regression analysis, and we end with contingency table analysis. Several industrial data sets are analyzed. The complete data, if too long to be printed, are given on the accompanying disk of computer applications.

3.1
Graphical Display and Analysis
3.1.1 Scatterplots

Suppose we are given a data set consisting of N records (elements). Each record contains observed values on k variables. Some of these variables might be qualitative (categorical) and some quantitative. Scatterplots display the values of pairwise quantitative variables in two-dimensional plots.

EXAMPLE **3.1** Consider the data set PLACE.DAT. The observations are the displacements (position errors) of electronic components on printed circuit boards. The data were collected by a large U.S. manufacturer of automatic insertion machines used in the mass production of electronic devices. The components are fed to the machine on reels. A robot arm picks up the components and places them in a prescribed location on a printed circuit board. The placement of the component is controlled by a computer built into the insertion machine. There are 26 boards, with 16 components on each. Each component has to be placed at a specific location (x, y) on the board and with a correct orientation θ. Due to mechanical and other design or environmental factors, placement errors are sometimes made. It is interesting to analyze whether these errors are within the specified tolerances. There are $k = 4$ variables in the data set. The first is categorical and gives the board number. The other three variables are continuous. The variable x-dev represents a placement error along the x-axis of the system. The variable y-dev stands for a placement error along the y-axis. Tet-dev is the variable representing an error in orientation.

Figure 3.1 is a scatterplot of y-dev versus x-dev for each record. The picture immediately reveals certain unexpected clustering of the data points. The y-dev placement variable does not depend on the x-dev variable. The scatterplot of Figure 3.1 shows three distinct clusters of points, which will be investigated later.

In a similar manner we can plot the values of tet-dev against those of x-dev or y-dev. We do this by creating a multiple scatterplot, or **matrix scatterplot**. Figure 3.2 illustrates the matrix scatterplot of x-dev, y-dev, and tet-dev.

The matrix scatterplot gives us a general picture of the relationships among the three variables. Although this scatterplot does not provide a great deal of detail (more information is given in the individual scatterplots—compare Figure 3.1 to Figure 3.2), it does provide enough information to direct us into further investigation. For example, we see in Figure 3.2, that the variable tet-dev is highly concentrated at

F I G U R E 3.1

Scatterplot of y-dev vs. x-dev

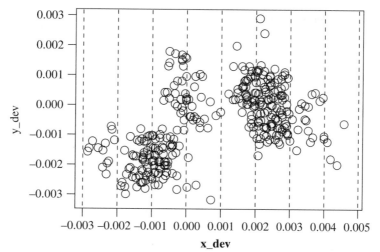

F I G U R E   3.2

Matrix scatterplot

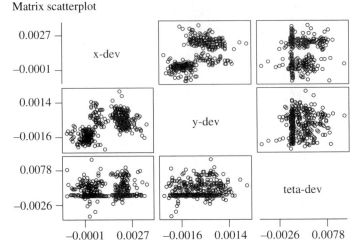

positive values. The frequency distribution of tet-dev, which is shown in Figure 3.3, reinforces this conclusion. Indeed, close to 50% of the tet-dev values are close to zero. The other values tend to be positive. The histogram in Figure 3.3 is skewed toward positive values.

A graph called a **3D scatterplot** shows the three dimensions simultaneously. An example of a 3D scatterplot is given in Figure 3.4 for the three variables x-dev (X direction), y-dev (Y direction), and tet-dev (Z direction). This plot expands the 2-dimensional scatterplot by adding a third variable horizontally. ■

F I G U R E   3.3

Frequency histogram of tet-dev

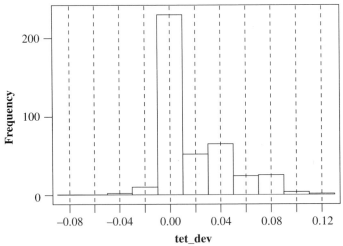

FIGURE **3.4**
3D scatterplot

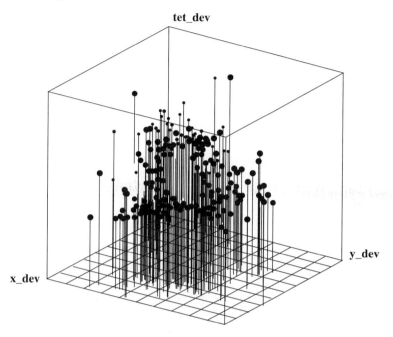

tet_dev

x_dev

y_dev

3.1.2 Multiple boxplots

The **multiple boxplot**, or side-by-side boxplot, is another graphical technique by which we can show distributions of a quantitative variable at different categories of a categorical variable.

EXAMPLE **3.2** Returning to the data set PLACE.DAT, we wish to investigate further the apparent clusters shown in Figure 3.1. As mentioned before, the data were collected in an experiment in which components were placed on 26 boards in a successive manner. The board number, board_n, is in the first column of the data set. We would like to examine whether the deviations in x, y or θ tend to change with time. To do this, we can plot the x-dev, y-dev, or tet-dev against board_n, or, for a more concise presentation, we can graph multiple boxplots by board number. In Figure 3.5, we present multiple boxplots of the x-dev against board_n. This figure presents an interesting picture. Boards 1–9 yield similar boxplots, boards 10–12 have boxplots significantly above those of the first group, and boards 13–26 have boxplots that are even higher. The groups seem to correspond to the three clusters in Figure 3.1. To verify this, we introduce a **code variable** to the data set, which assumes the value 1 if board_$n \leq 9$, the value 2 if $10 \leq$ board_$n \leq 12$, and the value 3 if board_$n \geq 13$. We then plot y-dev against x-dev again, denoting the points in the scatterplot by the code variable symbols $0, +, \times$.

F I G U R E   **3.5**

Multiple boxplots of *x*-dev versus board number

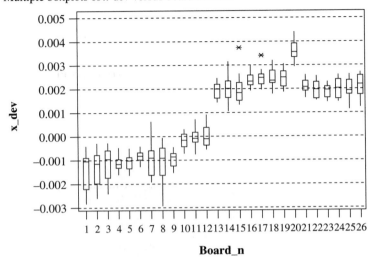

In Figure 3.6, we see this coded scatterplot. It is now clear that the three clusters are indeed formed by those three groups of boards. The differences among these groups might be due to a deficiency in the placement machine that caused the apparent time-related drift in the errors. Other possible reasons could be the composition of the printed circuit boards or different batches of raw material, such as the glue used for placing the components. ■

F I G U R E   **3.6**

Scatterplot of *y*-dev vs. *x*-dev by code variables

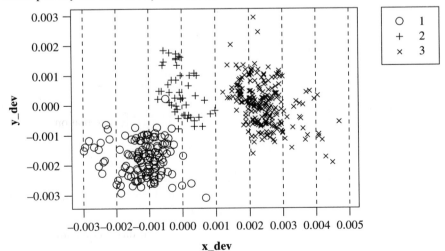

3.2
Dynamic Graphics

The application of advanced methods of graphical analysis depends on the type of software available. In this section, we discuss several methods that are available in software packages such as EXECUSTAT$^©$, S-PLUS$^©$, STATGRAPHICS$^©$, and others.

3.2.1 3D Spinning

The **3D spinning method** is a way of making stepwise or continuous changes to the orientation of the x-y-z axes in a 3D scatterplot. By changing the orientation, we spin the 3D scatterplot in various directions until a better picture is obtained of the three-dimensional scatter. We saw an example of a 3D scatterplot in Figure 3.4. The three clusters of points, which were discussed in the previous section, are not seen in Figure 3.4. By spinning the 3D scatterplot, the three-dimensional structure of these clusters becomes evident.

3.2.2 Brushing, Coding, and Transformations

Brushing is a procedure by which a scatterplot can be changed, usually by changing the color of the points to indicate the values of another variable. In brushing, the points of the scatterplot have two colors—say, red and yellow. When a color is changed, all points involved then take on the values of a specified (third) variable, satisfying a certain inequality.

EXAMPLE **3.3**   We can try to brush the y-dev versus x-dev of the data set PLACE.DAT, according to the value of tet-dev. We have already seen that the number of points having a tet-dev value smaller than zero is small. We can further see, using the brushing method, that only seven points have a tet-dev of less than -0.02 and that these seven points are distributed over the three clusters. Notice that only one point has a tet-dev value less than -0.003—in the second cluster (boards 10–12). In each of the first (boards 1–9) and third (boards 13–26) clusters, there are 11 such points. Note also that only two points in the first cluster and two in the second have a tet-dev value larger than 0.0712, whereas there are many such points in the third cluster. This shows some kind of dependence of the extremes of the distribution of tet-dev on the cluster. Because it is not possible to reproduce in print the dynamics of brushing and 3D spinning, readers should go through this section sitting at a computer terminal and attempting to confirm these observations.   ∎

We saw the **coding** option in the previous section (see Figure 3.6). EXECUSTAT and S-PLUS provide additional options, such as jitter, pt. identification, and transforms. The **transformation** option provides the possibility of changing the scale of a scatterplot by changing one of the two axes from arithmetic to logarithmic. Such transformations allow us to determine visually whether a change of scale will make the distribution of a variable less skewed. Notice that a transformation like this can be used only if all the values are positive.

Although spinning and brushing are not available in MINITAB, we can nevertheless obtain a great deal of information by 3D plotting by group variables, in which we can specify a column of code values. This is illustrated in Figure 3.7, in which a three-dimensional plotting of the placement data using x-dev, y-dev, and tet-dev, is coded by clusters, as in Figure 3.6. The three-dimensional plot reveals much more than a two-dimensional plot does. We see that clusters 2 and 3 tend to be below cluster 1 with respect to tet-dev, although almost all values of tet-dev are greater than or equal to zero.

FIGURE 3.7
3D plotting of placement deviations by cluster

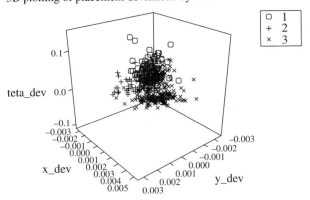

3.3
Frequency Distributions in Several Dimensions

In Chapter 2, we studied how to construct frequency distributions of single variables, either categorical or continuous. In this section, we extend those concepts to several variables simultaneously. For the sake of simplification we restrict the discussion to two variables. However, the methods presented have can be easily generalized to larger numbers of variables.

In order to enrich the examples, we introduce two additional data sets: ALMPIN.DAT and HADPAS.DAT. The ALMPIN.DAT set consists of 70 records on 6 variables based on measuring aluminum pins used in airplanes. The aluminum pins are inserted with air guns into predrilled holes in order to combine critical airplane parts such as wings, engine supports, and doors. Typical lot sizes consist of at least 1000 units, making this a prime example of the discrete mass production operations mentioned in Section 1.1. The main role of the aluminum pins is to reliably and securely connect two metal parts. The surface area where contact is established between the aluminum pins and the connected part determines the strength required to disconnect the part. A critical feature of the aluminum pin is that it fits perfectly the predrilled holes. Fixed correct diameter of the aluminum pin is therefore essential, and the pin's diameter is measured in three different locations, producing three measurements of width. Diameters 1, 2, and 3 should be all equal. Any deviation indicates lack of parallelism and therefore potential reliability problems because the surface area with actual contact will not be uniform. The measurements were taken in a computerized numerically controlled (CNC) metal-cutting operation. The six variables are Diameter 1, Diameter 2, Diameter 3, Cap Diameter, Lengthncp, and Lengthwcp. All measurements are in millimeters. The first three variables give the pin diameter at three specified locations. Cap Diameter is the diameter of the cap on top of the pin. The last two variables are the length of the pin, without the cap and with the cap, respectively.

Data set HADPAS.DAT provides several resistance measurements (ohms) of five types of resistances (Res 3, Res 18, Res 14, Res 7, and Res 20) located in six hybrid microcircuits (3 rows and 2 columns) manufactured simultaneously on ceramic substrates. Altogether there are 192 records for 32 ceramic plates.

3.3.1 Bivariate Joint Frequency Distributions

Joint frequency distribution is a function that provides the frequencies of elements (records) in the data set having values in specified intervals. More specifically, consider two variables X and Y. We assume that both variables are continuous. We partition the x-axis into k_1 subintervals (ξ_{i-1}, ξ_i), $i = 1, \ldots, k_1$, and the y-axis into k_2 subintervals (η_{j-1}, η_j), $j = 1, \ldots, k_2$. We denote by f_{ij} the number (count) of elements in the data set (sample) having X values in (ξ_{i-1}, ξ_i) and Y values in (η_{j-1}, η_j) simultaneously. The value f_{ij} is called the **joint frequency** of the rectangle $(\xi_{i-1}, \xi_i) \times (\eta_{j-1}, \eta_j)$. If N denotes the total number of elements in the data set, then obviously

$$\sum_i \sum_j f_{ij} = N \tag{3.3.1}$$

The frequencies f_{ij} can be represented in a table called a *table of the frequency distribution*. These row and column totals are called **marginal frequencies**. Generally, the marginal frequencies are

$$f_{i.} = \sum_{j=1}^{k_2} f_{ij} \qquad i = 1, \ldots, k_1 \tag{3.3.2}$$

and

$$f_{\cdot j} = \sum_{i=1}^{k_1} f_{ij} \qquad j = 1, \ldots, k_2 \tag{3.3.3}$$

These are the sums of the frequencies in a given row or a given column.

EXAMPLE **3.4**   In Table 3.1, we present the joint frequency distribution of Lengthncp and Lengthwcp of the data set ALMPIN.DAT

The row totals are the frequency distribution of Lengthncp. The column totals provide the frequency distribution of the variable Lengthwcp. Similar tabulations can be done on data set HADPAS.DAT for the frequency distributions of resistances. Table 3.2 shows the joint frequency distribution of Res 3 and Res 7.

Figure 3.8 presents the corresponding 3D histogram.   ■

TABLE   3.1
Joint frequency distribution

	Lenthwcp			
Lengthncp	**59.9–60.0**	**60.0–60.1**	**60.1–60.2**	**Row total**
49.8–49.9	8	18	0	26
49.9–50.0	5	33	2	40
50.0–50.1	0	0	4	4
Column total	13	51	6	70

TABLE   3.2
Joint frequency distribution of Res 3 and Res 7 (in ohms)

	Res 7					
Res 3	**1300–1500**	**1500–1700**	**1700–1900**	**1900–2100**	**2100–2300**	**Row Totals**
1500–1700	1	12	1	0	0	14
1700–1900	0	15	28	3	0	46
1900–2100	0	2	43	41	2	88
2100–2300	0	0	5	32	6	43
2300–2500	0	0	0	0	1	1
Column total	1	29	77	76	9	192

FIGURE   3.8
3D histogram of Res 3 and Res 7

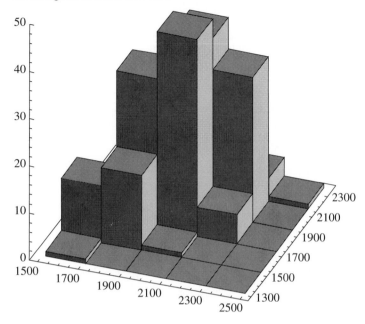

The bivariate frequency distribution also provides information on the association, or dependence, between the two variables. In Table 3.2, we see that resistance values of Res 3 tend to be similar to those of Res 7. For example, if the resistance value of Res 3 is in the interval (1500, 1700), 12 of the 14 resistance values of Res 7 are in the same interval. This association can be better illustrated by using a box and whiskers plot of the variable Res 3 by the categories (intervals) of the variable Res 7. To obtain these plots, we first partition the 192 cases into five subgroups, according to the resistance values of Res 7. The single case of Res 7 in the interval (1300, 1500) belongs to subgroup 1. The 29 cases of Res 7 values in (1500, 1700) belong to subgroup 2, and so on. We then perform an analysis by subgroups, yielding the information given in Table 3.3.

TABLE   3.3
Means and standard deviations of Res 3

Subgroup	Interval of Res 7	Sample size	Mean	Standard deviation
1	1300–1500	1	1600.0	——
2	1500–1700	29	1718.9	80.27
3	1700–1900	77	1932.9	101.17
4	1900–2100	76	2068.5	99.73
5	2100–2300	9	2204.0	115.49

We see in the table that the subgroup means grow steadily with the values of Res 7. The standard deviations do not change much. (There is no estimate of the standard deviation of subgroup 1.) A better picture of the dependence of Res 3 on the intervals of Res 7 is given by Figure 3.9, in which the boxplots of the Res 3 values are presented by subgroup.

F I G U R E   3.9

Boxplots of Res 3 by intervals of Res 7

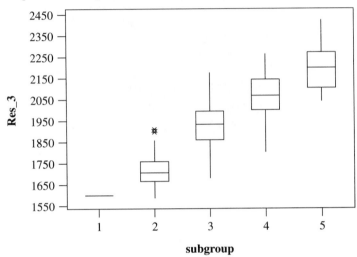

3.3.2   Conditional Distributions

Consider a population (or sample) of elements. Each element assumes random values of two (or more) variables $X, Y, Z, \ldots$. The distribution of X over elements whose Y value is restricted to a given interval (or set) A is called the **conditional distribution** of X, given Y is in A. If the conditional distributions of X given Y are different from the marginal distribution of X, we say that the variables X and Y are **statistically dependent**. We will learn later how to test whether the differences between the conditional distributions and the marginal ones are significant, and not just due to randomness in small samples.

EXAMPLE **3.5**   If we divide the frequencies in Table 3.2 by their column sums, we obtain the proportional frequency distributions of Res 3, given the intervals of Res 7. In Table 3.4, we compare these conditional frequency distributions with the marginal frequency distribution of Res 3. We see in Table 3.4 that the proportional frequencies of the conditional distributions of Res 3 depend strongly on the intervals of Res 7 to which they are restricted.   ■

TABLE 3.4

Conditional and marginal frequency distributions of Res 3

Res 3	Res7					Marginal distribution
	1300–1500	1500–1700	1700–1900	1900–2100	2100–2300	
1500–1700	100.0	44.8	1.2	0	0	7.8
1700–1900	0	51.7	38.3	1.4	0	24.5
1900–2100	0	3.4	54.3	55.6	22.2	45.3
2100–2300	0	0	6.2	43.0	66.7	21.9
2300–2500	0	0	0	0	11.1	0.5
Column Sums	100.0	100.0	100.0	100.0	100.0	100.0

3.4
Correlation and Regression Analysis

In the previous sections, we presented various graphical procedures for analyzing multivariate data. In particular, we examined multivariate scatterplots, 3-dimensional histograms, conditional boxplots, and so on. In this section, we begin numerical analysis of multivariate data.

3.4.1 Covariances and Correlations

The statistic that summarizes the simultaneous variability of two variables is called the **sample covariance**. This is a generalization of the sample variance statistic, S_x^2, of one variable, X. We denote the sample covariance of two variables, X and Y, by S_{xy}. The formula of S_{xy} is

$$S_{xy} = \frac{1}{n-1} \sum_{i=1}^{n} (X_i - \overline{X})(Y_i - \overline{Y}) \tag{3.4.1}$$

where $\overline{X}$ and $\overline{Y}$ are the sample means of X and Y, respectively. Notice that S_{xx} is the sample variance S_x^2, and S_{yy} is S_y^2. The sample covariance can assume positive or negative values. If one of the variables, say X, assumes a constant value c for all X_i ($i = 1, \ldots, n$), then $S_{xy} = 0$. This can be immediately verified, because $\overline{X} = c$ and $X_i - \overline{X} = 0$ for all $i = 1, \ldots, n$.

It can be proven that, for any variables X and Y,

$$S_{xy}^2 \leq S_x^2 \cdot S_y^2 \tag{3.4.2}$$

This inequality is the celebrated *Schwarz inequality*. Although a simple proof of this inequality can be presented here, we will deduce the inequality indirectly from a later result. By dividing S_{xy} by $S_x \cdot S_y$, we obtain a standardized index of dependence called the **sample correlation** (Pearson's product-moment correlation)—namely

$$R_{xy} = \frac{S_{xy}}{S_x \cdot S_y} \tag{3.4.3}$$

From the Schwarz inequality, the sample correlation always assumes values between -1 and $+1$. In Table 3.5 we present the sample covariances of the six variables measured on the aluminum pins. Because $S_{xy} = S_{yx}$ (symmetric statistic), it is sufficient to present the values at the bottom half of the table (on and below the diagonal).

EXAMPLE 3.6   Tables 3.5 and 3.6 show the sample covariances and sample correlations in the data file ALMPIN.DAT.

TABLE 3.5
Sample covariances of aluminum pin variables $\times$ 100

			Y			
X	**Diameter 1**	**Diameter 2**	**Diameter3**	**Cap diameter**	**Lengthncp**	**Lengthwcp**
Diameter 1	0.0270					
Diameter 2	0.0283	0.0324				
Diameter 3	0.0255	0.0284	0.0275			
Cap diameter	0.0291	0.0311	0.0286	0.0361		
Lengthncp	−0.0165	−0.0204	−0.0144	−0.0138	0.1907	
Lengthwcp	−0.0326	−0.0409	−0.0333	−0.0322	0.1546	0.2307

TABLE 3.6
Sample Correlations of Aluminum Pins Variables

			Y			
X	**Diameter 1**	**Diameter 2**	**Diameter3**	**Cap diameter**	**Lengthncp**	**Lengthwcp**
Diameter 1	1.000					
Diameter 2	.957	1.000				
Diameter 3	.935	.949	1.000			
Cap diameter	.932	.909	.907	1.000		
Lengthncp	−.230	−0.260	−.179	−.166	1.000	
Lengthwcp	−.413	−0.473	−.417	−.353	.737	1.000

We see from the table that the sample correlations among Diameter 1, Diameter 2, Diameter 3, and Cap Diameter are all greater than .9. Figure 3.10 (the multivariate scatterplots) shows that the points of these variables are scattered close

F I G U R E   3.10

Multiple scatterplots of the aluminum pin measurements

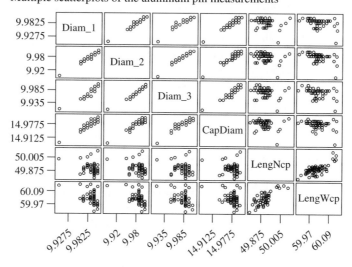

to straight lines. However, no clear relationship is evident between the first four variables and the length of the pin (with or without the cap). Negative correlations usually indicate that the points are scattered around a straight line having a negative slope. In this case, the magnitude of the negative correlations is due to the one outlier (pin 66). If we delete this outlier, the correlations are reduced in magnitude, as shown in Table 3.7.

T A B L E   3.7

Sample correlations of aluminum pin variables, after excluding outlying observation 66

X	Diameter 1	Diameter 2	Diameter3	Cap diameter	Lengthncp
			Y		
Diameter 2	.923				
Diameter 3	.922	.937			
Cap diameter	.875	.838	.874		
Lengthncp	−.103	−.150	−.093	.019	
Lengthwcp	−.313	−.396	−.328	−.229	.720

We see that the correlations between the three diameter variables and Length-nocp are much closer to zero after the outlier is excluded. Moreover, the correlation with Cap Diameter changed signs. This shows that the sample correlation, as just defined, is sensitive to the influence of extreme observations (outliers). More robust correlation statistics will be discussed in Section 3.4.2.3. ∎

An important question to ask is: How significant is the value of the correlation statistic? In other words, what is the effect on the correlation of the random components of the measurements? If $X_i = \xi_i + e_i$, $i = 1, \ldots, n$, where ξ_i are deterministic components and e_i are random, and if $Y_i = \alpha + \beta \xi_i + f_i$, $i = 1, \ldots, n$, where α and β are constants and f_i are random components, how large can the correlation between X and Y be if $\beta = 0$?

Questions that deal with assessing the significance of results will be discussed in Chapter 6.

3.4.2   Fitting Simple Regression Lines to Data

We have seen examples in which the relationship between two variables, X and Y, is close to linear. This is the case when the (x, y) points scatter along a straight line. Suppose we are given n pairs of observations $\{(x_i, y_i), i = 1, \ldots, n\}$. If the Y observations are related to those on X according to the **linear model**

$$y_i = \alpha + \beta x_i + e_i \qquad i = 1, \ldots, n \tag{3.4.4}$$

where α and β are constant coefficients and e_i are random components, with zero mean and constant variance, we say that Y relates to X according to a **simple linear regression**. The coefficients α and β, are called the *regression coefficients*. α is the *intercept* and β is the *slope coefficient*. Generally, the coefficients α and β are unknown. We fit a straight line, called the *estimated regression line*, or prediction line, to the data points.

3.4.2.1   The Least-Squares Method

The most common method of fitting a regression line is the **least-squares method**.

Suppose that $\hat{y} = a + bx$ is the straight line fitted to the data. To determine estimates of α and β, the *principle of least squares* requires a and b, which minimize the sum of squares of residuals around the line—that is,

$$SSE = \sum_{i=1}^{n} (y_i - a - bx_i)^2 \tag{3.4.5}$$

If we require that the regression line pass through the point $(\bar{x}, \bar{y})$, where $\bar{x}, \bar{y}$ are the sample means of the x's and y's, then

$$\bar{y} = a + b\bar{x}$$

or the coefficient a should be determined by the equation

$$a = \bar{y} - b\bar{x} \tag{3.4.6}$$

Substituting equation 3.4.6 into equation 3.4.5 we obtain

$$SSE = \sum_{i=1}^{n} (y_i - \bar{y} - b(x_i - \bar{x}))^2$$

$$= \sum_{i=1}^{n}(y_i - \bar{y})^2 - 2b\sum_{i=1}^{n}(x_i - \bar{x})(y_i - \bar{y}) + b^2\sum_{i=1}^{n}(x_i - \bar{x})^2$$

Dividing the two sides of the equation by $(n - 1)$, we obtain

$$\frac{SSE}{n-1} = S_y^2 - 2bS_{xy} + b^2 S_x^2$$

The coefficient b should be determined in order to minimize this quantity. We can write

$$\frac{SSE}{n-1} = S_y^2 + S_x^2\left(b^2 - 2b\frac{S_{xy}}{S_x^2} + \frac{S_{xy}^2}{S_x^4}\right) - \frac{S_{xy}^2}{S_x^2}$$

$$= S_y^2(1 - R_{xy}^2) + S_x^2\left(b - \frac{S_{xy}}{S_x^2}\right)^2$$

It is now clear that the least squares estimate of β is

$$b = \frac{S_{xy}}{S_x^2} = R_{xy}\frac{S_y}{S_x} \tag{3.4.7}$$

The value of $SSE/(n - 1)$, corresponding to the least-squares estimate, is

$$S_{y|x}^2 = S_y^2(1 - R_{xy}^2) \tag{3.4.8}$$

$S_{y|x}^2$ is the sample variance of the residuals around the least-squares regression line. By definition, $S_{y|x}^2 \geq 0$, and hence $R_{xy}^2 \leq 1$, or $-1 \leq R_{xy} \leq 1$. Note that $R_{xy} = \pm 1$ only if $S_{y|x}^2 = 0$. This is the case when all the points (x_i, y_i), $i = 1, \ldots, n$, lie on a straight line. If $R_{xy} = 0$, then the slope of the regression line is $b = 0$ and $S_{y|x}^2 = S_y^2$. Notice that

$$R_{xy}^2 = 1 - \frac{S_{y|x}^2}{S_y^2} \tag{3.4.9}$$

Thus, R_{xy}^2 is the proportion of variability in Y, which is explainable by the linear relationship $\hat{y} = a + bx$. For this reason, R_{xy}^2 is also called the **coefficient of determination**. The coefficient of correlation (squared) measures the extent of linear relationship in the data. The linear regression line, or prediction line, could be used to predict the values of Y corresponding to X values if R_{xy}^2 is not too small. To interpret the coefficient of determination—particularly when dealing with multiple regression models (see Chapter 8)—it is sometimes useful to consider an "adjusted" R^2. The adjustment accounts for the number of predictor or explanatory variables in the model and the sample size. In simple linear regression, we define

$$R_{xy}^2(\text{adjusted}) = 1 - \left[(1 - R_{xy}^2)\frac{n-1}{n-2}\right] \tag{3.4.10}$$

EXAMPLE **3.7**   Telecommunication satellites are powered while in orbit by solar cells. Tadicell, a solar cells producer that supplies several satellite manufacturers, was asked to provide data on the degradation of its solar cells over time. Tadicell engineers performed a simulated experiment in which solar cells were subjected to temperature and illumination changes similar to those in orbit, and the short-circuit current (ISC; in amperes) of solar cells at three different time periods were measured in order to determine their rate of degradation. Table 3.8 shows the ISC values of $n = 16$ solar cells, measured at three time epochs, one month apart. The data are given in file SOCELL.DAT. In Figure 3.11 we see the scatter of the ISC values at t_1, and at t_2.

T A B L E   3.8
ISC values of solar cells at three time epochs

		Time	
Cell	t_1	t_2	t_3
1	4.18	4.42	4.55
2	3.48	3.70	3.86
3	4.08	4.39	4.45
4	4.03	4.19	4.28
5	3.77	4.15	4.22
6	4.01	4.12	4.16
7	4.49	4.56	4.52
8	3.70	3.89	3.99
9	5.11	5.37	5.44
10	3.51	3.81	3.76
11	3.92	4.23	4.14
12	3.33	3.62	3.66
13	4.06	4.36	4.43
14	4.22	4.47	4.45
15	3.91	4.17	4.14
16	3.49	3.90	3.81

We now make a regression analysis of ISC at time t_2, Y, versus ISC at time t_1, X. The computations can be easily performed by MINITAB, as shown in the following exhibit:

```
MTB > regress c2 on 1 pred in c1

The regression equation is
C2 = 0.536 + 0.929 C1

Predictor    Coef     Stdev    t-ratio    P
Constant   0.5358    0.2031      2.64   0.019
C1         0.92870   0.05106    18.19   0.0000

s = 0.08709   R-sq = 95.9%   R-sq(adj) = 95.7%
```

Relationship of ISC values at t_1 and t_2

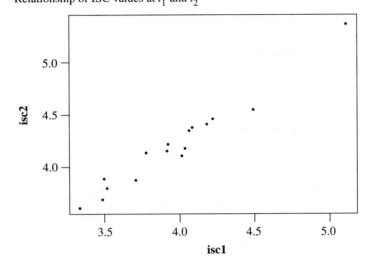

We see in this MINITAB analysis that the least-squares regression (prediction) line is $\hat{y} = 0.536 + 0.929x$ and the coefficient of determination is $R_{xy}^2 = 0.959$. This means that only 4% of the variability in the ISC values at time period t_2 are not explained by the linear regression on the ISC values at time t_1. Moreover, observation 9 is an "unusual observation." It has a relatively large influence on the regression line, as can be seen in Figure 3.11.

The MINITAB output also provides additional analysis. The Stdev corresponding to the least-squares regression coefficients are the square roots of the variances of these estimates, which are given by the formulas:

$$S_a^2 = S_e^2 \left[\frac{1}{n} + \frac{\bar{x}^2}{\sum\limits_{i=1}^{n}(x_i - \bar{x})^2} \right] \qquad (3.4.11)$$

and

$$S_b^2 = S_e^2 / \sum\limits_{i=1}^{n}(x_i - \bar{x})^2 \qquad (3.4.12)$$

where

$$S_e^2 = \frac{1 - R_{xy}^2}{n - 2} \sum\limits_{i=1}^{n}(y_i - \bar{y})^2 \qquad (3.4.13)$$

We see here that $S_e^2 = \frac{(n-1)}{(n-2)} S_{y|x}^2$. This modification is done for testing purposes. The value of S_e^2 in this analysis is 0.0076. The standard deviation of y is $S_y = 0.4175$. The standard deviation of the residuals around the regression line is $S_e = 0.08709$. This explains the high value of $R_{y|x}^2$.

In Table 3.9 we present the values of ISC at time t_2, y, and their predicted values at time t_1, $\hat{y}$. We also present a graph (Figure 3.12) of the residuals, $\hat{e} = y - \hat{y}$, versus the predicted values $\hat{y}$. If the simple linear regression explains the variability adequately, the residuals should be randomly distributed around zero, without any additional relationship to the regression x.

In Figure 3.12, the residuals $\hat{e} = y - \hat{y}$ are plotted against the predicted values $\hat{y}$ of the ISC values for time t_2. The residuals are randomly dispersed around zero. In Chapter 8, we will learn how to test whether this dispersion is indeed random.   ∎

TABLE   3.9

Observed and predicted values of ISC at time t_2

i	y_i	$\hat{y}_i$	$\hat{e}_i$
1	4.42	4.419	0.0008
2	3.70	3.769	−0.0689
3	4.39	4.326	0.0637
4	4.19	4.280	−0.0899
5	4.15	4.038	0.1117
6	4.12	4.261	−0.1413
7	4.56	4.707	−0.1472
8	3.89	3.973	−0.0833
9	5.37	5.283	0.0868
10	3.81	3.797	0.0132
11	4.23	4.178	0.0523
12	3.62	3.630	−0.0096
13	4.36	4.308	0.0523
14	4.47	4.456	0.0136
15	4.17	4.168	0.0016
16	3.90	3.778	0.1218

3.4.2.2   Regression and Prediction Intervals

Suppose that we wish to predict the possible outcomes of Y for some specific value of X, say x_0. If the true regression coefficients α and β are known, then the predicted value of Y is $\alpha + \beta x_0$. However, when α and β are unknown, we predict the outcome at x_0 to be $\hat{y}(x_0) = a + bx_0$. We know, however, that the actual, observed value of Y will not be exactly equal to $\hat{y}(x_0)$. We can determine a prediction interval around $\hat{y}(x_0)$ such that the likelihood of obtaining a Y value within this interval will be high. In Chapter 8, we will study the relationship between this likelihood and the distribution of the residuals. Generally, the prediction interval limits, given by the formula

$$\hat{y}(x_0) \pm 3 \cdot \left[1 + \frac{1}{n} + \frac{(x_0 - \bar{x})^2}{\sum_i (x_i - \bar{x})^2} \right]^{1/2} \tag{3.4.14}$$

F I G U R E   3.12

Residual vs. predicted ISC values

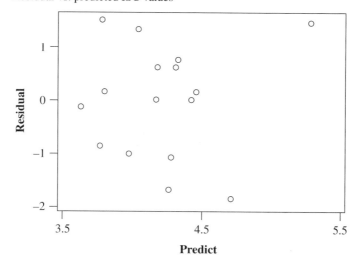

will yield good predictions. Table 3.10 gives the prediction intervals for the ISC values at time t_2, for selected ISC values at time t_1. Figure 3.13 shows the scatterplot, regression line, and prediction limits for Res 3 versus Res 7 of the HADPAS.DAT set.

T A B L E   3.10

Prediction intervals for ISC values at time t_2

x_0	$\hat{y}(x_0)$	Lower limit	Upper limit
4.0	4.251	3.987	4.514
4.4	4.622	4.350	4.893
4.8	4.993	4.701	5.285
5.2	5.364	5.042	5.687

3.4.2.3   Robust Methods of Correlation and Regression

As mentioned in Section 2.6.4, the sample mean, $\overline{X}$, and standard deviation, S, are sensitive to the extremes. To make more robust statistics, $\overline{X}$ and S are trimmed. The statistics $\overline{X}^*$ and S^* of the trimmed sample statistics, are called *α-trimmed mean* and *α-trimmed standard deviation*, where $α$ is the proportion of observations trimmed from the sample. Devlin and colleagues (1975) introduce a statistic of robust correlation, r^*, which is obtained in the following manner: Let $\overline{X}^*$ and S_X^* be robust statistics of the mean and the standard deviation of the X sample (like the $α$-trimmed statistics). Let $\overline{Y}^*$ and S_Y^* be the corresponding ones for the Y sample.

FIGURE　3.13

Prediction intervals for Res 3 values, given the Res 7 values

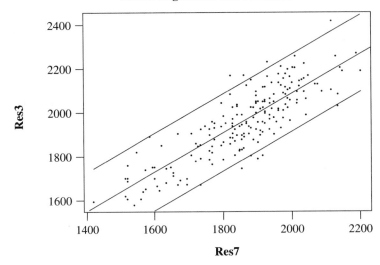

Linear Regression with Prediction Lines

From the X and Y samples compute

$$Z_{1i} = (X_i - \overline{X}^*)/S_X^* + (Y_i - \overline{Y}^*)/S_Y^* \qquad i = 1, \ldots, n$$
$$Z_{2i} = (X_i - \overline{X}^*)/S_X^* - (Y_i - \overline{Y}^*)/S_Y^* \qquad i = 1, \ldots, n$$

Let V_1^* be a robust variance of the Z_1 sample, and V_2^* a robust variance of the Z_2 sample. Then

$$r_{xy}^* = \frac{V_1^* - V_2^*}{V_1^* + V_2^*} \tag{3.4.15}$$

EXAMPLE　3.8　　As seen in previous examples, data set PLACE.DAT contains the (x, y, θ) deviations in the placement of components on 26 boards. If we plot side-by-side boxplots of the orientation deviations, θ, where each sample contains 16 observations on a board, we see that some samples contain outliers. In particular, sample 4 has an upper outlier, and sample 7 a lower outlier. Otherwise, the boxplots of samples 4 and 7 are similar. We therefore combine samples 4 and 7 and obtain the data presented in Table 3.11.

TABLE　3.11

Deviations of 32 placements (tet-dev)

0.01768	0.00317	0.07677	0.02990	0.05252	0.02959	0.12981
0.04797	−0.00288	−0.00285	−0.00101	−0.00079	−0.00091	−0.00211
−0.00201	0.00064	0.04642	0.09559	0.00114	0.05712	0.03536
0.04184	−0.07792	0.02970	−0.00332	−0.00098	0.00056	0.00062
−0.00166	−0.00360	0.00038	−0.00158			

We denote by X the variable tet-dev.

The mean and standard deviation of this combined sample are $\overline{X} = 0.0186$ and $S = 0.03742$. These statistics are affected by the extreme values, which are the smallest and the largest. If we delete these two extremes from the sample, we obtain a new mean and standard deviation: $\overline{X}^* = 0.0181$ and $S^* = 0.02736$. We see that by trimming the sample from the extremes, with $\alpha = 1/16$, the mean does not change much because the extremes are almost symmetric on its two sides; the standard deviation, however, is considerably reduced. If we trim two extreme values from each side—that is, $\alpha = 2/16$—we obtain $\overline{X}^* = 0.0161$ and $S^* = 0.02366$.

Using for the Y sample the corresponding data for the y deviations of samples 4 and 7 from PLACE.DAT, we obtain $\overline{Y}^* = -0.00209$ and $S_Y^* = 0.000503$. The values of X_i, Y_i, Z_{1i}, and Z_{2i} are given in Table 3.12.

TABLE   3.12

The $(X_i, Y_i, Z_{1i}, Z_{2i})$ from the placement data

i	X_i	Y_i	Z_{1i}	Z_{2i}
1	0.01768	−0.00240	−0.54675	0.67883
2	0.00317	−0.00209	−0.54357	−0.55067
3	0.07677	−0.00179	3.16309	1.96307
4	0.02990	−0.00104	2.67359	−1.50871
5	0.05252	−0.00201	1.70093	1.37571
6	0.02959	−0.00167	1.40793	−0.26925
7	0.12981	0.00025	9.46038	0.14853
8	0.04797	−0.00121	3.09920	−0.40712
9	−0.00288	−0.00217	−0.95829	−0.64728
10	−0.00285	−0.00189	−0.40033	−1.20271
11	−0.00101	−0.00160	0.25400	−1.70153
12	−0.00079	−0.00113	1.19775	−2.62668
13	−0.00091	−0.00250	−1.53115	0.09207
14	−0.00211	−0.00175	−0.09071	−1.44978
15	−0.00201	−0.00136	0.68891	−2.22095
16	0.00064	−0.00114	1.23830	−2.54637
17	0.04642	−0.00206	1.34374	1.21735
18	0.09559	−0.00266	2.22866	4.48810
19	0.00114	−0.00189	−0.23172	−1.03410
20	0.05712	−0.00211	1.69649	1.76892
21	0.03536	−0.00273	−0.45572	2.08206
22	0.04184	−0.00281	−0.34095	2.51495
23	−0.07792	−0.00264	−5.06380	−2.88388
24	0.02970	−0.00296	−1.15219	2.30016
25	−0.00332	−0.00197	−0.57925	−1.06351
26	−0.00098	−0.00234	−1.21599	−0.22900
27	0.00056	−0.00228	−1.03162	−0.28321
28	0.00062	−0.00259	−1.64543	0.33566
29	−0.00166	−0.00262	−1.80142	0.29896
30	−0.00360	−0.00316	−2.95703	1.29061
31	0.00038	−0.00244	−1.35734	0.02729
32	−0.00158	−0.00272	−1.99686	0.50116

The robust (α-trimmed) variances of Z_{1i} and Z_{2i} are $V_1^* = 2.0400$ and $V_2^* = 1.7032$. From this we obtain the robust correlation $r_{XY}^* = 0.0899$. The Pearson correlation of X and Y is $\rho_{XY} = .366$. This illustrates the effect of the outliers in the samples on the correlation. We expect a low correlation (close to zero) because there is no reason for y-dev and tet-dev to be dependent.   ∎

Another robust statistic of correlation is the so-called *Spearman rank-order correlation*. The X and Y data are assigned ranks. The smallest value in a sample gets the rank 1, the next smallest gets the rank 2, and so on. The largest value gets the rank n. We then compute the correlation between the ranks of the X's and Y's.

Table 3.13 gives the ranks of the (X, Y) values of Table 3.12. The correlation between the ranks of X, Y is $r_{R(X),R(Y)} = 0.148$. Notice that the rank correlation of this sample is also considerably smaller than r_{XY}. The rank correlation is a robust statistic.

T A B L E   3.13
The values of (X_i, Y_i) and their ranks $(R(X_i), R(Y_i))$

$R(X_i)$	X_i	Y_i	$R(Y_i)$	$R(X_i)$	X_i	Y_i	$R(Y_i)$
1	−0.07792	−0.00264	7	17	0.00064	−0.00114	29
2	−0.00360	−0.00316	1	18	0.00114	−0.00189	22
3	−0.00332	−0.00197	20	19	0.00317	−0.00209	17
4	−0.00288	−0.00217	15	20	0.01768	−0.00240	12
5	−0.00285	−0.00189	21	21	0.02959	−0.00167	25
6	−0.00211	−0.00175	24	22	0.02970	−0.00296	2
7	−0.00201	−0.00136	27	23	0.02990	−0.00104	31
8	−0.00166	−0.00262	8	24	0.03536	−0.00273	4
9	−0.00158	−0.00272	5	25	0.04184	−0.00281	3
10	−0.00101	0.00160	26	26	0.04642	−0.00206	18
11	−0.00098	−0.00234	13	27	0.04797	−0.00121	28
12	−0.00091	−0.00250	10	28	0.05252	−0.00201	19
13	−0.00079	−0.00113	30	29	0.05712	−0.00211	16
14	0.00038	−0.00244	11	30	0.07677	−0.00179	23
15	0.00056	−0.00228	14	31	0.09559	−0.00266	6
16	0.00062	−0.00259	9	32	0.12981	0.00025	32

The regression line fitted by the method of least squares is also sensitive to outliers. A robust regression line, called *resistant regression*, is fitted to the data in the following manner:

1 The (X, Y) array is first sorted by the variable X.

2 Then the data are partitioned into three parts. The first part consists of $[n/3]$ points (X, Y), with the smallest X values. The third part consists of $[n/3]$ points (X, Y), with the largest X values. The second part consists of all other points.

Let $(M_X^{(1)}, M_Y^{(1)})$ be the medians of X and Y in the first part. Let $(M_X^{(3)}, M_Y^{(3)})$ be the medians of X and Y in the third part. Finally, let (M_X, M_Y) be the medians of X and Y in the whole sample.

The coefficients of the resistant regression line are

$$b_R = \frac{M_Y^{(3)} - M_Y^{(1)}}{M_X^{(3)} - M_X^{(1)}} \tag{3.4.16}$$

and

$$a_R = M_Y - b_R M_X \tag{3.4.17}$$

For more advanced methods of robust regression lines, see Rosseeuw and Leroy (1987).

EXAMPLE 3.9 For the ISC data of Table 3.8, we saw that the least-squares regression of Y (ISC at t_2) on X (ISC at t_1) is $\hat{Y} = 0.536 + 0.929X$. The resistant line is $Y_R = 0.595 + 0.904X$. ∎

EXAMPLE 3.10 In the HADPAS data set, the least-squares regression of Res 3 on Res 7 is

$$\hat{Y} = 273.658 + 0.9105X$$

The corresponding resistant line is

$$\hat{Y}_R = -92.856 + 1.0957X$$

In Figure 3.14, we plot the least-squares and resistant lines on the scatterplot of (Res 7, Res 3). We see that the resistant has a negative intercept and a larger slope

F I G U R E   3.14

Least-squares and resistant regression lines for Res 3 vs. Res 7

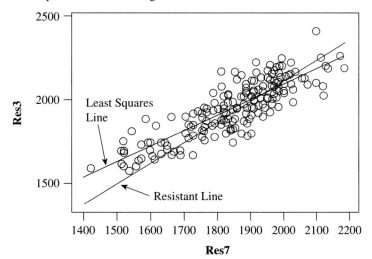

than that of the least-squares line. This is due to the negative skewness of the Res 7 values; those of Res 3 are more symmetric. Thus, the least-squares line is pulled upward on the left-hand side of the graph. ∎

A more detailed discussion of the theory of resistant lines is available in the literature. In MINITAB, the procedure called RLINE finds a regression line for which the median of the residuals in the lower third of the data is equal to that of the upper third. The procedure generally requires several iterations. The RLINE obtained for the Res 3 vs. Res 7 is $\hat{Y}_{RL} = 190.129 + 0.9513X$.

3.5
Contingency Tables
3.5.1 Structure of Contingency Tables

When data are categorical, we generally summarize them in a table that gives the frequency of each category by variable. Such a table is called a **contingency table**.

EXAMPLE **3.11**   Consider a test of a machine that inserts components onto a board. The displacement errors of such a machine were analyzed in Example 3.1. In this test, we perform a large number of insertions with $k = 9$ different components. The result of each trial (insertion) is either Success (no insertion error) or Failure (insertion error). In this test, there are two categorical variables: Component Type and Insertion Result. The first variable has nine categories:

$C1$: Diode

$C2$: 1/2-Watt canister

$C3$: Jump wire

$C4$: Small corning

$C5$: Large corning

$C6$: Small bullet

$C7$: 1/8-Watt dogbone

$C8$: 1/4-Watt dogbone

$C9$: 1/2-Watt dogbone

The second variable, Insertion Result, has two categories only (Success, Failure). The contingency table (Table 3.14) presents the frequencies of the various insertion results by component type.

The table shows that the proportional frequency of errors in insertions is very small ($190/1,056,764 = .0001798$)—about 180 FPM (failures per million). This may be judged to be in conformity with the industry standard. We see, however, that there are apparent differences among the failure proportions by component types.

TABLE   3.14

Contingency table of insertion results by component type

Component Type	Insertion result		Row Total
	Failure	Success	
C1	61	108058	108119
C2	34	136606	136640
C3	10	107328	107338
C4	23	105042	105065
C5	25	108829	108854
C6	9	96864	96873
C7	12	107379	107391
C8	3	105851	105854
C9	13	180617	180630
Column Total	190	1,056,574	1,056,764

Figure 3.15 shows the FPMs of the insertion failures by component type. The largest one is that of $C1$ (Diode), followed by those of components $C2$, $C4$, $C5$. A smaller proportion of failures is seen in $C3$, $C6$, $C7$, $C9$. The smallest error rate is that of $C8$.

FIGURE   3.15

Bar-chart of component error rates

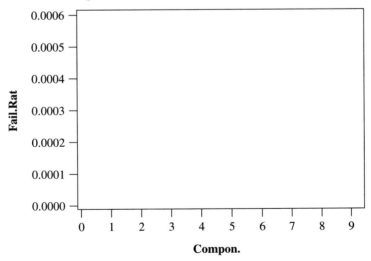

The differences in the component error rates can be shown to be very significant. Tests of significance of such differences will be discussed in Chapter 8. ∎

The structure of a contingency table can be considerably more complicated than that of Table 3.14. Example 3.12 illustrates a contingency table with three variables.

EXAMPLE **3.12**   The data are the placement errors of an OMNI 4621 automatic insertion machine. The variables are:

1   Machine Structure: Basic, EMI1, EMI2, EMI4

2   Components: 805, 1206, SOT_23

3   Placement Result: Error, No_Error

The contingency table is given in Table 3.15 and summarizes the results of 436,431 placements.

We see in the table that the total failure rate of this machine type is $124/436,307 = 284$ (FPM). The failure rates by machine structure (in FPMs), are 281, 275, 281, and 304, respectively. The first three structural types have almost the same FPMs, whereas the fourth one is slightly larger. The component failure rates are 378, 234, and 241 FPM, respectively. It remains to check the failure rates according to structure by component. These are given in Table 3.16.

We see that the effect of the structure is different on different components. Again, one should test whether the observed differences are statistically significant or due only to chance variability. Methods for testing this will be discussed in Chapter 8.   ■

TABLE   **3.15**
Contingency table of placement errors

Component Structure	805		1206		SOT_23		Total components		Total rows
	Err	N_Err	Err	N_Err	Err	N_Err	Err	N_Err	
Basic	11	40,279	7	40,283	16	40,274	34	120,836	120,870
EMI1	11	25,423	8	25,426	2	25,432	21	76,281	76,302
EMI2	19	54,526	15	54,530	12	54,533	46	163,589	163,635
EMI4	14	25,194	4	25,204	5	25,203	23	75,601	75,624
Total	55	145,422	34	145,443	35	145,442	124	436,307	436,431

TABLE   **3.16**
Failure rates (FPMs) by structure and component type

Structure	Component		
	805	1206	SOT_23
Basic	273	174	397
EMI1	433	315	79
EMI2	348	275	220
EMI4	555	159	198

Contingency tables can be constructed using MINITAB. We illustrate this on the data in file CAR.DAT. This file consists of information on 109 car models from 1989. The file contains 109 records on 5 variables: Number of Cylinders (4, 6, 8), Origin (United States = 1, Europe = 2, Asia = 3), Turn Diameter (in meters), Horsepower, and Number of Miles per Gallon in City Driving. One variable, Origin, is categorical; the other four are interval-scaled variables. One variable is discrete (Number of Cylinders) and the other three are continuous. By using the command

```
MTB> Table C1-C2;
SUBC> Counts.
```

we obtain the contingency table, which is illustrated in Table 3.17.

We can also prepare a contingency table from continuous data by selecting the number and length of interval for each variable, and counting the frequencies of each cell in the table.

For example, for the car data, if we wish to construct a contingency table of turn diameter versus miles/gallon, we obtain the contingency table shown in Table 3.18.

TABLE   **3.17**

Contingency table of number of cylinders against origin

	Origin			
Number of cylinders	**1**	**2**	**3**	**Total**
4	33	7	26	66
6	13	7	10	30
8	12	0	1	13
Total	58	14	37	109

TABLE   **3.18**

Contingency table of turn diameter versus miles/gallon city

	Miles/gallon city			
Turn diameter	**12–18**	**19–24**	**25–35**	**Total**
27–30.6	2	0	4	6
30.7–34.2	4	12	15	31
34.3–37.8	10	26	6	42
37.9–44.0	15	15	0	30
Total	31	53	25	109

3.5.2   Indices of Association for Contingency Tables

In this section, we construct several **measures of association**, which reflect the degree of dependence, or association, between variables. For the sake of simplicity, we consider here indices for two-way tables—that is, association between two variables.

3.5.2.1   **Two Interval-Scaled Variables**

If two variables are continuous, measured on an interval scale or on a transformation of one, we can use some of the dependence indices discussed earlier. For example, we can represent each interval by its midpoint and compute the correlation coefficient between the midpoints. As in Section 3.3.1, if variable X is classified into k intervals,

$$(\xi_0, \xi_1), (\xi_1, \xi_2), \ldots, (\xi_{k-1}, \xi_k)$$

and variable Y is classified into m intervals

$$(\eta_0, \eta_1), \ldots, (\eta_{m-1}, \eta_m)$$

let

$$\tilde{\xi}_i = \frac{1}{2}(\xi_{i-1} + \xi_i) \qquad i = 1, \ldots, k$$

$$\tilde{\eta}_j = \frac{1}{2}(\eta_{j-1} + \eta_j) \qquad j = 1, \ldots, m$$

Let $p_{ij} = f_{ij}/N$ denote the proportional frequency of the (i, j)th cell—that is, the X values in (ξ_{i-1}, ξ_i) and Y values in (η_{j-1}, η_j). Then, an estimate of the coefficient of correlation obtained from the contingency table is

$$\hat{\rho}_{XY} = \frac{\displaystyle\sum_{i=1}^{k}\sum_{j=1}^{m} p_{ij}(\tilde{\xi}_i - \overline{\xi})(\tilde{\eta}_j - \overline{\eta})}{\left[\displaystyle\sum_{i=1}^{k} p_{i.}(\tilde{\xi}_i - \overline{\xi})^2\right]^{1/2}\left[\displaystyle\sum_{j=1}^{m} p_{.j}(\tilde{\eta}_j - \overline{\eta})^2\right]^{1/2}} \qquad (3.5.1)$$

where

$$p_{i.} = \sum_{j=1}^{m} p_{ij} \qquad i = 1, \ldots, k$$

$$p_{.j} = \sum_{i=1}^{k} p_{ij} \qquad j = 1, \ldots, m$$

$$\overline{\xi} = \sum_{i=1}^{k} p_{i.}\tilde{\xi}_i$$

and

$$\overline{\eta} = \sum_{j=1}^{m} p_{.j}\tilde{\eta}_j$$

Notice that the sample correlation r_{XY} obtained from the sample data is different from $\hat{\rho}_{XY}$ due to the reduced information that is given by the contingency table. We illustrate this in the following example.

EXAMPLE **3.13**   Consider the data in file CAR.DAT. The sample correlation between the turn diameter, X, and the gas consumption (miles/gal) in a city is $r_{XY} = -.541$. If we compute this correlation on the basis of the data in Table 3.18, we obtain $\hat{\rho}_{XY} = -.478$. The approximation given by $\hat{\rho}_{XY}$ depends on the number of intervals, k and m; on the length of the intervals; and on the sample size, N.   ■

3.5.2.2   Indices of Association for Categorical Variables

If one of the variables or both are categorical, there is no meaning to the correlation coefficient. We need to devise another index of association. Such an index should not depend on the labeling or ordering of the categories. Common indices of association are based on a comparison of the observed frequencies of the cells ($i = 1, \ldots, k; j = 1, \ldots, m$), f_{ij}, to the expected ones if the events associated with the categories are independent. The concept of independence, in a probability sense, is defined in Chapter 4. We have already seen conditional frequency distributions. If $N_{i.} = \sum_{j=1}^{m} f_{ij}$, the conditional proportional frequency of the jth category of Y, given the ith category of X, is

$$p_{j|i} = \frac{f_{ij}}{N_{i.}} \qquad j = 1, \ldots, m$$

We say that X and Y are *not associated* if

$$p_{j|i} = p_{\cdot j} \qquad \text{for all } i = 1, \ldots, k$$

where

$$p_{\cdot j} = \frac{N_{\cdot j}}{N} \qquad j = 1, \ldots, m$$

and

$$N_{\cdot j} = \sum_{i=1}^{k} f_{ij}$$

Accordingly, the expected frequency of cell (i, j), if there is no association, is

$$\tilde{f}_{ij} = \frac{N_{i.} N_{\cdot j}}{N} \qquad i = 1, \ldots, k, j = 1, \ldots, m$$

A common index of discrepancy between f_{ij} and $\tilde{f}_{ij}$ is

$$X^2 = \sum_{i=1}^{k} \sum_{j=1}^{m} \frac{(f_{ij} - \tilde{f}_{ij})^2}{\tilde{f}_{ij}} \tag{3.5.2}$$

This index is called the **chi-square statistic**. We can compute this statistic using MINITAB, with the command

```
MTB> Table C1 C2;
SUBC> Counts;
SUBC> ChiSquare.
```

EXAMPLE **3.14**   For the CAR.DAT data, the chi-square statistic for the association between Origin and Number of Cylinders is $X^2 = 12.13$. In Chapter 8, we will study how to assess the statistical significance of such a magnitude of X^2.

Another option in MINITAB is to set the observed frequencies of the contingency table into columns, and use the command MTB> ChiSquare C_ − C_.

For example, let us set the frequencies of Table 3.18 into three columns—say, C6–C8. The preceding command yields Table 3.19, in which the expected frequencies, $\tilde{f}_{ij}$, appear below the observed ones.

The chi-square statistic is

$$X^2 = \frac{(2 - 1.71)^2}{1.71} + \cdots + \frac{6.88^2}{6.88} = 34.99 \quad \blacksquare$$

There are several association indices in the literature, based on the X^2. Three popular ones are:

Mean-squared contingency: $\Phi^2 = \dfrac{X^2}{N}$ (3.5.3)

Tschuprow's index: $T = \Phi / \sqrt{(k-1)(m-1)}$ (3.5.4)

Cramer's index: $C = \Phi / \sqrt{\min(k-1, m-1)}$ (3.5.5)

TABLE   **3.19**
Observed and expected frequencies of turn diameter by miles/gallon, CAR.DAT ($\tilde{f}_{ij}$ under $\tilde{f}_{ij}$)

Turn diameter	Miles/gallon city			Total
	12–18	19–24	25–35	
27–30.6	2	0	4	6
	1.71	2.92	1.38	
30.7–34.2	4	12	15	31
	8.82	15.07	7.11	
34.3–37.8	10	26	6	42
	11.94	20.42	9.63	
37.9–44.0	15	15	0	30
	8.53	14.59	6.88	
Total	31	53	25	109

No association corresponds to $\Phi^2 = T = C = 0$. The larger the index, the stronger the association. For the data of Table 3.20,

$$\Phi^2 = \frac{34.99}{109} = 0.321$$
$$T = 0.231$$
$$C = 0.401$$

The following is another example of contingency table analysis—this time using the Cramer index.

EXAMPLE **3.15** Compu Star, a service company providing technical support and sales of personal computers and printers, decided to investigate the various components of customer satisfaction specific to the company. A special questionnaire with 13 questions was designed and, after a pilot run, was mailed to a large sample of customers along with a self-addressed stamped envelope and a prize incentive. The prize was to be awarded by lottery among the customers who returned the questionnaire.

The customers were asked to rate, on a 1–6 ranking order, various aspects of the service, with a rating of 1 corresponding to *very poor* and the rating of 6 to *very good*. The areas covered included:

Q1: First impression of service representative

Q2: Friendliness of service representative

Q3: Speed in responding to service request

Q4: Technical knowledge of service representative

Q5: Professional level of service provided

Q6: Helpfulness of service representative

Q7: Additional information provided by service representative

Q8: Clarity of questions asked by service representative

Q9: Clarity of answers provided by service representative

Q10: Efficient use of time by service representative

Q11: Overall satisfaction with service

Q12: Overall satisfaction with product

Q13: Overall satisfaction with company

The response ranks were:

1   Very poor

2   Poor

3   Below average

4   Above average

5   Good

6   Very good

The responses were tallied and contingency tables computed linking the questions on overall satisfaction with questions on specific service dimensions. For example, Table 3.20 is a contingency table of responses to Q13 versus Q3.

TABLE 3.20

Two-by-two contingency table of customer responses for Q3 and Q13

	Q13					
Q3	1	2	3	4	5	6
1	0	1	0	0	3	1
2	0	2	0	1	0	0
3	0	0	4	2	3	0
4	0	1	1	10	7	5
5	0	0	0	10	71	38
6	0	0	0	1	30	134

In order to compute χ^2 we combine columns 1–3 of the table. The resulting statistic is $\chi^2 = 256.08$.

Cramer's index for Table 3.20 is:

$$C = 0.398$$

There were 10 detailed questions (Q1–Q10) and 3 questions on overall customer satisfaction (Q11–Q13). A table was constructed for every combination of the 3 overall customer satisfaction questions and the 10 specific questions. For each of these 30 tables, a Cramer's index was computed. These indices are given in Table 3.21, which uses a code of graphical symbols.

TABLE 3.21

Cramer's indices of Q1–Q10 by Q11–Q13

		Q1	Q2	Q3	Q4	Q5	Q6	Q7	Q8	Q9	Q10
Q11:	Overall satisfaction with service		•	++	•			+	•		
Q12:	Overall satisfaction with product		+	•	•		++	•	•		
Q13:	Overall satisfaction with company		•	+			+	++	•	•	

The indices are coded according to the following key:

Cramer's index	Code
0 - 0.2	
0.2 - 0.3	$\bullet$
0.3 - 0.4	$+$
0.4 - 0.5	$++$
0.5 -	$+++$

We can see from Table 3.21 that "Overall satisfaction with company" (Q13) is highly correlated with "Speed in responding to service requests" (Q3). However, "Efficient use of time" (Q10) is not associated with overall satisfaction.

We also notice that questions Q1, Q5, and Q10 show no correlation with overall satisfaction. For more examples of indices of association and graphical analysis of contingency tables, see Kenett (1983).   ∎

3.6
Chapter Highlights

Several techniques for graphical analysis of data in several dimensions are introduced and demonstrated using case studies. These include matrix scatterplots, 3D scatterplots, multiple boxplots, and dynamic graphical analysis such as brushing, linking, and spinning. Topics covered also include simple linear regression, multiple regression models, and contingency tables. Prediction intervals are constructed for currents of solar cells and resistances on hybrid circuits. Robust regression is used to analyze data on placement errors of components on circuit boards. A special section on indices of association for categorical variables includes an analysis of a customer satisfaction survey designed to identify the main components of customer satisfaction and dissatisfaction. The material covered by this chapter is best studied in front of a personal computer so the reader can reproduce and even expand the data analysis offered in the text.

The main concepts and tools introduced in this chapter include:

- Matrix scatterplot
- 3D scatterplot
- Multiple boxplot
- Code variable
- 3D spinning method
- Brushing
- Coding
- Transformations
- Joint frequency distribution
- Marginal frequency
- Conditional distribution
- Sample covariance
- Sample correlation
- Linear model
- Simple linear regression
- Least-squares method
- Coefficient of determination
- Contingency table
- Measures of association

Exercises

3.1.1 Use file CAR.DAT to prepare multiple or matrix scatterplots of Turn Diameter versus Horsepower versus Miles per Gallon. What can you learn from these plots?

3.1.2 Make a multiple (side-by-side) boxplot of Turn Diameter by car Origin for the data in file CAR.DAT. Can you infer that Turn Diameter depends on Origin?

3.1.3 Data file HADPAS.DAT contains the resistance values (in ohms) of five resistors placed in six hybrids on 32 ceramic substrates. The file contains eight columns. The variables in these columns are:

i	Record Number	v	Res 18
ii	Substrate Number	vi	Res 14
iii	Hybrid Number	vii	Res 7
iv	Res 3	viii	Res 20

a Make a multiple boxplot of the resistance in Res 3, by hybrid.

b Make a matrix plot of all the Res variables. What can you learn from the plots?

3.3.1 Construct a joint frequency distribution of the variables Horsepower and MPG/City for the data in file CAR.DAT.

3.3.2 Construct a joint frequency distribution for the resistance values of Res 3 and Res 14 in data file HADPAS.DAT. (Code the variables first; see the instructions in Exercise 3.3.4.)

3.3.3 Construct the conditional frequency distribu-

tion of Res 3 given that the resistance values of Res 14 are between 1300 and 1500 (ohms).

3.3.4 Compute the conditional means and standard deviations of one variable given another one using file HADPAS.DAT. First classify the data according to the values of Res 14 (Column C6) into five subgroups using the MINITAB command **CODE**, which is:

```
MTB> CODE(900:1200)1 (1201:1500)2
    (1501:1800)3 (1801:2100)4
       (2101:2700)5 C6 C9
```

Then use the command

```
MTB> DESC C4;
SUBC> By C9.
```

These give the conditional means and standard deviations of Res 3 given the subgroups of Res 7. Use these commands and write a report on the results obtained.

3.4.1 Following are four data sets of (X, Y) observations.

a Compute the least-squares regression coefficients of Y on X for the four data sets.

b Compute the coefficient of determination, R^2, for each set.

Data set 1		Data set 2		Data set 3		Data set 4	
$X^{(1)}$	$Y^{(1)}$	$X^{(2)}$	$Y^{(2)}$	$X^{(3)}$	$Y^{(3)}$	$X^{(4)}$	$Y^{(4)}$
10.0	8.04	10.0	9.14	10.0	7.46	8.0	6.68
8.0	6.95	8.0	8.14	8.0	6.67	8.0	5.76
13.0	7.58	13.0	8.74	13.0	12.74	8.0	7.71
9.0	8.81	9.0	8.77	9.0	7.11	8.0	8.84
11.0	8.33	11.0	9.26	11.0	7.81	8.0	8.47
14.0	9.96	14.0	8.1	14.0	8.84	8.0	7.04
6.0	7.24	6.0	6.13	6.0	6.08	8.0	5.25
4.0	4.26	4.0	3.1	4.0	5.39	19.0	12.5
12.0	10.84	11.0	9.13	12.0	8.16	8.0	5.56
7.0	4.82	7.0	7.26	7.0	6.42	8.0	7.91
5.0	5.68	5.0	4.74	5.0	5.73	8.0	6.89

3.4.2 Compute the correlation matrix of the variables Turn Diameter, Horsepower, and Miles per Gallon/City for the data in file CAR.DAT.

3.4.3

a Compute the robust correlations, r_{XY}^*, among the three continuous variables of data file CAR.DAT.

b Compute the Spearman rank-order correlations among these variables.

3.4.4 Compute the resistant regression lines for each data set given in Exercise 3.4.1. Compare the regression coefficients of the resistant lines to those of the least-squares regression lines. [Use the MINITAB procedure.]

3.5.1 In a customer satisfaction survey, several questions were asked regarding specific services and products provided to customers. The answers were on a 1–5 scale, where 5 means "Very satisfied with the service or product" and 1 means "Very dissatisfied." Compute the mean-squared contingency, Tschuprow's index, and Cramer's index for both contingency tables. Since frequencies in rows 1–2 and columns 1–2 are small, collapse the tables by combining rows 1–3 and columns 1–3.

	Question 1				
Question 3	**1**	**2**	**3**	**4**	**5**
1	0	0	0	1	0
2	1	0	2	0	0
3	1	2	6	5	1
4	2	1	10	23	13
5	0	1	1	15	100

	Question 2				
Question 3	**1**	**2**	**3**	**4**	**5**
1	1	0	0	3	1
2	2	0	1	0	0
3	0	4	2	3	0
4	1	1	10	7	5
5	0	0	1	30	134

3.5.2 From data file CAR.DAT, prepare a contingency table for Origin vs. Miles/Gallon City. Compute the value of X^2 and the Cramer index for this table.

4

Basic Models of Probability and Distribution Functions

4.1

Basic Probability

4.1.1 Events and Sample Spaces:
Formal Presentation of Random Measurements

Experiments or trials of interest are those that may yield varying results, with outcomes that are not known with certainty ahead of time. In the previous chapters, we saw a large number of examples in which outcomes of measurements vary. Before conducting a particular experiment, we may find it of interest to figure out the chances of obtaining results in a certain range. To provide a quantitative answer to such a problem, we have to formalize the framework of the discussion so there is no ambiguity.

In the general sense, when we say a "trial" or "experiment," we mean a well-defined process of measuring certain characteristic(s) or variable(s). For example, if an experiment is to measure the compressive strength of concrete cubes, we must specify exactly how the concrete mixture was prepared—that is, proportions of cement, sand, aggregate, and water in the batch; the length of mixing time; dimensions of the mold; number of days during which the concrete has hardened; the temperature and humidity during preparation and storage of the concrete cubes; and so on. All these factors influence the resulting compressive strength. A well-documented protocol of an experiment enables us to replicate it as many times as needed. In a well-controlled experiment, we can assume that variability in the measured variables is due to randomness. We can think of random experimental results as sample values from a hypothetical population. The set of all possible sample values is called the **sample space**. In other words, the sample space is the set of all possible outcomes of a specified experiment. The outcomes do not have to be numerical. They can be names, categorical values, functions, or collections of items. An individual outcome of an experiment is called an **elementary event**, or a *sample point* (element). Following are some examples.

EXAMPLE **4.1** The experiment consists of choosing 10 names (without replacement) from a list of 400 undergraduate students at a given university. The outcome of such an experiment is a sublist of 10 names. The sample space is the collection of *all* possible such sublists that can be drawn from the original list of 400 students. ∎

EXAMPLE **4.2** The experiment is to produce 20 concrete cubes under identical manufacturing conditions and count the number of cubes with compressive strength above 200 kg/cm^2. The sample space is the set $S = \{0, 1, 2, \ldots, 20\}$. The elementary events, or sample points, are the elements of S. ∎

EXAMPLE **4.3** The experiment is to choose a steel bar from a specific production process and measure its weight. The sample space S is the interval (ω_0, ω_1) of possible weights. The weight of a particular bar is a sample point. ∎

Thus, sample spaces can be finite sets of sample points or countable or noncountable infinite sets.

Any subset of the sample space, S, is called an **event**. S itself is called the *sure event*. The empty set, $\emptyset$, is called the **null event**. We denote events by the letters $A, B, C, \ldots$ or by $E_1, E_2, \ldots$. All events under consideration are subsets of the same sample space S. Thus, events are sets of sample points.

For any event $A \subseteq S$, we denote by A^c the **complementary event**—that is, the set of all points of S that are not in A.

An event A is said to *imply* an event B if all elements of A are elements of B. We denote this **inclusion relationship** by $A \subset B$. If $A \subset B$ and $B \subset A$, then the two events are **equivalent**, $A \equiv B$.

EXAMPLE **4.4** The experiment is to select a sequence of five letters for transmission of a code in a money transfer operation. Let $A_1, A_2, \ldots, A_5$ denote the first, second,..., fifth letter chosen, respectively. The sample space is the set of all possible sequences of five letters. Formally,

$$S = \{(A_1 A_2 A_3 A_4 A_5) : A_i \in \{a, b, c, \ldots, z\}, \quad i = 1, \ldots, 5\}$$

This is a finite sample space containing 26^5 possible sequences of five letters. Any such sequence is a sample point.

Let E be the event that all five letters in the sequence are the same. Thus

$$E = \{aaaaa, bbbbb, \ldots, zzzzz\}$$

This event contains 26 sample points. The complement of E, E^c, is the event that at least one letter in the sequence is different from the others. ∎

4.1.2   Basic Rules of Operations with Events: Unions, Intersections

Given events $A, B, \ldots$ of a sample space S, we can generate new events by the **operations with events**: union, intersection, and complementation.

The **union** of two events A and B, denoted $A \cup B$, is an event having elements that belong *either* to A *or* to B.

The **intersection** of two events, $A \cap B$, is an event whose elements belong both to A *and* to B. By pairwise union or intersection, we immediately extend the definition to a finite number of events $A_1, A_2, \ldots, A_n$. That is,

$$\bigcup_{i=1}^{n} A_i = A_1 \cup A_2 \cup \cdots \cup A_n$$

and

$$\bigcap_{i=1}^{n} A_i = A_1 \cap A_2 \cap \cdots \cap A_n$$

A *finite union*, $\bigcup_{i=1}^{n} A_i$, is an event whose elements belong to *at least one* of the n events. A *finite intersection*, $\bigcap_{i=1}^{n} A_i$, is an event whose elements belong to *all* the n events.

Any two events, A and B, are said to be *mutually exclusive*, or **disjoint**, if $A \cap B = \emptyset$—that is, they do not contain common elements. Obviously, by definition, any event is disjoint of its complement—that is, $A \cap A^c = \emptyset$. The operations of union and intersection are:

1   *Commutative*

$$A \cup B = B \cup A$$
$$A \cap B = B \cap A$$

2   *Associative*

$$(A \cup B) \cup C = A \cup (B \cup C)$$
$$= A \cup B \cup C$$
$$(A \cap B) \cap C = A \cap (B \cap C)$$
$$= A \cap B \cap C$$

(4.1.1)

3   *Distributive*

$$A \cap (B \cup C) = (A \cap B) \cup (A \cap C)$$
$$A \cup (B \cap C) = (A \cup B) \cap (A \cup C)$$

(4.1.2)

The intersection of events is sometimes denoted as a product:

$$A_1 \cap A_2 \cap \cdots \cap A_n \equiv A_1 A_2 A_3 \cdots A_n$$

The following law, called *DeMorgan's rule*, is fundamental to the algebra of events and yields the **complement** of the union, or intersection, of two events—namely:

1 $(A \cup B)^c = A^c \cap B^c$ (4.1.3)

2 $(A \cap B)^c = A^c \cup B^c$

Finally, a collection of n events $E_1, \ldots, E_n$ is called a *partition* of the sample space S if

1 $\quad \bigcup\limits_{i=1}^{n} E_i = S$

2 $\quad E_i \cap E_j = \emptyset \quad$ for all $i \neq j$ $(i, j = 1, \ldots, n)$

That is, the events in any partition are mutually disjoint, and their union exhausts all sample space.

EXAMPLE 4.5 The experiment is to generate on the computer a random number, U, in the interval $(0, 1)$. A random number in $(0, 1)$ can be obtained as

$$U = \sum_{j=1}^{\infty} I_j 2^{-j}$$

where I_j is the random result of tossing a coin:

$$I_j = \begin{cases} 1 & \text{if Head} \\ 0 & \text{if Tail} \end{cases}$$

For generating random numbers from a set of integers, the summation index j is bounded by a finite number N. For generating random numbers on a continuous interval, this method is not practical. Computer programs generate "pseudorandom" numbers.

Methods for generating random numbers are described in various books on simulation (see, for example, Bratley, Fox, and Schrage, 1983). The most commonly applied method is *linear congruential generator*. This method is based on the recursive equation

$$U_i = (aU_{i-1} + c)\bmod m \qquad i = 1, 2, \ldots$$

The parameters a, c, and m depend on the computer's architecture. In many programs, $a = 65539$, $c = 0$, and $m = 2^{31} - 1$. The first integer U_0 is called the *seed*. Different choices of the parameters a, c, and m yield pseudorandom sequences with different statistical properties.

The sample space of this experiment is

$$S = \{u : 0 \leq u \leq 1\}$$

Let E_1 and E_2 be the events

$$E_1 = \{u : 0 \leq u < 0.5\}$$
$$E_2 = \{u : 0.35 \leq u \leq 1\}$$

The union of these two events is

$$E_3 = E_1 \cup E_2 = \{u : 0 \leq u \leq 1\} = S$$

The intersection of these events is

$$E_4 = E_1 \cap E_2 = \{u : 0.35 \leq u < 0.5\}$$

Thus, E_1 and E_2 are *not* disjoint.

The complementary events are

$$E_1^c = \{u : 0.5 \le u \le 1\} \qquad \text{and} \qquad E_2^c = \{u : 0 \le u < 0.35\}$$

$E_1^c \cap E_2^c = \emptyset$; that is, the complementary events are disjoint. By DeMorgan's law

$$(E_1 \cap E_2)^c = E_1^c \cup E_2^c$$
$$= \{u : 0 \le u < 0.35 \qquad \text{or} \qquad 0.5 \le u \le 1.0\}$$

However,

$$\emptyset = \mathcal{S}^c = (E_1 \cup E_2)^c = E_1^c \cap E_2^c$$

Finally, the following is a partition of $\mathcal{S}$:

$$B_1 = \{u : 0 \le u < 0.1\}, \qquad B_2 = \{u : 0.1 \le u < 0.2\},$$
$$B_3 = \{u : 0.2 \le u < 0.5\}, \qquad B_4 = \{u : 0.5 \le u \le 1\}$$

Notice that $B_4 = E_1^c$. ∎

Different identities can be derived by these rules of operations on events; a few will be given as exercises.

4.1.3 Probabilities of Events

A probability function $\Pr\{\cdot\}$ assigns to events of $\mathcal{S}$, real numbers, following these basic axioms:

1 $\Pr\{E\} \ge 0$

2 $\Pr\{\mathcal{S}\} = 1$

3 If $E_1, \cdots, E_n$ $(n \ge 1)$ are mutually disjoint events, then

$$\Pr\left\{\bigcup_{i=1}^{n} E_i\right\} = \sum_{i=1}^{n} \Pr\{E_i\}$$

From these three basic axioms, we deduce the following results:

Result 1. If $A \subset B$, then

$$\Pr\{A\} \le \Pr\{B\}$$

Indeed, because $A \subset B$, $B = A \cup (A^c \cap B)$. Moreover, $A \cap A^c \cap B = \emptyset$. Hence, by axioms 1 and 3, $\Pr\{B\} = \Pr\{A\} + \Pr\{A^c \cap B\} \ge \Pr\{A\}$.

Thus, if E is any event, because $E \subset \mathcal{S}$, $0 \le \Pr\{E\} \le 1$

Result 2. For any event E,

$$\Pr\{E^c\} = 1 - \Pr\{E\}$$

Indeed, $S = E \cup E^c$. Because $E \cap E^c = \emptyset$,

$$1 = \Pr\{S\} = \Pr\{E\} + \Pr\{E^c\}$$

This implies the result.

Result 3. For any events A, B

$$\Pr\{A \cup B\} = \Pr\{A\} + \Pr\{B\} - \Pr\{A \cap B\} \tag{4.1.4}$$

Indeed, we can write

$$A \cup B = A \cup A^c \cap B$$

where $A \cap (A^c \cap B) = \emptyset$. Thus, by the third axiom,

$$\Pr\{A \cup B\} = \Pr\{A\} + \Pr\{A^c \cap B\}.$$

Moreover, $B = A^c \cap B \cup A \cap B$, where again $A^c \cap B$ and $A \cap B$ are disjoint. Thus, $\Pr\{B\} = \Pr\{A^c \cap B\} + \Pr\{A \cap B\}$, or $\Pr\{A^c \cap B\} = \Pr\{B\} - \Pr\{A \cap B\}$. Substituting this into the preceding equation, we obtain the result.

Result 4. If $B_1, \cdots, B_n$ $(n \geq 1)$ **is a partition of** S, **then for any event** E,

$$\Pr\{E\} = \sum_{i=1}^{n} \Pr\{E \cap B_i\}$$

Indeed, by the distributive law,

$$E = E \cap S = E \cap \left(\bigcup_{i=1}^{n} B_i \right)$$

$$= \bigcup_{i=1}^{n} EB_i$$

Finally, because $B_1, \ldots, B_n$ are mutually disjoint, $(EB_i) \cap (EB_j) = E \cap B_i \cap B_j = \emptyset$ for all $i \neq j$. Therefore, by the third axiom

$$\Pr\{E\} = \Pr\left\{ \bigcup_{i=1}^{n} EB_i \right\} = \sum_{i=1}^{n} \Pr\{EB_i\} \tag{4.1.5}$$

EXAMPLE **4.6**   Fuses are used to protect electronic devices from unexpected power surges. Modern fuses are produced on glass plates through processes of metal deposition and photographic lithography. On each plate, several hundred fuses are simultaneously produced. At the end of the process, the plates undergo precise cutting with special saws. A certain fuse is handled on one of three alternative cutting machines. Machine M_1 yields 200 fuses per hour, machine M_2 yields 250 fuses per hour, and machine M_3 yields 350 fuses per hour. The fuses are then mixed together. The proportions of defective parts that are typically produced on these machines are .01, .02, and .005, respectively. A fuse is chosen at random from the production of a given hour. What is the probability that it complies with the amperage requirements (nondefective)?

Let E_i be the event that the chosen fuse is from machine M_i ($i = 1, 2, 3$). Because the choice of fuse is random, each fuse has the same probability, $\frac{1}{800}$, of being chosen. Hence, $\Pr\{E_1\} = \frac{1}{4}$, $\Pr\{E_2\} = \frac{5}{16}$, and $\Pr\{E_3\} = \frac{7}{16}$

Let G denote the event that the selected fuse is nondefective. For example, for machine M_1, $\Pr\{G\} = 1 - .01 = .99$. We can assign $\Pr\{G \cap M_1\} = .99 \times .25 = .2475$, $\Pr\{G \cap M_2\} = .98 \times \frac{5}{16} = .3062$, and $\Pr\{G \cap M_3\} = .995 \times \frac{7}{16} = .4353$. Hence, the probability of selecting a nondefective fuse is, according to result 4,

$$\Pr\{G\} = \Pr\{G \cap M_1\} + \Pr\{G \cap M_2\} + \Pr\{G \cap M_3\} = .989 \quad \blacksquare$$

EXAMPLE **4.7**   Consider the problem of generating random numbers, which was discussed in Example 4.5. Suppose the probability function assigns any interval $I(a, b) = \{u : a < u < b\}$, $0 \le a < b \le 1$, the probability

$$\Pr\{I(a, b)\} = b - a$$

Let $E_3 = I(.1, .4)$ and $E_4 = I(.2, .5)$. $C = E_3 \cup E_4 = I(.1, .5)$. Hence,

$$\Pr\{C\} = .5 - .1 = .4$$

However, $\Pr\{E_3 \cap E_4\} = .4 - .2 = .2$.

$$\begin{aligned} \Pr\{E_3 \cup E_4\} &= \Pr\{E_3\} + \Pr\{E_4\} - \Pr\{E_3 \cap E_4\} \\ &= (.4 - .1) + (.5 - .2) - .2 = .4 \end{aligned}$$

This illustrates result 3. $\blacksquare$

4.1.4 Probability Functions for Random Sampling

Consider a finite population $\mathcal{P}$, and suppose that the random experiment is to select a random sample from $\mathcal{P}$, with or without replacement. More specifically, let $\mathcal{L}_N = \{w_1, w_2, \ldots, w_N\}$ be a list of the elements of $\mathcal{P}$, where N is its size. w_j ($j = 1, \cdots, N$) is an identification number of the jth element.

Suppose that a sample of size n is drawn from $\mathcal{L}_N$ (respectively, $\mathcal{P}$) *with* replacement. Let W_1 denote the first element selected from $\mathcal{L}_N$. If j_1 is the index of this element, then $W_1 = w_{j_1}$. Similarly, let W_i ($i = 1, \ldots, n$) denote the ith element of the sample. The corresponding sample space is the collection

$$\mathcal{S} = \{(W_1, \ldots, W_n) : W_i \in \mathcal{L}_N, \quad i = 1, \ldots, n\}$$

of all samples, with replacement from $\mathcal{L}_N$. The total number of possible samples is N^n. Indeed, w_{j_1} could be any one of the elements of $\mathcal{L}_N$, and so could $w_{j_2}, \ldots, w_{j_n}$. With each of the N possible choices of w_{j_1}, we should combine the N possible choices of w_{j_2} and so on. Thus, there are N^n possible ways of selecting a sample of size n with replacement. The sample points are the elements of $\mathcal{S}$ (possible samples). Recall from Chapter 2 that the sample is called *random with replacement* (RSWR) if each one of these N^n possible samples is assigned the same probability, $1/N^n$, for being selected.

Let $M(i)$ $(i = 1, \ldots, N)$ be the number of samples in $\mathcal{S}$ that contain the ith element of $\mathcal{L}_N$ (at least once). Because sampling is *with* replacement

$$M(i) = N^n - (N - 1)^n$$

Indeed, $(N - 1)^n$ is the number of samples with replacement that do *not* include w_i. Because all samples are equally probable, the probability that a RSWR, $\mathbf{S}_n$, includes w_i $(i = 1, \ldots, N)$ is

$$\Pr\{w_i \in \mathbf{S}_n\} = \frac{N^n - (N - 1)^n}{N^n}$$

$$= 1 - \left(1 - \frac{1}{N}\right)^n$$

If $n > 1$, then this probability is larger than $1/N$, which is the probability of selecting the element W_i in any given trial, but smaller than n/N. Notice also that this probability does not depend on i; that is, all elements of $\mathcal{L}_N$ have the same probability of being included in a RSWR. It can be shown that the probability that w_i is included in the sample exactly once is $n/N[1 - (1/N)]^{n-1}$. If sampling is *without* replacement, the number of sample points in $\mathcal{S}$ is $N(N - 1) \cdots (N - n + 1)/n!$ because the order of selection is immaterial. $n! = 1 \cdot 2 \cdot \cdots \cdot n$ is the product of the first n positive integers, $0! = 1$. The number of sample points that includes w_i is $M(i) = (N - 1)(N - 2) \cdots (N - n + 1)/(n - 1)!$. Recall from Chapter 2 that a sample $\mathbf{S}_n$ is called *random without replacement* (RSWOR) if all possible samples are equally probable. Thus, under RSWOR,

$$\Pr\{w_i \in \mathbf{S}_n\} = \frac{n!M(i)}{N(N - 1) \cdots (N - n + 1)} = \frac{n}{N}$$

for all $i = 1, \ldots, N$.

We now consider events that depend on the attributes of the elements of a population. Suppose that we sample to obtain information on the number of defective (nonstandard) elements in a population. The attribute in this case is "the element complies with the requirements of the standard." Suppose that M out of N elements in $\mathcal{L}_N$ are nondefective (have the attribute). Let E_j be the event that j out of the n elements in the sample are nondefective. Notice that $E_0, \ldots, E_n$ is a partition of the sample space. What is the probability, under RSWR, of E_j? Let K_j^n denote the number of sample points in which j out of n are G elements (nondefective) and $(n - j)$ elements are D (defective). To determine K_j^n, we can proceed in the following way.

First choose j Gs and $(n - j)$ Ds from the population. This can be done in $M^j(N - M)^{n-j}$ different ways. We now have to assign the j Gs into j out of n components of the vector $(w_1, \ldots, w_n)$. This can be done in $n(n - 1) \cdots (n - j + 1)/j!$ possible ways. This is known as the number of combinations of j out of n:

$$\binom{n}{j} = \frac{n!}{j!(n - j)!} \qquad j = 0, 1, \ldots, n \tag{4.1.6}$$

Hence, $K_j^n = \binom{n}{j}M^j(N - M)^{n-j}$. Because every sample is equally probable under RSWR,

$$\Pr\{E_j\} = K_j^n/N^n = \binom{n}{j}P^j(1 - P)^{n-j} \qquad j = 0, \ldots, n \tag{4.1.7}$$

where $P = M/N$. If sampling is without replacement, then

$$K_j^n = \binom{M}{j}\binom{N-M}{n-j}$$

and

$$\Pr\{E_j\} = \frac{\binom{M}{j}\binom{N-M}{n-j}}{\binom{N}{n}} \tag{4.1.8}$$

These results are valid because the order of selection is immaterial for the event E_j.

These probabilities of E_j under RSWR and RSWOR are called, respectively, the *binomial* and *hypergeometric probabilities*.

EXAMPLE **4.8**   The experiment consists of randomly transmitting a sequence of binary signals, 0 or 1. What is the probability that 3 out of 6 signals are 1s? Let E_3 denote this event.

The sample space of 6 signals consists of 2^6 points. Each point is equally probable. The probability of E_3 is

$$\Pr\{E_3\} = \binom{6}{3}\frac{1}{2^6} = \frac{6\cdot5\cdot4}{1\cdot2\cdot3\cdot64}$$
$$= \frac{20}{64} = \frac{5}{16} = .3125 \quad\blacksquare$$

EXAMPLE **4.9**   Two of 10 television sets are defective. A RSWOR of $n = 2$ sets is chosen. What is the probability that the two sets in the sample are good (nondefective)? This is the hypergeometric probability of E_0 when $M = 2, N = 10, n = 2$:

$$\Pr\{E_0\} = \frac{\binom{8}{2}}{\binom{10}{2}} = \frac{8\cdot7}{10\cdot9} = .622 \quad\blacksquare$$

4.1.5   Conditional Probabilities and Independence of Events

In Chapter 3, we introduced the notion of conditional frequency distributions. In this section, we discuss a similar notion of conditional probabilities. When different events are related, the realization of one event may provide us with relevant information that may help us improve our probability assessment of the other event(s). In Section 4.1.3, we gave an example of three machines that manufacture the same part but with different production rates and different proportions of defective parts in their output. The experiment was to choose a part at random from the mixed yield of the three machines. We saw earlier that the probability that the chosen part is nondefective is .989. If we could identify, before the quality test, the machine from which the part came, the probabilities of nondefective part would be conditional on this information.

The probability of choosing a nondefective part at random from machine M_1 is .99. If we are given the information that the machine is M_2, the probability is .98, and given machine M_3, the probability is .995. These probabilities are called **conditional probabilities**. The information given changes our probabilities.

We now formally define the concept of conditional probability.

Let A and B be two events such that $\Pr\{B\} > 0$. The conditional probability of A, given B, is

$$\Pr\{A \mid B\} = \frac{\Pr\{A \cap B\}}{\Pr\{B\}} \tag{4.1.9}$$

EXAMPLE **4.10** The random experiment is to measure the length of a steel bar. The sample space is $S = (19.5, 20.5)$ (in cm). The probability function assigns any subinterval a probability equal to its length. Let $A = (19.5, 20.1)$ and $B = (19.8, 20.5)$. $\Pr\{B\} = .7$. Suppose we are told that the length belongs to the interval B, and we have to guess whether it belongs to A. We compute the conditional probability

$$\Pr\{A \mid B\} = \frac{\Pr\{A \cap B\}}{\Pr\{B\}} = \frac{.3}{.7} = .4286$$

However, if the information that the length belongs to B is not given, then $\Pr\{A\} = .6$. Thus, there is a difference between conditional and nonconditional probabilities. This indicates that the two events A and B are dependent. ∎

Two events A, B are called **independent** if

$$\Pr\{A \mid B\} = \Pr\{A\}$$

If A and B are independent events, then

$$\Pr\{A\} = \Pr\{A \mid B\} = \frac{\Pr\{A \cap B\}}{\Pr\{B\}}$$

or, equivalently,

$$\Pr\{A \cap B\} = \Pr\{A\}\Pr\{B\}$$

If there are more than two events, $A_1, A_2, \ldots, A_n$, we say that the events are *pairwise independent* if

$$\Pr\{A_i \cap A_j\} = \Pr\{A_i\}\Pr\{A_j\} \quad \text{for all} \quad i \neq j, \ i,j = 1, \ldots, n$$

The n events are said to be *mutually independent* if, for any subset of k events, $k = 2, \ldots, n$, indexed by $A_{i_1}, \ldots, A_{i_k}$:

$$\Pr\{A_{i_1} \cap A_{i_2} \cap \cdots \cap A_{i_k}\} = \Pr\{A_{i_1}\} \cdots \Pr\{A_{i_n}\}$$

In particular, if n events are mutually independent, then

$$\Pr\left\{\bigcap_{i=1}^{n} A_i\right\} = \prod_{i=1}^{n} \Pr\{A_i\} \tag{4.1.10}$$

One can show examples of events that are pairwise independent but *not* mutually independent.

One can further show that if two events are independent, then the corresponding complementary events are independent (see the exercises). Furthermore, if n events are mutually independent, then any pair of events is pairwise independent, every three events are triplewise independent, and so on.

EXAMPLE 4.11   Five identical parts are manufactured in a given production process. Let $E_1, \ldots, E_5$ be the events that these five parts comply with the quality specifications (nondefective). Under the model of mutual independence, the probability that all five parts are indeed nondefective is

$$\Pr\{E_1 \cap E_2 \cap \cdots \cap E_5\} = \Pr\{E_1\}\Pr\{E_2\} \cdots \Pr\{E_5\}$$

Because these parts come from the same production process, we can assume that $\Pr\{E_i\} = p$, all $i = 1, \ldots, 5$. Thus, the probability that all five parts are nondefective is p^5.

What is the probability that one part is defective and the other four are nondefective? Let A_1 be the event that one of five parts is defective. To simplify the notation, we write the intersection of events as their product. Thus,

$$A_1 = E_1^c E_2 E_3 E_4 E_5 \cup E_1 E_2^c E_3 E_4 E_5 \cup E_1 E_2 E_3^c E_4 E_5$$
$$\cup E_1 E_2 E_3 E_4^c E_5 \cup E_1 E_2 E_3 E_4 E_5^c$$

A_1 is the union of five disjoint events. Therefore

$$\Pr\{A_1\} = \Pr\{E_1^c E_2 \cdots E_5\} + \cdots + \Pr\{E_1 E_2 \cdots E_5^c\}$$
$$= 5p^4(1 - p)$$

Indeed, because $E_1, \ldots, E_5$ are mutually independent events

$$\Pr\{E_1^c E_2 \cdots E_5\} = \Pr\{E_1^c\}\Pr\{E_2\} \cdots \Pr\{E_5\} = (1 - p)p^4$$

Also,

$$\Pr\{E_1 E_2^c E_3 E_4 E_5\} = (1 - p)p^4$$

and so on. Generally, if J_5 denotes the number of defective parts among the five, then

$$\Pr\{J_5 = i\} = \binom{5}{i}p^{(5-i)}(1 - p)^i \quad i = 0, 1, 2, \ldots, 5 \quad \blacksquare$$

4.1.6   Bayes' Formula and Its Applications

Bayes' formula, which is derived in this section, provides us with a fundamental formula for weighing the evidence in the data concerning unknown parameters or some unobservable events.

Suppose the results of a random experiment depend on some event(s) that is (are) not directly observable. The observable event is related to the unobservable one(s) via conditional probabilities. More specifically, suppose that $\{B_1, \ldots, B_m\}$

$(m \geq 2)$ is a partition of the sample space. The events $B_1, \ldots, B_m$ are not directly observable, or verifiable. The random experiment results in an event A (or its complement). The conditional probabilities $\Pr\{A \mid B_i\}$, $i = 1, \ldots, m$ are known. The question is whether, after observing the event A, we can assign probabilities to the events $B_1, \ldots, B_m$. To weigh the evidence that A has on $B_1, \ldots, B_m$, we first assume some probabilities, $\Pr\{B_i\}$, $i = 1, \ldots, m$, called **prior probabilities**. Prior probabilities express our degree of belief in the occurrence of the events B_i $(i = 1, \ldots, m)$. After observing the event A, we convert the prior probabilities of B_i $(i = 1, \ldots, m)$ into **posterior probabilities**, $\Pr\{B_i \mid A\}$, $i = 1, \ldots, m$, by using Bayes' formula:

$$\Pr\{B_i \mid A\} = \frac{\Pr\{B_i\}\Pr\{A \mid B_i\}}{\displaystyle\sum_{j=1}^{m}\Pr\{B_j\}\Pr\{A \mid B_j\}} \qquad i = 1, \ldots, m \qquad \text{(4.1.11)}$$

These posterior probabilities reflect the weight of evidence that the event A has concerning $B_1, \ldots, B_m$.

Bayes' formula can be obtained from the basic rules of probability. Indeed, assuming that $\Pr\{A\} > 0$,

$$\Pr\{B_i \mid A\} = \frac{\Pr\{A \cap B_i\}}{\Pr\{A\}}$$
$$= \frac{\Pr\{B_i\}\Pr\{A \mid B_i\}}{\Pr\{A\}}$$

Furthermore, because $\{B_1, \ldots, B_m\}$ is a partition of the sample space,

$$\Pr\{A\} = \sum_{j=1}^{m}\Pr\{B_j\}\Pr\{A \mid B_j\}$$

Substituting this expression into the preceding one, we obtain Bayes' formula.

The following example illustrates the applicability of Bayes' formula to a problem of decision making.

EXAMPLE **4.12**   Two vendors, B_1 and B_2, produce ceramic plates for a given production process of hybrid microcircuits. The parts of vendor B_1 have probability $p_1 = .10$ of being defective. The parts of vendor B_2 have probability $p_2 = .05$ of being defective. A delivery of $n = 20$ parts arrives, but the label identifying the vendor is missing. We wish to apply Bayes' formula to assign a probability that the package came from vendor B_1.

Suppose that it is a priori equally likely that the package was mailed by vendor B_1 or vendor B_2. In this case, the prior probabilities are $\Pr\{B_1\} = \Pr\{B_2\} = .5$. We inspect the 20 parts in the package and find $J_{20} = 3$ defective items. A is the event $\{J_{20} = 3\}$. The conditional probabilities of A, given B_i $(i = 1, 2)$ are

$$\Pr\{A \mid B_1\} = \binom{20}{3}p_1^3(1 - p_1)^{17}$$
$$= .1901$$

Similarly,

$$\Pr\{A \mid B_2\} = \binom{20}{3} p_2^3 (1 - p_2)^{17}$$
$$= .0596$$

According to Bayes' formula

$$\Pr\{B_1 \mid A\} = \frac{.5 \times .1901}{0.5 \times .1901 + .5 \times .0596} = .7613$$

$$\Pr\{B_2 \mid A\} = 1 - \Pr\{B_1 \mid A\} = .2387$$

Thus, after observing three defective parts in a sample of $n = 20$, we believe that the delivery came from vendor B_1. The posterior probability of B_1, given A, is more than three times higher than that of B_2 given A. The a priori odds of B_1 against B_2 were 1:1. The a posteriori odds are 19:6.   ■

Random Variables and Their Distributions

Random variables are formally defined as real-valued functions, $X(w)$, over the sample space, S, such that, events $\{w : X(w) \leq x\}$ can be assigned probabilities, for all $-\infty < x < \infty$, where w are the elements of S.

EXAMPLE **4.13**   Suppose that S is the sample space of all RSWOR of size n, from a finite population, $\mathcal{P}$, of size N, $1 \leq n < N$. The elements w of S are subsets of distinct elements of the population $\mathcal{P}$. A random variable $X(w)$ is some function that assigns w a finite real number—for example, the number of defective elements of w. In this example, $X(w) = 0, 1, \ldots, n$ and

$$\Pr\{X(w) = j\} = \frac{\binom{M}{j}\binom{N-M}{n-j}}{\binom{N}{n}} \qquad j = 0, \ldots, n$$

where M is the number of defective elements of $\mathcal{P}$.   ■

EXAMPLE **4.14**   Another example of a random variable is the compressive strength of a concrete cube of a certain dimension. In this example, the random experiment is to manufacture a concrete cube according to a specified process. The sample space S is the space of all cubes, w, that can be manufactured by this process. $X(w)$ is the compressive strength of w. The probability function assigns each event $\{w : X(w) \leq \xi\}$ a probability, according to some mathematical model that satisfies the laws of probability. Any continuous nondecreasing function $F(x)$, such that $\lim_{x \to -\infty} F(x) = 0$ and

$\lim_{x \to \infty} F(x) = 1$ will do the job. For example, for compressive strength of concrete cubes, the following model has been shown to fit experimental results:

$$
\Pr\{X(w) \le x\} =
\begin{cases}
0 & x \le 0 \\
\dfrac{1}{\sqrt{2\pi}\sigma} \displaystyle\int_0^x \dfrac{1}{y} \exp\left\{ -\dfrac{(\ln y - \mu)^2}{2\sigma^2} \right\} dy & 0 < x < \infty
\end{cases}
$$

The constants μ and σ, $-\infty < \mu < \infty$ and $0 < \sigma < \infty$, are called **parameters** of the model. Such parameters characterize the manufacturing process. ▪

In Chapter 2, we distinguished between *two types* of random variables: discrete and continuous.

Discrete random variables, $X(w)$, are random variables having a finite or countable range. For example, the number of defective elements in a random sample is a discrete random variable. The number of blemishes on a ceramic plate is a discrete random variable. A *continuous random variable* is one whose range consists of whole intervals of possible values. Weight, length, compressive strength, tensile strength, cycle time, output voltage, and so on are continuous random variables.

4.2.1   Discrete and Continuous Distributions

4.2.1.1   Discrete Random Variables

Suppose that a discrete random variable can assume the distinct values $x_0, x_1, x_2, \ldots, x_k$ (k is finite or infinite). The function

$$
p(x) = \Pr\{X(w) = x\} \qquad -\infty < x < \infty \tag{4.2.1}
$$

is called the **probability distribution function (p.d.f.)** of X.

Notice that if x is not one of the values in the specified range $S_X = \{x_j; j = 0, 1, \ldots, k\}$ then $\{X(w) = x\} = \phi$ and $p(x) = 0$. Thus, $p(x)$ assumes positive values only on the specified sequence S_X (S_X is also called the sample space of X), such that

1 $p(x_j) \ge 0, \quad j = 0, \ldots, k$ \hfill (4.2.2)

2 $\displaystyle\sum_{j=0}^{k} p(x_j) = 1$

EXAMPLE **4.15**   Suppose that the random experiment is to cast a die once. The sample points are six possible faces of the die, $\{w_1, \ldots, w_6\}$. Let $X(w_j) = j, j = 1, \ldots, 6$, be the random variable representing the face number. The probability model yields

$$
p(x) =
\begin{cases}
\dfrac{1}{6} & \text{if } x = 1, 2, \ldots, 6 \\
0 & \text{otherwise}
\end{cases}
\quad ▪
$$

EXAMPLE 4.16   Consider the example of Section 4.1.5, of drawing independently $n = 5$ parts from a production process, and counting the number of "defective" parts in this sample. The random variable is $X(w) = J_5$. $\mathcal{S}_X = \{0, 1, \ldots, 5\}$ and the p.d.f. is

$$p(x) = \begin{cases} \binom{5}{x} p^{5-x}(1-p)^x & x = 0, 1, \ldots, 5 \\ 0 & \text{otherwise} \end{cases} \quad \blacksquare$$

The probability of the event $\{X(w) \leq x\}$, for any $-\infty < x < \infty$, can be computed by summing the probabilities of the values in $\mathcal{S}_X$, which belong to the interval $(-\infty, x]$. This sum is called the **cumulative distribution function** (c.d.f.) of X and is denoted by

$$\begin{aligned} P(x) &= \Pr\{X(w) \leq x\} \\ &= \sum_{\{x_j \leq x\}} p(x_j) \end{aligned} \tag{4.2.3}$$

where $x_j \in \mathcal{S}_X$.

The c.d.f. corresponding to Example 4.16 is

$$P(x) = \begin{cases} 0 & x < 0 \\ \sum_{j=0}^{[x]} \binom{5}{j} p^{5-j}(1-p)^j & 0 \leq x < 5 \\ 1 & 5 \leq x \end{cases}$$

where $[x]$ denotes the integer part of x—that is, the *largest integer* smaller than or equal to x.

Generally, the graph of the p.d.f. of a discrete variable is a bar chart (see Figure 4.1a). The corresponding c.d.f. is a step function, as shown in Figure 4.1b.

4.2.1.2   Continuous Random Variables

In the case of continuous random variables, the model assigns the variable under consideration a function $F(x)$, which is:

1   Continuous

2   Nondecreasing—that is, if $x_1 < x_2$, then $F(x_1) \leq F(x_2)$

and

3   $\lim_{x \to -\infty} F(x) = 0$ and $\lim_{x \to \infty} F(x) = 1$

Such a function can serve as a c.d.f. for X.

An example of a c.d.f. for a continuous random variable that assumes nonnegative values—for example, the total operation time until a part fails—is

$$F(x) = \begin{cases} 0 & \text{if } x \leq 0 \\ 1 - e^{-x} & \text{if } x > 0 \end{cases}$$

FIGURE   4.1

The graph of the p.d.f. (a) and c.d.f. (b) of $P(x) = \sum_{j=0}^{[x]} \frac{\binom{5}{j}}{2^5}$, a random variable

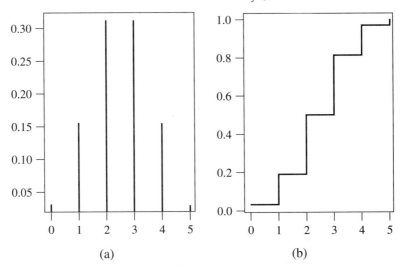

(a)   (b)

This function (see Figure 4.2) is continuous and monotonically increasing, and $\lim_{x \to \infty} F(x) = 1 - \lim_{x \to \infty} e^{-x} = 1$. If the c.d.f. of a continuous random variable can be represented as

$$F(x) = \int_{-\infty}^{x} f(y)dy \qquad \text{(4.2.4)}$$

FIGURE   4.2

The c.d.f. of $F(x) = 1 - e^{-x}$

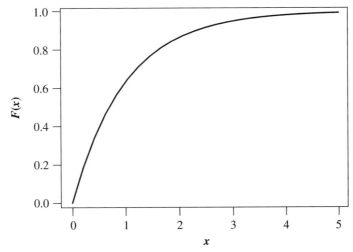

for some $f(y) \geq 0$, then we say that $F(x)$ is **absolutely continuous** and $f(x) = (d/dx)F(x)$. (The derivative $f(x)$ may not exist on a finite number of x values in any finite interval.) The function $f(x)$ is called the **probability density function** (p.d.f.) of X.

In the preceding example of total operational time, the p.d.f. is

$$f(x) = \begin{cases} 0 & \text{if } x < 0 \\ e^{-x} & \text{if } x \geq 0 \end{cases}$$

Thus, as in the discrete case, we have $F(x) = \Pr\{X \leq x\}$. It is now possible to write

$$\Pr\{a \leq X < b\} = \int_a^b f(t)dt = F(b) - F(a) \tag{4.2.5}$$

or

$$\Pr\{X \geq b\} = \int_b^\infty f(t)dt = 1 - F(b) \tag{4.2.6}$$

Thus, if X has the above c.d.f., then

$$\Pr\{1 \leq X \leq 2\} = F(2) - F(1) = e^{-1} - e^{-2} = .2325$$

Certain phenomena need more complicated modeling. The random variables under consideration may not have purely discrete or purely absolutely continuous distribution. Many random variables have c.d.f.'s that are absolutely continuous within certain intervals, and have jump points (points of discontinuity) at the endpoints of the intervals, as shown in Figure 4.3. Distributions of such random variables can be expressed as mixtures of purely discrete c.d.f., $F_d(x)$, and of absolutely continuous

FIGURE　4.3

The c.d.f. of the mixture distribution $F(x) = 0.5 \, (1 - e^{-x}) + 0.5 \times e^{-1} \sum_{j=0}^{[x]} \frac{1}{j!}$

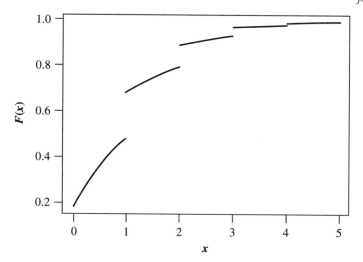

c.d.f., $F_{ac}(x)$. That is,

$$F(x) = pF_d(x) + (1-p)F_{ac}(x) \qquad -\infty < x < \infty \qquad \text{(4.2.7)}$$

where $0 \le p \le 1$.

EXAMPLE **4.17**   A distribution that is a mixture of discrete and continuous distributions is obtained, for example, when a measuring instrument is not sensitive enough to measure small quantities or large quantities that are outside its range. This could be the case for a weighing instrument that assigns the value 0 mg to any weight smaller than 1 mg, the value 1 g to any weight greater than 1 g, and the correct weight to values in between.

Another example is the total number of minutes within a given working hour that a service station is busy serving customers. In this case, the c.d.f. has a jump at 0, of height p, which is the probability that the service station is idle at the beginning of the hour, and no customer arrives during that hour. In this case,

$$F(x) = p + (1-p)G(x) \qquad 0 \le x < \infty$$

where $G(x)$ is the c.d.f. of the total service time, $G(0) = 0$.   ∎

4.2.2 Expected Values and Moments of Distributions

The **expected value** of a function $g(X)$, under the distribution $F(x)$, is

$$E_F\{g(X)\} = \begin{cases} \displaystyle\int_{-\infty}^{\infty} g(x)f(x)dx & \text{if } X \text{ is continuous} \\[2mm] \displaystyle\sum_{j=0}^{k} g(x_j)p(x_j) & \text{if } X \text{ is discrete} \end{cases}$$

In particular,

$$\mu_l(F) = E_F\{X^l\} \qquad l = 1, 2, \ldots \qquad \text{(4.2.8)}$$

is called the **lth moment** of $F(x)$. $\mu_1(F) = E_F\{X\}$ is the expected value of X, or the *population mean*, according to the model $F(x)$.

Moments around $\mu_1(F)$ are called **central moments**:

$$\mu_l^*(F) = E\{(X - \mu_1(F))^l\} \qquad l = 1, 2, 3, \ldots \qquad \text{(4.2.9)}$$

Obviously, $\mu_1^*(F) = 0$. The second central moment is called the **variance** of $F(x)$, $V_F\{X\}$.

In the following discussion, as long as there is no room for confusion, the notation $\mu_l(F)$ will be simplified to μ_l.

Expected values of a function $g(X)$, and in particular the moments, may not exist because an integral $\displaystyle\int_{-\infty}^{\infty} x^l f(x)dx$ may not be well defined. An example of such a case is the distribution called the *Cauchy distribution*, with p.d.f.

$$f(x) = \frac{1}{\pi} \cdot \frac{1}{1+x^2} \qquad -\infty < x < \infty$$

Notice that under this model, moments do not exist for any $l = 1, 2, \ldots$. Indeed, the integral

$$\frac{1}{\pi} \int_{-\infty}^{\infty} \frac{x}{1+x^2} dx$$

does not exist. If the second moment exists, then

$$V\{X\} = \mu_2 - \mu_1^2$$

EXAMPLE 4.18   Consider the random experiment of casting a die once. The random variable, X, is the face number. Thus $p(x) = \frac{1}{6}, x = 1, \ldots, 6$ and

$$\mu_1 = E\{X\} = \frac{1}{6} \sum_{j=1}^{6} j = \frac{6(6+1)}{2 \times 6} = \frac{7}{2} = 3.5$$

$$\mu_2 = \frac{1}{6} \sum_{j=1}^{6} j^2 = \frac{6(6+1)(2 \times 6+1)}{6 \times 6} = \frac{7 \times 13}{6} = \frac{91}{6} = 15.167$$

The variance is

$$V\{X\} = \frac{91}{6} - \left(\frac{7}{2}\right)^2 = \frac{182 - 147}{12}$$

$$= \frac{35}{12} \quad \blacksquare$$

EXAMPLE 4.19 X has a continuous distribution with p.d.f.

$$f(x) = \begin{cases} 0 & \text{otherwise} \\ 1 & \text{if } 1 \le x \le 2 \end{cases}$$

Thus,

$$\mu_1 = \int_1^2 x dx = \frac{1}{2} \left(x^2 \Big|_1^2\right) = \frac{1}{2}(4-1) = 1.5$$

$$\mu_2 = \int_1^2 x^2 dx = \frac{1}{3} \left(x^3 \Big|_1^2\right) = \frac{7}{3}$$

$$V\{X\} = \mu_2 - \mu_1^2 = \frac{7}{3} - \frac{9}{4} = \frac{28 - 27}{12} = \frac{1}{12} \quad \blacksquare$$

The following is a useful formula when X assumes only positive values—that is, $F(x) = 0$ for all $x \le 0$,

$$\mu_1 = \int_0^\infty (1 - F(x))dx \qquad \text{(4.2.10)}$$

for continuous c.d.f. $F(x)$. Indeed,

$$\mu_1 = \int_0^\infty xf(x)dx$$

$$= \int_0^\infty \left(\int_0^x dy \right) f(x)dx$$

$$= \int_0^\infty \left(\int_y^\infty f(x)dx \right) dy$$

$$= \int_0^\infty (1 - F(y))dy$$

For example, suppose that $f(x) = \mu e^{-\mu x}$, for $x \geq 0$. Then $F(x) = 1 - e^{-\mu x}$ and

$$\int_0^\infty (1 - F(x))dx = \int_0^\infty e^{-\mu x}dx = \frac{1}{\mu}.$$

When X is discrete, assuming the values $\{1, 2, 3, \ldots\}$, then we have a similar formula

$$E\{X\} = 1 + \sum_{i=1}^\infty (1 - F(i))$$

4.2.3 The Standard Deviation, Quantiles, Measures of Skewness and Kurtosis

The **standard deviation** of a distribution $F(x)$ is $\sigma = (V\{X\})^{1/2}$. Recall that the standard deviation is used as a measure of dispersion of a distribution. An important theorem in probability theory, called the **Chebyshev inequality**, relates the standard deviation to the probability of deviation from the mean. More formally, the theorem states that if σ exists, then

$$\Pr\{|X - \mu_1| > \lambda\sigma\} \leq \frac{1}{\lambda^2} \qquad \text{(4.2.11)}$$

Thus, by this theorem, the probability that a random variable will deviate from its expected value by more than 3 standard deviations is less than or equal to 1/9, whatever the distribution is. This theorem has important implications, which will be highlighted later.

The pth quantile of a distribution $F(x)$ is the smallest value of x, ξ_p such that $F(x) \geq p$. We also write $\xi_p = F^{-1}(p)$.

For example, if $F(x) = 1 - e^{-\lambda x}, 0 \leq x < \infty$, where $0 < \lambda < \infty$, then ξ_p is such that

$$F(\xi_p) = 1 - e^{-\lambda \xi_p} = p$$

Solving for ξ_p, we get

$$\xi_p = -\frac{1}{\lambda} \cdot \ln(1 - p)$$

The *median* of $F(x)$ is $F^{-1}(.5) = \xi_{.5}$. Similarly, $\xi_{.25}$ and $\xi_{.75}$ are the first and third quartiles of F.

A distribution $F(x)$ is **symmetric** about the mean, $\mu_1(F)$, if

$$F(\mu_1 + \delta) = 1 - F(\mu_1 - \delta)$$

for all $\delta \geq 0$.

In particular, if F is symmetric, then $F(\mu_1) = 1 - F(\mu_1)$ or $\mu_1 = F^{-1}(.5) = \xi_{.5}$. Accordingly, the mean and median of a symmetric distribution coincide. In terms of the p.d.f., a distribution is symmetric about its mean if

$$f(\mu_1 + \delta) = f(\mu_1 - \delta) \quad \text{for all} \quad \delta \geq 0$$

A commonly used index of *skewness* (asymmetry) is

$$\beta_3 = \frac{\mu_3^*}{\sigma^3} \tag{4.2.12}$$

where μ_3^* is the third central moment of F. One can prove that **if $F(x)$ is symmetric, then $\beta_3 = 0$**. If $\beta_3 > 0$, we say that $F(x)$ is positively skewed; otherwise, it is negatively skewed.

EXAMPLE **4.20**   Consider the binomial distribution, with p.d.f.

$$p(x) = \binom{n}{x} p^x (1-p)^{n-x} \qquad x = 0, 1, \ldots, n$$

In this case

$$\mu_1 = \sum_{x=0}^{n} x \binom{n}{x} p^x (1-p)^{n-x}$$

$$= np \sum_{x=1}^{n} \binom{n-1}{x-1} p^{x-1} (1-p)^{n-1-(x-1)}$$

$$= np \sum_{j=0}^{n-1} \binom{n-1}{j} p^j (1-p)^{n-1-j}$$

$$= np$$

Indeed,

$$x\binom{n}{x} = x \frac{n!}{x!(n-x)!} = \frac{n!}{(x-1)!((n-1)-(x-1))!}$$

$$= n\binom{n-1}{x-1}$$

Similarly, we can show that

$$\mu_2 = n^2 p^2 + np(1-p)$$

and

$$\mu_3 = np[n(n-3)p^2 + 3(n-1)p + 1 + 2p^2]$$

The third central moment is

$$\mu_3^* = \mu_3 - 3\mu_2\mu_1 + 2\mu_1^3$$
$$= np(1-p)(1-2p)$$

Furthermore,

$$V\{X\} = \mu_2 - \mu_1^2$$
$$= np(1-p)$$

Hence,

$$\sigma = \sqrt{np(1-p)}$$

and the index of asymmetry is

$$\beta_3 = \frac{\mu_3^*}{\sigma^3} = \frac{np(1-p)(1-2p)}{(np(1-p))^{3/2}}$$

$$= \frac{1-2p}{\sqrt{np(1-p)}}$$

Thus, if $p = \frac{1}{2}$, then $\beta_3 = 0$ and the distribution is symmetric. If $p < \frac{1}{2}$, the distribution is positively skewed, and it is negatively skewed if $p > \frac{1}{2}$. ∎

In Chapter 2 we mentioned also the index of *kurtosis* (steepness). This is given by

$$\beta_4 = \frac{\mu_4^*}{\sigma^4} \qquad\qquad\qquad (4.2.13)$$

EXAMPLE **4.21**   Consider the exponential c.d.f.

$$F(x) = \begin{cases} 0 & \text{if } x < 0 \\ 1 - e^{-x} & \text{if } x \geq 0 \end{cases}$$

The p.d.f. is $f(x) = e^{-x}$, $x \geq 0$. Thus, for this distribution

$$\mu_1 = \int_0^\infty xe^{-x}dx = 1$$

$$\mu_2 = \int_0^\infty x^2e^{-x}dx = 2$$

$$\mu_3 = \int_0^\infty x^3e^{-x}dx = 6$$

$$\mu_4 = \int_0^\infty x^4e^{-x}dx = 24$$

Therefore,

$$V\{X\} = \mu_2 - \mu_1^2 = 1$$
$$\sigma = 1$$
$$\mu_4^* = \mu_4 - 4\mu_3 \cdot \mu_1 + 6\mu_2\mu_1^2 - 3\mu_1^4$$
$$= 24 - 4 \times 6 \times 1 + 6 \times 2 \times 1 - 3 = 9$$

Finally, the index of kurtosis is

$$\beta_4 = 9 \quad \blacksquare$$

4.2.4 Moment-Generating Functions

The **moment-generating function (m.g.f.)** of a distribution of X is defined as a function of a real variable t,

$$M(t) = E\{e^{tX}\} \tag{4.2.14}$$

$M(0) = 1$ for all distributions. $M(t)$, however, may not exist for some $t \neq 0$. To be useful, it is sufficient that $M(t)$ exists in some interval containing $t = 0$.

For example, if X has a continuous distribution, with p.d.f.

$$f(x) = \begin{cases} \dfrac{1}{b-a} & \text{if } a \leq x \leq b, \quad a < b \\ 0 & \text{otherwise} \end{cases}$$

then

$$M(t) = \frac{1}{b-a} \int_a^b e^{tx} dx = \frac{1}{t(b-a)}(e^{tb} - e^{ta})$$

This is a differentiable function of t, for all t, $-\infty < t < \infty$.

However, if for $0 < \lambda < \infty$,

$$f(x) = \begin{cases} \lambda e^{-\lambda x} & 0 \leq x < \infty \\ 0 & x < 0 \end{cases}$$

then

$$M(t) = \lambda \int_0^\infty e^{tx - \lambda x} dx$$

$$= \frac{\lambda}{\lambda - t} \quad t < \lambda$$

This m.g.f. exists only for $t < \lambda$. The m.g.f. $M(t)$ is a transform of the distribution $F(x)$, and the correspondence between $M(t)$ and $F(x)$ is one-to-one. In this example, $M(t)$ is the **Laplace transform** of the p.d.f. $\lambda e^{-\lambda x}$. This correspondence is often useful in identifying the distributions of some statistics, as will be shown later.

Another useful property of the m.g.f. $M(t)$ is that often we can obtain the moments of $F(x)$ by differentiating $M(t)$. More specifically, consider the rth order derivative of $M(t)$. Assuming that this derivative exists, and differentiation can be interchanged with integration (or summation), then

$$M^{(r)}(t) = \frac{d^r}{dt^r} \int e^{tx} f(x) dx = \int \left(\frac{d^r}{dt^r} e^{tx} \right) f(x) dx$$

$$= \int x^r e^{tx} f(x) dx$$

Thus, if these operations are justified, then

$$M^{(r)}(t)\Big|_{t=0} = \int x^r f(x)dx = \mu_r \qquad (4.2.15)$$

In the following sections we will illustrate the usefulness of the m.g.f.

4.3 Families of Discrete Distributions

In this section, we discuss several families of discrete distributions, and illustrate possible applications in modeling industrial phenomena.

4.3.1 The Binomial Distribution

Consider n identical independent trials. In each trial, the probability of success is fixed at some value p, and successive events of "success" or "failure" are *independent*. Such trials are called **Bernoulli trials**. The distribution of the number of successes, J_n, is binomial, with p.d.f.

$$b(j; n, p) = \binom{n}{j} p^j (1-p)^{n-j} \qquad j = 0, 1, \ldots, n \qquad (4.3.1)$$

This p.d.f. was derived in Example 4.11 as a special case.

A binomial random variable, with parameters (n, p) will be designated as $B(n, p)$. Parameter n is a given integer, and p belongs to the interval $(0, 1)$. The collection of all such binomial distributions is called the **binomial family**.

The binomial distribution is a proper model whenever we have a sequence of independent binary events (0–1, or success and failure) with the same probability of success.

EXAMPLE **4.22**   We draw a random sample of $n = 10$ items from a mass production line of lightbulbs. Each lightbulb undergoes an inspection, and if it complies with the production specifications, we say that the bulb is compliant (successful event). Let $X_i = 1$ if the ith bulb is compliant and $X_i = 0$ otherwise. If we can assume that the probability of $\{X_i = 1\}$ is the same, p, for all bulbs, and if the n events are mutually independent, then the number of bulbs in the sample that comply with the specifications—that is,

$$J_n = \sum_{i=1}^{n} X_i$$—has the binomial p.d.f., $b(i; n, p)$. Notice that if we draw a sample at random **with** replacement, RSWR, from a lot of size N, which contains M compliant units, then J_n is $B(n, M/N)$.

Indeed, if sampling is with replacement, the probability that the ith item selected is compliant is $p = M/N$ for all $i = 1, \ldots, n$. Furthermore, selections are independent of each other.   ∎

The binomial c.d.f. will be denoted by $B(i; n, p)$. Recall that

$$B(i; n, p) = \sum_{j=0}^{i} b(j; n, p) \tag{4.3.2}$$

$i = 0, 1, \ldots, n$. The m.g.f. of $B(n, p)$ is

$$
\begin{aligned}
M(t) &= E\{e^{tX}\} \\
&= \sum_{j=0}^{n} \binom{n}{j} (pe^t)^j (1-p)^{n-j} \\
&= (pe^t + (1-p))^n \qquad -\infty < t < \infty
\end{aligned}
\tag{4.3.3}
$$

Notice that

$$M'(t) = n(pe^t + (1-p))^{n-1}pe^t$$

and

$$M''(t) = n(n-1)p^2 e^{2t}(pe^t + (1-p))^{n-2} + npe^t(pe^t + (1-p))^{n-1}$$

The expected value and variance of $B(n, p)$ are

$$E\{J_n\} = np \tag{4.3.4}$$

and

$$V\{J_n\} = np(1-p) \tag{4.3.5}$$

This was shown in Example 4.20 and can be verified directly by the above formulas of $M'(t)$ and $M''(t)$. To obtain the values of $b(i; n, p)$, we can use MINITAB. For example, suppose we wish to tabulate the values of the p.d.f., $b(i; n, p)$, and those of the c.d.f., $B(i; n, p)$, for $n = 30$ and $p = .60$. We first put in column C1 the integers $0, 1, \ldots, 30$ and put the values of $b(i; 30, .60)$ in column C2, and those of $B(i; 30, .60)$ in C3. This is done with the following commands:

```
MTB> Set C1
DATA> 1(0 : 30/1)1
DATA> End.
MTB> PDF C1 C2;
SUBC> Binomial 30 .60.
MTB> CDF C1 C3;
SUBC> Binomial 30 .60.
```

In Table 4.1 we present some of these values.

After tabulating the values of the c.d.f., we can obtain the quantiles (or fractiles) of the distribution. Recall that in the discrete case, the pth quantile of a random variable X is

$$x_p = \text{smallest } x \quad \text{such that} \quad F(x) \geq p$$

Thus, from Table 4.1, we find that the lower quartile, the median, and the upper

TABLE **4.1**

Values of the p.d.f. and c.d.f. of $B(30, .6)$

i	$b(i; 30, .6)$	$B(i; 30, .6)$
8	.0002	.0002
9	.0006	.0009
10	.0020	.0029
11	.0054	.0083
12	.0129	.0212
13	.0269	.0481
14	.0489	.0971
15	.0783	.1754
16	.1101	.2855
17	.1360	.4215
18	.1474	.5689
19	.1396	.7085
20	.1152	.8237
21	.0823	.9060
22	.0505	.9565
23	.0263	.9828
24	.0115	.9943
25	.0041	.9985
26	.0012	.9997
27	.0003	1.0000

quartile of $B(30, .6)$ are $Q_1 = 16$, $M_e = 18$, and $Q_3 = 20$. These values can also be obtained directly with the following MINITAB command:

```
MTB> Set C1
DATA> 1(.24:.75/.25)1
DATA> End.
MTB> InvCDF C1 C2;
SUBC> Binomial 30 .6.
```

The value of Q_1, M_e and Q_3 are stored in C2.

Figure 4.4 shows the p.d.f. of three binomial distributions, with $n = 50$ and $p = .25, .50$, and $.75$. We see that if $p = .25$, the p.d.f. is positively skewed. When $p = .5$, it is symmetric, and when $p = .75$ it is negatively skewed. This is in accordance with the index of skewness β_3, which was presented in Example 4.20.

4.3.2 The Hypergeometric Distribution

Let J_n denote the number of units in a RSWOR of size n, from a population of size N, having a certain property. The number of population units before sampling that have this property is M. The distribution of J_n is called the **hypergeometric**

FIGURE   4.4

The p.d.f. of $B(50, p)$, $p = .25, .50, .75$

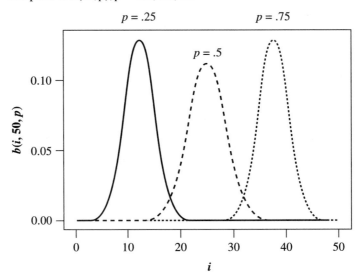

distribution. We denote a random variable having such a distribution by $H(N, M, n)$. The p.d.f. of J_n is

$$h(j; N, M, n) = \frac{\binom{M}{j}\binom{N-M}{n-j}}{\binom{N}{n}} \qquad j = 0, \ldots, n \tag{4.3.6}$$

This formula was shown in Section 4.1.4.

The c.d.f. of $H(N, M, n)$ will be designated by $H(j; N, M, n)$. In Table 4.2, we present the p.d.f. and c.d.f. of $H(75, 15, 10)$.

We cannot compute the p.d.f. or c.d.f. directly by MINITAB. Instead, we furnish a special executable program that computes the p.d.f. and c.d.f. of $H(N, M, n)$. The name of the program is HYPERG.EXE, and it is stored in subdirectory CHA4. The

TABLE   4.2

The p.d.f. and c.d.f. of $H(75, 15, 10)$

j	$h(j; 75, 15, 10)$	$H(j; 75, 15, 10)$
0	.0910	.0910
1	.2675	.3585
2	.3241	.6826
3	.2120	.8946
4	.0824	.9770
5	.0198	.9968
6	.0029	.9997
7	.0003	1.0000

output is stored in C:\ISTAT\DATA\HYPERG.DAT. For S-PLUS functions, see Appendix VI.

In Figure 4.5 we show the p.d.f. of $H(500, 350, 100)$, which was computed with this program.

FIGURE 4.5
The p.d.f. $h(i; 500, 350, 100)$.

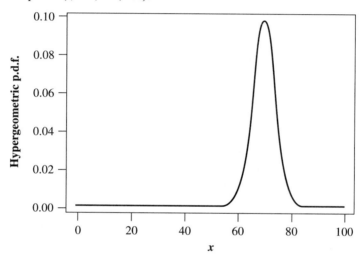

The expected value and variance of $H(N, M, n)$ are:

$$E\{J_n\} = n \cdot \frac{M}{N} \qquad (4.3.7)$$

and

$$V\{J_n\} = n \cdot \frac{M}{N} \cdot \left(1 - \frac{M}{N}\right)\left(1 - \frac{n-1}{N-1}\right) \qquad (4.3.8)$$

Notice that when $n = N$, the variance of J_n is $V\{J_N\} = 0$. Indeed, if $n = N$, $J_N = M$, which is not a random quantity. Derivation of these formulas is given in Section 5.2.2. There is no simple expression for the m.g.f.

If the sample size n is small relative to N—that is, if $n/N << 0.1$—the hypergeometric p.d.f. can be approximated by that of the binomial $B(n, M/N)$. In Table 4.3, we compare the p.d.f. of $H(500, 350, 20)$ to that of $B(20, .7)$.

The expected value and variance of the binomial and the hypergeometric distributions are compared in Table 4.4. We see that the expected values have the same formula, but that the variance formulas differ by the correction factor $(N - n)/(N - 1)$, which becomes 1 when $n = 1$ and 0 when $n = N$.

EXAMPLE 4.23   At the end of a production day, printed circuit boards (PCBs) soldered by a wave-soldering process are subjected to a sampling audit. A RSWOR of size n is drawn

TABLE   4.3

The p.d.f. of $H(500, 350, 20)$ and $B(20, .7)$

i	$h(i; 500, 350, 20)$	$b(i; 20, .7)$
5	.00003	.00004
6	.00016	.00022
7	.00082	.00102
8	.00333	.00386
9	.01093	.01202
10	.02928	.03082
11	.06418	.06537
12	.11491	.11440
13	.16715	.16426
14	.19559	.19164
15	.18129	.17886
16	.12999	.13042
17	.06949	.07160
18	.02606	.02785
19	.00611	.00684
20	.00067	.00080

TABLE   4.4

The expected value and variance of the hypergeometric and binomial distribution

	$H(N, M, n)$ **Hypergeometric**	$B\left(n, \dfrac{M}{N}\right)$ **Binomial**
Expected Value	$n\dfrac{M}{N}$	$n\dfrac{M}{N}$
Variance	$n\dfrac{M}{N}\left(1 - \dfrac{M}{N}\right)\left(1 - \dfrac{n-1}{N-1}\right)$	$n\dfrac{M}{N}\left(1 - \dfrac{M}{N}\right)$

from the lot, which consists of all the PCBs produced on that day. If the sample has any defective PCB, another RSWOR of size $2n$ is drawn from the lot. If there are more than three defective boards in the combined sample, the lot is sent for rectification, a process in which every PCB is inspected. If the lot consists of $N = 100$ PCBs, and the number of defective ones is $M = 5$, what is the probability that the lot will be rectified, when $n = 10$?

Let J_1 be the number of defective items in the first sample. If $J_1 > 3$, then the lot is rectified without taking a second sample. If $J_1 = 1$, 2, or 3, a second sample is drawn. Thus, if R denotes the event "the lot is sent for rectification,"

$$\Pr\{R\} = 1 - H(3; 100, 5, 10)$$

$$+ \sum_{i=1}^{3} h(i; 100, 5, 10) \cdot [1 - H(3 - i; 90, 5 - i, 20)]$$

$$= .00025 + .33939 \times .03313$$
$$+ .07022 \times .12291$$
$$+ .00638 \times .397 = .0227 \quad \blacksquare$$

4.3.3 The Poisson Distribution

A third discrete distribution that plays an important role in quality control is the **Poisson distribution**, denoted by $P(\lambda)$. This is sometimes called the distribution of rare events because it is used as an approximation to the binomial distribution when the sample size, n, is large and the proportion of defectives, p, is small. The parameter λ represents the "rate" at which defectives occur—that is, the expected number of defectives per time interval or per sample. The Poisson probability distribution function is given by the formula

$$p(j; \lambda) = \frac{e^{-\lambda}\lambda^j}{j!} \qquad j = 0, 1, 2, \ldots \tag{4.3.9}$$

and the corresponding c.d.f. is

$$P(j; \lambda) = \sum_{i=0}^{j} p(i; \lambda) \qquad j = 0, 1, 2, \ldots \tag{4.3.10}$$

EXAMPLE **4.24**   Suppose that a machine produces aluminum pins for airplanes. The probability p that a single pin emerges defective is small—say, $p = .002$. In one hour, the machine makes $n = 1000$ pins (considered here to be a random sample of pins). The number of defective pins produced by the machine in one hour has a binomial distribution with a mean of $\mu = np = 1000(.002) = 2$, so the rate of defective pins for the machine is $\lambda = 2$ pins per hour. In this case, the binomial probabilities are very close to the Poisson probabilities. This approximation is illustrated in Table 4.5 by considering processes that produce defective items at a rate of $\lambda = 2$ parts per hour, based on various sample sizes. In Exercise 4.3.5 the student is asked to prove that the binomial p.d.f. converges to that of the Poisson with mean λ when $n \to \infty$, $p \to 0$, but $np \to \lambda$. $\blacksquare$

The m.g.f. of the Poisson distribution is

$$M(t) = e^{-\lambda} \sum_{j=0}^{\infty} e^{tj} \frac{\lambda^j}{j!} \tag{4.3.11}$$

$$= e^{-\lambda} \cdot e^{\lambda e^t} = e^{-\lambda(1-e^t)} \qquad -\infty < t < \infty$$

Thus,

$$M'(t) = \lambda M(t)e^t$$
$$M''(t) = \lambda^2 M(t)e^{2t} + \lambda M(t)e^t$$
$$= (\lambda^2 e^{2t} + \lambda e^t)M(t)$$

TABLE **4.5**

Binomial distributions for $np = 2$ and the Poisson distribution with $\lambda = 2$

	Binomial				Poisson
k	$n = 20$ $p = .1$	$n = 40$ $p = .05$	$n = 100$ $p = .02$	$n = 1000$ $p = .002$	$\lambda = 2$
0	.121576	.128512	.132619	.135065	.135335
1	.270170	.270552	.270651	.270670	.270671
2	.285179	.277671	.273413	.270942	.270671
3	.190119	.185114	.182275	.180628	.180447
4	.089779	.090121	.090208	.090223	.090223
5	.031921	.034151	.035347	.036017	.036089
6	.008867	.010485	.011422	.011970	.012030
7	.001970	.002680	.003130	.003406	.003437
8	.000356	.000582	.000743	.000847	.000859
9	.000053	.000109	.000155	.000187	.000191

Hence, the mean and variance of the Poisson distribution are

$$\mu = E\{X\} = \lambda$$

and

$$\sigma^2 = V\{X\} = \lambda$$

(4.3.12)

The Poisson distribution is used not only as an approximation to the Binomial. It is a useful model for describing the number of "events" occurring in a unit of time (or area, volume, and so on) when those events occur "at random." The rate at which these events occur is denoted by λ. An example of a Poisson random variable is the number of decaying atoms from a radioactive substance detected by a Geiger counter in a fixed period of time. If the rate of detection is 5 per second, then the number of atoms detected in a second has a Poisson distribution with mean $\lambda = 5$. The number detected in 5 seconds, however, will have a Poisson distribution with $\lambda = 25$. A rate of 5 per second equals a rate of 25 per 5 seconds. Other examples of Poisson random variables include:

1 The number of blemishes found in a unit area of a finished surface (ceramic plate)

2 The number of customers arriving at a store in one hour

3 The number of defective soldering points found on a circuit board

The p.d.f., c.d.f., and quantiles of the Poisson distribution can be computed using MINITAB. In Figure 4.6, we illustrate the p.d.f. for three values of λ.

4.3.4 The Geometric and Negative Binomial Distributions

Consider a sequence of independent trials, each having the same probability for success—say, p. Let N be a random variable that counts the number of trials until

FIGURE   4.6
Poisson p.d.f. λ = 5, 10, 15

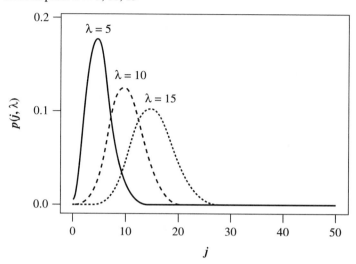

the first success is realized, including the successful trial. N may assume nonnegative integer values with probabilities

$$\Pr\{N = n\} = p(1 - p)^{n-1} \qquad n = 1, 2 \ldots \tag{4.3.13}$$

This probability function is the p.d.f. of the **geometric distribution**.

Let $g(n; p)$ designate the p.d.f. The corresponding c.d.f. is

$$G(n; p) = 1 - (1 - p)^n \qquad n = 1, 2, \ldots$$

From this, we see that the α-quantile $(0 < \alpha < 1)$ is given by

$$N_\alpha = \left[\frac{\log(1 - \alpha)}{\log(1 - p)}\right] + 1$$

where $[x]$ designates the integer part of x.

The expected value and variance of the geometric distribution are

$$E\{N\} = \frac{1}{p}$$

and

$$V\{N\} = \frac{1 - p}{p^2}$$

(4.3.14)

Indeed, the m.g.f. of the geometric distribution is

$$M(t) = pe^t \sum_{j=0}^{\infty} (e^t(1 - p))^j$$

$$= \frac{pe^t}{1 - e^t(1 - p)} \qquad \text{if } t < -\log(1 - p)$$

(4.3.15)

Thus, for $t < -\log(1-p)$,

$$M'(t) = \frac{pe^t}{(1 - e^t(1-p))^2}$$

and

$$M''(t) = \frac{pe^t}{(1 - e^t(1-p))^2} + \frac{2p(1-p)e^{2t}}{(1 - e^t(1-p))^3}$$

Hence,

$$\mu_1 = M'(0) = \frac{1}{p}$$

$$\mu_2 = M''(0) = \frac{2-p}{p^2}$$

(4.3.16)

and the above formulas of $E\{X\}$ and $V\{X\}$ are obtained.

The geometric distribution is applicable in many problems. We illustrate one such application in Example 4.25.

EXAMPLE **4.25**   An insertion machine stops automatically if a failure occurs in the handling of a component during an insertion cycle. A cycle starts immediately after the insertion of a component and ends at the insertion of the next component. Suppose that the probability of stopping is $p = 10^{-3}$ per cycle. Let N be the number of cycles until the machine stops. It is assumed that events at different cycles are mutually independent. Thus, N has a geometric distribution and $E\{N\} = 1000$. We expect a run of 1000 cycles between consecutive stopping. The number of cycles, N, however is a random variable with standard deviation of $\sigma = [(1-p)/p^2]^{1/2} = 999.5$. This high value of σ indicates that we may see very short runs and also long ones. Indeed, for $\alpha = .05, .95$, the quantiles of N are, $N_{.05} = 52$ and $N_{.95} = 2995$. ■

The number of failures until the first success, $N - 1$, has a shifted geometric distribution, which is a special case of the family of **Negative-Binomial distributions**.

We say that a nonnegative, integer-valued random variable X has a negative-binomial distribution, with parameters (p, k), where $0 < p < 1$ and $k = 1, 2, \ldots$, if its p.d.f. is

$$g(j; p, k) = \binom{j + k - 1}{k - 1} p^k (1-p)^j$$

(4.3.17)

$j = 0, 1, \ldots$. The shifted geometric distribution is the special case of $k = 1$.

A more general version of the negative-binomial distribution can be formulated in which $k - 1$ is replaced by a positive real parameter. A random variable having this negative binomial will be designated by $\text{NB}(p, k)$. The $\text{NB}(p, k)$ represents the number of failures observed until the kth success. The expected value and variance of $\text{NB}(p, k)$ are:

$$E\{X\} = k\frac{1-p}{p}$$

and   **(4.3.18)**

$$V\{X\} = k\frac{1-p}{p^2}$$

In Figure 4.7, we present the p.d.f. of $NB(p, k)$. The negative-binomial distribution has been used as a model of the distribution for the periodic demand of parts in inventory theory. An example of such an application will be shown in Chapter 6.

FIGURE 4.7
p.d.f. of $NB(p, 5)$ with $p = 0.10, 0.20$

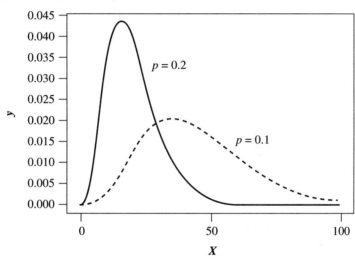

4.4
Continuous Distributions
4.4.1 The Uniform Distribution on the Interval (a, b), $a < b$

We denote a random variable having a **uniform distribution** on the interval (a, b), $a < b$, by $U(a, b)$. The p.d.f. is given by

$$f(x; a, b) = \begin{cases} 1/(b-a) & a \le x \le b \\ 0 & \text{elsewhere} \end{cases} \qquad \textbf{(4.4.1)}$$

and the c.d.f. is

$$F(x; a, b) = \begin{cases} 0 & \text{if } x < a \\ (x-a)/(b-a) & \text{if } a \le x < b \\ 1 & \text{if } b \le x \end{cases} \qquad \textbf{(4.4.2)}$$

The expected value and variance of $U(a, b)$ are

$$\mu = (a + b)/2,$$

and

$$\sigma^2 = (b - a)^2/12$$

(4.4.3)

The pth fractile is $x_p = a + p(b - a)$.

To verify the formula for μ, we write:

$$\mu_1 = \frac{1}{b - a} \int_a^b x dx = \frac{1}{b - a} \left. \frac{1}{2} x^2 \right|_a^b = \frac{1}{2(b - a)} (b^2 - a^2)$$

$$= \frac{a + b}{2}$$

Similarly,

$$\mu_2 = \frac{1}{b - a} \int_a^b x^2 dx = \frac{1}{b - a} \left. \frac{1}{3} x^3 \right|_a^b$$

$$= \frac{1}{3(b - a)} (b^3 - a^3) = \frac{1}{3} (a^2 + ab + b^2)$$

Thus,

$$\sigma^2 = \mu_2 - \mu_1^2 = \frac{1}{3} (a^2 + ab + b^2) - \frac{1}{4} (a^2 + 2ab + b^2)$$

$$= \frac{1}{12} (4a^2 + 4ab + 4b^2 - 3a^2 - 6ab - 3b^2)$$

$$= \frac{1}{12} (b - a)^2$$

We can get these moments also from the m.g.f., which is

$$M(t) = \frac{1}{t(b - a)} (e^{tb} - e^{ta}) \qquad -\infty < t < \infty$$

Moreover, for values of t close to 0

$$M(t) = 1 + \frac{1}{2} t(b + a) + \frac{1}{6} t^2 (b^2 + ab + a^2) + \cdots.$$

4.4.2   The Normal and Log-Normal Distributions

4.4.2.1   The Normal Distribution

The normal, or Gaussian, distribution, denoted by $N(\mu, \sigma)$, occupies a central role in statistical theory. Its density function (p.d.f.) is given by the formula

$$n(x; \mu, \sigma) = \frac{1}{\sigma \sqrt{2\pi}} \exp \left\{ -\frac{1}{2\sigma^2} (x - \mu)^2 \right\}$$

(4.4.4)

This p.d.f. is symmetric around the location parameter, μ. σ is a scale parameter. The m.g.f. of $N(0, 1)$ is

$$M(t) = \frac{1}{\sqrt{2\pi}} e^{tx - \frac{1}{2}x^2} dx$$

$$= \frac{e^{t^2/2}}{\sqrt{2\pi}} \int_{-\infty}^{\infty} e^{-\frac{1}{2}(x^2 - 2tx + t^2)} dx \tag{4.4.5}$$

$$= e^{t^2/2}$$

Indeed, $1/\sqrt{2\pi} \exp\left\{-\frac{1}{2}(x - t)^2\right\}$ is the p.d.f. of $N(t, 1)$. Furthermore,

$$M'(t) = tM(t)$$

$$M''(t) = t^2 M(t) + M(t) = (1 + t^2)M(t)$$

$$M'''(t) = (t + t^3)M(t) + 2tM(t)$$

$$= (3t + t^3)M(t)$$

$$M^{(4)}(t) = (3 + 6t^2 + t^4)M(t)$$

Thus, by substituting $t = 0$, we obtain that

$$E\{N(0, 1)\} = 0$$

$$V\{N(0, 1)\} = 1$$

$$\mu_3^* = 0 \tag{4.4.6}$$

$$\mu_4^* = 3$$

To obtain the moments in the general case of $N(\mu, \sigma)$, we write $X = \mu + \sigma N(0, 1)$. Then

$$E\{X\} = E\{\mu + \sigma N(0, 1)\}$$

$$= \mu + \sigma E\{N(0, 1)\} = \mu$$

$$V\{X\} = E\{(X - \mu)^2\} = \sigma^2 E\{N^2(0, 1)\} = \sigma^2$$

$$\mu_3^* = E\{(X - \mu)^3\} = \sigma^3 E\{N^3(0, 1)\} = 0$$

$$\mu_4^* = E\{(X - \mu)^4\} = \sigma^4 E\{N^4(0, 1)\} = 3\sigma^4$$

Thus, the index of kurtosis in the normal case is $\beta_4 = 3$.

The graph of the p.d.f. $n(x; \mu, \sigma)$ is a symmetric bell-shaped curve that is centered at μ (shown in Figure 4.8). The spread of the density is determined by the variance σ^2 in the sense that most of the area under the curve (in fact, 99.7% of the area) lies between $\mu - 3\sigma$ and $\mu + 3\sigma$. Thus, if X has a normal distribution with mean $\mu = 25$ and standard deviation $\sigma = 2$, the probability is .997 that the observed value of X will fall between 19 and 31.

Areas (that is, probabilities) under the normal p.d.f. are found in practice using a table or appropriate software like MINITAB. Because it is not practical to have a table for each pair of parameters, μ and σ, we use the standardized form of the normal random variable. A random variable Z is said to have a **standard normal distribution** if it has a normal distribution with mean 0 and variance 1. The standard

FIGURE   4.8

The p.d.f. of $N(\mu, \sigma)$, $\mu = 10$, $\sigma = 1, 2, 3$

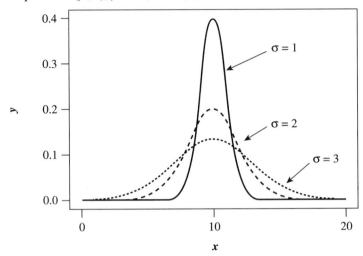

normal density function is $\phi(x) = n(x; 0, 1)$ and the standard cumulative distribution function is denoted by $\Phi(x)$. This function is also called the **standard normal integral**—that is,

$$\Phi(x) = \int_{-\infty}^{x} \phi(t)dt = \int_{-\infty}^{x} \frac{1}{\sqrt{2\pi}} e^{-\frac{1}{2}t^2} dt \tag{4.4.7}$$

The c.d.f., $\Phi(x)$, represents the area over the x-axis under the standard normal p.d.f. to the left of the value x. This c.d.f. is plotted in Figure 4.9.

FIGURE   4.9

Standard normal c.d.f.

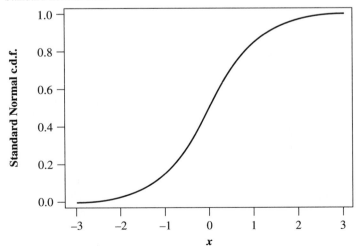

If we wish to determine the probability that a standard normal random variable is less than 1.5, for example, we use the following MINITAB command:

```
MTB> CDF 1.5;
SUBC> NORMAL 0 1.
```

We find that $\Pr\{Z \le 1.5\} = \Phi(1.5) = .9332$. To obtain the probability that Z lies between .5 and 1.5, we first find the probability that Z is less than 1.5, then subtract from this number the probability that Z is less than .5. This yields

$$\Pr\{.5 < Z < 1.5\} = \Pr\{Z < 1.5\} - \Pr\{Z < .5\}$$
$$= \Phi(1.5) - \Phi(.5) = .9332 - .6915 = .2417$$

Many tables of the normal distribution do not list values of $\Phi(x)$ for $x < 0$. This is because the normal density is symmetric about $x = 0$, and we have the relation

$$\Phi(-x) = 1 - \Phi(x) \qquad \text{for all } x \tag{4.4.8}$$

This is illustrated in Figure 4.10. Thus, to compute the probability that Z is less than -1, for example, we write

$$\Pr\{Z < -1\} = \Phi(-1) = 1 - \Phi(1) = 1 - .8413 = .1587$$

FIGURE 4.10
The symmetry of the normal distribution

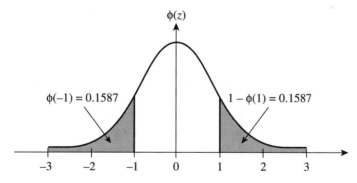

The **pth quantile (fractile)** of the standard normal distribution is the number z_p that satisfies the statement

$$\Phi(z_p) = \Pr\{Z \le z_p\} = p \tag{4.4.9}$$

If X has a normal distribution with mean μ and standard deviation σ, we denote the pth fractile of the distribution by x_p. We can show that x_p is related to the standard normal fractile by

$$x_p = \mu + z_p \sigma$$

The *p*th fractile of the normal distribution can be obtained by using the MINITAB command

```
MTB> InvCDF 0.95;
SUBC> Normal 0.0 1.0.
```

In this command we used $p = .95$. The printed result is $z_{.95} = 1.6449$. We can use any value of μ and σ in the subcommand. Thus, for $\mu = 10$ and $\sigma = 1.5$

$$x_{.95} = 10 + z_{.95} \times \sigma = 12.4673$$

Now suppose that X is a random variable having a normal distribution with mean μ and variance σ^2. That is, X has a $N(\mu, \sigma)$ distribution. We define the *standardized form* of X as

$$Z = \frac{X - \mu}{\sigma}$$

By subtracting the mean from X and then dividing by the standard deviation, we transform X to a standard normal random variable. (That is, Z has expected value 0 and standard deviation 1.) This will allow us to use the standard normal table to compute probabilities involving X. Thus, to compute the probability that X is less than a, we write

$$\Pr\{X \le a\} = \Pr\left\{\frac{X - \mu}{\sigma} < \frac{a - \mu}{\sigma}\right\}$$
$$= \Pr\left\{Z < \frac{a - \mu}{\sigma}\right\} = \Phi\left(\frac{a - \mu}{\sigma}\right)$$

EXAMPLE **4.26**   Let X represent the length (with cap) of a randomly selected aluminum pin. Suppose we know that X has a normal distribution with mean $\mu = 60.02$ and standard deviation $\sigma = 0.048$ (in mm). What is the probability that the length with cap of a randomly selected pin will be less than 60.1 (mm)? Using the MINITAB command

```
MTB> CDF 60.1;
SUBC> Normal 60.02 0.048.
```

we obtain $\Pr\{X \le 60.1\} = .9522$. If we have to use the table of $\Phi(Z)$, we write

$$\Pr\{X \le 60.1\} = \Phi\left(\frac{60.1 - 60.02}{.048}\right)$$
$$= \Phi(1.667) = .9522$$

Continuing with the example, consider the following question: If a pin is considered "acceptable" when its length is between 59.9 and 60.1 mm, what proportion of pins can we expect to be rejected? To answer this question, we first compute the probability of accepting a single pin. This is the probability that X lies between 59.9 and 60.1—that is,

$$\Pr\{59.9 < X < 60.1\} = \Phi\left(\frac{60.1 - 60.02}{.048}\right) - \Phi\left(\frac{59.9 - 60.02}{.048}\right)$$
$$= \Phi(1.667) - \Phi(-2.5)$$
$$= .9522 - .0062 = .946$$

Thus, we expect that 94.6% of the pins will be accepted, and that 5.4% of them will be rejected. ∎

4.4.2.2 The Log-Normal Distribution

A random variable X is said to have a **log-normal distribution**, $LN(\mu, \sigma)$, if $Y = \log X$ has the normal distribution $N(\mu, \sigma^2)$.

The log-normal distribution has been used for modeling distributions of strength variables, like the tensile strength of fibers (see Chapter 2), the compressive strength of concrete cubes, and so on. It has also been used for random quantities of pollutants in water or air and for other phenomena with skewed distributions.

The p.d.f. of $LN(\mu, \sigma)$ is given by the formula

$$f(x; \mu, \sigma^2) = \begin{cases} \dfrac{1}{\sqrt{2\pi}\sigma x} \exp\left\{-\dfrac{1}{2\sigma^2}(\log x - \mu)^2\right\} & 0 < x < \infty \\ 0 & x \le 0 \end{cases} \tag{4.4.10}$$

The c.d.f. is expressed in terms of the standard normal integral as

$$F(x) = \begin{cases} 0 & x \le 0 \\ \Phi\left(\dfrac{\log x - \mu}{\sigma}\right) & 0 < x < \infty \end{cases} \tag{4.4.11}$$

The expected value and variance of $LN(\mu, \sigma)$ are

$$E\{X\} = e^{\mu + \sigma^2/2}$$

and $\tag{4.4.12}$

$$V\{X\} = e^{2\mu + \sigma^2}(e^{\sigma^2} - 1)$$

We can show that the third central moment of $LN(\mu, \sigma^2)$ is

$$\mu_3^* = e^{3\mu + \frac{3}{2}\sigma^2}(e^{3\sigma^2} - 3e^{\sigma^2} + 2)$$

Hence, the *index of skewness* of this distribution is

$$\beta_3 = \frac{\mu_3^*}{(V\{X\}^{3/2})} = \frac{e^{3\sigma^2} - 3e^{\sigma^2} + 2}{(e^{\sigma^2} - 1)^{3/2}} \tag{4.4.13}$$

It is interesting that the index of skewness does not depend on μ, and is positive for all $\sigma^2 > 0$. This index of skewness grows very fast as σ^2 increases. This is shown in Figure 4.11.

FIGURE   4.11
The index of skewness of $LN(\mu, \sigma)$

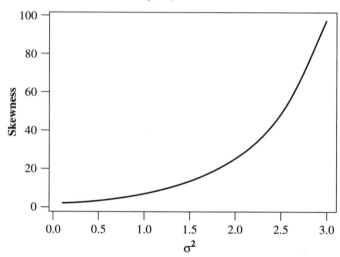

4.4.3   The Exponential Distribution

We designate the exponential distribution by $E(\beta)$. The p.d.f. of $E(\beta)$ is given by the formula

$$f(x; \beta) = \begin{cases} 0 & \text{if } x < 0 \\ (1/\beta)e^{-x/\beta} & \text{if } x \geq 0 \end{cases} \tag{4.4.14}$$

where β is a positive parameter—that is, $0 < \beta < \infty$. In Figure 4.12, we present these p.d.f.'s for various values of β.

The corresponding c.d.f. is

$$F(x; \beta) = \begin{cases} 0 & \text{if } x < 0 \\ 1 - e^{-x/\beta} & \text{if } x \geq 0 \end{cases} \tag{4.4.15}$$

The expected value and the variance of $E(\beta)$ are

$$\mu = \beta$$

and

$$\sigma^2 = \beta^2$$

Indeed,

$$\mu = \frac{1}{\beta} \int_0^\infty x e^{-x/\beta} dx$$

F I G U R E   4.12

The p.d.f. of $E(\beta)$, $\beta = 1, 2, 3$

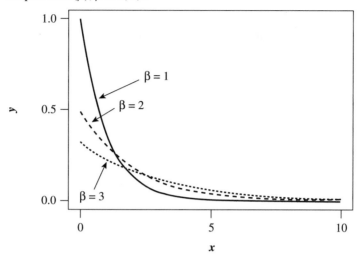

Making the change of variable to $y = x/\beta$, $dx = \beta dy$, we obtain

$$\mu = \beta \int_0^\infty y e^{-y} dy$$
$$= \beta$$

Similarly,

$$\mu_2 = \frac{1}{\beta} \int_0^\infty x^2 e^{-x/\beta} dx = \beta^2 \int_0^\infty y^2 e^{-y} dy$$
$$= 2\beta^2$$

Hence,

$$\sigma^2 = \beta^2$$

The pth quantile is $x_p = -\beta \ln(1 - p)$.

The exponential distribution is related to the Poisson model in the following way: If the number of events occurring in a period of time follows a Poisson distribution with rate λ, then the time between occurrences of events has an exponential distribution with parameter $\beta = 1/\lambda$. The exponential model can also be used to describe the lifetime (that is, time to failure) of certain electronic systems. For example, if the mean life of a system is 200 hours, then the probability that it will work at least 300 hours without failure is

$$Pr\{X \geq 300\} = 1 - Pr\{X < 300\}$$
$$= 1 - F(300) = 1 - (1 - e^{-300/200}) = .223$$

The exponential distribution is positively skewed, and its index of skewness is

$$\beta_3 = \frac{\mu_3^*}{\sigma^3} = 2$$

irrespective of the value of β. We have seen before that the kurtosis index is $\beta_4 = 9$.

4.4.4 The Gamma and Weibull Distributions

Two important distributions for studying the reliability and failure rates of systems are the **gamma** and the **Weibull distributions**. We will need these distributions in our study of reliability methods (Chapter 14). We discuss them here as additional examples of continuous distributions.

Suppose we use in a manufacturing process a machine that mass produces a particular part. In a random manner, it produces defective parts at a rate of λ per hour. The number of defective parts produced by this machine in a time period $[0, t]$ is a random variable $X(t)$ having a Poisson distribution with mean λt—that is,

$$\Pr\{X(t) = j\} = (\lambda t)^j e^{-(\lambda t)}/j! \qquad j = 0, 1, 2, \ldots \tag{4.4.16}$$

Suppose we wish to study the distribution of the time until the kth defective part is produced. Call this continuous random variable Y_k. We use the fact that the kth defect will occur before time t (that is, $Y_k \leq t$) if and only if at least k defects occur up to time t (that is, $X(t) \geq k$). Thus, the c.d.f. for Y_k is

$$
\begin{aligned}
G(t; k, \lambda) &= \Pr\{Y_k \leq t\} \\
&= \Pr\{X(t) \geq k\} \\
&= 1 - \sum_{j=0}^{k-1} (\lambda t)^j e^{-\lambda t}/j!
\end{aligned}
\tag{4.4.17}
$$

The corresponding p.d.f. for Y_k is

$$g(t; k, \lambda) = \frac{\lambda^k}{(k-1)!} t^{k-1} e^{-\lambda t} \quad \text{for } t \geq 0 \tag{4.4.18}$$

This p.d.f. is a member of a general family of distributions that depend on two parameters, ν and β, and are called the **gamma distributions** $G(\nu, \beta)$. The p.d.f. of a gamma distribution $G(\nu, \beta)$ is

$$g(x; \nu, \beta) = \begin{cases} \dfrac{1}{\beta^\nu \Gamma(\nu)} x^{\nu-1} e^{-x/\beta} & x \geq 0 \\ 0 & x < 0 \end{cases} \tag{4.4.19}$$

where $0 < \nu, \beta < \infty$, and $\Gamma(\nu)$ is called the *gamma function* of ν and is defined as the integral

$$\Gamma(\nu) = \int_0^\infty x^{\nu-1} e^{-x} dx \qquad \nu > 0 \tag{4.4.20}$$

The gamma function satisfies the relationship

$$\Gamma(\nu) = (\nu - 1)\Gamma(\nu - 1) \qquad \text{for all } \nu > 1 \tag{4.4.21}$$

Hence, for every positive integer k, $\Gamma(k) = (k-1)!$. Also, $\Gamma\left(\frac{1}{2}\right) = \sqrt{\pi}$. We note also that the exponential distribution, $E(\beta)$, is a special case of the gamma distribution with $\nu = 1$. Some gamma p.d.f.'s are presented in Figure 4.13. The value

FIGURE 4.13

The gamma densities, with $\beta = 1$ and $v = .5, 1, 2$

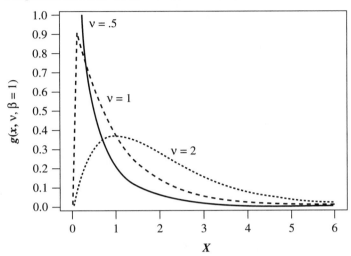

of $\Gamma(v)$ can be computed in MINITAB by the following commands, which compute $\Gamma(5)$. Generally, replace 5 in line 2 by v.

```
MTB> PDF 1 K1;
SUBC> GAMMA 5 1.
MTB> Let K2 = EXPO(−1)/K1
MTB> PRINT K2
```

The expected value and variance of the gamma distribution, $G(v, \beta)$, are, respectively,

$$\mu = v\beta$$

and   **(4.4.22)**

$$\sigma^2 = v\beta^2$$

To verify these formulas we write

$$\mu = \frac{1}{\beta^v \Gamma(v)} \int_0^\infty x \cdot x^{v-1} e^{-x/\beta} \, dx$$

$$= \frac{\beta^{v+1}}{\beta^v \Gamma(v)} \int_0^\infty y^v e^{-y} \, dy$$

$$= \beta \frac{\Gamma(v+1)}{\Gamma(v)} = v\beta$$

Similarly,

$$\mu_2 = \frac{1}{\beta^\nu \Gamma(\nu)} \int_0^\infty x^2 \cdot x^{\nu-1} e^{-x/\beta} dx$$

$$= \frac{\beta^{\nu+2}}{\beta^\nu \Gamma(\nu)} \int_0^\infty y^{\nu+1} e^{-y} dy$$

$$= \beta^2 \frac{\Gamma(\nu+2)}{\Gamma(\nu)} = (\nu+1)\nu\beta^2$$

Hence,

$$\sigma^2 = \mu_2 - \mu_1^2 = \nu\beta^2$$

An alternative way is to differentiate the m.g.f.:

$$M(t) = (1 - t\beta)^{-\nu} \qquad t < \frac{1}{\beta} \tag{4.4.23}$$

Weibull distributions are often used in reliability models in which the system either "ages" with time or becomes "younger" (see Chapter 14). The Weibull family of distributions will be denoted by $W(\alpha, \beta)$. The parameters α and β, $\alpha, \beta > 0$, are called the shape and the scale parameters, respectively. The p.d.f. of $W(\alpha, \beta)$ is given by

$$w(t; \alpha, \beta) = \begin{cases} \dfrac{\alpha t^{\alpha-1}}{\beta^\alpha} e^{-(t/\beta)^\alpha} & t \geq 0 \\ 0 & t < 0 \end{cases} \tag{4.4.24}$$

The corresponding c.d.f. is

$$W(t; \alpha, \beta) = \begin{cases} 1 - e^{-(t/\beta)^\alpha} & t \geq 0 \\ 0 & t < 0 \end{cases} \tag{4.4.25}$$

Notice that $W(1, \beta) = E(\beta)$. The mean and variance of this distribution are

$$\mu = \beta \cdot \Gamma\left(1 + \frac{1}{\alpha}\right) \tag{4.4.26}$$

and

$$\sigma^2 = \beta^2 \left\{ \Gamma\left(1 + \frac{2}{\alpha}\right) - \Gamma^2\left(1 + \frac{1}{\alpha}\right) \right\} \tag{4.4.27}$$

respectively. The values of $\Gamma(1 + (1/\alpha))$ and $\Gamma(1 + (2/\alpha))$ can be computed by MINITAB. If, for example, $\alpha = 2$, then

$$\mu = \beta\sqrt{\pi}/2 = .8862\beta$$
$$\sigma^2 = \beta^2(1 - \pi/4) = .2145\beta^2$$

because

$$\Gamma\left(1 + \frac{1}{2}\right) = \frac{1}{2} \cdot \Gamma\left(\frac{1}{2}\right) = \frac{1}{2}\sqrt{\pi}$$

and

$$\Gamma\left(1 + \frac{2}{2}\right) = \Gamma(2) = 1$$

Figure 4.14 shows two p.d.f.'s of $W(\alpha, \beta)$ for $\alpha = 1.5, 2.0$, and $\beta = 1$.

FIGURE   4.14
Weibull density functions, $\alpha = 1.5, 2$

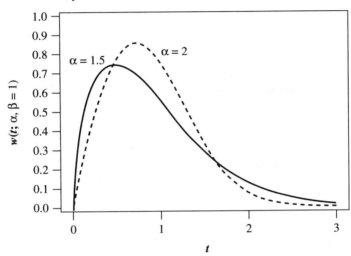

4.4.5   The Beta Distribution

Distributions having p.d.f. of the form

$$f(x; \nu_1, \nu_2) = \begin{cases} \dfrac{1}{B(\nu_1, \nu_2)} x^{\nu_1 - 1}(1 - x)^{\nu_2 - 1} & 0 < x < 1 \\ 0 & \text{otherwise} \end{cases} \qquad \textbf{(4.4.28)}$$

where, for ν_1, ν_2 positive,

$$B(\nu_1, \nu_2) = \int_0^1 x^{\nu_1 - 1}(1 - x)^{\nu_2 - 1} dx \qquad \textbf{(4.4.29)}$$

are called **beta distributions**. The function $B(\nu_1, \nu_2)$ is called the *beta integral*. One can prove that

$$B(\nu_1, \nu_2) = \frac{\Gamma(\nu_1)\Gamma(\nu_2)}{\Gamma(\nu_1 + \nu_2)} \qquad \textbf{(4.4.30)}$$

The parameters ν_1 and ν_2 are shape parameters. Notice that when $\nu_1 = 1$ and $\nu_2 = 1$, the beta reduces to $U(0, 1)$. We designate distributions of this family by beta(ν_1, ν_2).

The c.d.f. of beta(v_1, v_2) is denoted also by $I_x(v_1, v_2)$, which is known as the *incomplete beta function ratio*—that is,

$$I_x(v_1, v_2) = \frac{1}{B(v_1, v_2)} \int_0^x u^{v_1-1}(1-u)^{v_2-1} du \tag{4.4.31}$$

for $0 \le x \le 1$. Notice that $I_x(v_1, v_2) = 1 - I_{1-x}(v_2, v_1)$. The density functions of the p.d.f. beta$(2.5, 5.0)$ and beta$(2.5, 2.5)$ are plotted in Figure 4.15. Notice that if $v_1 = v_2$, then the p.d.f. is symmetric around $\mu = 1/2$. There is no simple formula for the m.g.f. of beta(v_1, v_2). However, the mth moment is equal to

$$
\begin{aligned}
\mu_m &= \frac{1}{B(v_1, v_2)} \int_0^1 u^{m+v_1-1}(1-u)^{v_2-1} du \\
&= \frac{B(v_1+m, v_2)}{B(v_1, v_2)} \\
&= \frac{v_1(v_1+1)\cdots(v_1+m-1)}{(v_1+v_2)(v_1+v_2+1)\cdots(v_1+v_2+m-1)}
\end{aligned}
\tag{4.4.32}
$$

Hence,

$$
\begin{aligned}
E\{\text{beta}(v_1, v_2)\} &= \frac{v_1}{v_1+v_2} \\
V\{\text{beta}(v_1, v_2)\} &= \frac{v_1 v_2}{(v_1+v_2)^2(v_1+v_2+1)}
\end{aligned}
\tag{4.4.33}
$$

The beta distribution has an important role in the theory of statistics. As will be seen later, many methods of statistical inference are based on the order statistics (see Section 4.7). The distribution of the order statistics is related to the beta distribution. Moreover, because the beta distribution can have a variety of shapes, it has been applied in many cases in which the variable has a distribution on a finite domain. By introducing a location and a scale parameter, one can fit a shifted-scaled beta distribution to various frequency distributions.

4.5
Joint, Marginal, and Conditional Distributions
4.5.1 Joint and Marginal Distributions

Let $X_1, \ldots, X_k$ be random variables that are jointly observed at the same experiments. In Chapter 3, we presented various examples of bivariate and multivariate frequency distributions. In this section, we present only the fundamentals of the theory, mainly for future reference, and we focus on continuous random variables. The theory also holds generally for discrete and for a mixture of continuous and discrete random variables. We introduce now the **joint distribution** of several random variables.

A function $F(x_1, \ldots, x_k)$ is called the *joint c.d.f.* of $X_1, \ldots, X_k$ if

$$F(x_1, \ldots, x_k) = \Pr\{X_1 \le x_1, \ldots, X_k \le x_k\} \tag{4.5.1}$$

FIGURE 4.15
Beta densities, $v_1 = 2.5$, $v_2 = 2.5$; $v_1 = 2.5$, $v_2 = 5.00$

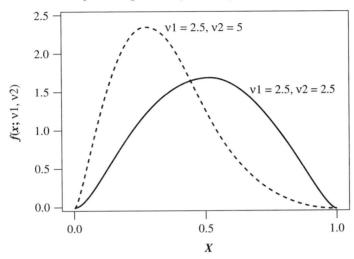

for all $(x_1, \ldots, x_k) \in \mathbb{R}^k$ (the Euclidean k-space). By letting one or more variables tend to infinity, we obtain the joint c.d.f. of the remaining variables. For example,

$$F(x_1, \infty) = \Pr\{X_1 \le x_1, X_2 \le \infty\}$$
$$= \Pr\{X_1 \le x_1\} = F_1(x_1) \tag{4.5.2}$$

The c.d.f.'s of the individual variables are called the **marginal distributions**. $F_1(x_1)$ is the *marginal c.d.f.* of X_1.

A nonnegative function $f(x_1, \ldots, x_k)$ is called the *joint p.d.f.* of $X_1, \ldots, X_k$, if

1 $f(x_1, \ldots, x_k) \ge 0$ for all $(x_1, \ldots, x_k)$, where $-\infty < x_i < \infty$ $(i = 1, \ldots, k)$

2 $\int_{-\infty}^{\infty} \cdots \int_{-\infty}^{\infty} f(x_1, \ldots, x_k) dx_1, \ldots, dx_k = 1$

and

3 $F(x_1, \ldots, x_k) = \int_{-\infty}^{x_1} \cdots \int_{-\infty}^{x_k} f(y_1, \ldots, y_k) dy_1 \ldots dy_k$

The *marginal p.d.f.* of X_i $(i = 1, \ldots, k)$ can be obtained from the joint p.d.f. $f(x_1, \ldots, x_k)$, by integrating the joint p.d.f. with respect to all x_j, $j \ne i$. For example, if $k = 2$, $f(x_1, x_2)$ is the joint p.d.f. of X_1, X_2. The marginal p.d.f. of X_1 is

$$f_1(x_1) = \int_{-\infty}^{\infty} f(x_1, x_2) dx_2$$

Similarly, the marginal p.d.f. of X_2 is

$$f_2(x_2) = \int_{-\infty}^{\infty} f(x_1, x_2) dx_1$$

Indeed, the marginal c.d.f. of X is

$$F(x_1) = \int_{-\infty}^{x_1} \int_{-\infty}^{\infty} f(y_1, y_2) dy_1 dy_2$$

Differentiating $F(x_1)$ with respect to x_1, we obtain the marginal p.d.f. of X_1:

$$f(x_1) = \frac{d}{dx_1} \int_{-\infty}^{x_1} \int_{-\infty}^{\infty} f(y_1, y_2) dy_1 dy_2$$

$$= \int_{-\infty}^{\infty} f(x_1, y_2) dy_2$$

If $k = 3$, we can obtain the marginal joint p.d.f. of a pair of random variables by integrating with respect to a third variable. For example, the joint marginal p.d.f. of (X_1, X_2) can be obtained from that of (X_1, X_2, X_3) as

$$f_{1,2}(x_1, x_2) = \int_{-\infty}^{\infty} f(x_1, x_2, x_3) dx_3$$

Similarly,

$$f_{1,3}(x_1, x_3) = \int_{-\infty}^{\infty} f(x_1, x_2, x_3) dx_2$$

and

$$f_{2,3}(x_2, x_3) = \int_{-\infty}^{\infty} f(x_1, x_2, x_3) dx_1$$

EXAMPLE **4.27**   This example is theoretical and is designed to illustrate the preceding concepts.

Let (X, Y) be a pair of random variables having a joint uniform distribution on the region

$$T = \{(x, y) : 0 \leq x, y, \ x + y \leq 1\}$$

T is a triangle in the (x, y)-plane with vertices at $(0, 0)$, $(1, 0)$, and $(0, 1)$. According to the assumption of uniform distribution, the joint p.d.f. of (X, Y) is

$$f(x, y) = \begin{cases} 2 & \text{if } (x, y) \in T \\ 0 & \text{otherwise} \end{cases}$$

The marginal p.d.f. of X is

$$f_1(x) = 2 \int_0^{1-x} dy = 2(1 - x) \qquad 0 \leq x \leq 1$$

Obviously, $f_1(x) = 0$ for x outside the interval $[0, 1]$. Similarly, the marginal p.d.f. of Y is

$$f_2(y) = \begin{cases} 2(1 - y) & 0 \leq y \leq 1 \\ 0 & \text{otherwise} \end{cases}$$

Both X and Y have the same marginal beta $(1, 2)$ distribution. Thus,

$$E\{X\} = E\{Y\} = \frac{1}{3}$$

and

$$V\{X\} = V\{Y\} = \frac{1}{18} \quad \blacksquare$$

4.5.2 Covariance and Correlation

Given any two random variables (X_1, X_2) having a joint distribution with p.d.f. $f(x_1, x_2)$, the **covariance** of X_1 and X_2 is defined as

$$\text{cov}(X_1, X_2) = \int_{-\infty}^{\infty} \int_{-\infty}^{\infty} (x_1 - \mu_1)(x_2 - \mu_2) f(x_1, x_2) dx_1 dx_2 \qquad \textbf{(4.5.3)}$$

where

$$\mu_i = \int_{-\infty}^{\infty} x f_i(x) dx \qquad i = 1, 2$$

is the expected value of X_i. Notice that

$$\text{cov}(X_1, X_2) = E\{(X_1 - \mu_1)(X_2 - \mu_2)\}$$
$$= E\{X_1 X_2\} - \mu_1 \mu_2$$

The **correlation** between X_1 and X_2 is defined as

$$\rho_{12} = \frac{\text{cov}(X_1, X_2)}{\sigma_1 \sigma_2} \qquad \textbf{(4.5.4)}$$

where σ_i $(i = 1, 2)$ is the standard deviation of X_i.

EXAMPLE 4.28 In this example, we compute $\text{cov}(X, Y)$.

We have seen that $E\{X\} = E\{Y\} = \frac{1}{3}$. We now compute the expected value of their product

$$E\{XY\} = 2 \int_0^1 x \int_0^{1-x} y \, dy$$

$$= 2 \int_0^1 x \cdot \frac{1}{2} (1-x)^2 dx$$

$$= B(2, 3) = \frac{\Gamma(2)\Gamma(3)}{\Gamma(5)} = \frac{1}{12}$$

Hence,

$$\text{cov}(X, Y) = E\{XY\} - \mu_1 \mu_2 = \frac{1}{12} - \frac{1}{9}$$

$$= -\frac{1}{36}$$

Finally, the correlation between X, Y is

$$\rho_{XY} = -\frac{1/36}{1/18} = -\frac{1}{2} \quad \blacksquare$$

The following are some properties of covariance

i

$$|\text{cov}(X_1, X_2)| \leq \sigma_1 \sigma_2$$

where σ_1 and σ_2 are the standard deviations of X_1 and X_2, respectively.

ii If c is any constant, then

$$\text{cov}(X, c) = 0 \tag{4.5.5}$$

iii For any constants a_1 and a_2,

$$\text{cov}(a_1 X_1, a_2 X_2) = a_1 a_2 \text{cov}(X_1, X_2) \tag{4.5.6}$$

iv For any constants a, b, c, and d,

$$\text{cov}(aX_1 + bX_2, cX_3 + dX_4) = ac\,\text{cov}(X_1, X_3) + ad\,\text{cov}(X_1, X_4)$$
$$+ bc\,\text{cov}(X_2, X_3) + bd\,\text{cov}(X_2, X_4)$$

Property (iv) can be generalized to

$$\text{cov}\left(\sum_{i=1}^{m} a_i X_i, \sum_{j=1}^{n} b_j Y_j\right) = \sum_{i=1}^{m}\sum_{j=1}^{n} a_i b_j \,\text{cov}(X_i, Y_j) \tag{4.5.7}$$

From property (i), we deduce that $-1 \leq \rho_{12} \leq 1$. The correlation obtains the values ± 1 only if the two variables are linearly dependent.

Random variables $X_1, \ldots, X_k$ are said to be **mutually independent** if, for every $(x_1, \ldots, x_k)$,

$$f(x_1, \ldots, x_k) = \prod_{i=1}^{k} f_i(x_i) \tag{4.5.8}$$

where $f_i(x_i)$ is the marginal p.d.f. of X_i. The variables X, Y of Example 4.27 are dependent, because $f(x, y) \neq f_1(x)f_2(y)$.

If two random variables are independent, then their correlation (or covariance) is zero. The converse is generally not true. Zero correlation *does not* imply independence.

We illustrate this in the following example.

EXAMPLE 4.29   Let (X, Y) be discrete random variables having the following joint p.d.f.

$$p(x, y) = \begin{cases} \dfrac{1}{3} & \text{if } X = -1, Y = 0 \text{ or } X = 0, Y = 0 \quad \text{or} \quad X = 1, Y = 1 \\ 0 & \text{elsewhere} \end{cases}$$

In this case, the marginal p.d.f.'s are

$$p_1(x) = \begin{cases} \dfrac{1}{3} & x = -1, 0, 1 \\ 0 & \text{otherwise} \end{cases}$$

$$p_2(y) = \begin{cases} \dfrac{1}{3} & y = 0 \\[2mm] \dfrac{2}{3} & y = 1 \end{cases}$$

For example, $p(x, y) \neq p_1(x)p_2(y)$ if $X = 1$, $Y = 1$. Thus, X and Y are dependent. However, $E\{X\} = 0$ and $E\{XY\} = 0$. Hence, $\text{cov}(X, Y) = 0$. ∎

The following result is very important for independent random variables.

If $X_1, X_2, \ldots, X_k$ are mutually independent, then, for any integrable functions $g_1(X_1), \ldots, g_k(X_k)$,

$$E\left\{ \prod_{i=1}^{k} g_i(X_i) \right\} = \prod_{i=1}^{k} E\{g_i(X_i)\} \tag{4.5.9}$$

Indeed,

$$E\left\{ \prod_{i=1}^{k} g_i(X_i) \right\} = \int \cdots \int g_1(x_1) \cdots g_k(x_k)$$

$$f(x_1, \ldots, x_k)dx_1, \ldots, dx_k$$

$$= \int \cdots \int g_1(x_1) \cdots g_k(x_k)f_1(x_1) \cdots f_k(x_k)dx_1 \cdots dx_k$$

$$= \int g_1(x_1)f_1(x_1)dx_1 \cdot \int g_2(x_2)f_2(x_2)dx_2 \cdots \int g_k(x_k)f_k(x_k)dx_k$$

$$= \prod_{i=1}^{k} E\{g_i(X_i)\}$$

4.5.3 Conditional Distributions

If (X_1, X_2) are two random variables having a joint p.d.f. $f(x_1, x_2)$ and marginal p.d.f.'s, $f_1(\cdot)$ and $f_2(\cdot)$, respectively, then the **conditional** p.d.f. of X_2, given $\{X_1 = x_1\}$, where $f_1(x_1) > 0$, is defined to be

$$f_{2 \cdot 1}(x_2 \mid x_1) = \frac{f(x_1, x_2)}{f_1(x_1)} \tag{4.5.10}$$

Notice that $f_{2 \cdot 1}(x_2 \mid x_1)$ is a p.d.f. Indeed, $f_{2 \cdot 1}(x_2 \mid x_1) \geq 0$ for all x_2, and

$$\int_{-\infty}^{\infty} f_{2 \cdot 1}(x_2 \mid x_1)dx_2 = \frac{\displaystyle\int_{-\infty}^{\infty} f(x_1, x_2)dx_2}{f_1(x_1)}$$

$$= \frac{f_1(x_1)}{f_1(x_1)} = 1$$

The **conditional expectation** of X_2, given $\{X_1 = x_1\}$ such that $f_1(x_1) > 0$, is the expected value of X_2 with respect to the conditional p.d.f. $f_{2 \cdot 1}(x_2 \mid x_1)$—that is,

$$E\{X_2 \mid X_1 = x_1\} = \int_{-\infty}^{\infty} x f_{2 \cdot 1}(x \mid x_1)dx.$$

Similarly, we can define the **conditional variance** of X_2, given $\{X_1 = x_1\}$, as the variance of X_2 with respect to the conditional p.d.f. $f_{2 \cdot 1}(x_2 \mid x_1)$. If X_1 and X_2 are independent, then, by substituting $f(x_1, x_2) = f_1(x_1)f_2(x_2)$, we obtain

$$f_{2 \cdot 1}(x_2 \mid x_1) = f_2(x_2)$$

and

$$f_{1 \cdot 2}(x_1 \mid x_2) = f_1(x_1)$$

EXAMPLE **4.30**   Returning to Example 4.27, we compute the conditional distribution of Y, given $\{X = x\}$, for $0 < x < 1$.

According to the preceding definition, the conditional p.d.f. of Y, given $\{X = x\}$, for $0 < x < 1$, is

$$f_{Y\mid X}(y \mid x) = \begin{cases} \dfrac{1}{1-x} & \text{if } 0 < y < (1-x) \\ 0 & \text{otherwise} \end{cases}$$

Notice that this is a uniform distribution over $(0, 1 - x)$, $0 < x < 1$. If $x \notin (0, 1)$, then the conditional p.d.f. does not exist. This is, however, an event of zero probability. From this result, the conditional expectation of Y, given $X = x$, $0 < x < 1$, is

$$E\{Y \mid X = x\} = \frac{1-x}{2}$$

The conditional variance is

$$V\{Y \mid X = x\} = \frac{(1-x)^2}{12}.$$

In a similar fashion we show that the conditional distribution of X, given $Y = y$, $0 < y < 1$, is uniform on $(0, 1 - y)$ ∎

One can immediately prove that if X_1 and X_2 are independent, then the conditional distribution of X_1 given $\{X_2 = x_2\}$, when $f_2(x_2) > 0$, is just the marginal distribution of X_1. Thus, X_1 and X_2 are independent if and only if,

$$f_{2 \cdot 1}(x_2 \mid x_1) = f_2(x_2) \quad \text{for all } x_2$$

and

$$f_{1 \cdot 2}(x_1 \mid x_2) = f_1(x_1) \quad \text{for all } x_1$$

provided that the conditional p.d.f. are well defined.

Notice that for a pair of random variables (X, Y), $E\{Y \mid X = x\}$ changes with x, as shown in Example 4.30, if X and Y are dependent. Thus, we can consider $E\{Y \mid X\}$

to be a random variable, which is a function of X. It is interesting to compute the expected value of this function of X:

$$E\{E\{Y \mid X\}\} = \int E\{Y \mid X = x\}f_1(x)dx$$

$$= \int \left\{\int yf_{Y \cdot X}(y \mid x)dy\right\}f_1(x)dx$$

$$= \int \int y\frac{f(x, y)}{f_1(x)}f_1(x)dydx$$

If we can interchange the order of integration (whenever $\int |y|f_2(y)dy < \infty$), then

$$E\{E\{Y \mid X\}\} = \int y\left\{\int f(x, y)dx\right\}dy$$

$$= \int yf_2(y)dy \hspace{3cm} \textbf{(4.5.11)}$$

$$= E\{Y\}$$

This result, known as the **law of iterated expectation**, is often very useful. An example of the use of the law of iterated expectation is the following.

EXAMPLE 4.31 Let (J, N) be a pair of random variables. The conditional distribution of J, given $\{N = n\}$, is the binomial $B(n, p)$. The marginal distribution of N is Poisson with mean λ. What is the expected value of J?

By the law of iterated expectation,

$$E\{J\} = E\{E\{J \mid N\}\}$$

$$= E\{Np\} = pE\{N\} = p\lambda$$

We can show that the marginal distribution of J is Poisson, with mean $p\lambda$. ∎

Another important result relates variances and conditional variances. That is, if (X, Y) is a pair of random variables having finite variances, then

$$V\{Y\} = E\{V\{Y \mid X\}\} + V\{E\{Y \mid X\}\} \hspace{2cm} \textbf{(4.5.12)}$$

We call this relationship the **law of total variance**.

EXAMPLE 4.32 Let (X, Y) be a pair of independent random variables having finite variances σ_X^2 and σ_Y^2 and expected values μ_X, μ_Y. Determine the variance of $W = XY$. By the law of total variance,

$$V\{W\} = E\{V\{W \mid X\}\} + V\{E\{W \mid X\}\}$$

Because X and Y are independent

$$V\{W \mid X\} = V\{XY \mid X\} = X^2 V\{Y \mid X\}$$

$$= X^2\sigma_Y^2$$

Similarly,

$$E\{W \mid X\} = X\mu_Y$$

Hence,

$$
\begin{aligned}
V\{W\} &= \sigma_Y^2 E\{X^2\} + \mu_Y^2 \sigma_X^2 \\
&= \sigma_Y^2 (\sigma_X^2 + \mu_X^2) + \mu_Y^2 \sigma_X^2 \\
&= \sigma_X^2 \sigma_Y^2 + \mu_X^2 \sigma_Y^2 + \mu_Y^2 \sigma_X^2 \quad \blacksquare
\end{aligned}
$$

4.6
Some Multivariate Distributions
4.6.1   The Multinomial Distribution

The multinomial distribution is a generalization of the binomial distribution to cases of n independent trials in which the results are classified to k possible categories (for example, Excellent, Good, Average, Poor). The random variables $(J_1, J_2, \ldots, J_k)$ are the number of trials yielding results in each one of the k categories. These random variables are dependent, because $J_1 + J_2 + \ldots + J_k = n$. Furthermore, let $p_1, p_2, \ldots, p_k; p_i \geq 0, \sum_{i=1}^{k} p_i = 1$, be the probabilities of the k categories. The binomial distribution is the special case of $k = 2$. Because $J_k = n - (J_1 + \ldots + J_{k-1})$, the joint probability function is written as a function of $k - 1$ arguments, and its formula is

$$p(j_1, \ldots, j_{k-1}) = \binom{n}{j_1, \ldots, j_{k-1}} p_1^{j_1} \ldots p_{k-1}^{j_{k-1}} p_k^{j_k} \tag{4.6.1}$$

for $j_1, \ldots, j_{k-1} \geq 0$ such that $\sum_{i=1}^{k-1} j_i \leq n$. In this formula,

$$\binom{n}{j_1, \ldots, j_{k-1}} = \frac{n!}{j_1! j_2! \ldots j_k!} \tag{4.6.2}$$

and $j_k = n - (j_1 + \ldots + j_{k-1})$. For example, if $n = 10$, $k = 3$, $p_1 = .3$, $p_2 = .4$, $p_3 = .3$,

$$
\begin{aligned}
p(5, 2) &= \frac{10!}{5!2!3!} (.3)^5 (.4)^2 (.3)^3 \\
&= .02645
\end{aligned}
$$

The marginal distribution of each one of the k variables is binomial, with parameters n and p_i $(i = 1, \ldots, k)$. The joint marginal distribution of (J_1, J_2) is trinomial, with parameters n, p_1, p_2 and $(1 - p_1 - p_2)$. Finally, the conditional distribution of $(J_1, \ldots, J_r)$, $1 \le r < k$, given $\{J_{r+1} = j_{r+1}, \ldots, J_k = j_k\}$ is $(r+1)$-nomial, with parameters $n_r = n - (j_{r+1} + \ldots + j_k)$ and $p'_1, \ldots, p'_r, p'_{r+1}$, where

$$p'_i = \frac{p_i}{(1 - p_{r+1} - \ldots - p_k)} \qquad i = 1, \ldots, r$$

and

$$p'_{r+1} = 1 - \sum_{i=1}^{r} p'_i$$

Finally, we can show that, for $i \ne j$,

$$\text{cov}(J_i, J_j) = -n p_i p_j \tag{4.6.3}$$

EXAMPLE 4.33 An insertion machine is designed to insert components onto computer-printed circuit boards. Every component inserted on a board is scanned optically. An insertion is either error free or its error is classified as being in one of two main categories: misinsertion (broken lead, off pad, and so on) or wrong component. Thus, we have altogether three general categories. Let

$$J_1 = \text{Number of error free components}$$
$$J_2 = \text{Number of misinsertions}$$
$$J_3 = \text{Number of wrong components}$$

The probabilities that an insertion belongs to one of these categories is $p_1 = .995$, $p_2 = .001$, and $p_2 = .004$.

The insertion rate of this machine is $n = 3500$ components per hour of operation. Thus, we expect during one hour of operation $n \times (p_2 + p_3) = 175$ insertion errors.

Given that there are 16 insertion errors during a particular hour of operation, the conditional distribution of the number of misinsertions is binomial $B(16, .01/.05)$.

Thus,

$$E\{J_2 \mid J_2 + J_3 = 16\} = 16 \times 0.2 = 3.2$$

However,

$$E\{J_2\} = 3500 \times 0.001 = 3.5$$

We see that the information concerning the total number of insertion errors makes a difference.

Finally

$$\text{cov}(J_2, J_3) = -3500 \times 0.001 \times 0.004$$
$$= -0.014$$

$$V\{J_2\} = 3500 \times 0.001 \times 0.999 = 3.4965$$

and

$$V\{J_3\} = 3500 \times 0.004 \times 0.996 = 13.944$$

Hence, the correlation between J_2 and J_3 is

$$\rho_{2,3} = \frac{-0.014}{\sqrt{3.4965 \times 13.944}} = -0.0020$$

This correlation is quite small. ∎

4.6.2 The Multi-hypergeometric Distribution

Suppose we draw a RSWOR of size n from a population of size N. Each one of the n units in the sample is classified as belonging to one of k categories. Let $J_1, J_2, \ldots, J_k$ be the number of sample units belonging to each one of these categories; $J_1 + \ldots + J_k = n$. The distribution of $J_1, \ldots, J_k$ is k-variate hypergeometric. If $M_1, \ldots, M_k$ are the number of units in the population in these categories before the sample is drawn, then the joint p.d.f. of $J_1, \ldots, J_k$ is

$$p(j_1, \ldots, j_{k-1}) = \frac{\binom{M_1}{j_1}\binom{M_2}{j_2} \cdots \binom{M_k}{j_k}}{\binom{N}{n}} \tag{4.6.4}$$

where $j_k = n - (j_1 + \ldots + j_{k-1})$. This distribution is a generalization of the hypergeometric distribution $H(N, M, n)$. The hypergeometric distribution $H(N, M_i, n)$ is the marginal distribution of J_i $(i = 1, \ldots, k)$. Thus,

$$E\{J_i\} = n\frac{M_i}{N} \qquad i = 1, \ldots, k$$

$$V\{J_i\} = n\frac{M_i}{N}\left(1 - \frac{M_i}{N}\right)\left(1 - \frac{n-1}{N-1}\right), \qquad i = 1, \ldots, k \tag{4.6.5}$$

and for $i \neq j$

$$\mathrm{cov}(J_i, J_j) = -n\frac{M_i}{N} \cdot \frac{M_j}{N}\left(1 - \frac{n-1}{N-1}\right)$$

EXAMPLE **4.34**   A lot of 100 spark plugs contains 20 plugs from vendor V_1, 50 plugs from vendor V_2, and 30 plugs from vendor V_3.

A random sample of $n = 20$ plugs is drawn from the lot without replacement.

Let J_i be the number of plugs in the sample from the vendor V_i, $i = 1, 2, 3$. Accordingly,

$$\Pr\{J_1 = 5, J_2 = 10\} = \frac{\binom{20}{5}\binom{50}{10}\binom{30}{5}}{\binom{100}{20}}$$

$$= .00096$$

If we are told that 5 of the 20 plugs in the sample are from vendor V_3, then the conditional distribution of J_1 is

$$\Pr\{J_1 = j_1 \mid J_3 = 5\} = \frac{\binom{20}{j_1}\binom{50}{15-j_1}}{\binom{70}{15}} \qquad j_1 = 0, \ldots, 15$$

Indeed, given $J_3 = 5$, then J_1 can assume only the values $0, 1, \ldots, 15$. The conditional probability that j_1 of the 15 remaining plugs in the sample are from vendor V_1 is the same as that of choosing a RSWOR of size 15 from a lot of size $70 = 20 + 50$, with 20 plugs from vendor V_1. ∎

4.6.3 The Bivariate Normal Distribution

The bivariate normal distribution is the joint distribution of two continuous random variables (X, Y) having a joint p.d.f.

$$f(x, y; \mu, \eta, \sigma_X, \sigma_Y, \rho) = \frac{1}{2\pi \sigma_X \sigma_Y \sqrt{1 - \rho^2}} \cdot \qquad \text{(4.6.6)}$$

$$\exp\left\{-\frac{1}{2(1-\rho^2)}\left[\left(\frac{x-\mu}{\sigma_X}\right)^2 - 2\rho\frac{x-\mu}{\sigma_X}\cdot\frac{y-\eta}{\sigma_Y} + \left(\frac{y-\eta}{\sigma_Y}\right)^2\right]\right\}$$

$$-\infty < x, y < \infty$$

$\mu, \eta, \sigma_X, \sigma_Y$, and ρ are parameters of this distribution.

Integration of y yields that the marginal distribution of X is $N(\mu, \sigma_x^2)$. Similarly, the marginal distribution of Y is $N(\eta, \sigma_Y^2)$. Furthermore, ρ is the correlation between X and Y. Notice that if $\rho = 0$, then the joint p.d.f. becomes the product of the two marginal ones:

$$f(x, y; \mu, \eta, \sigma_X, \sigma_Y, 0) = \frac{1}{\sqrt{2\pi}\sigma_X}\exp\left\{-\frac{1}{2}\left(\frac{x-\mu}{\sigma_X}\right)^2\right\} \cdot$$

$$\frac{1}{\sqrt{2\pi}\sigma_Y}\exp\left\{-\frac{1}{2}\left(\frac{y-\eta}{\sigma_Y}\right)^2\right\}$$

$$\text{for all} -\infty < x, y < \infty$$

Hence, if $\rho = 0$, then X and Y are independent. However, if $\rho \neq 0$, then $f(x, y; \mu, \eta, \sigma_X, \sigma_Y, \rho) \neq f_1(x; \mu, \sigma_X)f_2(y; \eta, \sigma_Y)$, and the two random variables are dependent.

Figure 4.16 shows the bivariate p.d.f. for $\mu = \eta = 0, \sigma_X = \sigma_Y = 1$, and $\rho = 0.5$.

One can verify also that the conditional distribution of Y, given $\{X = x\}$, is normal with mean

$$\mu_{Y\cdot x} = \eta + \rho\frac{\sigma_Y}{\sigma_X}(x - \mu) \qquad \text{(4.6.7)}$$

and variance

$$\sigma_{Y\cdot x}^2 = \sigma_Y^2(1 - \rho^2) \qquad \text{(4.6.8)}$$

It is interesting to see that $\mu_{Y\cdot x}$ is a linear function of x. The formula is analogous to that of the simple linear regression of Y on x, where η replaces $\overline{Y}$, μ replaces $\overline{X}$, and $\rho(\sigma_Y/\sigma_X)$ replaces the regression coefficient b. We can say that $\mu_{Y\cdot x} = E\{Y \mid X = x\}$ is, in the bivariate normal case, the theoretical (linear) regression of Y on X.

F I G U R E   4.16
Bivariate normal p.d.f.

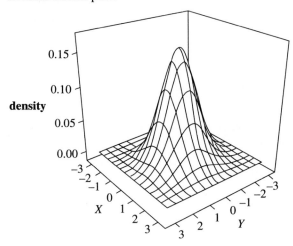

Similarly,

$$\mu_{X \cdot y} = \mu + \rho \frac{\sigma_X}{\sigma_Y}(y - \eta)$$

and

$$\sigma^2_{X \cdot y} = \sigma^2_X(1 - \rho^2)$$

If $\mu = \eta = 0$ and $\sigma_X = \sigma_Y = 1$, we have the standard bivariate normal distribution. The joint c.d.f. in the standard case is denoted by $\Phi_2(x, y; \rho)$, and its formula is

$$\Phi_2(x, y; \rho) = \frac{1}{2\pi\sqrt{1 - \rho^2}} \qquad (4.6.9)$$

$$\int_{-\infty}^{x} \int_{-\infty}^{y} \exp\left\{ -\frac{1}{2(1 - \rho^2)}(z_1^2 - 2\rho z_1 z_2 + z_2^2) \right\} dz_1 dz_2$$

$$= \int_{-\infty}^{x} \phi(z_1) \Phi\left(\frac{y - \rho z_1}{\sqrt{1 - \rho^2}} \right) dz_1$$

Values of $\Phi_2(x, y; \rho)$ can be obtained by numerical integration. If one has to compute the bivariate c.d.f. in the general case, the following formula is useful:

$$F(x, y; \mu, \eta, \sigma_X, \sigma_Y, \rho) = \Phi_2\left(\frac{x - \mu}{\sigma_X}, \frac{y - \eta}{\sigma_Y}; \rho \right)$$

For computing $\Pr\{a \leq X \leq b, c \leq Y \leq d\}$, we use the formula

$$\Pr\{a \leq X \leq b, c \leq Y \leq d\} = F(b, d; -) - F(a, d; -)$$
$$- F(b, c; -) + F(a, c; -)$$

EXAMPLE **4.35** Suppose that (X, Y) deviations in component placement on PCBs by an automatic machine have a bivariate normal distribution with means $\mu = \eta = 0$, standard deviations $\sigma_X = 0.00075$ and $\sigma_Y = 0.00046$ (in inches), and $\rho = 0.160$. The placement errors are within the specifications if $|X| < 0.001$ (in inches) and $|Y| < 0.001$ (in inches). What proportion of components is expected to have X, Y deviations in line with the specifications? The standardized version of the spec limits are $Z_1 = 0.001/0.00075 = 1.33$ and $Z_2 = 0.001/0.00046 = 2.174$. We compute

$$
\begin{aligned}
\Pr\{|X| < 0.001, |Y| < 0.001\} = {} & \Phi_2(1.33, 2.174, .16) \\
& - \Phi_2(-1.33, 2.174, .16) \\
& - \Phi_2(1.33, -2.174; .16) \\
& + \Phi_2(-1.33, -2.174; .16) \\
= {} & .793
\end{aligned}
$$

This is the expected proportion of good placements. ∎

4.7
Distribution of Order Statistics

As defined in Chapter 2, the order statistics of the sample are the sorted data. More specifically, let $X_1, \ldots, X_n$ be independent and identically distributed (i.i.d.) random variables. The order statistics are $X_{(i)}$, $i = 1, \ldots, n$, where

$$X_{(1)} \le X_{(2)} \le \cdots \le X_{(n)}$$

In this section, we discuss the distributions of these order statistics when $F(x)$ is (absolutely) continuous, having a p.d.f. $f(x)$.

We start with the extreme statistics $X_{(1)}$ and $X_{(n)}$.

Because the random variables X_i $(i = 1, \ldots, n)$ are i.i.d., the c.d.f. of $X_{(1)}$ is

$$
\begin{aligned}
F_{(1)}(x) &= \Pr\{X_{(1)} \le x\} \\
&= 1 - \Pr\{X_{(1)} \ge x\} = 1 - \prod_{i=1}^{n} \Pr\{X_i \ge x\} \\
&= 1 - (1 - F(x))^n
\end{aligned}
$$

By differentiation, we see that the p.d.f. of $X_{(1)}$ is

$$f_{(1)}(x) = nf(x)[1 - F(x)]^{n-1} \tag{4.7.1}$$

Similarly, the c.d.f. of the sample maximum $X_{(n)}$ is

$$
\begin{aligned}
F_{(n)}(x) &= \prod_{i=1}^{n} \Pr\{X_i \le x\} \\
&= (F(x))^n
\end{aligned}
$$

The p.d.f. of $X_{(n)}$ is

$$f_{(n)}(x) = nf(x)(F(x))^{n-1} \tag{4.7.2}$$

EXAMPLE 4.36 a. A switching circuit consists of n modules that operate independently and are connected in series (see Figure 4.17). Let X_i be the time till failure of the ith module. The system fails when any module fails. Thus, the time till failure of the system is $X_{(1)}$. If all X_i are exponentially distributed with mean life β, then the c.d.f. of $X_{(1)}$ is

$$F_{(1)}(x) = 1 - e^{-nx/\beta} \qquad x \geq 0$$

Thus, $X_{(1)}$ is distributed like $E(\beta/n)$. It follows that the expected time till failure of the circuit is $E\{X_{(1)}\} = \beta/n$.

b. If the modules are connected in parallel, then the circuit fails at the instant the last of the n modules fails, which is $X_{(n)}$. Thus, if X_i is $E(\beta)$, the c.d.f. of $X_{(n)}$ is

$$F_{(n)}(x) = (1 - e^{-(x/\beta)})^n$$

The expected value of $X_{(n)}$ is

$$
\begin{aligned}
E\{X_{(n)}\} &= \frac{n}{\beta} \int_0^\infty x e^{-x/\beta}(1 - e^{-x/\beta})^{n-1} dx \\
&= n\beta \int_0^\infty y e^{-y}(1 - e^{-y})^{n-1} dy \\
&= n\beta \sum_{j=0}^{n-1} (-1)^j \binom{n-1}{j} \int_0^\infty y e^{-(1+j)y} dy \\
&= n\beta \sum_{j=1}^{n} (-1)^{j-1} \binom{n-1}{j-1} \frac{1}{j^2}
\end{aligned}
$$

Furthermore, because $n\binom{n-1}{j-1} = j\binom{n}{j}$, we find that

$$E\{X_{(n)}\} = \beta \sum_{j=1}^{n} (-1)^{j-1} \binom{n}{j} \frac{1}{j}$$

FIGURE 4.17
Series and parallel systems

Components in Series

Components in Parallel

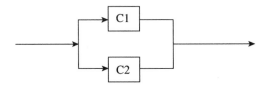

We can also show that this formula is equivalent to

$$E\{X_{(n)}\} = \beta \sum_{j=1}^{n} \frac{1}{j}$$

Accordingly, if the parallel circuit consists of 3 modules, and the time till failure of each module is exponential with $\beta = 1000$ hr, the expected time till failure of the system is 1833.3 hr. ∎

Generally, the distribution of $X_{(i)}$ $(i = 1, \ldots, n)$ can be obtained by the following argument. The event $\{X_{(i)} \leq x\}$ is equivalent to the event that the number of X_i values in the random sample that are smaller or equal to x is at least i.

Consider n independent and identical trials in which success is that $\{X_i \leq x\}$ $(i = 1, \ldots, n)$. The probability of success is $F(x)$. The distribution of the number of successes is $B(n, F(x))$. Thus, the c.d.f. of $X_{(i)}$ is

$$F_{(i)}(x) = \Pr\{X_{(i)} \leq x\} = 1 - B(i - 1; n, F(x))$$
$$= \sum_{j=i}^{n} \binom{n}{j} (F(x))^j (1 - F(x))^{n-j}$$

Differentiating this c.d.f. with respect to x yields the p.d.f. of $X_{(i)}$—namely:

$$f_{(i)}(x) = \frac{n!}{(i-1)!(n-i)!} f(x)(F(x))^{i-1}(1 - F(x))^{n-i} \tag{4.7.3}$$

Notice that if X has a uniform distribution on $(0, 1)$, then the distribution of $X_{(i)}$ is like that of beta$(i, n - i + 1)$, $i = 1, \ldots, n$. In a similar manner, we can derive the joint p.d.f. of $(X_{(i)}, X_{(j)})$, $1 \leq i < j \leq n$, and so on. This joint p.d.f. is given by

$$f_{(i),(j)}(x, y) = \frac{n!}{(i-1)!(j-1-i)!(n-j)!} f(x)f(y) \cdot$$
$$(F(x))^{i-1}[F(y) - F(x)]^{j-i-1}(1 - F(y))^{n-j} \tag{4.7.4}$$

for $-\infty < x < y < \infty$.

4.8
Linear Combinations of Random Variables

Let $X_1, X_2, \ldots, X_n$ be random variables having a joint distribution, with joint p.d.f. $f(x_1, \ldots, x_n)$. Let $\alpha_1, \ldots, \alpha_n$ be given constants. Then

$$W = \sum_{i=1}^{n} \alpha_i X_i$$

is a linear combination of the X's. The p.d.f. of W can generally be derived using various methods. In this section, we discuss only the formulas of the expected value and variance of W.

It is easily shown that

$$E\{W\} = \sum_{i=1}^{n} \alpha_i E\{X_i\} \tag{4.8.1}$$

That is, the expected value of a linear combination is the same linear combination of the expectations.

The formula for the variance is somewhat more complicated and is given by

$$V\{W\} = \sum_{i=1}^{n} \alpha_i^2 V\{X_i\} + \sum\sum_{i \neq j} \alpha_i \alpha_j \text{cov}(X_i, X_j) \tag{4.8.2}$$

EXAMPLE 4.37   Let $X_1, X_2, \ldots, X_n$ be i.i.d. random variables, with common expectations μ and common finite variances σ^2. The sample mean $\overline{X}_n = 1/n \sum_{i=1}^{n} X_i$ is a particular linear combination, with

$$\alpha_1 = \alpha_2 = \ldots = \alpha_n = \frac{1}{n}$$

Hence,

$$E\{\overline{X}_n\} = \frac{1}{n} \sum_{i=1}^{n} E\{X_i\} = \mu$$

and, because $X_1, X_2, \ldots, X_n$ are mutually independent, $\text{cov}(X_i, X_j) = 0$, all $i \neq j$. Hence

$$V\{\overline{X}_n\} = \frac{1}{n^2} \sum_{i=1}^{n} V\{X_i\} = \frac{\sigma^2}{n}$$

Thus, we have shown that in a random sample of n i.i.d. random variables, the sample mean has the same expectation as that of the individual variables, but its sample variance is reduced by a factor of $1/n$.

Moreover, from Chebyshev's inequality, for any $\epsilon > 0$

$$\Pr\{|\overline{X}_n - \mu| > \epsilon\} < \frac{\sigma^2}{n\epsilon^2}$$

Therefore, because $\lim_{n \to \infty} \sigma^2/n\epsilon^2 = 0$,

$$\lim_{n \to \infty} \Pr\{|\overline{X}_n - \mu| > \epsilon\} = 0$$

This property is called the **convergence in probability** of $\overline{X}_n$ to μ.   ∎

EXAMPLE 4.38   Let U_1, U_2, U_3 be three i.i.d. random variables having uniform distributions on $(0, 1)$. We consider the statistic

$$W = \frac{1}{4}U_{(1)} + \frac{1}{2}U_{(2)} + \frac{1}{4}U_{(3)}$$

where $0 < U_{(1)} < U_{(2)} < U_{(3)} < 1$ are the order statistics. We saw in Section 4.7 that the distribution of $U_{(i)}$ is like that of beta$(i, n - i + 1)$. Hence

$$E\{U_{(1)}\} = E\{\text{beta}(1, 3)\} = \frac{1}{4}$$

$$E\{U_{(2)}\} = E\{\text{beta}(2, 2)\} = \frac{1}{2}$$

$$E\{U_{(3)}\} = E\{\text{beta}(3, 1)\} = \frac{3}{4}$$

It follows that

$$E\{W\} = \frac{1}{4} \cdot \frac{1}{4} + \frac{1}{2} \cdot \frac{1}{2} + \frac{1}{4} \cdot \frac{1}{4} = \frac{1}{2}$$

To find the variance of W, we need more derivations.

First

$$V\{U_{(1)}\} = V\{\text{beta}(1, 3)\} = \frac{3}{4^2 \times 5} = \frac{3}{80}$$

$$V\{U_{(2)}\} = V\{\text{beta}(2, 2)\} = \frac{4}{4^2 \times 5} = \frac{1}{20}$$

$$V\{U_{(3)}\} = V\{\text{beta}(3, 1)\} = \frac{3}{4^2 \times 5} = \frac{3}{80}$$

We need to find $\text{cov}(U_{(1)}, U_{(2)})$, $\text{cov}(U_{(1)}, U_{(3)})$, and $\text{cov}(U_{(2)}, U_{(3)})$. From the joint p.d.f. formula of order statistics, the joint p.d.f. of $(U_{(1)}, U_{(2)})$ is

$$f_{(1),(2)}(x, y) = 6(1 - y) \qquad 0 < x \le y < 1$$

Hence

$$E\{U_{(1)} U_{(2)}\} = 6 \int_0^1 x \left(\int_0^1 y(1 - y) dy \right) dx$$

$$= \frac{6}{40}$$

Thus,

$$\text{cov}(U_{(1)}, U_{(2)}) = \frac{6}{40} - \frac{1}{4} \cdot \frac{1}{2}$$

$$= \frac{1}{40}$$

Similarly, the p.d.f. of $(U_{(1)}, U_{(3)})$ is

$$f_{(1),(3)}(x, y) = 6(y - x) \qquad 0 < x \le y < 1$$

Thus,

$$E\{U_{(1)} U_{(3)}\} = 6 \int_0^1 x \left(\int_x^1 y(y - x) dy \right) dx$$

$$= 6 \int_0^1 x \left(\frac{1}{3} \left(1 - x^3 \right) - \frac{x}{2}(1 - x^2) \right) dx$$

$$= \frac{1}{4}$$

and

$$\text{cov}(U_{(1)}, U_{(3)}) = \frac{1}{5} - \frac{1}{4} \cdot \frac{3}{4} = \frac{1}{80}$$

The p.d.f. of $(U_{(2)}, U_{(3)})$ is

$$f_{(2),(3)}(x, y) = 6x \qquad 0 < x \le y \le 1$$

and

$$\text{cov}(U_{(2)}, U_{(3)}) = \frac{1}{40}$$

Finally,

$$
\begin{aligned}
V\{W\} &= \frac{1}{16} \cdot \frac{3}{80} + \frac{1}{4} \cdot \frac{1}{20} + \frac{1}{16} \cdot \frac{3}{80} \\
&\quad + 2 \cdot \frac{1}{4} \cdot \frac{1}{2} \cdot \frac{1}{40} + 2 \cdot \frac{1}{4} \cdot \frac{1}{4} \cdot \frac{1}{80} \\
&\quad + 2 \cdot \frac{1}{2} \cdot \frac{1}{4} \cdot \frac{1}{40} \\
&= \frac{1}{32} = 0.03125 \quad \blacksquare
\end{aligned}
$$

The following is a useful result:

If $X_1, X_2, \ldots, X_n$ are mutually independent, then the m.g.f. of $T_n = \sum\limits_{i=1}^{n} X_i$ is

$$M_{T_n}(t) = \prod_{i=1}^{n} M_{X_i}(t) \tag{4.8.3}$$

Indeed, as shown in Section 4.5.2, when $X_1, \ldots, X_n$ are independent, the expected value of the product of functions is the product of their expectations. Therefore,

$$
\begin{aligned}
M_{T_n}(t) &= E\{e^{t \sum\limits_{i=1}^{n} X_i}\} \\
&= E\left\{\prod_{i=1}^{n} e^{tX_i}\right\} \\
&= \prod_{i=1}^{n} E\{e^{tX_i}\} \\
&= \prod_{i=1}^{n} M_{X_i}(t)
\end{aligned}
$$

The expected value of the product is equal to the product of the expectations because $X_1, \ldots, X_n$ are mutually independent.

EXAMPLE 4.39   In this example, we illustrate some applications of the last result.

a. If $X_1, X_2, \ldots, X_k$ be independent random variables having binomial distributions like $B(n_i, p)$, $i = 1, \ldots, k$, then their sum T_k has the binomial distribution. To show this,

$$M_{T_k}(t) = \prod_{i=1}^{k} M_{X_i}(t)$$

$$= [e^t p + (1 - p)]^{\sum_{i=1}^{k} n_i}$$

That is, T_k is distributed like $B\left(\sum_{i=1}^{k} n_i, p\right)$. This result is intuitively clear.

b. If $X_1, \ldots, X_n$ are independent random variables having Poisson distributions with parameters λ_i $(i = 1, \ldots, n)$, then the distribution of $T_n = \sum_{i=1}^{n} X_i$ is Poisson with parameter $\mu_n = \sum_{i=1}^{n} \lambda_i$. Indeed,

$$M_{T_n}(t) = \prod_{j=1}^{n} \exp\{-\lambda_j(1 - e^t)\}$$

$$= \exp\left\{-\sum_{j=1}^{n} \lambda_j(1 - e^t)\right\}$$

$$= \exp\{-\mu_n(1 - e^t)\}$$

c. Suppose $X_1, \ldots, X_n$ are independent random variables, and the distribution of X_i is normal $N(\mu_i, \sigma_i^2)$; then the distribution of $W = \sum_{i=1}^{n} \alpha_i X_i$ is normal, like that of

$$N\left(\sum_{i=1}^{n} \alpha_i \mu_i, \sum_{i=1}^{n} \alpha_i^2 \sigma_i^2\right)$$

To verify this, we recall that $X_i = \mu_i + \sigma_i Z_i$, where Z_i is $N(0, 1)$ $(i = 1, \ldots, n)$. Thus

$$M_{\alpha_i X_i}(t) = E\{e^{t(\alpha_i \mu_i + \alpha_i \sigma_i Z_i)}\}$$

$$= e^{t\alpha_i \mu_i} M_{Z_i}(\alpha_i \sigma_i t)$$

We derived before that $M_{Z_i}(u) = e^{u^2/2}$. Hence,

$$M_{\alpha_i X_i}(t) = \exp\left\{\alpha_i \mu_i t + \frac{\alpha_i^2 \sigma_i^2}{2} t^2\right\}$$

Finally,

$$M_W(t) = \prod_{i=1}^{n} M_{\alpha_i X_i}(t)$$

$$= \exp\left\{ \left(\sum_{i=1}^{n} \alpha_i \mu_i\right) t + \frac{\sum_{i=1}^{n} \alpha_i^2 \sigma_i^2}{2} t^2 \right\}$$

This implies that the distribution of W is normal, with

$$E\{W\} = \sum_{i=1}^{n} \alpha_i \mu_i$$

and

$$V\{W\} = \sum_{i=1}^{n} \alpha_i^2 \sigma_i^2$$

d. If $X_1, X_2, \ldots, X_n$ are independent random variables having gamma distributions like $G(\nu_i, \beta)$, respectively, $i = 1, \ldots, n$, then the distribution of $T_n = \sum_{i=1}^{n} X_i$ is gamma, like that of $G\left(\sum_{i=1}^{n} \nu_i, \beta\right)$. Indeed,

$$M_{T_n}(t) = \prod_{i=1}^{n} (1 - t\beta)^{-\nu_i}$$

$$= (1 - t\beta)^{-\sum_{i=1}^{n} \nu_i} \quad \blacksquare$$

4.9
Large-Sample Approximations
4.9.1   The Law of Large Numbers

We saw in Example 4.37 that the mean of a random sample, $\overline{X}_n$, converges in probability to the expected value of X, μ (the population mean). This is the **law of large numbers** (LLN), which states that, if $X_1, X_2, \ldots$ are i.i.d. random variables and $E\{|X_1|\} < \infty$, then, for any $\epsilon > 0$,

$$\lim_{n \to \infty} \Pr\{|\overline{X}_n - \mu| > \epsilon\} = 0$$

We also write,

$$\lim_{n \to \infty} \overline{X}_n = \mu \quad \text{in probability}$$

This is known as the *weak LLN* (WLLN). There is a stronger law which states that, under the preceding conditions,

$$\Pr\{\lim_{n \to \infty} \overline{X}_n = \mu\} = 1$$

It is beyond the scope of the book to discuss the meaning of the strong LLN.

4.9.2 The Central Limit Theorem

The **central limit theorem** (CLT), is one of the most important theorems in probability theory. We formulate here the simplest version of this theorem, which is often sufficient for applications. The theorem states that if $\overline{X}_n$ is the sample mean of n i.i.d. random variables, then, if the population variance σ^2 is positive and finite, the sampling distribution of $\overline{X}_n$ is approximately normal as $n \to \infty$. More precisely, **If** $X_1, X_2, \ldots$ **is a sequence of i.i.d. random variables, with**

$$E\{X_1\} = \mu \text{ and } V\{X_1\} = \sigma^2, 0 < \sigma^2 < \infty, \text{ then}$$

$$\lim_{n \to \infty} \Pr \left\{ \frac{(\overline{X}_n - \mu)\sqrt{n}}{\sigma} \leq z \right\} = \Phi(z) \tag{4.9.1}$$

where $\Phi(z)$ is the c.d.f. of $N(0, 1)$.

The proof of this basic version of the CLT is based on a result in probability theory, that states that if $X_1, X_2, \ldots$ is a sequence of random variables having m.g.f.'s, $M_n(T)$, $n = 1, 2, \ldots$, and if $\lim_{n \to \infty} M_n(t) = M(t)$ is the m.g.f. of a random variable X^* having a c.d.f. $F^*(x)$, then $\lim_{n \to \infty} F_n(x) = F^*(x)$, where $F_n(x)$ is the c.d.f. of X_n.

The m.g.f. of

$$Z_n = \frac{\sqrt{n}(\overline{X}_n - \mu)}{\sigma}$$

can be written as

$$M_{Z_n}(t) = E\left\{ \exp\left\{ \frac{t}{\sqrt{n}\sigma} \sum_{i=1}^{n} (X_i - \mu) \right\} \right\}$$

$$= \left(E\left\{ \exp\left\{ \frac{t}{\sqrt{n}\sigma} (X_1 - \mu) \right\} \right\} \right)^n$$

because the random variables are independent. Furthermore, Taylor expansion of $\exp\left\{ \frac{t}{\sqrt{n}\sigma} (X_1 - \mu) \right\}$ is

$$1 + \frac{t}{\sqrt{n}\sigma}(X_1 - \mu) + \frac{t^2}{2n\sigma^2}(X_1 - \mu)^2 + o\left(\frac{1}{n}\right)$$

for n large. Here $o(\frac{1}{n})$ indicates a remainder R_n such that $\lim_{n \to \infty} nR_n = 0$. Hence, as $n \to \infty$

$$E\left\{ \exp\left\{ \frac{t}{\sqrt{n}\sigma}(X_1 - \mu) \right\} \right\} = 1 + \frac{t^2}{2n} + o\left(\frac{1}{n}\right)$$

Hence,

$$\lim_{n \to \infty} M_{Z_n}(t) = \lim_{n \to \infty} \left(1 + \frac{t^2}{2n} + o\left(\frac{1}{n}\right) \right)^n$$

$$= e^{t^2/2}$$

which is the m.g.f. of $N(0, 1)$. This is a sketch of the proof. For rigorous proofs and extensions, see textbooks on probability theory.

4.9.3　Some Normal Approximations

The CLT can be applied to provide an approximation to the distribution of the sum of n i.i.d. random variables, by a standard normal distribution, when n is large. Following are a few such useful approximations.

1　Binomial distributions

When n is large, then the c.d.f. of $B(n, p)$ can be approximated by

$$B(k; n, p) \cong \Phi\left(\frac{k + \frac{1}{2} - np}{\sqrt{np(1 - p)}}\right) \tag{4.9.2}$$

We add $\frac{1}{2}$ to k in the argument of $\Phi(\cdot)$ to obtain a better approximation when n is not too large. This modification is called **correction for discontinuity**.

How large should n be to get a "good" approximation? A general rule is

$$n > \frac{9}{p(1 - p)} \tag{4.9.3}$$

2　Poisson distributions

The c.d.f. of Poisson with parameter λ can be approximated by

$$P(k; \lambda) \cong \Phi\left(\frac{k + \frac{1}{2} - \lambda}{\sqrt{\lambda}}\right) \tag{4.9.4}$$

if λ is large (greater than 30).

3　Gamma distribution

The c.d.f. of $G(\nu, \beta)$ can be approximated by

$$G(x; \nu, \beta) \cong \Phi\left(\frac{x - \nu\beta}{\beta\sqrt{\nu}}\right) \tag{4.9.5}$$

for large values of ν.

EXAMPLE **4.40**　a. A lot consists of $n = 10,000$ screws. The probability that a screw is defective is $p = .01$. What is the probability that there are more than 120 defective screws in the lot?

The number of defective screws in the lot, J_n, has a distribution like $B(10000, .01)$. Hence,

$$\Pr\{J_{10000} > 120\} = 1 - B(120; 10000, .01)$$

$$\cong 1 - \Phi\left(\frac{120.5 - 100}{\sqrt{99}}\right)$$

$$= 1 - \Phi(2.06) = .0197$$

b. In the production of industrial film, we find, on average, 1 defect per 100 ft^2 of film. What is the probability that fewer than 100 defects will be found on 12,000 ft^2 of film?

We assume that the number of defects per unit area of film is a Poisson random variable. Thus, our model is that the number of defects, X, per 12,000 ft^2 has a Poisson distribution with parameter $\lambda = 120$. Thus,

$$\Pr\{X < 100\} \cong \Phi\left(\frac{99.5 - 120}{\sqrt{120}}\right)$$
$$= .0306$$

c. The time till failure, T, of radar equipment is exponentially distributed with mean time till failure (MTTF) of $\beta = 100$ hr.

A sample of $n = 50$ units is put to the test. Let $\overline{T}_{50}$ be the sample mean. What is the probability that $\overline{T}_{50}$ will fall in the interval (95, 105) hr?

We have seen that $\sum_{i=1}^{50} T_i$ is distributed like $G(50, 100)$ becauuse $E(\beta)$ is distributed like $G(1, \beta)$. Hence, $\overline{T}_{50}$ is distributed like $\frac{1}{50}G(50, 100)$, which is $G(50, 2)$. By the normal approximation

$$\Pr\{95 < \overline{T}_{50} < 105\} \cong \Phi\left(\frac{105 - 100}{2\sqrt{50}}\right)$$
$$-\Phi\left(\frac{95 - 100}{2\sqrt{50}}\right) = 2\Phi(0.3536) - 1 = .2764 \quad \blacksquare$$

4.10
Additional Distributions of Statistics of Normal Samples

In this section, we assume that $X_1, X_2, \ldots, X_n$ are i.i.d. $N(\mu, \sigma^2)$ random variables. We present the chi-square, t and F distributions, which play an important role in the theory of statistical inference (see Chapter 6).

4.10.1 Distribution of the Sample Variance

Writing $X_i = \mu + \sigma Z_i$, where $Z_1, \ldots, Z_n$ are i.i.d. $N(0, 1)$, we see that the sample variance S^2 is distributed like

$$S^2 = \frac{1}{n-1}\sum_{i=1}^{n}(X_i - \overline{X}_n)^2$$
$$= \frac{1}{n-1}\sum_{i=1}^{n}(\mu + \sigma Z_i - (\mu + \sigma \overline{Z}_n))^2$$
$$= \frac{\sigma^2}{n-1}\sum_{i=1}^{n}(Z_i - \overline{Z}_n)^2$$

We can show that $\sum_{i=1}^{n}(Z_i - \bar{Z}_n)^2$ is distributed like $\chi^2[n-1]$, where $\chi^2[\nu]$ is called a **chi-square random variable with ν degrees of freedom**. Moreover, $\chi^2[\nu]$ is distributed like $G\left(\frac{\nu}{2}, 2\right)$.

The αth quantile of $\chi^2[\nu]$ is denoted by $\chi^2_\alpha[\nu]$. Accordingly, the c.d.f. of the sample variance is

$$
\begin{aligned}
H_{S^2}(x; \sigma^2) &= \Pr\left\{ \frac{\sigma^2}{n-1} \chi^2[n-1] \le x \right\} \\
&= \Pr\left\{ \chi^2[n-1] \le \frac{(n-1)x}{\sigma^2} \right\} \\
&= \Pr\left\{ G\left(\frac{n-1}{2}, 2\right) \le \frac{(n-1)x}{\sigma^2} \right\} \\
&= G\left(\frac{(n-1)x}{2\sigma^2}; \frac{n-1}{2}, 1 \right)
\end{aligned}
\tag{4.10.1}
$$

The probability values of the distribution of $\chi^2[\nu]$, as well as the α-quantiles, can be computed by MINITAB, or read from appropriate tables.

The expected value and variance of the sample variance are

$$
\begin{aligned}
E\{S^2\} &= \frac{\sigma^2}{n-1} E\{\chi^2[n-1]\} \\
&= \frac{\sigma^2}{n-1} E\left\{ G\left(\frac{n-1}{2}, 2\right) \right\} \\
&= \frac{\sigma^2}{n-1} \cdot (n-1) = \sigma^2
\end{aligned}
$$

Similarly,

$$
\begin{aligned}
V\{S^2\} &= \frac{\sigma^4}{(n-1)^2} V\{\chi^2[n-1]\} \\
&= \frac{\sigma^4}{(n-1)^2} V\left\{ G\left(\frac{n-1}{2}, 2\right) \right\} \\
&= \frac{\sigma^4}{(n-1)^2} \cdot 2(n-1) \\
&= \frac{2\sigma^4}{n-1}
\end{aligned}
\tag{4.10.2}
$$

Thus, applying the Chebyshev's inequality, for any given $\epsilon > 0$,

$$
\Pr\{|S^2 - \sigma^2| > \epsilon\} < \frac{2\sigma^4}{(n-1)\epsilon^2}
$$

Hence, S^2 converges in probability to σ^2. Moreover,

$$\lim_{n\to\infty} \text{Pr}\left\{ \frac{(S^2 - \sigma^2)}{\sigma^2\sqrt{2}}\sqrt{n-1} \le z \right\} = \Phi(z) \tag{4.10.3}$$

That is, the distribution of S^2 can be approximated by the normal distributions in large samples.

4.10.2 The Student's *t* statistic

We have seen that

$$Z_n = \frac{\sqrt{n}(\overline{X}_n - \mu)}{\sigma}$$

has a $N(0, 1)$ distribution. As we will see in Chapter 6, when σ is unknown, we test hypotheses concerning μ by the statistic

$$t = \frac{\sqrt{n}(\overline{X}_n - \mu_0)}{S},$$

where S is the sample standard deviation. If $X_1, \ldots, X_n$ are i.i.d., like $N(\mu_0, \sigma^2)$, then the distribution of t is called the **Student's t distribution with $v = n - 1$ degrees of freedom**. The corresponding random variable is denoted by $t[v]$.

The p.d.f. of $t[v]$ is symmetric about 0 (see Figure 4.18). Thus,

$$E\{t[v]\} = 0 \quad \text{for} \quad v \ge 2 \tag{4.10.4}$$

FIGURE 4.18

Density functions of $t[v]$, $v = 5, 50$

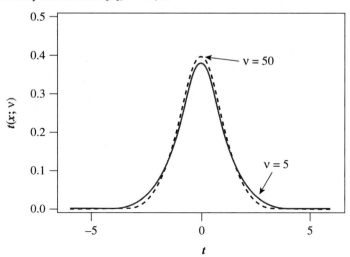

and

$$V\{t[\nu]\} = \frac{\nu}{\nu - 2} \qquad \nu \geq 3 \tag{4.10.5}$$

The α-quantile of $t[\nu]$ is denoted by $t_\alpha[\nu]$. It can be read from a table or determined by MINITAB.

4.10.3   Distribution of the Variance Ratio

Consider now two independent samples of size n_1 and n_2, respectively, which have been taken from normal populations having variances σ_1^2 and σ_2^2. Let

$$S_1^2 = \frac{1}{n_1 - 1} \sum_{i=1}^{n_1} (X_{1i} - \overline{X}_1)^2$$

and

$$S_2^2 = \frac{1}{n_2 - 1} \sum_{i=1}^{n_2} (X_{2i} - \overline{X}_2)^2$$

be the variances of the two samples where $\overline{X}_1$ and $\overline{X}_2$ are the corresponding sample means. The F-ratio

$$F = \frac{S_1^2 \sigma_2^2}{S_2^2 \sigma_1^2}$$

has a distribution denoted by $F[\nu_1, \nu_2]$, with $\nu_1 = n_1 - 1$ and $\nu_2 = n_2 - 1$. This distribution is called the **F distribution with ν_1 and ν_2 degrees of freedom**. A graph of the densities of $F[\nu_1, \nu_2]$ is given in Figure 4.19.

F I G U R E   4.19
Density function of $F(\nu_1, \nu_2)$

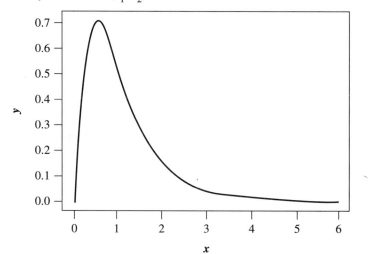

The expected value and the variance of $F[v_1, v_2]$ are:

$$E\{F[v_1, v_2]\} = v_2/(v_2 - 2) \qquad v_2 > 2 \tag{4.10.6}$$

and

$$V\{F[v_1, v_2]\} = \frac{2v_2^2(v_1 + v_2 - 2)}{v_1(v_2 - 2)^2(v_2 - 4)} \qquad v_2 > 4 \tag{4.10.7}$$

The $(1 - \alpha)$th quantile of $F[v_1, v_2]$—that is, $F_{1-\alpha}[v_1, v_2]$—can be computed by MINITAB. If we wish to obtain the αth fractile, $F_{\alpha}[v_1, v_2]$, for values of $\alpha < .5$, we can apply the relationship:

$$F_{1-\alpha}[v_1, v_2] = \frac{1}{F_{\alpha}[v_2, v_1]} \tag{4.10.8}$$

Thus, for example, to compute $F_{.05}[15, 10]$, we write

$$F_{.05}[15, 10] = 1/F_{.95}[10, 15] = 1/2.54 = .3937$$

4.11
Chapter Highlights

The chapter provides the basics of probability theory and of the theory of distribution functions. The probability model for random sampling is discussed. This is fundamental for sampling procedures to be discussed in Chapters 5, 7, and 9. Bayes' theorem has important ramifications in statistical inference, as discussed in Chapter 6. The distribution theory presented is parallel to the empirical concepts discussed in Chapters 2 and 3. MINITAB calculations of distribution functions were shown (see Appendix VI for the corresponding S-PLUS functions). The concepts and definitions introduced are:

- Sample space
- Elementary event
- Event
- Operations with events
- Union
- Intersection
- Disjoint event
- Complement
- Probability function
- Conditional probabilities
- Independent event

- Bayes' formula
- Prior probability
- Posterior probability
- Probability distribution function
- Cumulative distribution function
- Expected value
- Central moments
- Chebyshev inequality
- Moment-generating function
- Independent trials
- pth quantile
- Joint distribution
- Marginal distribution
- Covariance
- Correlation
- Mutual independence
- Conditional distribution
- Law of iterated expectation
- Law of total variance
- Convergence in probability
- Law of large numbers
- Central limit theorem

4.12

Exercises

4.1.1   An experiment consists of making 20 observations on the quality of chips. Each observation is recorded as G or D.

a   What is the sample space, S, corresponding to this experiment?

b   How many elementary events in S?

c   Let A_n, $n = 0, \ldots, 20$, be the event that exactly n G observations are made. Write the events A_n formally. How many elementary events belong to A_n?

4.1.2   An experiment consists of 10 measurements, $w_1, \ldots, w_{10}$, of the weights of packages. All packages under consideration have weights between 10 and 20 pounds. What is the sample space S? Let $A = \{(w_1, w_2, \ldots, w_{10}) : w_1 + w_2 = 25\}$. Let $B = \{(w_1, \ldots, w_{10}) : w_1 + w_2 \leq 25\}$. Describe the events A and B graphically. Show that $A \subset B$.

4.1.3   Strings of 30 binary $(0, 1)$ signals are transmitted.

a   Describe the sample space, S.

b   Let A_{10} be the event that the first 10 signals transmitted are all 1s. How many elementary events belong to A_{10}?

c   Let B_{10} be the event that exactly 10 signals out of 30 transmitted are 1s. How many elementary events belong to B_{10}? Is $A_{10} \subset B_{10}$?

4.1.4   Prove DeMorgan's laws:

a $(A \cup B)^c = A^c \cap B^c$

b $(A \cap B)^c = A^c \cup B^c$

4.1.5.   Consider Exercise 4.1.1. Show that the events $A_0, A_1, \ldots, A_{20}$ are a partition of the sample space, S.

4.1.6   Let $A_1, \ldots, A_n$ be a partition of S. Let B be an event. Show that $B = \bigcup_{i=1}^{n} A_i B$, where $A_i B = A_i \cap B$.

4.1.7   Develop a formula for the probability $Pr\{A \cup B \cup C\}$, where A, B, C are arbitrary events.

4.1.8   Show that if $A_1, \ldots, A_n$ is a partition, then for any event B, $P\{B\} = \sum_{i=1}^{n} P\{A_i B\}$. [Use the result of Exercise 4.1.6.]

4.1.9   An unbiased die has the numbers $1, 2, \ldots, 6$ written on its faces. The die is thrown twice. What is the probability that the two numbers shown on its upper face add up to 10?

4.1.10   The time till failure, T, of electronic equipment is a random quantity. The event $A_t = \{T > t\}$ is assigned the probability $Pr\{A_t\} = \exp\{-t/200\}$, $t \geq 0$. What is the probability of the event $B = \{150 < T < 280\}$?

4.1.11   A box contains 40 parts, 10 of type A, 10 of type B, 15 of type C, and 5 of type D. A random sample of 8 parts is drawn without replacement. What is the probability that 2 parts of each type will be found in the sample?

4.1.12   How many samples of size $n = 5$ can be drawn from a population of size $N = 100$

i   with replacement?

ii   without replacement?

4.1.13   A lot of 1000 items contains $M = 900$ "good" items and 100 "defective" items. A random sample of size $n = 10$ is drawn from the lot. What is the probability that at least 8 good items will be observed in the sample.

i   when sampling is with replacement?

ii   when sampling is without replacement?

4.1.14   Continuing with the previous exercise, what is the probability of observing in an RSWR at least one defective item?

4.1.15   Consider the problem of Exercise 4.1.10. What is the conditional probability $Pr\{T > 300 \mid T > 200\}$?

4.1.16   A point (X, Y) is chosen at random within the unit square—that is,

$$S = \{(x, y) : 0 \leq x, y \leq 1\}$$

Any set A contained in S having an area given by

$$\text{Area}\{A\} = \iint_A dxdy$$

is an event whose probability is the area of A. Define the events

$$B = \left\{(x, y) : x > \frac{1}{2}\right\}$$
$$C = \{(x, y) : x^2 + y^2 \leq 1\}$$
$$D = \{(x, y) : (x + y) \leq 1\}$$

a   Compute the conditional probability $\Pr\{D \mid B\}$.

b   Compute the conditional probability $\Pr\{C \mid D\}$.

4.1.17   Show that if A and B are independent events, then A^c and B^c are also independent events.

4.1.18   Show that if A and B are disjoint events, then A and B are dependent events.

4.1.19   Show that if A and B are independent events, then

$$\Pr\{A \cup B\} = \Pr\{A\}(1 - \Pr\{B\}) + \Pr\{B\}$$
$$= \Pr\{A\} + \Pr\{B\}(1 - \Pr\{A\})$$

4.1.20   A machine that tests whether a part is defective, D, or good, G, may err. The probabilities of errors are given by

$$\Pr\{A \mid G\} = .95$$
$$\Pr\{A \mid D\} = .10$$

where A is the event "the part is considered G after testing." If $\Pr\{G\} = .99$, what is the probability of D given A?

Additional problems in combinatorial and geometric probabilities

4.1.21   Assume 365 days in a year. If there are 10 people at a party, what is the probability that their birthdays fall on different days? Show that if there are more than 22 people in the party, the probability is greater than $\frac{1}{2}$ that at least 2 will have birthdays on the same day.

4.1.22   A number is constructed at random by choosing 10 digits from $\{0, 1, \ldots, 9\}$ with replacement. We allow the digit 0 at any position. What is the probability that the number does not contain 3 specific digits?

4.1.23   A caller remembers all seven digits of a telephone number, but is uncertain about the order of the last four. He keeps dialing the last four digits at random, without repeating the same number, until he reaches the right number. What is the probability that he will dial at least 10 wrong numbers?

4.1.24   One hundred lottery tickets are sold. There are 4 prizes and 10 consolation prizes. If you buy 5 tickets, what is the probability that you win

i   one prize?

ii   a prize and a consolation prize?

iii something?

4.1.25   Ten PCBs are in a bin, two defective. The boards are chosen at random, one by one, without replacement. What is the probability that exactly five good boards will be found between the drawing of the first and second defective PCB?

4.1.26   A random sample of 11 integers is drawn without replacement from the set $\{1, 2, \ldots, 20\}$. What is the probability that the sample median, Me, is equal to the integer k? ($6 \leq k \leq 15$).

4.1.27   A stick is broken at random into three pieces. What is the probability that these pieces can be the sides of a triangle?

4.1.28   A particle is moving at a uniform speed on a circle of unit radius and is released at a random point on the circumference. Draw a line segment of length $2h$ ($h < 1$) centered at a point A of distance $a > 1$ from the center of the circle, O. The line segment is perpendicular to the line connecting O with A. What is the probability that the particle will hit the line segment? [The particle flies along a straight line tangential to the circle.]

4.1.29   A block of 100 bits is transmitted over a binary channel, with probability $p = 10^{-3}$ of bit error. Errors occur independently. Find the probability that the block contains at least three errors.

4.1.30   A coin is tossed repeatedly until 2 "heads" occur. What is the probability that 4 tosses are required?

4.2.1   Consider the sample space, S, of all sequences of 10 binary numbers (0–1 signals). Define on this sample space two random variables and derive their probability distribution function, assuming the model that all sequences are equally probable.

4.2.2   The number of blemishes on a ceramic plate is a discrete random variable. Assume the probability model, with p.d.f.

$$p(x) = e^{-5}\frac{5^x}{x!} \qquad x = 0, 1, \ldots$$

a   Show that $\displaystyle\sum_{x=0}^{\infty} p(x) = 1$

b   What is the probability of at most 1 blemish on a plate?

c   What is the probability of no more than 7 blemishes on a plate?

4.2.3   Consider a distribution function of a mixed type, with c.d.f.

$$F_x(x) = \begin{cases} 0 & \text{if } x < -1 \\ .3 + .2(x + 1) & \text{if } -1 \leq x < 0 \\ .7 + .3x & \text{if } 0 \leq x < 1 \\ 1 & \text{if } 1 \leq x \end{cases}$$

a What is $\Pr\{X = -1\}$?

b What is $\Pr\{-.5 < X < 0\}$?

c What is $\Pr\{0 \le X < .75\}$?

d What is $\Pr\{X = 1\}$?

e Compute the expected value, $E\{X\}$, and variance, $V\{X\}$.

4.2.4 A random variable has the Rayleigh distribution, with c.d.f.

$$F(x) = \begin{cases} 0 & x < 0 \\ 1 - e^{-x^2/2\sigma^2} & x \ge 0 \end{cases}$$

where σ^2 is a positive parameter. Find the expected value $E\{X\}$.

4.2.5 A random variable X has a discrete distribution over the integers $\{1, 2, \ldots, N\}$ with equal probabilities. Find $E\{X\}$ and $V\{X\}$.

4.2.6 A random variable has expectation $\mu = 10$ and standard deviation $\sigma = 0.5$. Use Chebyshev's inequality to find a lower bound to the probability

$$\Pr\{8 < X < 12\}$$

4.2.7 Consider the random variable X with c.d.f.

$$F(x) = \frac{1}{2} + \frac{1}{\pi} \tan^{-1}(x) \qquad -\infty < x < \infty$$

Find the .25th, .50th, and .75th quantiles of this distribution.

4.2.8 Show that the central moments μ_l^* relate to the moments μ_l around the origin by the formula

$$\mu_l^* = \sum_{j=0}^{l-2} (-1)^j \binom{l}{j} \mu_{l-j} \mu_1^j + (-1)^{l-1}(l-1)\mu_1^l$$

4.2.9 Find the expected value, μ_1, and the second moment, μ_2, of the random variable whose c.d.f. is given in Exercise 4.2.3.

4.2.10 A random variable X has a continuous uniform distribution over the interval (a, b):

$$f(x) = \begin{cases} \dfrac{1}{b-a} & \text{if } a \le x \le b \\ 0 & \text{otherwise} \end{cases}$$

Find the moment-generating function of X. Find the mean and variance by differentiating the m.g.f.

4.2.11 Consider the moment-generating function, m.g.f., of the exponential distribution:

$$M(t) = \frac{\lambda}{\lambda - t} \qquad t < \lambda$$

a Find the first four moments of the distribution by differentiating $M(t)$.

b Convert the moments to central moments.

c What is the index of kurtosis β_4?

4.3.1 Using MINITAB, prepare a table of the p.d.f. and c.d.f. of the binomial distribution $B(20, .17)$.

4.3.2 What are the 1st quantile (Q_1), median (M_e), and 3rd quantile (Q_3) of $B(20, .17)$?

4.3.3 Compute the mean $E\{X\}$ and standard deviation, σ, of $B(45, .35)$.

4.3.4 A PCB is populated by 50 chips randomly chosen from a lot. The probability that an individual chip is nondefective is p. What should be the value of p so that the probability of no defective chip on the board is $\gamma = .99$? [The answer to this question shows why industry standards are so stringent.]

4.3.5 Let $b(j; n, p)$ be the p.d.f. of the binomial distribution. Show that if $n \to \infty$, and $p \to 0$ so that $np \to \lambda$, $0 < \lambda < \infty$, then

$$\lim_{\substack{n \to \infty \\ p \to 0 \\ np \to \lambda}} b(j; n, p) = e^{-\lambda} \frac{\lambda^j}{j!} \qquad j = 0, 1, \ldots$$

4.3.6 Use the result of the previous exercise to find the probability that a block of 1000 bits in a binary communication channel will have fewer than 4 errors, when the probability of a bit error is $p = 10^{-3}$.

4.3.7 Compute $E\{X\}$ and $V\{X\}$ of the hypergeometric distribution $H(500, 350, 20)$.

4.3.8 A lot of size $N = 500$ items contains $M = 5$ defective ones. A random sample of size $n = 50$ is drawn from the lot without replacement (RSWOR). What is the probability that more than 1 defective item in the sample will be observed?

4.3.9 Consider Example 4.23. What is the probability that the lot will be rectified if $M = 10$ and $n = 20$?

4.3.10 Use the m.g.f. to compute the third and fourth central moments of the Poisson distribution $P(10)$. What is the index of skewness and kurtosis of this distribution?

4.3.11 The number of blemishes on ceramic plates has a Poisson distribution with mean $\lambda = 1.5$. What is the probability that more than 2 blemishes will be observed on a plate?

4.3.12 The error rate of an insertion machine is 380 ppm (per 10^6 parts inserted). What is the probability of observing more than 6 insertion errors in 2 hours of operation, when the insertion rate is 4000 parts per hour?

4.3.13 Continuing with the previous exercise, let N be the number of parts inserted until an error occurs. What is the distribution of N? Compute the expected value and the standard deviation of N.

4.3.14 What are Q_1, M_e, and Q_3 of the negative-binomial $NB(p, k)$ with $p = .01$ and $k = 3$?

4.3.15 Derive the m.g.f. of N.B. (p, k).

4.3.16 Differentiate the m.g.f. of the geometric distribution—that is,

$$M(t) = \frac{pe^t}{(1 - e^t(1 - p))} \qquad t < -\log(1 - p)$$

to obtain its first four moments, and then derive the indices of skewness and kurtosis.

4.3.17 The proportion of defective RAM chips is $p = .002$. You have to install 50 chips on a board. Each chip is tested before its installation. How many chips should you order so that, with probability greater than $\gamma = .95$, you will have at least 50 good chips to install?

4.3.18 The random variable X assumes the values $\{1, 2, \ldots\}$ with probabilities of a geometric distribution, with parameter p, $0 < p < 1$. Prove the "memoryless" property of the geometric distribution—namely,

$$P[X > n + m \mid X > m] = P[X > n]$$

for all $n, m = 1, 2, \ldots$

4.4.1 Let X be a random variable having a continuous c.d.f., $F(x)$. Let $Y = F(X)$. Show that Y has a uniform distribution on $(0, 1)$. Conversely, if U has a uniform distribution on $(0, 1)$, show that $X = F^{-1}(U)$ has the c.d.f. $F(x)$.

4.4.2 Compute the expected value and the standard deviation of a uniform distribution $U(10, 50)$.

4.4.3 Show that if U is uniform on $(0, 1)$, then $X = -\log(U)$ has an exponential distribution $E(1)$.

4.4.4 Use MINITAB to compute the probabilities, for $N(100, 15)$, of

i $92 < X < 108$

ii $X > 105$

iii $2X + 5 < 200$

4.4.5 The .9 quantile of $N(\mu, \sigma)$ is 15 and its .99 quantile is 20. Find the mean μ and standard deviation σ.

4.4.6 A communication channel accepts an arbitrary voltage input v and outputs a voltage $v + E$, where E is distributed like $N(0, 1)$. The channel is used to transmit binary information as follows:

i To transmit 0, input $-v$.

ii To transmit 1, input v.

The receiver decides 0 if the voltage Y is negative, and 1 otherwise. What should be the value of v so that the receiver's probability of bit error is $\alpha = .01$?

4.4.7 Aluminum pins manufactured for the aviation industry have a random diameter whose distribution is (approximately) normal with mean of $\mu = 10$ (in mm) and standard deviation $\sigma = 0.02$ (in mm). Holes are automatically drilled on aluminum plates, with diameters having a normal distribution with mean μ_d (in mm) and $\sigma = 0.02$ (in mm). What should be the value of μ_d so that the probability that a pin will not enter a hole (too wide) is $\alpha = .01$?

4.4.8 Let $X_1, \ldots, X_n$ be a random sample (i.i.d.) from a normal distribution $N(\mu, \sigma^2)$. Find the expected value and variance of $Y = \sum_{i=1}^{n} iX_i$.

4.4.9 Concrete cubes have compressive strength with log-normal distribution LN(5, 1). Find the probability that the compressive strength X of a random concrete cube will be greater than 300 kg/cm^2.

4.4.10 Using the m.g.f. of $N(\mu, \sigma)$, derive the expected value and variance of LN(μ, σ). [Recall that X is distributed like $e^{N(\mu, \sigma)}$.]

4.4.11 What are the Q_1, M_e, and Q_3 of $E(\beta)$?

4.4.12 Show that if the life length of a chip is exponential, $E(\beta)$, then only 36.7% of the chips will function longer than the mean time till failure, β.

4.4.13 Show that the m.g.f. of $E(\beta)$ is $M(t) = (1 - \beta t)^{-1}$, for $t < \dfrac{1}{\beta}$.

4.4.14 Let X_1, X_2, X_3 be independent random variables having an identical exponential distribution, $E(\beta)$. Compute $\Pr\{X_1 + X_2 + X_3 \geq 3\beta\}$.

4.4.15 Establish the formula

$$G(t; k, \lambda) = 1 - e^{-\lambda t} \sum_{j=0}^{k-1} \frac{(\lambda t)^j}{j!}$$

by integrating in parts the p.d.f. of

$$G\left(k; \frac{1}{\lambda}\right)$$

4.4.16   Use MINITAB to compute $\Gamma(1.17)$, $\Gamma\left(\frac{1}{2}\right)$, $\Gamma\left(\frac{3}{2}\right)$.

4.4.17   Using m.g.f., show that the sum of k independent exponential random variables, $E(\beta)$, has the gamma distribution $G(k, \beta)$.

4.4.18   What is the expected value and variance of the Weibull distribution $W(2, 3.5)$?

4.4.19   The time till failure (days) of electronic equipment has the Weibull distribution $W(1.5, 500)$. What is the probability that the failure time will not be before 600 days?

4.4.20   Compute the expected value and standard deviation of a random variable having the beta distribution $\text{beta}\left(\frac{1}{2}, \frac{3}{2}\right)$.

4.4.21   Show that the index of kurtosis of $\text{beta}(v, v)$ is $\beta_2 = [3(1 + 2v)/(3 + 2v)]$.

4.5.1   The joint p.d.f. of two random variables (X, Y) is

$$f(x, y) = \begin{cases} \dfrac{1}{2} & \text{if } (x, y) \in S \\ 0 & \text{otherwise} \end{cases}$$

where S is a square of area 2 whose vertices are $(1, 0)$, $(0, 1)$, $(-1, 0)$, $(0, -1)$.

a   Find the marginal p.d.f. of X and of Y.

b   Find $E\{X\}$, $E\{Y\}$, $V\{X\}$, $V\{Y\}$.

4.5.2   Let (X, Y) have a joint p.d.f.

$$f(x, y) = \begin{cases} \dfrac{1}{y} \exp\left\{-y - \dfrac{x}{y}\right\} & \text{if } 0 < x, y < \infty \\ 0 & \text{otherwise} \end{cases}$$

Find $\text{cov}(X, Y)$ and the coefficient of correlation ρ_{XY}.

4.5.3   Show that the random variables (X, Y) whose joint distribution is defined in Example 4.26 are dependent. Find $\text{cov}(X, Y)$.

4.5.4   Find the correlation coefficient of J and N of Example 4.31.

4.5.5   Let X and Y be independent random variables, X is distributed like $G(2, 100)$ and Y is like $W(1.5, 500)$. Find the variance of XY.

4.6.1   Consider the trinomial distribution of Example 4.33.

a   What is the probability that during one hour of operation there will be no more than 20 errors?

b   What is the conditional distribution of wrong components, given that there are 15 misinsertions in a given hour of operation?

c   Approximating the conditional distribution of part b by a Poisson distribution, compute the conditional probability of no more than 15 wrong components?

4.6.2   Continuing Example 4.34, compute the correlation between J_1 and J_2.

4.6.3   In a bivariate normal distribution, the conditional variance of Y given X is 150 and the variance of Y is 200. What is the correlation ρ_{XY}?

4.7.1 $n = 10$ electronic devices start to operate at the same time. The times till failure of these devices are independent random variables having an identical $E(100)$ distribution.

a   What is the expected value of the first failure?

b   What is the expected value of the last failure?

4.7.2   A factory has $n = 10$ machines of a certain type. At each given day, the probability is $p = .95$ that a machine will be working. Let J denote the number of machines that work on a given day. The time it takes to produce an item on a given machine is $E(10)$—that is, exponentially distributed with mean $\mu = 10$ (min). The machines operate independently of each other. Let $X_{(1)}$ denote the minimal time for the first item to be produced. Determine:

a $Pr\{J = k, X_{(1)} \le x\}$, $k = 1, 2, \ldots$

b $Pr\{X_{(1)} \le x \mid J \ge 1\}$ Notice that when $J = 0$, no machine is working. The probability of this event is $(.05)^{10}$.

4.7.3   Let $X_1, X_2, \ldots, X_{11}$ be a random sample of exponentially distributed random variables with p.d.f. $f(x) = \lambda e^{-\lambda x}$, $x \ge 0$.

a   What is the p.d.f. of the median $M_e = X_{(6)}$?

b   What is the expected value of M_e?

4.8.1   Let X and Y be independent random variables having an $E(\beta)$ distribution. Let $T = X + Y$ and $W = X - Y$. Compute the variance of $T + \frac{1}{2}W$.

4.8.2   Let X and Y be independent random variables having a common variance σ^2. What is the covariance $\text{cov}(X, X + Y)$?

4.8.3   Let (X, Y) have a bivariate normal distribution. What is the variance of $\alpha X + \beta Y$?

4.8.4 Let X have a normal distribution $N(\mu, \sigma)$. Let $\Phi(z)$ be the standard normal c.d.f. Verify that $E\{\Phi(X)\} = P\{U < X\}$, where U is independent of X and U is distributed like $N(0, 1)$. Show that

$$E\{\Phi(X)\} = \Phi\left(\frac{\mu}{\sqrt{1 + \sigma^2}}\right)$$

4.8.5 Let X have a normal distribution $N(\mu, \sigma)$. Show that

$$E\{\Phi^2(X)\} = \Phi_2\left(\frac{\mu}{\sqrt{1 + \sigma^2}}, \frac{\mu}{\sqrt{1 + \sigma^2}}; \frac{\sigma^2}{1 + \sigma^2}\right)$$

4.8.6 X and Y are independent random variables having Poisson distributions with means $\lambda_1 = 5$ and $\lambda_2 = 7$, respectively. Compute the probability that $X + Y$ is greater than 15.

4.8.7 Let X_1 and X_2 be independent random variables having continuous distributions with p.d.f. $f_1(x)$ and $f_2(x)$, respectively. Let $Y = X_1 + X_2$. Show that the p.d.f. of Y is

$$g(y) = \int_{-\infty}^{\infty} f_1(x) f_2(y - x)\, dx$$

(This integral transform is called the convolution of $f_1(x)$ with $f_2(x)$. The convolution operation is denoted by $f_1 \times f_2$.)

4.8.8 Let X_1 and X_2 be independent random variables having uniform distributions on $(0, 1)$. Apply the convolution operation to find the p.d.f. of $Y = X_1 + X_2$.

4.8.9 Let X_1 and X_2 be independent random variables having a common exponential distribution $E(1)$. Determine the p.d.f. of $U = X_1 - X_2$. (The distribution of U is called bi-exponential or Laplace and its p.d.f. is $f(u) = \frac{1}{2} e^{-|u|}$.)

4.9.1 Apply the central limit theorem to approximate $P\{X_1 + \cdots + X_{20} \leq 50\}$, where $X_1, \ldots, X_{20}$ are independent random variables having a common mean $\mu = 2$ and a common standard deviation $\sigma = 10$.

4.9.2 Let X have a binomial distribution $B(200, .15)$. Find the normal approximation to $\Pr\{25 < X < 35\}$.

4.9.3 Let X have a Poisson distribution with mean $\lambda = 200$. Find, approximately, $\Pr\{190 < X < 210\}$.

4.9.4 $X_1, X_2, \ldots, X_{200}$ are 200 independent random variables with a common beta distribution, $B(3, 5)$. Approximate the probability $\Pr\{|\overline{X}_{200} - .375| < .2282\}$, where

$$\overline{X}_n = \frac{1}{n} \sum_{i=1}^{n} X_i \qquad n = 200$$

4.10.1 Use MINITAB to compute the .95 quantiles of $t[10]$, $t[15]$, $t[20]$.

4.10.2 Use MINITAB to compute the .95 quantiles of $F[10, 30]$, $F[15, 30]$, $F[20, 30]$.

4.10.3 Show that, for each $0 < \alpha < 1$, $t_{1-\alpha/2}^2[n] = F_{1-\alpha}[1, n]$.

4.10.4 Verify the relationship

$$F_{1-\alpha}[v_1, v_2] = \frac{1}{F_\alpha[v_2, v_1]} \qquad 0 < \alpha < 1$$

$v_1, v_2 = 1, 2, \ldots$

4.10.5 Verify the formula

$$V\{t[v]\} = \frac{v}{v - 2} \qquad v > 2$$

4.10.6 Find the expected value and variance of $F[3, 10]$.

5

Sampling for Estimation of Finite Population Quantities

Sampling and the Estimation Problem

5.1.1 Basic Definitions

In this chapter, we consider the problem of estimating quantities (parameters) of a finite population. The problem of testing hypotheses concerning such quantities—in the context of conducting a sampling inspection of product quality—will be studied in Chapter 9. Estimating and testing the parameters of statistical models for infinite populations will be discussed in Chapter 6.

Let $\mathcal{P}$ designate a finite population of N units. Here, we assume that the population size, N, is known. We also assume that a list (or a **frame**) of the population units $\mathcal{L}_N = \{u_1, \ldots, u_N\}$ is available.

Let X be a variable of interest and $x_i = X(u_i)$, $i = 1, \ldots, N$, the value ascribed by X to the ith unit, u_i, of $\mathcal{P}$.

The population mean and population variance for the variable X—that is,

$$\mu_N = \frac{1}{N} \sum_{i=1}^{N} x_i$$

and

$$\sigma_N^2 = \frac{1}{N} \sum_{i=1}^{N} (x_i - \mu_N)^2$$

(5.1.1)

are called **population quantities**. In some books (Cochran, 1977) these quantities are called *population parameters*. However, we distinguish between population quantities and parameters of distributions, which represent variables in infinite populations. Parameters are not directly observable and can only be estimated, whereas finite population quantities can be determined exactly if the whole population is observed.

The population quantity μ_N is the expected value of the distribution of X in the population, whose c.d.f. is

$$F_N(x) = \frac{1}{N} \sum_{i=1}^{N} I(x; x_i)$$

where

$$I(x; x_i) = \begin{cases} 1 & \text{if } x_i \le x \\ 0 & \text{if } x_i > x \end{cases}$$

(5.1.2)

σ_N^2 is the variance of $F_N(x)$.

In this chapter, we focus on estimating the population mean, μ_N, when a sample of size n, $n < N$, is observed. The problem of estimating the population variance σ_N^2 will be discussed in context of estimating the standard errors of estimators of μ_N.

Two types of sampling strategies will be considered. One type consists of sampling randomly (with or without replacement) from the whole population. This strategy is called **simple random sampling**. The other type of sampling strategy is called **stratified random sampling**. In stratified random sampling, the population is first partitioned into strata (subpopulations) and then a simple random sample is drawn from each stratum independently. If the strata are determined so that the variability within strata is smaller relative to the general variability in the population, the precision in estimating the population mean μ_N using a stratified random sampling will generally be higher than that in simple random sampling. This will be shown in Section 5.3.

The following is an example of a case where stratification could be helpful. At the end of each production day, we draw a random sample from the lot of that day's products to estimate the proportion of defective items. Suppose that several machines operate in parallel and manufacture the same item. Stratification by machine will provide higher precision for the global estimate, as well as provide information on the level of quality of each machine. If we can stratify by shift, by vendor, or by other factors that may contribute to the variability, we may also increase the precision of our estimates.

5.1.2 Sample Estimates of Population Quantities and Their Sampling Distribution

So far, we have discussed the nature of variable phenomena and presented some methods of exploring and presenting the results of experiments. More specifically, the methods of analysis described in Chapters 2 and 3 explore the given data but do not assess what might happen in future experiments. For example, Table 2.2 gives the frequency distribution of the number of blemishes on 30 ceramic plates. We see from that table that 77% of the plates in that particular sample have at most one blemish. Can we say that this will be the case in future samples as well? After all, this result is based on a small sample of 30 plates. It should be clear that our statements about the number of blemishes expected to be observed in future samples are restricted to ceramic plates coming from the *same* production process. If the production process is improved, or deteriorates, the expected number of blemishes should either diminish or increase. Suppose now that we plan to ship a lot of 1000 ceramic plates to a customer, all manufactured in the same process as those in the sample of 30. Based on the given sample, what can we say about the proportion of plates in the lot to be shipped having at most one blemish? We know that the proportion in the lot will likely be different from .77, but how much different? This is a problem of **statistical inference**.

If we draw from the same population several different random samples of the same size, we generally find that statistics of interest assume different values in the different samples. This can be illustrated by using MINITAB. In MINITAB, we can draw samples with or without replacement from a collection of numbers (population) stored in some column. To show it, let us store in column $C1$ the integers 1,2,...,100. This can be done by using the command **SET PATTERNED DATA** ... under **CALC**. To sample at random with replacement (RSWR) a sample of size $n = 20$ from $C1$ and put the random sample at $C2$, we use

```
MTB> Sample 20 C1 C2;
SUBC> Replace.
```

This can be repeated four times, and each time we put the sample in a new column.

In Table 5.1, we present the results of the sampling, and for each sample we present its mean and standard deviation. The notions of random sampling with replacement (RSWR) and random sampling without replacement (RSWOR) were introduced and defined in Section 2.3.

Notice that the "population" mean (that of column $C1$) is 50.5 and its standard deviation is 28.87. The sample means and standard deviations are estimates of these population parameters, and, as seen previously, they differ from the parameters. The distribution of sample estimates of a parameter is called the **sampling distribution of an estimate**.

Theoretically (hypothetically), the number of possible random samples with replacement is either infinite, if the population is infinite, or of magnitude N^n, if the population is finite (n is the sample size and N is the population size). This number is practically too large, even if the population is finite (100^{20} in the preceding example). We can, however, approximate the distribution by drawing a large number, M, of such samples. Figure 5.1 shows the histogram of the sampling distribution of $\overline{X}_n$, for $M = 1000$ random samples with replacement of size $n = 20$, from the population $\{1, 2, \ldots, 100\}$ of the previous example. This can be done using MINITAB, by executing the following block of commands 1000 times:

```
MTB> Sample 20 C1 C2;
SUBC> Replace.
```

TABLE   5.1

Four random samples with replacement of size 20, from {1, 2, . . . , 100}

	Sample		
1	2	3	4
26	54	4	15
56	59	81	52
63	73	87	46
46	62	85	98
1	57	5	44
4	2	52	1
31	33	6	27
79	54	47	9
21	97	68	28
5	6	50	52
94	62	89	39
52	70	18	34
79	40	4	30
33	70	53	58
6	45	70	18
33	74	7	14
67	29	68	14
33	40	49	32
21	21	70	10
8	43	15	52

	Means		
37.9	49.6	46.4	33.6

	Standard deviations		
28.0	23.7	31.3	22.6

```
MTB> let k1 = mean (C2)
MTB> stack C3 k1 C3
MTB> end
```

Remember that column $C1$ has 100 rows with the integers 1, . . . , 100. Column $C2$ should be empty, and column $C3$ should initially have 1 row, with some number in it. Put the mean of $C1$ there. Such a block of commands, to be executed repeatedly, is called a **macro**.

FIGURE   5.1

Histogram of 1000 sample means

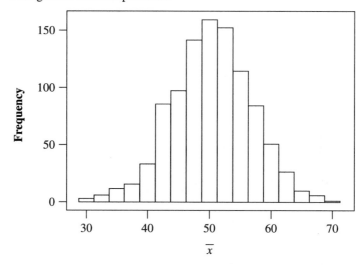

After executing the preceding macro 1000 times, we obtained the following frequency distribution of the sample means, $\overline{X}_{20}$:

Midpoint	Frequency
28	1
32	10
36	19
40	55
44	153
48	224
52	251
56	168
60	86
64	26
68	6
72	1

This frequency distribution is an approximation of the sampling distribution of $\overline{X}_{20}$. It is interesting to note that this distribution has mean $\overline{\overline{X}} = 50.42$ and standard deviation $\overline{S} = 6.412$. $\overline{\overline{X}}$ is quite close to the population mean 50.5, and $\overline{S}$ is approximately $\sigma/\sqrt{20}$, where σ is the population standard deviation. A proof of this is given in the following section.

   Our computer sampling procedure provided a close estimate of this standard error. Often we are interested in properties of statistics for which it is difficult to derive standard error formulas. Computer sampling techniques, like the one just illustrated, provide good approximations to the standard errors of sample statistics.

5.2
Estimation with Simple Random Samples

In this section, we investigate the properties of population quantity estimators when sampling is simple random.

The probability structure for simple random samples with or without replacement, RSWR and RSWOR, was studied in Section 4.1.4.

Let $X_1, \ldots, X_n$ denote the values of the variable $X(u)$ of the n elements in the random samples. The marginal distributions of X_i ($i = 1, \ldots, n$) if the sample is random, with or without replacement, is the distribution $\hat{F}_N(x)$. If the sample is random with replacement, then $X_1, \ldots, X_n$ are independent. If the sample is random without replacement, then $X_1, \ldots, X_n$ are correlated (dependent).

For an estimator of μ_N, we use the sample mean

$$\overline{X}_n = \frac{1}{n} \sum_{j=1}^{n} X_j$$

For an estimator of σ_N^2, we use the sample variance

$$S_n^2 = \frac{1}{n-1} \sum_{j=1}^{n} (X_j - \overline{X}_n)^2$$

Both estimators are random variables, which may change their values from one sample to another.

An estimator is called **unbiased** if its expected value is equal to the population value of the quantity it estimates. The **precision** of an estimator is the inverse of its sampling variance.

EXAMPLE 5.1 We illustrate the preceding with a numerical example. A population is of size $N = 100$. For simplicity, we take $X(u_i) = i$ ($i = 1, \ldots, 100$). For this simple population, $\mu_{100} = 50.5$ and $\sigma_{100}^2 = 833.25$.

Draw from this population 100 independent samples of size $n = 10$ with and without replacement. Do this using the MINITAB macro of Section 5.1. The means and standard deviations (Std) of the 100 sample estimates are given in Table 5.2.

TABLE 5.2
Statistics of sampling distributions

	RSWR		RSWOR	
Estimate	Mean	Std	Mean	Std
$\overline{X}_{10}$	49.85	9.2325	50.12	9.0919
S_{10}^2	774.47	257.96	782.90	244.36

As will be shown in the following section, the theoretical expected value of $\overline{X}_{10}$, both in RSWR and RSWOR, is $\mu = 50.5$. Note that the means of the sample estimates are close to the value of μ. The theoretical standard deviation of $\overline{X}_{10}$ is 9.128 for RSWR and 8.703 for RSWOR. The empirical standard deviations are also close to these values. The empirical means of S_{10}^2 are somewhat lower than their expected values of 833.25 and 841.67, for RSWR and RSWOR, respectively. But, as will be shown later, they are not significantly smaller than σ^2. ■

5.2.1   Properties of $\overline{X}_n$ and S_n^2 under RSWR

If the sampling is RSWR, the random variables $X_1, \ldots, X_n$ are independent, having the same c.d.f., $F_N(x)$. The corresponding p.d.f. is

$$p_N(x) = \begin{cases} \dfrac{1}{N} & \text{if } x = x_j, j = 1, \ldots, N \\ 0 & \text{otherwise} \end{cases} \tag{5.2.1}$$

Accordingly,

$$E\{X_j\} = \frac{1}{N} \sum_{j=1}^{N} x_j = \mu_N \qquad \text{all } j = 1, \ldots, N \tag{5.2.2}$$

It follows from the results of Section 4.8 that

$$E\{\overline{X}_n\} = \frac{1}{n} \sum_{j=1}^{n} E\{X_j\} \tag{5.2.3}$$

$$= \mu_N$$

Thus, the sample mean is an unbiased estimator of the population mean.

The variance of X_j is the variance associated with $F_N(x)$—that is,

$$V\{X_j\} = \frac{1}{N} \sum_{j=1}^{N} x_j^2 - \mu_N^2$$

$$= \frac{1}{N} \sum_{j=1}^{N} (x_j - \mu_N)^2$$

$$= \sigma_N^2$$

Moreover, because $X_1, X_2, \ldots, X_n$ are i.i.d.,

$$V\{\overline{X}_n\} = \frac{\sigma_N^2}{n} \tag{5.2.4}$$

Thus, as explained in Section 4.8, the sample mean converges in probability to the population mean as $n \to \infty$. An estimator having such a property is called **consistent**.

We show now that S_n^2 is an unbiased estimator of σ_N^2.

Indeed, if we write

$$S_n^2 = \frac{1}{n-1} \sum_{j=1}^{n} (X_j - \overline{X}_n)^2$$

$$= \frac{1}{n-1} \left(\sum_{j=1}^{n} X_j^2 - n\overline{X}_n^2 \right)$$

we obtain

$$E\{S_n^2\} = \frac{1}{n-1} \left(\sum_{j=1}^{n} E\{X_j^2\} - nE\{\overline{X}_n^2\} \right)$$

Moreover, because $X_1, \ldots, X_n$ are i.i.d.,

$$E\{X_j^2\} = \sigma_N^2 + \mu_N^2 \qquad j = 1, \ldots, n$$

and

$$E\{\overline{X}_n^2\} = \frac{\sigma_N^2}{n} + \mu_N^2$$

Substituting these into the expression for $E\{S_n^2\}$, we obtain

$$E\{S_n^2\} = \frac{1}{n-1} \left(n(\sigma_N^2 + \mu_N^2) - n \left(\frac{\sigma_N^2}{n} + \mu_N^2 \right) \right) \tag{5.2.5}$$

$$= \sigma_N^2$$

An estimator of the standard error of $\overline{X}_n$ is $S_n/\sqrt{n}$. This estimator is biased.

In large samples, the distribution of $\overline{X}_n$ is approximately normal, like $N\left(\mu_N, \sigma_N^2/n\right)$, as implied by the CLT. Therefore, the interval

$$\left(\overline{X}_n - z_{1-\alpha/2} \frac{S_n}{\sqrt{n}}, \overline{X}_n + z_{1-\alpha/2} \frac{S_n}{\sqrt{n}} \right)$$

has, in large samples, the property that $\Pr\{\overline{X}_n - z_{1-\alpha/2}(S_n/\sqrt{n}) < \mu_N < \overline{X}_n + z_{1-\alpha/2}(S_n/\sqrt{n})\} \cong 1 - \alpha$. An interval having this property is called a **confidence interval** for μ_N, with an approximate confidence level $(1 - \alpha)$. In the preceding formula, $z_{1-\alpha/2} = \Phi^{-1}(1 - (\alpha/2))$.

It is considerably more complicated to derive the formula for $V\{S_n^2\}$. An approximation for large samples is

$$V\{S_n^2\} \cong \frac{\mu_{4,N} - (\sigma_N^2)^2}{n} + \frac{2(\sigma_N^2)^2 - \mu_{3,N}}{n^2} + \frac{\mu_{4,N} - 3(\sigma_N^2)^2}{n^3} \tag{5.2.6}$$

where

$$\mu_{3,N} = \frac{1}{N} \sum_{j=1}^{N} (x_j - \mu_N)^3 \tag{5.2.7}$$

and

$$\mu_{4,N} = \frac{1}{N} \sum_{j=1}^{N} (x_j - \mu_N)^4 \tag{5.2.8}$$

EXAMPLE **5.2**   In file PLACE.DAT, we have data on x, y, and θ deviations of $N = 416$ component placements by automatic insertion on 26 PCBs.

Let us consider this record as a finite population. Suppose we are interested in the population quantities of the variable x-dev. Using MINITAB, we find that the population mean, variance, and third and fourth central moments are

$$\mu_N = 0.9124$$
$$\sigma_N^2 = 2.91999$$
$$\mu_{3,N} = -0.98326$$
$$\mu_{4,N} = 14.655$$

The unit of measurement of the x-dev is 10^{-3} (inches).

Thus, if we draw a simple RSWR of size $n = 50$, the variance of $\overline{X}_n$ will be $V\{\overline{X}_{50}\} = \sigma_N^2/50 = 0.0584$. The variance of S_{50}^2 will be

$$V\{S_{50}^2\} \cong \frac{14.655 - (2.9199)^2}{50} + \frac{2(2.9199)^2 + 0.9833}{2500}$$
$$+ \frac{14.655 - 3(2.9199)^2}{125,000}$$
$$= 0.1297 \quad \blacksquare$$

5.2.2 Properties of $\overline{X}_n$ and S_n^2 under RSWOR

Under RSWOR, $\overline{X}_n$ is an unbiased estimator of μ_N.

Let I_j be an indicator variable that assumes the value 1 if u_j belongs to the selected sample, $\mathbf{s}_n$, and equal to 0 otherwise. Then we can write

$$\overline{X}_n = \frac{1}{n} \sum_{j=1}^{N} I_j x_j \qquad (5.2.9)$$

Accordingly,

$$E\{\overline{X}_n\} = \frac{1}{n} \sum_{j=1}^{N} x_j E\{I_j\}$$

$$= \frac{1}{n} \sum_{j=1}^{N} x_j \Pr\{I_j = 1\}$$

As shown in Section 4.1.4,

$$\Pr\{I_j = 1\} = \frac{n}{N} \qquad \text{all } j = 1, \ldots, N$$

Substituting this into the preceding equation yields

$$E\{\overline{X}_n\} = \mu_N \qquad (5.2.10)$$

It is shown below that

$$V\{\overline{X}_n\} = \frac{\sigma_N^2}{n}\left(1 - \frac{n-1}{N-1}\right)$$

(5.2.11)

To derive the formula for the variance of $\overline{X}_n$ under RSWOR, we use the result of Section 4.8 on the variance of linear combinations of random variables. First, we write

$$V\{\overline{X}_n\} = V\left\{\frac{1}{n}\sum_{i=1}^N x_i I_i\right\}$$

$$= \frac{1}{n^2}V\left\{\sum_{i=1}^N x_i I_i\right\}$$

$\sum_{i=1}^N x_i I_i$ is a linear combination of the random variables $I_1, \ldots, I_N$.

Then we show that

$$V\{I_i\} = \frac{n}{N}\left(1 - \frac{n}{N}\right) \qquad i = 1, \ldots, N$$

Indeed, because $I_i^2 = I_i$,

$$V\{I_i\} = E\{I_i^2\} - (E\{I_i\})^2$$

$$= E\{I_i\}(1 - E\{I_i\})$$

$$= \frac{n}{N}\left(1 - \frac{n}{N}\right) \qquad i = 1, \ldots, N$$

Moreover, for $i \neq j$,

$$\text{cov}(I_i, I_j) = E\{I_i I_j\} - E\{I_i\}E\{I_j\}$$

But,

$$E\{I_i I_j\} = \Pr\{I_i = 1, I_j = 1\}$$

$$= \frac{n(n-1)}{N(N-1)}$$

Hence, for $i \neq j$,

$$\text{cov}(I_i, I_j) = -\frac{n}{N^2}\cdot\frac{N-n}{N-1}$$

Finally,

$$V\left\{\sum_{i=1}^N x_i I_i\right\} = \sum_{i=1}^N x_i^2 V\{I_i\} + \sum\sum_{i\neq j} x_i x_j \,\text{cov}(I_i, I_j)$$

Substituting these expressions into

$$V\{\overline{X}_n\} = \frac{1}{n^2}V\left\{\sum_{i=1}^N x_i I_i\right\}$$

we obtain

$$V\{\bar{X}_n\} = \frac{1}{n^2}\left\{\frac{n}{N}\left(1 - \frac{n}{N}\right)\sum_{i=1}^{N}x_i^2 - \frac{n(N-n)}{N^2(N-1)}\sum\sum_{i\neq j}x_i x_j\right\}$$

But, $\displaystyle\sum\sum_{i\neq j}x_i x_j = \left(\sum_{i=1}^{N}x_i\right)^2 - \sum_{i=1}^{N}x_i^2$. Hence,

$$V\{\bar{X}_n\} = \frac{N-n}{nN^2}\left\{\frac{N}{N-1}\sum_{i=1}^{N}x_i^2 - \frac{1}{N-1}\left(\sum_{i=1}^{N}x_i\right)^2\right\}$$

$$= \frac{N-n}{n(N-1)N}\sum_{i=1}^{N}(x_i - \mu_N)^2$$

$$= \frac{\sigma_N^2}{n}\left(1 - \frac{n-1}{N-1}\right)$$

We see that the variance of $\bar{X}_n$ is smaller under RSWOR than under RSWR by a factor of $[1 - (n-1)/(N-1)]$. This factor is called, the **finite population multiplier**, or **correction factor**.

The formula we saw in Section 4.3.2 for the variance of the hypergeometric distribution can be obtained from the preceding formula. In the hypergeometric model, we have a finite population of size N. M elements have a certain attribute. Let

$$x_i = \begin{cases} 1 & \text{if } w_i \text{ has the attribute} \\ 0 & \text{if } w_i \text{ does not have it} \end{cases}$$

Because $\displaystyle\sum_{i=1}^{N}x_i = M$ and $x_i^2 = x_i$,

$$\sigma_N^2 = \frac{M}{N}\left(1 - \frac{M}{N}\right)$$

If $J_n = \displaystyle\sum_{i=1}^{n}X_i$, we have

$$V\{J_n\} = n^2 V\{\bar{X}_n\}$$

$$= n\frac{M}{N}\left(1 - \frac{M}{N}\right)\left(1 - \frac{n-1}{N-1}\right) \tag{5.2.12}$$

This is the formula given in Section 4.3.2.

To estimate σ_N^2 we can again use the sample variance, S_n^2. The sample variance has, however, a slight positive bias. Indeed,

$$E\{S_n^2\} = \frac{1}{n-1}E\left\{\sum_{j=1}^{n}X_j^2 - n\bar{X}_n^2\right\}$$

$$= \frac{1}{n-1} \left(n(\sigma_N^2 + \mu_N^2) - n \left(\mu_N^2 + \frac{\sigma_N^2}{n} \left(1 - \frac{n-1}{N-1} \right) \right) \right)$$

$$= \sigma_N^2 \left(1 + \frac{1}{N-1} \right)$$

This bias is negligible if σ_N^2/N is small. Thus, the standard error of $\overline{X}_n$ can be estimated by

$$\text{SE}\{\overline{X}_n\} = \frac{S_n}{\sqrt{n}} \left(1 - \frac{n-1}{N-1} \right)^{1/2} \tag{5.2.13}$$

When sampling is RSWOR, the random variables $X_1, \ldots, X_n$ are not independent, and we cannot justify theoretically the use of the normal approximation to the sampling distribution of $\overline{X}_n$. However, if n/N is small, the normal approximation is expected to yield good results. Thus, if $(n/N) < 0.1$, we can approximate the confidence interval of level $(1 - \alpha)$ for μ_N by the interval with limits:

$$\overline{X}_n \pm z_{1-\alpha/2} \cdot \text{SE}\{\overline{X}_n\}$$

In order to estimate the coverage probability of this interval estimator, when $n/N = 0.3$, we perform the following simulation example.

EXAMPLE 5.3   We use MINITAB to select RSWOR of size $n = 30$ from the population $\mathcal{P} = \{1, 2, \ldots, 100\}$ of $N = 100$ units, whose values are $x_i = i$.

For this purpose, set the integers $1, \ldots, 100$ into column $C1$. Notice that when $n = 30, N = 100, \alpha = .05, z_{1-\alpha/2} = 1.96$, and $(1.96/\sqrt{n})[1 - (n-1)/(N-1)]^{1/2} = 0.301$. In $C2(1)$, we store $k1 = (\text{mean}(C1) - 0.301 * \text{stan}(C1))$, and in $C3(1)$, we store $k2 = (\text{mean}(C1) + 0.301 * \text{stan}(C1))$. The following macro was executed $M = 1000$ times:

```
Sample 30 C1 C4
let k1 = mean(C4) − .301 * stan(C4)
let k2 = mean(C4) + .301 * stan(C4)
stack C2 k1 C2
stack C3 k2 C3
end
```

The true population mean is $\mu_N = 50.5$. The estimated coverage probability is the proportion of cases for which $k_1 \leq \mu_N \leq k_2$. In this simulation, the proportion of coverage is .947. The nominal confidence level is $1 - \alpha = .95$. The estimated coverage probability is .947. Thus, the example shows that even in cases where $n/N > 0.1$, the approximate confidence limits are quite effective.   ∎

5.3
Estimating the Mean with Stratified RSWOR

Now we consider the problem of estimating the population mean, μ_N, with stratified RSWOR. Suppose the population $\mathcal{P}$ is partitioned into k strata (subpopulations) $\mathcal{P}_1, \mathcal{P}_2, \ldots, \mathcal{P}_k, k \geq 2$.

Let $N_1, N_2, \ldots, N_k$ denote the sizes; $\mu_{N_1}, \ldots, \mu_{N_k}$ the means; and $\sigma_{N_1}^2, \ldots, \sigma_{N_k}^2$ the variances of these strata, respectively. Notice that the population mean is

$$\mu_N = \frac{1}{N} \sum_{i=1}^{k} N_i \mu_{N_i} \tag{5.3.1}$$

and, according to the formula of total variance (see Section 4.8), the population variance is

$$\sigma_N^2 = \frac{1}{N} \sum_{i=1}^{k} N_i \sigma_{N_i}^2 + \frac{1}{N} \sum_{i=1}^{k} N_i (\mu_{N_i} - \mu_N)^2 \tag{5.3.2}$$

We see that if the means of the strata are not the same, the population variance is greater than the weighted average of the within-strata variances, $\sigma_{N_i}^2$ $(i = 1, \ldots, k)$.

A stratified RSWOR is a sampling procedure in which k independent random samples without replacement are drawn from the strata. Let n_i, $\overline{X}_{n_i}$, and $S_{n_i}^2$ be the size, mean, and variance of the RSWOR from the ith stratum, $\mathcal{P}_i$ $(i = 1, \ldots, k)$.

We showed in the previous section that $\overline{X}_{n_i}$ is an unbiased estimator of μ_{N_i}. Thus, an unbiased estimator of μ_N is the weighted average

$$\hat{\mu}_N = \sum_{i=1}^{k} W_i \overline{X}_{n_i} \tag{5.3.3}$$

where $W_i = N_i/N$, $i = 1, \ldots, k$. Indeed,

$$E\{\hat{\mu}_N\} = \sum_{i=1}^{k} W_i E\{\overline{X}_{n_i}\}$$

$$= \sum_{i=1}^{k} W_i \mu_{N_i} \tag{5.3.4}$$

$$= \mu_N$$

Because $\overline{X}_{n_1}, \overline{X}_{n_2}, \ldots, \overline{X}_{n_k}$ are independent random variables, the variance of $\hat{\mu}_N$ is

$$V\{\hat{\mu}_N\} = \sum_{i=1}^{k} W_i^2 V\{\overline{X}_{n_i}\}$$

$$= \sum_{i=1}^{k} W_i^2 \frac{\sigma_{N_i}^2}{n_i} \left(1 - \frac{n_i - 1}{N_i - 1}\right) \tag{5.3.5}$$

$$= \sum_{i=1}^{k} W_i^2 \frac{\tilde{\sigma}_{N_i}^2}{n_i} \left(1 - \frac{n_i}{N_i}\right)$$

where

$$\tilde{\sigma}_{N_i}^2 = \frac{N_i}{N_i - 1}\sigma_{N_i}^2$$

EXAMPLE 5.4   Returning to the data of Example 5.2 on deviations in the x direction of automatically inserted components, the units are partitioned to $k = 3$ strata: boards 1–10 in stratum 1, boards 11–13 in stratum 2, and boards 14–26 in stratum 3. The population characteristics of these strata are:

Stratum	Size	Mean	Variance
1	160	−0.966	0.4189
2	48	0.714	1.0161
3	208	2.403	0.3483

The relative sizes of the strata are $W_1 = 0.385$, $W_2 = 0.115$, and $W_3 = 0.5$. If we select a stratified RSWOR of sizes $n_1 = 19$, $n_2 = 6$, and $n_3 = 25$, the variance of $\hat{\mu}_N$ will be

$$V\{\hat{\mu}_N\} = (0.385)^2 \frac{0.4189}{19}\left(1 - \frac{18}{159}\right) + (0.115)^2 \frac{1.0161}{6}\left(1 - \frac{5}{47}\right)$$

$$+(0.5)^2 \frac{0.3483}{25}\left(1 - \frac{24}{207}\right)$$

$$= 0.00798$$

This variance is considerably smaller than the variance of $\overline{X}_{50}$ in a simple RSWOR, which is

$$V\{\overline{X}_{50}\} = \frac{2.9199}{50}\left(1 - \frac{49}{415}\right)$$

$$= 0.0515 \quad \blacksquare$$

5.4
Proportional and Optimal Allocation

An important question to consider in designing the stratified RSWOR is how to allocate the total number of observations, n, to the different strata—that is, how to determine $n_i \geq 0$ ($i = 1, \ldots, k$) so that $\sum_{i=1}^{k} n_i = n$ for a given n. This is called the **sample allocation**. One type of sample allocation is the so-called **proportional allocation**—that is,

$$n_i = nW_i \qquad i = 1, \ldots, k \tag{5.4.1}$$

The variance of the estimator $\hat{\mu}_N$ under proportional allocation is

$$V_{\text{prop}}\{\hat{\mu}_N\} = \frac{1}{n} \sum_{i=1}^{k} W_i \tilde{\sigma}_{N_i}^2 \left(1 - \frac{n}{N}\right)$$

$$= \frac{\bar{\sigma}_N^2}{n} \left(1 - \frac{n}{N}\right)$$

(5.4.2)

where

$$\bar{\sigma}_N^2 = \sum_{i=1}^{k} W_i \tilde{\sigma}_{N_i}^2$$

is the weighted average of the within-strata variances.

We showed in the previous section that if we take a simple RSWOR, the variance of $\bar{X}_n$ is

$$V_{\text{simple}}\{\bar{X}_n\} = \frac{\sigma_N^2}{n} \left(1 - \frac{n-1}{N-1}\right)$$

$$= \frac{\tilde{\sigma}_N^2}{n} \left(1 - \frac{n}{N}\right)$$

where

$$\tilde{\sigma}_N^2 = \frac{N}{N-1} \sigma_N^2$$

In large-sized populations, σ_N^2 and $\tilde{\sigma}_N^2$ are very close, and we can write

$$V_{\text{simple}}\{\bar{X}_n\} \cong \frac{\sigma_N^2}{N} \left(1 - \frac{n}{N}\right)$$

$$= \frac{1}{n} \left(1 - \frac{n}{N}\right) \left\{ \sum_{i=1}^{k} W_i \sigma_{N_i}^2 + \sum_{i=1}^{k} W_i (\mu_{N_i} - \mu_N)^2 \right\}$$

$$\cong V_{\text{prop}}\{\hat{\mu}_N\} + \frac{1}{n} \left(1 - \frac{n}{N}\right) \sum_{i=1}^{k} W_i (\mu_{N_i} - \mu_N)^2$$

This shows that $V_{\text{simple}}\{\bar{X}_n\} > V_{\text{prop}}\{\hat{\mu}_N\}$; that is, the estimator of the population mean, μ_N, under stratified RSWOR, with proportional allocation, generally has a smaller variance (is more precise) than the estimator under a simple RSWOR. The difference grows with the variance between the strata means, $\sum_{i=1}^{k} W_i (\mu_{N_i} - \mu_N)^2$.

Thus, effective stratification is one that partitions the population to strata that are homogeneous within (small values of $\sigma_{N_i}^2$) and heterogeneous between (large values of $\sum_{i=1}^{k} W_i (\mu_{N_i} - \mu_N)^2$). If sampling is stratified RSWR, then the variance $\hat{\mu}_N$, under proportional allocation, is

$$V_{\text{prop}}\{\hat{\mu}_N\} = \frac{1}{n} \sum_{i=1}^{k} W_i \sigma_{N_i}^2$$

(5.4.3)

This is strictly smaller than the variance of $\overline{X}_n$ in a simple RSWR. Indeed,

$$V_{\text{simple}}\{\overline{X}_n\} = \frac{\sigma_N^2}{n}$$

$$= V_{\text{prop}}\{\hat{\mu}_N\} + \frac{1}{n}\sum_{i=1}^{k} W_i(\mu_{N_i} - \mu_N)^2$$

EXAMPLE 5.5   Defective circuit breakers are a serious hazard because the function of circuit break-ers is to protect electronic systems from power surges or power drops. Variabil-ity in power supply voltage levels can cause major damage to electronic systems. Circuit breakers are used to shield electronic systems from such events. The pro-portion of potentially defective circuit breakers is a key parameter in designing redundancy levels of protection devices and preventive maintenance programs. A lot of $N = 10,000$ circuit breakers was put together by purchasing the products from $k = 3$ different vendors. We want to estimate the proportion of defective breakers by sampling and testing $n = 500$ breakers. Stratifying the lot by vendor, we have 3 strata of sizes $N_1 = 3000$, $N_2 = 5000$, and $N_3 = 2000$. Before installing the circuit breakers, we drew from the lot a stratified RSWOR with proportional allocation—that is, $n_1 = 150$, $n_2 = 250$, and $n_3 = 100$. After testing, we found in the first sample $J_1 = 3$ defective circuit breakers, in the second sample, $J_2 = 10$ defectives, and in the third sample, $J_3 = 2$ defectives. Testing is done with a special-purpose device, simulating intensive usage of the product.

In this case, we set $X = 1$ if the item is defective and $X = 0$ otherwise. Then μ_N is the proportion of defective items in the lot, and μ_{N_i} ($i = 1, 2, 3$) is the proportion of defectives in the ith stratum.

The unbiased estimator of μ_N is

$$\hat{\mu}_N = .3 \times \frac{J_1}{150} + .5 \times \frac{J_2}{250} + .2 \times \frac{J_3}{100}$$
$$= .03$$

The variance within each stratum is $\sigma_{N_i}^2 = P_{N_i}(1 - P_{N_i})$, $i = 1, 2, 3$, where P_{N_i} is the proportion in the ith stratum. Thus, the variance of $\hat{\mu}_N$ is

$$V_{\text{prop}}\{\hat{\mu}_N\} = \frac{1}{500}\overline{\sigma}_N^2\left(1 - \frac{500}{10,000}\right)$$

where

$$\overline{\sigma}_N^2 = .3\tilde{\sigma}_{N_1}^2 + .5\tilde{\sigma}_{N_2}^2 + .2\tilde{\sigma}_{N_3}^2$$

or

$$\overline{\sigma}_N^2 = .3 \times \frac{3000}{2999}P_{N_1}(1 - P_{N_1}) + .5\frac{5000}{4999}P_{N_2}(1 - P_{N_2})$$
$$+ .2\frac{2000}{1999}P_{N_3}(1 - P_{N_3})$$

Substituting $\frac{3}{150}$ for an estimate of P_{N_1}, $\frac{10}{250}$ for that of P_{N_2}, and $\frac{2}{100}$ for P_{N_3}, we obtain the estimate of $\bar{\sigma}_N^2$:

$$\bar{\sigma}_N^2 = .029008$$

Finally, an estimate of $V_{\text{prop}}\{\hat{\mu}_N\}$ is

$$\hat{V}_{\text{prop}}\{\hat{\mu}_N\} = \frac{.029008}{500}\left(1 - \frac{500}{10,000}\right)$$

$$= .00005511$$

The standard error of the estimator is .00742

Confidence limits for μ_N, at level $1 - \alpha = .95$, are given by

$$\hat{\mu}_N \pm 1.96 \times \text{SE}\{\hat{\mu}_N\} = \begin{cases} .0446 \\ .0154 \end{cases}$$

These limits can be used for determining the stock level of spare parts. ∎

When the variances $\tilde{\sigma}_N^2$ within strata are known, we can further reduce the variance of μ_N by an allocation called **optimal allocation**.

We wish to minimize

$$\sum_{i=1}^{k} W_i^2 \frac{\tilde{\sigma}_{N_i}^2}{n_i}\left(1 - \frac{n_i}{N_i}\right)$$

subject to the constraint:

$$n_1 + n_2 + \cdots + n_k = n$$

This can be done by minimizing

$$L(n_1, \ldots, n_k, \lambda) = \sum_{i=1}^{k} W_i^2 \frac{\tilde{\sigma}_{N_i}^2}{n_i} - \lambda\left(n - \sum_{i=1}^{k} n_i\right)$$

with respect to $n_1, \ldots, n_k$ and λ. This function is called the *Lagrangian* and λ is called the **Lagrange multiplier**.

The result is

$$n_i^0 = n\frac{W_i\tilde{\sigma}_{N_i}}{\sum_{j=1}^{k} W_j\tilde{\sigma}_{N_j}} \qquad i = 1, \ldots, k \tag{5.4.4}$$

We see that the proportional allocation is optimal when all $\tilde{\sigma}_{N_i}^2$ are equal.

The variance of $\hat{\mu}_N$ corresponding to the optimal allocation is

$$V_{\text{opt}}\{\hat{\mu}_N\} = \frac{1}{n}\left(\sum_{i=1}^{k} W_i\tilde{\sigma}_{N_i}\right)^2 - \frac{1}{N}\sum_{i=1}^{k} W_i\tilde{\sigma}_{N_i}^2 \tag{5.4.5}$$

5.5
Prediction Models with Known Covariates

In some problems of estimating the mean, μ_N, of a variable Y in a finite population, we may have information on variables $X_1, X_2, \ldots, X_k$ that are related to Y. The variables $X_1, \ldots, X_k$ are called **covariates**. The model relating Y to $X_1, \ldots, X_k$ is called a **prediction model**. If the values of Y are known only for the units in the sample, whereas the values of the covariates are known for all the units of the population, we can use the prediction model to improve the precision of the estimator. This method can be useful when, for example, the measurements of Y are destructive but the covariates can be measured without destroying the units. There are many such examples, like the case of measuring the compressive strength of a concrete cube. This measurement is destructive. The compressive strength Y is related to the ratio of cement to water in the mix, which is a covariate that can be known for all units. We will develop the idea with a simple prediction model.

Let $\{u_1, u_2, \ldots, u_N\}$ be a finite population, $\mathcal{P}$. The values of $x_i = X(u_i)$, $i = 1, \ldots, N$ are known for all the units of $\mathcal{P}$. Suppose that $Y(u_i)$ is related linearly to $X(u_i)$ according to the prediction model

$$y_i = \beta x_i + e_i \qquad i = 1, \ldots, N \tag{5.5.1}$$

where β is an unknown regression coefficient and $e_1, \ldots, e_N$ are i.i.d. random variables such that

$$E\{e_i\} = 0 \qquad i = 1, \ldots, N$$
$$V\{e_i\} = \sigma^2 \qquad i = 1, \ldots, N$$

The random variable e_i in the prediction model is there because the linear relationship between Y and X is not perfect, but subject to random deviations.

We are interested in the population quantity $\bar{y}_N = (1/N)\sum_{i=1}^{N} y_i$. We cannot, however, measure all the Y values. Even if we know the regression coefficient β, we can only predict $\bar{y}_N$ by $\beta\bar{x}_N$, where $\bar{x}_N = (1/N)\sum_{j=1}^{N} x_j$. Indeed, according to the prediction model, $\bar{y}_N = \beta\bar{x}_N + \bar{e}_N$, and $\bar{e}_N$ is a random variable with

$$E\{\bar{e}_N\} = 0 \qquad V\{\bar{e}_N\} = \frac{\sigma^2}{N} \tag{5.5.2}$$

Thus, because $\bar{y}_N$ has a random component, and because $E\{\bar{y}_N\} = \beta\bar{x}_N$, we say that a predictor of $\bar{y}_N$—say, $\hat{Y}_N$—is *unbiased* if $E\{\hat{Y}_N\} = \beta\bar{x}_N$. Generally, β is unknown. Thus, we draw a sample of units from $\mathcal{P}$ by a simple RSWOR of size n, $1 < n < N$, and measure their Y values in order to estimate β.

Let $(X_1, Y_1), \ldots, (X_n, Y_n)$ be the values of X and Y in the random sample. A predictor of $\bar{y}_N$ is some function of the observed sample values. Notice that after drawing a random sample, we have two sources of variability: one due to the random error components, $e_1, \ldots, e_n$, associated with the sample values, and the other due to the random sampling of the n units of $\mathcal{P}$. Notice that the error variables, $e_1, \ldots, e_n$, are independent of the X values and thus independent of $X_1, X_2, \ldots, X_n$, randomly chosen to the sample. We now examine a few alternative predictors of $\bar{y}_N$. Expectation and variances are taken with respect to the errors model and with respect to the sampling procedure.

First is the **sample mean**, $\bar{Y}_n$. Because

$$\bar{Y}_n = \beta \bar{X}_n + \bar{e}_n$$

we see that

$$E\{\bar{Y}_n\} = \beta E\{\bar{X}_n\} + E\{\bar{e}_n\}$$

$E\{\bar{e}_n\} = 0$ and because the sampling is RSWOR, $E\{\bar{X}_n\} = \bar{x}_N$. Thus, $E\{\bar{Y}_n\} = \beta \bar{x}_N$, and the predictor is unbiased. Because $\bar{e}_n$ is independent of $\bar{X}_n$, the variance of the predictor is

$$V\{\bar{Y}_n\} = \beta^2 V\{\bar{X}_n\} + \frac{\sigma^2}{n}$$

$$= \frac{\sigma^2}{n} + \frac{\beta^2 \sigma_x^2}{n}\left(1 - \frac{n-1}{N-1}\right) \tag{5.5.3}$$

where

$$\sigma_x^2 = \frac{1}{N}\sum_{j=1}^{N}(x_j - \bar{x}_N)^2$$

A second is the **ratio predictor**:

$$\hat{Y}_R = \bar{x}_N \frac{\bar{Y}_n}{\bar{X}_n} \tag{5.5.4}$$

The ratio predictor is used when all $x_i > 0$. In this case, $\bar{X}_n > 0$ in every possible sample. Substituting $\bar{Y}_n = \beta \bar{X}_n + \bar{e}_n$, we obtain

$$E\{\hat{Y}_R\} = \beta \bar{x}_N + \bar{x}_N E\left\{\frac{\bar{e}_n}{\bar{X}_n}\right\}$$

Again, because $\bar{e}_n$ and $\bar{X}_n$ are independent, $E\{\bar{e}_n/\bar{X}_n\} = 0$ and $\hat{Y}_R$ is an **unbiased predictor**. The variance of $\hat{Y}_R$ is

$$V\{\hat{Y}_R\} = (\bar{x}_N)^2 V\left\{\frac{\bar{e}_n}{\bar{X}_n}\right\}$$

Because $\bar{e}_n$ and $\bar{X}_n$ are independent, and $E\{\bar{e}_n\} = 0$, the law of the total variance implies that

$$V\{\hat{Y}_R\} = \frac{\sigma^2}{n}\bar{x}_N^2 E\left\{\frac{1}{\bar{X}_n^2}\right\}$$

$$= \frac{\sigma^2}{n}E\left\{\left(1 + \frac{(\bar{X}_n - \bar{x}_N)}{\bar{x}_N}\right)^{-2}\right\} \tag{5.5.5}$$

$$= \frac{\sigma^2}{n} E \left\{ 1 - \frac{2}{\bar{x}_N} (\bar{X}_n - \bar{x}_N) + \frac{3}{\bar{x}_N^2} (\bar{X}_n - \bar{x}_N)^2 - \cdots \right\}$$

$$\cong \frac{\sigma^2}{n} \left(1 + \frac{3\gamma_x^2}{n} \left(1 - \frac{n-1}{N-1} \right) \right)$$

where $\gamma_x = \sigma_x / \bar{x}_N$ is the coefficient of variation of X. This approximation is effective in large samples.

Using the large-sample approximation, we see that the ratio predictor $\hat{Y}_R$ has a smaller variance than $\bar{Y}_n$ if

$$\frac{3\sigma^2 \gamma_x^2}{n^2} \left(1 - \frac{n-1}{N-1} \right) < \frac{\beta^2 \sigma_x^2}{n} \left(1 - \frac{n-1}{N-1} \right)$$

or if

$$n > \frac{3\sigma^2}{(\beta \bar{x}_N)^2}$$

Other possible predictors for this model are

$$\hat{Y}_{RA} = \bar{x}_N \cdot \frac{1}{n} \sum_{i=1}^{n} \frac{X_i}{Y_i} \tag{5.5.6}$$

and

$$\hat{Y}_{RG} = \bar{x}_N \cdot \frac{\sum_{i=1}^{n} X_i Y_i}{\sum_{i=1}^{n} X_i^2}. \tag{5.5.7}$$

We leave it as an exercise to prove that both $\hat{Y}_{RA}$ and $\hat{Y}_{RG}$ are unbiased predictors, and to derive their variances.

What happens under this prediction model if the sample drawn is not random? Suppose that a nonrandom sample $(x_1, y_1), \ldots, (x_n, y_n)$ is chosen. Then

$$E\{\bar{y}_n\} = \beta \bar{x}_n$$

and

$$V\{\bar{y}_n\} = \frac{\sigma^2}{n}$$

The predictor $\bar{y}_n$ is biased unless $\bar{x}_n = \bar{x}_N$. A sample that satisfies this property is called a *balanced sample* with respect to X. Generally, the **prediction mean-squared error (MSE)** of a predictor $\hat{\theta}_n$ of a population quantity θ_N is

$$\text{MSE}\{\hat{\theta}_n\} = E\{(\hat{\theta}_n - \theta_N)^2\}$$

Under nonrandom sampling, the prediction MSE of $\bar{y}_n$ is

$$\text{MSE}\{\bar{y}_n\} = E\{(\bar{y}_n - \beta \bar{x}_N)^2\}$$

$$= \frac{\sigma^2}{n} + \beta^2 (\bar{x}_n - \bar{x}_N)^2 \tag{5.5.8}$$

Thus, if the sample is balanced with respect to X, then $\bar{y}_n$ is a more precise predictor than all the preceding, which are based on simple random samples.

EXAMPLE **5.6**   Electronic systems such as television sets, radios, and computers contain printed circuit boards with electronic components positioned in patterns determined by design engineers. After assembly (either by automatic insertion machines or manually) the components are soldered to the board. In the relatively new surface mount technology, minute components are simultaneously positioned and soldered to the boards. The occurrence of defective soldering points impacts the assembly plant productivity and is therefore closely monitored. In file PRED.DAT, we find 1000 records on variable X and Y. X is the number of soldering points on a board, and Y is the number of defective soldering points. The mean of Y is $\bar{y}_{1000} = 7.495$ and that of X is $\bar{x}_{1000} = 148.58$. Moreover, $\sigma_x^2 = 824.562$ and the coefficient of variation is $\gamma_x = .19326$. The relationship between X and Y is $y_i = \beta x_i + e_i$, where $E\{e_i\} = 0$ and $V\{e_i\} = 7.5$, $\beta = .05$. Thus, if we have to predict $\bar{y}_{1000}$ by a predictor based on a RSWR of size $n = 100$, the variances of $\overline{Y}_{100}$ and $\hat{Y}_R = \bar{x}_{1000}(\overline{Y}_{100}/\overline{X}_{100})$ are

$$V\{\overline{Y}_{100}\} = \frac{7.5}{100} + \frac{0.0025 \times 824.562}{100} = 0.0956$$

However, the large-sample approximation yields

$$V\{\hat{Y}_R\} = \frac{7.5}{100}\left(1 + \frac{3 \times 0.037351}{100}\right)$$
$$= 0.07508$$

We see that if we have to predict $\bar{y}_{1000}$ on the basis of a RSWR of size $n = 100$, the ratio predictor, $\hat{Y}_R$, is more precise.   ∎

In Figures 5.2 and 5.3, we present the histograms of 500 predictors $\overline{Y}_{100}$ and 500 $\hat{Y}_R$ based on a RSWR of size 100 from this population.

5.6
Chapter Highlights

Techniques for sampling finite populations and estimating population parameters are presented. Formulas are given for the expected value and variance of the sample mean and sample variance of simple random samples with and without replacement. Stratification is studied as a method to increase the precision of estimators. Formulas for proportional and optimal allocation are provided and demonstrated with case studies. The chapter concludes with a section on prediction models with known covariates.

The main concepts and definitions introduced in this chapter include:

- Population quantities
- Simple random sampling
- Stratified random sampling
- Statistical inference
- Unbiased (estimators)
- Precision (of an estimator)
- Finite population multiplier

FIGURE **5.2**
Sampling distribution of $\overline{Y}$

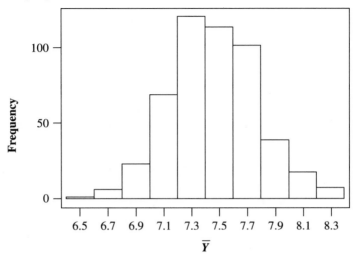

FIGURE **5.3**
Sampling distribution of $\hat{Y}_R$

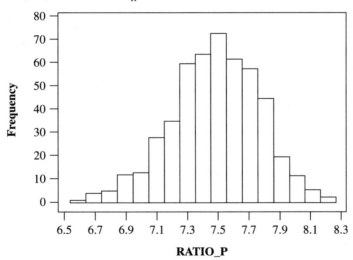

- Sample allocation
- Proportional allocation
- Optimal allocation
- Covariates
- Prediction model

- Sample mean
- Ratio predictor
- Unbiased predictor
- Prediction MSE
- Balanced sample

5.7
Exercises

5.2.1   Consider a finite population of size N, whose elements have values $x_1, \ldots, x_N$. Let $F_N(x)$ be the c.d.f.,—that is,

$$F_N(x) = \frac{1}{N} \sum_{i=1}^{N} I\{x_i \le x\}$$

Let $X_1, \ldots, X_n$ be the values of a RSWR. Show that $X_1, \ldots, X_n$ are independent, having a common distribution $F_N(x)$.

5.2.2   Show that if $\overline{X}_n$ is the mean of a RSWR, then $\overline{X}_n \to \mu_N$ as $n \to \infty$ in probability.

5.2.3   What is the large-sample approximation to $\Pr\{\sqrt{n} \mid \overline{X}_n - \mu_N| < \delta\}$ in RSWR?

5.2.4   Use MINITAB to draw random samples with or without replacement from data file PLACE.DAT. Write a macro that computes the sample correlation between the x-dev and y-dev in the sample values. Execute this macro 100 times and make a histogram of the sample correlations.

5.2.5   Use file CAR.DAT and MINITAB. Construct a macro that samples 50 records at random, without replacement (RSWOR). Stack the medians of the variables turn diameter, horsepower, and mpg (3, 4, 5). Execute the macro 200 times and show the histograms of the sampling distributions of the medians.

5.2.6   Continuing Example 5.4, how large should the sample be from the three strata so that the SE $\{\overline{X}_i\}$ ($i = 1, \ldots, 3$) will be smaller than $\delta = 0.005$?

5.2.7   The proportion of defective chips in a lot of $N = 10,000$ chips is $\mathbf{P} = 5 \times 10^{-4}$. How large should a RSWOR be so that the width of the confidence interval for P, with confidence level $1 - \alpha = .95$, will be 0.002?

5.3.1   Use MINITAB to perform stratified random samples from the three strata of the data file PLACE.DAT (see Example 5.4). Allocate 200 observations to the three samples proportionally. Estimate the population mean (of x-dev). Repeat this 100 times and estimate the standard error of your estimates. Compare the estimated standard error to the exact one.

5.4.1   Derive the formula for n_i^0 ($i = 1, \ldots, k$) in the optimal allocation by differentiating $L(n_1, \ldots, n_k, \lambda)$ and solving the equations.

5.5.1   Consider the prediction model

$$y_i = \beta + e_i \qquad i = 1, \ldots, N$$

where $E\{e_i\} = 0$, $V\{e_i\} = \sigma^2$, and $\text{cov}(e_i, e_j) = 0$ for $i \ne j$. We wish to predict the population mean $\mu_N = (1/N)\sum_{i=1}^{N} y_i$. Show that the sample mean, $\overline{Y}_n$, is prediction unbiased. What is the prediction MSE of $\overline{Y}_n$?

5.5.2   Consider the prediction model

$$y_i = \beta_0 + \beta_1 x_i + e_i \qquad i = 1, \ldots, N$$

where $e_1, \ldots, e_N$ are independent random variables with $E\{e_i\} = 0$ and $V\{e_i\} = \sigma^2 x_i$ ($i = 1, \ldots, n$). We wish to predict $\mu_N = (1/N)\sum_{i=1}^{N} y_i$. What should be a good predictor for μ_N?

5.5.3   Prove that $\hat{Y}_{RA}$ and $\hat{Y}_{RG}$ are unbiased predictors and derive their prediction variances.

6

Parametric Statistical Inference

In this chapter, we introduce the basic concepts and methods of **statistical inference**. The focus is on estimating the parameters of statistical distributions and testing hypotheses about them. Testing whether or not certain distributions fit observed data is considered as well. We begin with some basic problems of estimation theory.

A statistical population is represented by the distribution function(s) of the observable random variable(s) associated with its elements. The actual distributions representing the population under consideration are generally unspecified or only partially specified. Based on theoretical considerations and/or practical experience, we often assume that a distribution belongs to a particular family—normal, Poisson, Weibull, and so on. Such assumptions are called the *statistical model*. If the model assumes a specific distribution with known parameters, there is no need to estimate the parameters. We may, however, use sample data to test whether the hypothesis concerning the specific distribution in the model is valid. This is a *goodness-of-fit* testing problem. If the model assumes only the family to which the distribution belongs, and the specific values of the parameters are unknown, the problem is to estimate the unknown parameters. In Chapter 5 we studied the problem of estimating or predicting characteristics of a finite population.

Sampling Characteristics of Estimators

The means and the variances of random samples vary randomly around the true values of the parameters. In practice, we usually take one sample of data and then construct a single estimate for each population parameter. To illustrate the concept of error in estimation, consider what happens if we take many samples from the same population. The collection of estimates (one from each sample) can itself be thought of as a sample taken from a hypothetical population of all possible estimates. The distribution of all possible estimates is called the **sampling distribution**. The sampling distributions of the estimates may be of different types than the distribution of the original observations. Figures 6.1 and 6.2 show the frequency distributions of $\overline{X}_{10}$ and of S_{10}^2 for 100 random samples of size $n = 10$ drawn from the uniform distribution over the integers $\{1, \ldots, 100\}$.

We see in Figure 6.1 that the frequency distribution of sample means does not resemble a uniform distribution but seems to be close to normal. Moreover, the spread of the sample means is from 25 to 75 and not the original spread of 1 to 100. In Chapter 4, we discussed the CLT, which states that when the sample size is large,

FIGURE   6.1
Histogram of 100 Sample Means

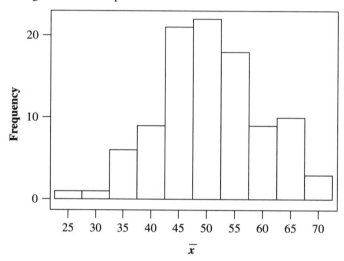

FIGURE   6.2
Histogram of 100 sample variances

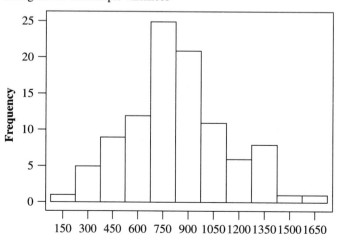

the sampling distribution of the sample mean of a simple random sample, $\overline{X}_n$, for any population having a finite positive variance, σ^2, is approximately normal with mean

$$E\{\overline{X}\} = \mu \tag{6.1.1}$$

and variance

$$V\{\overline{X}_n\} = \frac{\sigma^2}{n} \tag{6.1.2}$$

Notice that

$$\lim_{n \to \infty} V\{\overline{X}_n\} = 0$$

This means that the precision of the sample mean as an estimator of the population mean, μ, grows with the sample size.

Generally, if a function of the sample values $X_1, \ldots, X_n$, $\hat{\theta}(X_1, \ldots, X_n)$, is an estimator of a parameter θ of a distribution, then $\hat{\theta}_n$ is called an **unbiased estimator** if

$$E\{\hat{\theta}_n\} = \theta \quad \text{for all } \theta \tag{6.1.3}$$

Furthermore, $\hat{\theta}_n$ is called a **consistent estimator** of θ if for any $\epsilon > 0$, $\lim_{n \to \infty} \Pr\{|\hat{\theta}_n - \theta| > \epsilon\} = 0$. Applying the Chebyshev inequality, we see that a sufficient condition for consistency is to have $\lim_{n \to \infty} V\{\hat{\theta}_n\} = 0$. The sample mean is generally a consistent estimator. The standard deviation of the sampling distribution of $\hat{\theta}_n$ is called the **standard error** of $\hat{\theta}_n$; that is, SE $\{\hat{\theta}_n\} = (V\{\hat{\theta}_n\})^{1/2}$.

6.2
Some Methods of Point Estimation

Consider a statistical model that specifies the family $\mathcal{F}$ of possible distributions of the observed random variable. The family $\mathcal{F}$ is called a **parametric family** if the distributions in $\mathcal{F}$ are of the same functional type and differ only by the values of their parameters. For example, the family of all exponential distributions, $E(\beta)$, when $0 < \beta < \infty$, is a parametric family. In this case, we can write

$$\mathcal{F} = \{E(\beta) : 0 < \beta < \infty\}$$

Another example of a parametric family is $\mathcal{F} = \{N(\mu, \sigma); -\infty < \mu < \infty, 0 < \sigma < \infty\}$, which is the family of all normal distributions. The range Θ of the parameter(s) $\boldsymbol{\theta}$, is called the **parameter space**. Thus, a parametric statistical model specifies the parametric family $\mathcal{F}$. This specification gives both the functional form of the distribution and its parameter(s) space Θ.

We observe a random sample from the infinite population, which consists of the values of i.i.d. random variables $X_1, X_2, \ldots, X_n$, whose common distribution $F(x; \boldsymbol{\theta})$ is an element of $\mathcal{F}$.

A function of the observable random variables is called a **statistic**. A statistic cannot depend on unknown parameters. A statistic is thus a random variable whose value can be determined from the sample values $(X_1, \ldots, X_n)$. In particular, a statistic $\hat{\theta}(X_1, \ldots, X_n)$, which yields values in the parameter space is called a **point**

estimator of θ. If the distributions in $\mathcal{F}$ depend on several parameters, we have to determine point estimators for each parameter, or for a function of the parameters. For example, the pth quantile of a normal distribution is $\xi_p = \mu + z_p\sigma$, where μ and σ are the parameters and $z_p = \Phi^{-1}(p)$. This is a function of two parameters. An important problem in quality control is to estimate such quantiles. In this section, we discuss a few methods for deriving point estimators.

6.2.1   Moment Equation Estimators

If $X_1, X_2, \ldots, X_n$ are i.i.d. random variables (a random sample), then the sample lth moment ($l = 1, 2, \ldots$) is

$$M_l = \frac{1}{n}\sum_{i=1}^{n} X_i^l \tag{6.2.1}$$

The law of large numbers (strong) says that if $E\{|X|^l\} < \infty$, then M_l converges with probability 1 to the population lth moment $\mu_l(F)$. Accordingly, we know that if the sample size n is large, then, with probability close to 1, M_l is close to $\mu_l(F)$. The method of moments for parametric models equates M_l to μ_l, which is a function of θ and solves for θ. Generally, if $F(x; \boldsymbol{\theta})$ depends on k parameters $\theta_1, \theta_2, \ldots, \theta_k$, then we set up k equations

$$\begin{aligned}
M_1 &= \mu_1(\theta_1, \ldots, \theta_k) \\
M_2 &= \mu_2(\theta_1, \ldots, \theta_k) \\
&\;\;\vdots \\
M_k &= \mu_k(\theta_1, \ldots, \theta_k)
\end{aligned} \tag{6.2.2}$$

and solve for $\theta_1, \ldots, \theta_k$. The solutions are functions of the sample statistics $M_1, \ldots, M_k$, and are therefore estimators. This method does not always yield simple or good estimators. Following are two examples in which the estimators obtained by this method are reasonable.

EXAMPLE 6.1   Consider the family $\mathcal{F}$ of Poisson distributions—that is,

$$\mathcal{F} = \{P(x; \theta); 0 < \theta < \infty\}$$

The parameter space is $\Theta = (0, \infty)$. The distributions depend on one parameter, and

$$\mu_1(\theta) = E_\theta\{X\} = \theta$$

Thus, the method of moments yields the estimator

$$\hat{\theta}_n = \overline{X}_n$$

This is an unbiased estimator with $V\{\hat{\theta}_n\} = \theta/n$. ∎

EXAMPLE **6.2**   Consider a random sample of $X_1, X_2, \ldots, X_n$ from a log-normal distribution $LN(\mu, \sigma)$. The distributions depend on $k = 2$ parameters.

We have seen that

$$\mu_1(\mu, \sigma^2) = \exp\{\mu + \sigma^2/2\}$$

$$\mu_2(\mu, \sigma^2) = \exp\{2\mu + 2\sigma^2\}$$

Thus, let $\theta_1 = \mu$ and $\theta_2 = \sigma^2$ and set the equations

$$\exp\{\theta_1 + \theta_2/2\} = M_1$$

$$\exp\{2\theta_1 + 2\theta_2\} = M_2$$

The solutions $\hat{\theta}_1$ and $\hat{\theta}_2$ of this system of equations are

$$\hat{\theta}_1 = 2 \log M_1 - \frac{1}{2} \log M_2$$

and

$$\hat{\theta}_2 = \log \left(\frac{M_2}{M_1^2} \right)$$

The estimators obtained are biased, but we can show they are consistent. Simple formulas for $V\{\hat{\theta}_1\}$, $V\{\hat{\theta}_2\}$, and $\text{cov}(\hat{\theta}_1, \hat{\theta}_2)$ do not exist. We can derive large-sample approximations to these characteristics, or approximate them by a method of resampling called *bootstrapping*, which is discussed in Chapter 7.   ■

6.2.2   The Method of Least Squares

If $\mu = E\{X\}$, then the method of least squares chooses the estimator $\hat{\mu}$, which minimizes

$$Q(\mu) = \sum_{i=1}^{n} (X_i - \mu)^2 \tag{6.2.3}$$

It is easy to show that the **least-squares estimator (LSE)** is the sample mean—that is,

$$\hat{\mu} = \overline{X}_n$$

Indeed, we can write

$$Q(\mu) = \sum_{i=1}^{n} (X_i - \overline{X}_n + \overline{X}_n - \mu)^2$$

$$= \sum_{i=1}^{n} (X_i - \overline{X}_n)^2 + n(\overline{X}_n - \mu)^2$$

Thus, $Q(\hat{\mu}) \geq Q(\bar{X}_n)$ for all μ and $Q(\hat{\mu})$ is minimized only if $\hat{\mu} = \bar{X}_n$. This estimator is unbiased and consistent. Indeed,

$$V\{\hat{\mu}\} = \frac{\sigma^2}{n}$$

provided that $\sigma^2 < \infty$.

The LSE is more interesting in the case of linear regression (see Chapter 3).

In the simple linear regression case, we have n independent random variables $Y_1, \ldots, Y_n$ with equal variances, σ^2, but expected values that depend linearly on known regressors (predictors) $x_1, \ldots, x_n$. That is,

$$E\{Y_i\} = \beta_0 + \beta_1 x_i \qquad i = 1, \ldots, n \tag{6.2.4}$$

The least-squares estimators of the regression coefficients β_0 and β_1 are the values that minimize

$$Q(\beta_0, \beta_1) = \sum_{i=1}^{n} (Y_i - \beta_0 - \beta_1 x_i)^2 \tag{6.2.5}$$

We showed in Chapter 3 that these LSEs are

$$\hat{\beta}_0 = \bar{Y}_n - \hat{\beta}_1 \bar{x}_n \tag{6.2.6}$$

and

$$\hat{\beta}_1 = \frac{\sum_{i=1}^{n} Y_i (x_i - \bar{x}_n)}{\sum_{i=1}^{n} (x_i - \bar{x}_n)^2} \tag{6.2.7}$$

where $\bar{x}_n$ and $\bar{Y}_n$ are the sample means of the x's and the Ys, respectively. Thus, $\hat{\beta}_0$ and $\hat{\beta}_1$ are linear combinations of the Ys with known coefficients. From the results of Section 4.8, we see that

$$E\{\hat{\beta}_1\} = \sum_{i=1}^{n} \frac{(x_i - \bar{x}_n)}{\mathrm{SS}_x} E\{Y_i\}$$

$$= \sum_{i=1}^{n} \frac{(x_i - \bar{x}_n)}{\mathrm{SS}_x} (\beta_0 + \beta_1 x_i)$$

$$= \beta_0 \sum_{i=1}^{n} \frac{(x_i - \bar{x}_n)}{\mathrm{SS}_x} + \beta_1 \sum_{i=1}^{n} \frac{(x_i - \bar{x}_n) x_i}{\mathrm{SS}_x}$$

where $\mathrm{SS}_x = \sum_{i=1}^{n} (x_i - \bar{x}_n)^2$. Furthermore,

$$\sum_{i=1}^{n} \frac{x_i - \bar{x}_n}{\mathrm{SS}_x} = 0$$

and

$$\sum_{i=1}^{n} \frac{(x_i - \bar{x}_n)x_i}{SS_x} = 1$$

Hence, $E\{\hat{\beta}_1\} = \beta_1$. Also,

$$E\{\hat{\beta}_0\} = E\{\bar{Y}_n\} - \bar{x}_n E\{\hat{\beta}_1\}$$
$$= (\beta_0 + \beta_1 \bar{x}_n) - \beta_1 \bar{x}_n$$
$$= \beta_0$$

Thus, $\hat{\beta}_0$ and $\hat{\beta}_1$ are both unbiased. The variances of these LSEs are given by

$$V\{\hat{\beta}_1\} = \frac{\sigma^2}{SS_x}$$

$$V\{\hat{\beta}_0\} = \frac{\sigma^2}{n} + \frac{\sigma^2 \bar{x}_n^2}{SS_x} \tag{6.2.8}$$

and

$$\text{cov}(\hat{\beta}_0, \hat{\beta}_1) = -\frac{\sigma^2 \bar{x}_n}{SS_x} \tag{6.2.9}$$

Thus, $\hat{\beta}_0$ and $\hat{\beta}_1$ are not independent. A hint for deriving these formulas is given in Exercise 6.2.5.

The correlation between $\hat{\beta}_0$ and $\hat{\beta}_1$ is

$$\rho = -\frac{\bar{x}_n}{\left(\frac{1}{n}\sum_{i=1}^{n} x_i^2\right)^{1/2}} \tag{6.2.10}$$

6.2.3 Maximum Likelihood Estimators

Let $X_1, X_2, \ldots, X_n$ be i.i.d. random variables having a common distribution belonging to a parametric family $\mathcal{F}$. Let $f(x; \theta)$ be the p.d.f. of X, $\theta \in \Theta$. This is either a density function or a probability distribution function of a discrete random variable. Because $X_1, \ldots, X_n$ are independent, their joint p.d.f. is

$$f(x_1, \ldots, x_n; \theta) = \prod_{i=1}^{n} f(x_i; \theta)$$

The **likelihood function** of θ over Θ is defined as

$$L(\theta; x_1, \ldots, x_n) = \prod_{i=1}^{n} f(x_i; \theta) \tag{6.2.11}$$

The likelihood of θ is thus the probability in the discrete case, or the joint density in the continuous case, of the observed sample values under θ. In the likelihood function $L(\theta; x_1, \ldots, x_n)$, the sample values $(x_1, \ldots, x_n)$ play the role of parameters. A **maximum likelihood estimator (MLE)** of θ is a point in the parameter space, $\hat{\theta}_n$, for which $L(\theta; X_1, \ldots, X_n)$ is maximized. The notion of maximum is taken in a general sense. For example, the function

$$f(x; \lambda) = \begin{cases} \lambda e^{-\lambda x} & x \geq 0 \\ 0 & x < 0 \end{cases}$$

as a function of λ, $0 < \lambda < \infty$, attains a maximum at $\lambda = 1/x$.

However, the function

$$f(x; \theta) = \begin{cases} \dfrac{1}{\theta} & 0 \leq x \leq \theta \\ 0 & \text{otherwise} \end{cases}$$

as a function of θ over $(0, \infty)$ attains a lowest upper bound (supremum) at $\theta = x$, which is $1/x$. We say it is maximized at $\theta = x$. Notice that it is equal to zero for $\theta < x$. Following are two examples.

EXAMPLE 6.3   Suppose that $X_1, X_2, \ldots, X_n$ is a random sample from a normal distribution. Then the likelihood function of (μ, σ^2) is

$$L(\mu, \sigma^2; X_1, \ldots, X_n)$$

$$= \frac{1}{(2\pi)^{n/2}(\sigma)^n} \exp\left\{ -\frac{1}{2\sigma^2} \sum_{i=1}^{n} (X_i - \mu)^2 \right\}$$

$$= \frac{1}{(2\pi)^{n/2}(\sigma^2)^{n/2}} \exp\left\{ -\frac{1}{2\sigma^2} \sum_{i=1}^{n} (X_i - \overline{X}_n)^2 - \frac{n}{2\sigma^2} (\overline{X}_n - \mu)^2 \right\}$$

Notice that the likelihood function of (μ, σ^2) depends on the sample variables only through the statistics $(\overline{X}_n, Q_n)$, where $Q_n = \sum_{i=1}^{n} (X_i - \overline{X}_n)^2$. These statistics are called **likelihood statistics**. To maximize the likelihood, we can maximize the log-likelihood

$$l(\mu, \sigma^2; \overline{X}_n, Q_n) = -\frac{n}{2} \log(2\pi) - \frac{n}{2} \log(\sigma^2) - \frac{Q_n}{2\sigma^2} - \frac{n(\overline{X}_n - \mu)^2}{2\sigma^2}$$

With respect to μ, we maximize by $\hat{\mu}_n = \overline{X}_n$. With respect to σ^2, differentiate

$$l(\hat{\mu}_n, \sigma^2; \overline{X}_n, Q_n) = -\frac{n}{2} \log(2\pi) - \frac{n}{2} \log(\sigma^2) - \frac{Q_n}{2\sigma^2}$$

This is

$$\frac{\partial}{\partial \sigma^2} l(\hat{\mu}, \sigma^2; \overline{X}_n, Q_n) = -\frac{n}{2\sigma^2} + \frac{Q_n}{2\sigma^4}$$

Equating the derivative to zero and solving yields the MLE

$$\hat{\sigma}_n^2 = \frac{Q_n}{n}$$

Thus, the MLEs are

$$\hat{\mu}_n = \overline{X}_n \qquad \text{and} \qquad \hat{\sigma}_n^2 = \frac{n-1}{n} S_n^2$$

$\hat{\sigma}_n^2$ is biased, but the bias goes to zero as $n \to \infty$. ∎

EXAMPLE 6.4 Let X have a negative-binomial distribution $NB(k, p)$. Suppose that k is known, and $0 < p < 1$. The likelihood function of p is

$$L(p; X, k) = \binom{X+k-1}{k-1} p^k (1-p)^X$$

Thus, the log-likelihood is

$$l(p; X, k) = \log \binom{X+k-1}{k-1} + k \log p + X \log(1-p)$$

The MLE of p is

$$\hat{p} = \frac{k}{X+k}$$

We can show that $\hat{p}$ has a positive bias—that is, that $E\{\hat{p}\} > p$. For large values of k, the bias is approximately

$$\begin{aligned} \text{bias}(\hat{p}; k) &= E\{\hat{p}; k\} - p \\ &\cong \frac{3p(1-p)}{2k} \quad \text{large } k \end{aligned}$$

The variance of $\hat{p}$ for large k is approximately

$$V\{\hat{p}; k\} \cong \frac{p^2(1-p)}{k} \quad ∎$$

6.3
Comparison of Sample Estimates with Specified Standards— Testing Statistical Hypotheses

6.3.1 Basic Concepts

Statistical hypotheses are statements concerning the parameters or characteristics of the distribution that represents a certain random variable (or variables) in a population. For example, consider a manufacturing process. The parameter of interest may be the proportion, p, of nonconforming items. If $p \leq p_0$, the process is considered to be acceptable. If $p > p_0$, the process should be corrected.

Suppose that 20 items are randomly selected from the process and inspected. Let X be the number of nonconforming items in the sample. Then X has a binomial distribution $B(20, p)$. On the basis of the observed value of X, we have to decide whether the process should be stopped and adjusted. In the statistical formulation of the problem, we are testing the hypothesis

$$H_0 : p \leq p_0$$

against the hypothesis

$$H_1 : p > p_0$$

Hypothesis H_0 is called the **null hypothesis**; H_1 is called the **alternative hypothesis**. Only when the data provide significant evidence that the null hypothesis is wrong do we reject it in favor of the alternative. It may not be justifiable to disrupt a production process unless we have ample evidence that the proportion of nonconforming items is too high. It is important to distinguish between *statistical significance* and *technological significance*. The statistical level of significance is the probability of rejecting H_0 when it is true. If we reject H_0 at a low level of significance, the probability of committing an error is small, and we are confident that our conclusion is correct. Rejecting H_0 might not be technologically significant if the true value of p is not greater than $p_0 + \delta$, where δ is some acceptable level of indifference. If $p_0 < p < p_0 + \delta$, H_1 is true, but there is no technological significance to the difference $p - p_0$.

To construct a statistical test procedure based on a **test statistic**, X, consider first all possible values that could be observed. In our example, X can assume the values $0, 1, 2, \ldots, 20$. Determine a critical region, or **rejection region**, so that whenever the observed value of X belongs to this region, the null hypothesis H_0 is rejected. For example, if we were testing $H_0 : P \leq 0.10$ against $H_1 : P > 0.10$, we might reject H_0 if $X > 4$. The complement of this region, $X \leq 3$, is called the **acceptance region**.

There are two possible errors that can be committed. If the true proportion of nonconforming items were only .05 (unknown to us), for example, and our sample happened to produce four items, we would incorrectly decide to reject H_0, thus shutting down the process, which is actually performing acceptably. This is called a **Type I error**. However, if the true proportion were .15, and only three nonconforming items were found in the sample, we would incorrectly allow the process to continue, even though defectives ran higher than 10% (**a Type II error**).

We denote the probability of committing a type I error by $\alpha(p)$, for $p \leq p_0$, and the probability of committing a Type II error by $\beta(p)$, for $p > p_0$.

In most problems, the critical region is constructed in such a way that the probability of committing a type I error will not exceed a preassigned value called the **significance level** of the test. Let α denote the significance level. In our example, the significance level is

$$\alpha = \Pr\{X \geq 4; p = .1\} = 1 - B(3; 20, .1) = .133$$

Notice that the significance level is computed with $p = .10$, which is the largest p value for which the null hypothesis is true.

To further evaluate the test procedure, we would like to know the probability of accepting the null hypothesis for various values of p. Such a function is called the **operating characteristic function** and is denoted by $OC(p)$. The graph of $OC(p)$ vs. p is called the **OC curve**. Ideally, we would like to see $OC(p) = 1$ whenever $H_0 : p \le p_0$ is true, and $OC(p) = 0$ when $H_1 : p > p_0$ is true. This, however, cannot be obtained when the decision is based on a random sample of items.

In our example, we can compute the OC function as

$$OC(p) = \Pr\{X \le 3; p\} = B(3; 20, p)$$

From Table 6.1 we find that

$$OC(.10) = .8670$$
$$OC(.15) = .6477$$
$$OC(.20) = .4114$$
$$OC(.25) = .2252$$

Notice that the significance level α is the maximum probability of rejecting H_0 when it is true. Accordingly, $OC(p_0) = 1 - \alpha$. The OC curve for this example is shown in Figure 6.3.

TABLE 6.1

The binomial c.d.f. $B(x; n, p)$ for $n = 20$, $p = .10(.05).25$

x	$p = .10$	$p = .15$	$p = .20$	$p = .25$
0	.1216	.0388	.0115	.0032
1	.3917	.1756	.0692	.0243
2	.6769	.4049	.2061	.0913
3	.8670	.6477	.4114	.2252
4	.9568	.8298	.6296	.4148
5	.9887	.9327	.8042	.6172
6	.9976	.9781	.9133	.7858
7	.9996	.9941	.9679	.8982
8	.9999	.9987	.9900	.9591
9	1.0000	.9998	.9974	.9861
10	1.0000	1.0000	.9994	.9961

We see that as p grows, the value of $OC(p)$ decreases because the probability of observing at least four nonconforming items out of 20 is growing with p.

Suppose that the significance level of the test is decreased in order to reduce the probability of incorrectly interfering with a good process. We may choose the critical region to be $X \ge 5$. The new OC function is

$$OC(p) = \Pr\{X \le 4; p\} = B(4; 20, p)$$

FIGURE   6.3

The OC curve for testing $H_0 : p \leq .1$ against $H_1 : p > .1$ with a sample of size $n = 20$ and rejection region $X \geq 4$

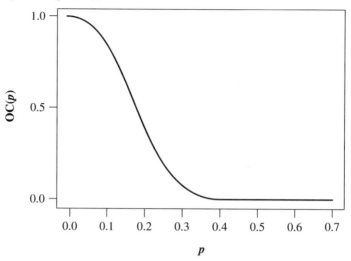

For this new critical region,

$$OC(.10) = .9568$$
$$OC(.15) = .8298$$
$$OC(.20) = .6296$$
$$OC(.25) = .4148$$

The new significance level is $\alpha = 1 - OC(.1) = .0432$. Notice that, although we reduced the risk of committing a type I error, we increased the risk of committing a type II error. Only with a larger sample size can we reduce simultaneously the risks of both type I and type II errors.

Instead of the OC function, we may want to consider the **power function** for evaluating the sensitivity of a test procedure. The power function, denoted by $\psi(p)$, is the probability of *rejecting* the null hypothesis when the alternative is true. Thus, $\psi(p) = 1 - OC(p)$.

Finally, we consider an alternative method of performing a test. Rather than specifying in advance the desired significance level—say, $\alpha = .05$—we can compute the probability of observing X_0 or more nonconforming items in a random sample if $p = p_0$. This probability is called the *attained significance level* or the **P-value** of the test. If the P-value is small—say, $\leq .05$—we consider the results to be significant and we reject the null hypothesis. For example, suppose we observed $X_0 = 6$ nonconforming items in a sample of size 20. The P-value is $\Pr\{X \geq 6; p = .10\} = 1 - B(5; 20, .10) = .0113$. This small probability suggests that we could reject H_0 in favor of H_1 without much of a risk.

The term P-value should not be confused with the parameter p of the binomial distribution.

6.3.2 Some Common One-Sample Tests of Hypotheses

A. The Z-test: Testing the Mean of a Normal Distribution, σ^2 Known

One-Sided Test   The hypothesis for a one-sided test on the mean of a normal distribution is:

$$H_0 : \mu \leq \mu_0$$

against

$$H_1 : \mu > \mu_0$$

where μ_0 is a specified value. Given a sample $X_1, \ldots, X_n$, we first compute the sample mean $\overline{X}_n$. Because large values of $\overline{X}_n$ relative to μ_0 would indicate that H_0 is possibly not true, the critical region should be of the form $\overline{X} \geq C$, where C is chosen so that the probability of committing a type I error is equal to α. (In many problems, we use $\alpha = .01$ or $.05$ depending on the consequences of a type I error.) For convenience, we use a modified form of the test statistic given by the Z statistic:

$$Z = \sqrt{n}(\overline{X}_n - \mu_0)/\sigma \qquad (6.3.1)$$

The critical region, in terms of Z, is given by

$$\{Z : Z \geq z_{1-\alpha}\}$$

where $z_{1-\alpha}$ is the $(1 - \alpha)$th quantile of the standard normal distribution. This critical region is equivalent to the region

$$\{\overline{X}_n : \overline{X}_n \geq \mu_0 + z_{1-\alpha}\sigma/\sqrt{n}\}$$

These regions are illustrated in Figure 6.4.

FIGURE 6.4
Critical regions for the one-sided Z-test

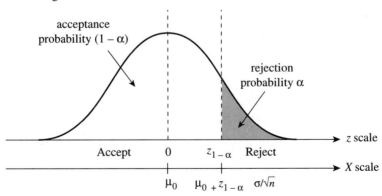

The operating characteristic function of this test is given by

$$OC(\mu) = \Phi(z_{1-\alpha} - \delta\sqrt{n}) \qquad (6.3.2)$$

where

$$\delta = (\mu - \mu_0)/\sigma \qquad\qquad (6.3.3)$$

EXAMPLE 6.5   Suppose we are testing the hypothesis $H_0 : \mu \le 5$ against $H_1 : \mu > 5$, with a sample of size $n = 100$ from a normal distribution with known standard deviation $\sigma = 0.2$. With a significance level of size $\alpha = .05$, we reject H_0 if

$$Z \ge z_{.95} = 1.645$$

The values of the OC function are computed in Table 6.2. In this table, $z = z_{1-\alpha} - \delta\sqrt{n}$ and $OC(\mu) = \Phi(z)$.

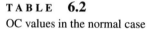

TABLE 6.2
OC values in the normal case

μ	$\delta\sqrt{n}$	z	$OC(\mu)$
5.	0	1.645	.9500
5.01	0.5	1.145	.8739
5.02	1.0	.645	.7405
5.03	1.5	.145	.5576
5.04	2.0	−.355	.3613
5.05	2.5	−.855	.1963

If the null hypothesis is $H_0 : \mu \ge \mu_0$ against the alternative $H_1 : \mu < \mu_0$, we reverse the direction of the test and reject H_0 if $Z \le -z_{1-\alpha}$. ∎

Two-Sided Test   The two-sided test has the form

$$H_0 : \mu = \mu_0$$

against

$$H_1 : \mu \ne \mu_0$$

The corresponding critical region is given by

$$\{Z : Z \ge z_{1-\alpha/2}\} \cup \{Z : Z \le -z_{1-\alpha/2}\}$$

The operating characteristic function is

$$OC(\mu) = \Phi(z_{1-\alpha/2} + \delta\sqrt{n}) - \Phi(-z_{1-\alpha/2} - \delta\sqrt{n}) \qquad (6.3.4)$$

The P-value of the two-sided test can be determined in the following manner. First compute

$$|Z_0| = \sqrt{n}|\overline{X}_n - \mu_0|/\sigma$$

and then compute the P-value

$$P = \Pr\{Z \geq |Z_0|\} + P\{Z \leq -|Z_0|\}$$
$$= 2(1 - \Phi(|Z_0|)) \tag{6.3.5}$$

B. The t-Test: Testing the Mean of a Normal Distribution, σ^2 Unknown In this case, we replace σ in the Z-test with the sample standard deviation, S, and $z_{1-\alpha}$ (or $z_{1-\alpha/2}$) with $t_{1-\alpha}[n-1]$ (or $t_{1-\alpha/2}[n-1]$). Thus, the critical region for the two-sided test becomes

$$\{t : |t| \geq t_{1-\alpha/2}[n-1]\}$$

where

$$t = (\overline{X}_n - \mu_0)\sqrt{n}/S \tag{6.3.6}$$

The operating characteristic function of the one-sided test is given approximately by

$$OC(\mu) \cong 1 - \Phi\left(\frac{\delta\sqrt{n} - t_{1-\alpha}[n-1](1 - 1/8(n-1))}{(1 + t_{1-\alpha}^2[n-1]/2(n-1))^{1/2}}\right) \tag{6.3.7}$$

where $\delta = |\mu - \mu_0|/\sigma$. (This is a good approximation to the exact formula, which is based on the complicated noncentral t distribution.)

Table 6.3 gives some numerical comparisons of the power of the one-sided test for the cases of σ^2 known and σ^2 unknown, when $n = 20$ and $\alpha = .05$. Notice that when σ is unknown, the power of the test is somewhat smaller than when it is known.

TABLE 6.3
Power functions of Z- and t-tests

δ	σ known	σ unknown
0.0	0.050	0.050
0.1	0.116	0.111
0.2	0.226	0.214
0.3	0.381	0.359
0.4	0.557	0.527
0.5	0.723	0.691

EXAMPLE 6.6 The cooling system of a large computer consists of metal plates that are attached together to create an internal cavity that allows for the circulation of special-purpose cooling liquids. The metal plates are attached with steel pins designed to measure 0.5 mm in diameter. Experience with the process of manufacturing similar steel pins has shown that the diameters of the pins are normally distributed, with mean μ and standard deviation σ. The process is aimed at maintaining a mean of $\mu_0 = 0.5$ (in mm). To control this process, we want to test $H_0 : \mu = 0.5$ against $H_1 : \mu \neq 0.5$. If we have prior information that the process standard deviation is constant at

$\sigma = 0.02$, we can use the Z-test to test these hypotheses. If we apply a significance level of $\alpha = .05$, then we will reject H_0 if $Z \geq z_{1-\alpha/2} = 1.96$.

Suppose that the following data were observed:

$$.53, .54, .48, .50, .50, .49, .52$$

The sample size is $n = 7$ with a sample mean of $\overline{X} = .509$. Therefore

$$Z = |.509 - .5|\sqrt{7}/.02 = 1.191$$

Because this value of Z does not exceed the critical value of 1.96, we do not reject the null hypothesis.

If there is no prior information about σ, use the sample standard deviation, S, and perform a t-test, rejecting H_0 if $|t| > t_{1-\alpha/2}[6]$. In this example, $S = 0.022$, and $t = 1.082$. Because $|t| < t_{.975}[6] = 2.447$, we reach the same conclusion.

C. The Chi-Square Test: Testing the Variance of a Normal Distribution   Consider a one-sided test of the hypothesis:

$$H_0 : \sigma^2 \leq \sigma_0^2$$

against

$$H_1 : \sigma^2 > \sigma_0^2$$

The test statistic corresponding to this hypothesis is:

$$Q^2 = (n-1)S^2/\sigma_0^2 \tag{6.3.8}$$

with a critical region

$$\{Q^2 : Q^2 \geq \chi_{1-\alpha}^2[n-1]\}$$

The operating characteristic function for this test is given by

$$OC(\sigma^2) = \Pr\{\chi^2[n-1] \leq \frac{\sigma_0^2}{\sigma^2}\chi_{1-\alpha}^2[n-1]\} \tag{6.3.9}$$

where $\chi^2[n-1]$ is a chi-square random variable with $n-1$ degrees of freedom.

Continuing the previous example, let us test the hypothesis

$$H_0 : \sigma^2 \leq 0.0004$$

against

$$H_1 : \sigma^2 > 0.0004$$

Because the sample standard deviation is $S = 0.022$, we find

$$Q^2 = (7-1)(0.022)^2/0.0004 = 7.26$$

H_0 is rejected at level $\alpha = .05$ if

$$Q^2 \geq \chi_{.95}^2[6] = 12.59$$

Because $Q^2 < \chi_{.95}^2[6]$, H_0 is not rejected. Whenever n is odd—that is, when $n = 2m + 1$ ($m = 0, 1, \ldots$), the c.d.f. of $\chi^2[n-1]$ can be computed according to the formula:

$$\Pr\{\chi^2[2m] \le x\} = 1 - P\left(m - 1; \frac{x}{2}\right)$$

where $P(a; \lambda)$ is the c.d.f. of the Poisson distribution with mean λ. For example, if $n = 21$, $m = 10$ and $\chi^2_{.95}[20] = 31.41$. Thus, the value of the OC function at $\sigma^2 = 1.5\,\sigma_0^2$ is

$$\text{OC}(1.5\sigma_0^2) = \Pr\left\{\chi^2[20] \le \frac{31.41}{1.5}\right\}$$
$$= 1 - P(9; 10.47) = 1 - .4007$$
$$= .5993$$

If n is even—that is, if $n = 2m$—we can compute the OC values for $n = 2m - 1$ and for $n = 2m + 1$ and take the average of these OC values. This will yield a good approximation.

The power function of the test is obtained by subtracting the OC function from 1.

Table 6.4 gives a few numerical values of the power function for $n = 20$, 30, 40 and for $\alpha = .05$. Here we let $\rho = \sigma^2/\sigma_0^2$ and used the values $\chi^2_{.95}[19] = 30.1$, $\chi^2_{.95}[29] = 42.6$, and $\chi^2_{.95}[39] = 54.6$.

As illustrated in Table 6.4, the power function changes more rapidly as n grows.

TABLE 6.4
Power of the χ^2-test, $\alpha = .05$, $\rho = \sigma^2/\sigma_0^2$

	n		
ρ	20	30	40
1.00	.050	.050	.050
1.25	.193	.236	.279
1.50	.391	.497	.589
1.75	.576	.712	.809
2.00	.719	.848	.920

D. Testing Hypotheses About the Success Probability, p, in Binomial Trials   Consider one-sided tests for which

the null hypothesis is $H_0 : p \le p_0$
the alternative hypothesis is $H_1 : p > p_0$
the critical region is $\{X : X > c_\alpha(n, p_0)\}$

where X is the number of successes among n trials and $c_\alpha(n, p_0)$ is the first value of k for which the binomial c.d.f., $B(k; n, p_0)$, exceeds $1 - \alpha$.

The operating characteristic function is

$$\text{OC}(p) = B(c_\alpha(n, p_0); n, p) \tag{6.3.10}$$

Notice that $c_\alpha(n, p_0) = B^{-1}(1 - \alpha; n, p_0)$ is the $(1 - \alpha)$-quantile of the binomial distribution $B(n, p_0)$. In order to determine $c(n, p_0)$, we can use the MINITAB command that—for $\alpha = .05$, $n = 20$, and $p_0 = .20$, is

```
MTB> INVCDF .95;
SUBC> BINOM 20 .2.
```

Table 6.5 is an output for the binomial distribution with $n = 20$ and $p = .2$.

The smallest value of k for which $B(k; 20, .2) = \Pr\{X \leq k\} \geq .95$ is 7. Thus, we set $c_{.05}(20, .20) = 7$. H_0 is rejected whenever $X > 7$. The level of significance of this test is actually .032, which is due to the discrete nature of the binomial distribution. The OC function of the test for $n = 20$ can be easily determined from the corresponding distribution of $B(20, p)$. For example, the $B(n, p)$ distribution for $n = 20$ and $p = .25$ is presented in Table 6.6.

We see that $B(7; 20, .25) = .8982$. Hence, the probability of accepting H_0 when $p = .25$ is $OC(.25) = .8982$.

A large-sample test in the binomial case can be based on the normal approximation to the binomial distribution. If the sample is indeed large, we can use the test statistic

$$Z = \frac{\hat{p} - p_0}{\sqrt{p_0 q_0}} \sqrt{n} \qquad\qquad (6.3.11)$$

TABLE   6.5
p.d.f. and c.d.f. of $B(20, .2)$

Binomial distribution: $n = 20, p = .2$

k	$\Pr(X = k)$	$\Pr(X \leq k)$
0	.0115	.0115
1	.0576	.0692
2	.1369	.2061
3	.2054	.4114
4	.2182	.6296
5	.1746	.8042
6	.1091	.9133
7	.0545	.9679
8	.0222	.9900
9	.0074	.9974
10	.0020	.9994
11	.0005	.9999
12	.0001	1.0000

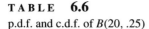

TABLE 6.6
p.d.f. and c.d.f. of $B(20, .25)$

Binomial distribution: $n = 20, p = .25$		
k	$\Pr(X = k)$	$\Pr(X \leq k)$
0	.0032	.0032
1	.0211	.0243
2	.0669	.0913
3	.1339	.2252
4	.1897	.4148
5	.2023	.6172
6	.1686	.7858
7	.1124	.8982
8	.0609	.9591
9	.0271	.9861
10	.0099	.9961
11	.0030	.9991
12	.0003	.9998
13	.0002	1.0000

with the critical region

$$\{Z : Z \geq z_{1-\alpha}\}$$

where $q_0 = 1 - p_0$. Here $\hat{p}$ is the sample proportion of successes. The operating characteristic function takes the form:

$$OC(p) = 1 - \Phi\left(\frac{(p - p_0)\sqrt{n}}{\sqrt{pq}} - z_{1-\alpha}\sqrt{\frac{p_0 q_0}{pq}}\right) \tag{6.3.12}$$

where $q = 1 - p$ and $q_0 = 1 - p_0$.

For example, suppose that $n = 450$ and the hypotheses are $H_0 : p \leq .1$ against $H_1 : p > .1$. The critical region, for $\alpha = .05$, is

$$\{\hat{p} : \hat{p} \geq .10 + 1.645\sqrt{(.1)(.9)/450}\} = \{\hat{p} : \hat{p} \geq .1233\}$$

Thus, H_0 is rejected whenever $\hat{p} \geq .1233$. The OC value of this test, at $p = .15$, is approximately

$$OC(.15) \cong 1 - \Phi\left(\frac{.05\sqrt{450}}{\sqrt{(.15)(.85)}} - 1.645\sqrt{\frac{(.1)(.9)}{(.15)(.85)}}\right)$$

$$= 1 - \Phi(2.970 - 1.382)$$

$$= 1 - .944 = .056$$

The corresponding value of the power function is .944. Notice that the power of rejecting H_0 for H_1 when $p = .15$ is so high because of the large sample size. ∎

6.4

Confidence Intervals

Confidence intervals for unknown parameters are intervals determined around the sample estimates of the parameters. They have the property that—whatever the true value of the parameter—in repetitive sampling, a prescribed proportion of the intervals (say, $1 - \alpha$) will contain the true value of the parameter. The prescribed proportion, $1 - \alpha$, is called the **confidence level** of the interval. In Figure 6.5, we illustrate 50 simulated confidence intervals, which correspond to independent samples. All of these intervals are designed to estimate the mean of the population from which the samples were drawn. In this particular simulation, the population was normally distributed with mean $\mu = 10$. We see from the figure that all of these 50 random intervals cover the true value of μ.

FIGURE   6.5

Simulated confidence intervals for the mean of a normal distribution, samples of size $n = 10$ from $N(10, 1)$

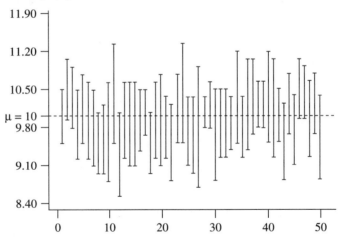

If the sampling distribution of the estimator, $\hat{\theta}_n$, is approximately normal, one can use, as a rule of thumb, the interval estimator with limits

$$\hat{\theta}_n \pm 2 \, SE\{\hat{\theta}_n\}$$

The confidence level of such an interval will be close to .95 for all θ.

Generally, if we have a powerful test procedure for testing the hypothesis $H_0 : \theta = \theta_0$ versus $H_1 : \theta \neq \theta_0$, we can obtain good confidence intervals for θ by the following method.

Let $T = T(\mathbf{X})$ be a test statistic for testing $H_0 : \theta = \theta_0$. Suppose that H_0 is rejected if $T \geq \overline{K}_\alpha(\theta_0)$ or if $T \leq \underline{K}_\alpha(\theta_0)$, where α is the significance level. The interval $(\underline{K}_\alpha(\theta_0), \overline{K}_\alpha(\theta_0))$ is the acceptance region for H_0. We can now consider the family of acceptance regions $\mathcal{A} = \{(\underline{K}_\alpha(\theta), \overline{K}_\alpha(\theta)), \; \theta \in \Theta\}$, where Θ is the parameter space. The interval $(L_\alpha(T), U_\alpha(T))$, defined as

$$L_\alpha(T) = \inf\{\theta : T \leq \overline{K}_\alpha(\theta)\}$$
$$U_\alpha(T) = \sup\{\theta : T \geq \underline{K}_\alpha(\theta)\}$$

(6.4.1)

is a confidence interval for θ at level of confidence $1 - \alpha$. Indeed, the hypothesis $H_0 : \alpha = \alpha_0$ with $L_\alpha(T) < \theta_0 < U_\alpha(T)$ is accepted with the observed value of the test statistic, and the probability of accepting such hypothesis is $1 - \alpha$. That is, if θ_0 is the true value of θ, the probability that H_0 is accepted is $1 - \alpha$. But H_0 is accepted if, and only if, θ_0 is covered by the interval $(L_\alpha(T), U_\alpha(T))$.

6.4.1 Confidence Intervals for μ; σ Known

When σ is known, the sample mean $\overline{X}$ is used as an estimator of μ or as a test statistic for the hypothesis $H_0 : \mu = \mu_0$. H_0 is rejected at level of significance α if $\overline{X} \geq \mu_0 + z_{1-\alpha/2}\sigma/\sqrt{n}$ or $\overline{X} \leq \mu_0 + z_{1-\alpha/2}\sigma/\sqrt{n}$, where $z_{1-\alpha/2} = \Phi^{-1}(1 - \alpha/2)$. Thus, $\overline{K}_\alpha(\mu) = \mu + z_{1-\alpha/2}\sigma/\sqrt{n}$ and $\underline{K}_\alpha(\mu) = \mu - z_{1-\alpha/2}\sigma/\sqrt{n}$. The limits of the confidence interval are, accordingly, the roots μ of the equations

$$\overline{K}_\alpha(\mu) = \overline{X}$$

and

$$\underline{K}_\alpha(\mu) = \overline{X}$$

These equations yield the confidence interval for μ:

$$\left(\overline{X} - z_{1-\alpha/2}\frac{\sigma}{\sqrt{n}}, \overline{X} + z_{1-\alpha/2}\frac{\sigma}{\sqrt{n}}\right)$$

(6.4.2)

6.4.2 Confidence Intervals for μ; σ Unknown

A confidence interval for μ, at level $1 - \alpha$, when σ is unknown is obtained from the corresponding t-test. The confidence interval is

$$\left(\overline{X} - t_{1-\alpha/2}[n-1]\frac{S}{\sqrt{n}}, \overline{X} + t_{1-\alpha/2}[n-1]\frac{S}{\sqrt{n}}\right)$$

(6.4.3)

where $\overline{X}$ and S are the sample mean and standard deviation, respectively. $t_{1-\alpha/2}[n-1]$ is the $(1 - \alpha/2)$th quantile of the t distribution with $n - 1$ degrees of freedom.

6.4.3 Confidence Intervals for σ^2

We have seen that, in the normal case, the hypothesis $H_0 : \sigma = \sigma_0$, is rejected at level of significance α if

$$S^2 \geq \frac{\sigma_0^2}{n-1}\chi^2_{1-\alpha/2}[n-1]$$

or

$$S^2 \leq \frac{\sigma_0^2}{n-1} \chi_{\alpha/2}^2[n-1]$$

where S^2 is the sample variance and $\chi_{\alpha/2}^2[n-1]$ and $\chi_{1-\alpha/2}^2[n-1]$ are the $\alpha/2$-th and $(1-\alpha/2)$th quantiles of χ^2, with $n-1$ degrees of freedom. The corresponding confidence interval for σ^2, at confidence level $1-\alpha$, is

$$\left(\frac{(n-1)S^2}{\chi_{1-\alpha/2}^2[n-1]}, \frac{(n-1)S^2}{\chi_{\alpha/2}^2[n-1]} \right) \tag{6.4.4}$$

EXAMPLE 6.7   Consider a normal distribution with unknown mean μ and unknown standard deviation σ. Suppose we draw a random sample of size $n = 16$ from this population, and the sample values are:

16.16	9.33	12.96	11.49
12.31	8.93	6.02	10.66
7.75	15.55	3.58	11.34
11.38	6.53	9.75	9.47

The mean and variance of this sample are: $\overline{X} = 10.20$ and $S^2 = 10.977$. The sample standard deviation is $S = 3.313$. For a confidence level of $1 - \alpha = 0.95$ we find

$$t_{.975}[15] = 2.131$$
$$\chi_{.975}^2[15] = 27.50$$
$$\chi_{.025}^2[15] = 6.26$$

Thus, the confidence interval for μ is $(8.435, 11.965)$. The confidence interval for σ^2 is $(5.987, 26.303)$.   ∎

6.4.4 Confidence Intervals for p

Let X be the number of successes in n independent trials, with unknown probability of successes, p. The sample proportion, $\hat{p} = X/n$, is an unbiased estimator of p. To construct a confidence interval for p using $\hat{p}$, we must find limits, $p_L(\hat{p})$ and $p_U(\hat{p})$, that satisfy

$$\Pr\{p_L(\hat{p}) < p < p_U(\hat{p})\} = 1 - \alpha$$

The null hypothesis $H_0 : p = p_0$ is rejected if $\hat{p} \geq \overline{K}_\alpha(p_0)$ or $\hat{p} \leq \underline{K}_\alpha(p_0)$, where

$$\overline{K}_\alpha(p_0) = \frac{1}{n} B^{-1}(1 - \alpha/2; n, p_0)$$

and

$$\underline{K}_\alpha(p_0) = \frac{1}{n} B^{-1}(\alpha/2; n, p_0) \tag{6.4.5}$$

$B^{-1}(\gamma; n, p)$ is the γth quantile of the binomial distribution $B(n, p)$. Thus, if $X = n\hat{p}$, the upper confidence limit for p, $p_U(\hat{p})$, is the largest value of p satisfying the inequality

$$B(X; n, p) \geq \alpha/2$$

The lower confidence limit for p is the smallest value of p satisfying the inequality

$$B(X; n, p) \leq 1 - \alpha/2$$

From the relationship between the F distribution, the beta distribution, and the binomial distribution, the lower and upper limits are given by the formulas:

$$p_L = \frac{X}{X + (n - X + 1)F_1} \tag{6.4.6}$$

and

$$p_U = \frac{(X + 1)F_2}{n - X + (X + 1)F_2} \tag{6.4.7}$$

where

$$F_1 = F_{1-\alpha/2}[2(n - X + 1), 2X] \tag{6.4.8}$$

and

$$F_2 = F_{1-\alpha/2}[2(X + 1), 2(n - X)] \tag{6.4.9}$$

are the $(1 - \alpha/2)$th quantiles of the F distribution with the indicated degrees of freedom.

EXAMPLE **6.8**   Suppose that among $n = 30$ Bernoulli trials, we find $X = 8$ successes. For level of confidence $1 - \alpha = .95$, the confidence limits are $p_L = .123$ and $p_U = .459$. Indeed,

$$B(7; 30, .123) = .975$$

and

$$B(8; 30, .459) = .025$$

Moreover,

$$F_1 = F_{.975}[46, 16] = 2.49$$

and

$$F_2 = F_{.975}[18, 44] = 2.07$$

Hence,

$$p_L = 8/(8 + 23(2.49)) = .123$$

and

$$p_U = 9(2.07)/(22 + 9(2.07)) = .459 \quad \blacksquare$$

When the sample size n is large, we may use the normal approximation to the binomial distribution. This approximation yields the following formula for a $1 - \alpha$ confidence interval:

$$\left(\hat{p} - z_{1-\alpha/2}\sqrt{\hat{p}\hat{q}/n}, \hat{p} + z_{1-\alpha/2}\sqrt{\hat{p}\hat{q}/n}\right) \tag{6.4.10}$$

where $\hat{q} = 1 - \hat{p}$. Applying this large-sample approximation to our previous example, in which $n = 30$, we obtain the approximate .95 confidence interval (.108, .425). This interval is slightly different from the interval obtained with the exact formulas. The difference is due to the inaccuracy of the normal approximation.

It is sometimes reasonable to use only a one-sided confidence interval—for example, when $\hat{p}$ is the estimated proportion of nonconforming items in a population. Obviously, the true value of p is always greater than 0, and we may wish to determine only an upper confidence limit. In this case, we apply the formula given earlier, but replace $\alpha/2$ by α. For example, in the case of $n = 30$ and $X = 8$, the upper confidence limit for p, in a one-sided confidence interval, is

$$p_U = \frac{(X+1)F_2}{n - X + (X+1)F_2}$$

where $F_2 = F_{1-\alpha}[2(X+1), 2(n-X)] = F_{.95}[18, 44] = 1.855$. Thus, the upper confidence limit of a .95 one-sided interval is $p_U = .431$. This limit is smaller than the upper limit of the two-sided interval.

6.5
Tolerance Intervals

Technological specifications for a given characteristic X may require that a specified proportion of elements of a statistical population satisfy certain constraints. For example, in the production of concrete, we may have the requirement that at least 90% of all concrete cubes of a certain size have a compressive strength of at least 240 kg/cm^2. As another example, suppose that, in the production of washers, it is required that at least 99% of the washers produced have a thickness between 0.121 and 0.129 inch. In both examples we want to be able to determine whether or not the requirements are satisfied. If the distributions of strength and thickness are completely known, we can determine, without data, if the requirements are met. However, if the distributions are not completely known, we can make these determinations only with a certain level of confidence and not with certainty.

6.5.1   Tolerance Intervals for the Normal Distributions

In order to construct confidence intervals, we first consider what happens when the distribution of the characteristic X is completely known. Suppose, for example, that the compressive strength X of the concrete cubes is such that $Y = \ln X$ has a normal distribution with mean $\mu = 5.75$ and standard deviation $\sigma = 0.2$. The proportion of concrete cubes exceeding the specification of 240 kg/cm^2 is

$$
\begin{aligned}
\Pr\{X \geq 240\} &= \Pr\{Y \geq \log 240\} \\
&= 1 - \Phi((5.481 - 5.75)/.2) \\
&= \Phi(1.345) = .911
\end{aligned}
$$

Because this probability is greater than the specified proportion of 90%, the requirement is satisfied.

We can also solve this problem by determining the compressive strength that is exceeded by 90% of the concrete cubes. Because 90% of the Y values are greater than the .1th fractile of the $N(5.75, .04)$ distribution,

$$
\begin{aligned}
Y_{0.1} &= \mu + z_{0.1}\sigma \\
&= 5.75 - 1.28(0.2) \\
&= 5.494
\end{aligned}
$$

Accordingly, 90% of the compressive strength values should exceed $e^{5.494} = 243.2$ kg/cm^2. Once again we see that the requirement is satisfied, more than 90% of the cubes have strength values that exceed the specification of 240 kg/cm^2. Notice that no sample values are required because the distribution of X is known. Furthermore, we are *certain* that the requirement is met.

Consider the situation in which we have only partial information on the distribution of Y. Suppose we know that Y is normally distributed with standard deviation $\sigma = 0.2$, but the mean μ is unknown. The .1th fractile of the distribution, $y_{0.1} = \mu + z_{0.1}\sigma$, cannot be determined exactly. Let $Y_1, \ldots, Y_n$ be a random sample from this distribution and let $\overline{Y}_n$ represent the sample mean. From the previous section we know that

$$
L(\overline{Y}_n) = \overline{Y}_n - z_{1-\alpha}\sigma/\sqrt{n}
$$

is a $1 - \alpha$ lower confidence limit for the population mean. That is,

$$
\Pr\{\overline{Y}_n - z_{1-\alpha}\sigma/\sqrt{n} < \mu\} = 1 - \alpha
$$

Substituting this lower bound for μ in the expression for the .1th fractile, we obtain a *lower tolerance limit* for 90% of the log-compressive strengths, with confidence level $1 - \alpha$. More specifically, the lower tolerance limit at level of confidence $1 - \alpha$ is

$$
L_{\alpha,.1}(\overline{Y}_n) = \overline{Y}_n - (z_{1-\alpha}/\sqrt{n} + z_{.9})\sigma
$$

In general, we say that, with confidence level of $1 - \alpha$, the proportion of population values exceeding the lower tolerance limit is at least $1 - \beta$. This lower tolerance limit is

$$
L_{\alpha,\beta}(\overline{Y}_n) = \overline{Y}_n - (z_{1-\alpha}/\sqrt{n} + z_{1-\beta})\sigma \tag{6.5.1}
$$

It can also be shown that the **upper tolerance limit** for a proportion $1 - \beta$ of the values, with confidence level $1 - \alpha$, is

$$
U_{\alpha,\beta}(\overline{Y}_n) = \overline{Y}_n + (z_{1-\alpha}/\sqrt{n} + z_{1-\beta})\sigma \tag{6.5.2}
$$

and a **tolerance interval** containing a proportion $1 - \beta$ of the values, with confidence $1 - \alpha$, is

$$(\bar{Y}_n - (z_{1-\alpha/2}/\sqrt{n} + z_{1-\beta/2})\sigma, \bar{Y}_n + (z_{1-\alpha/2}/\sqrt{n} + z_{1-\beta/2})\sigma)$$

When the standard deviation σ is unknown, we should use the sample standard deviation S to construct the tolerance limits and interval. The lower tolerance limits will be of the form $\bar{Y}_n - kS_n$, where the factor $k = k(\alpha, \beta, n)$ is determined so that, with confidence level $1 - \alpha$, we can state that a proportion $1 - \beta$ of the population values will exceed this limit. The corresponding upper limit is given by $\bar{Y}_n + kS_n$ and the tolerance interval is given by

$$(\bar{Y}_n - k'S_n, \bar{Y}_n + k'S_n)$$

The "two-sided" factor $k' = k'(\alpha, \beta, n)$ is determined so that the interval will contain a proportion $1 - \beta$ of the population with confidence $1 - \alpha$. Approximate solutions, for large values of n, are given by

$$k(\alpha, \beta, n) \doteq t(\alpha, \beta, n) \tag{6.5.3}$$

and

$$k'(\alpha, \beta, n) \doteq t(\alpha/2, \beta/2, n) \tag{6.5.4}$$

where

$$t(\alpha, \beta, n) = \frac{z_{1-\beta}}{1 - z_{1-\alpha}^2/2n} + \frac{z_{1-\alpha}(1 + z_\beta^2/2 - z_{1-\alpha}^2/2n)^{1/2}}{\sqrt{n}(1 - z_{1-\alpha}^2/2n)} \tag{6.5.5}$$

EXAMPLE **6.9**   The following data represent a sample of 20 compressive strength measurements (kg/cm^2) of concrete cubes at 7 days of age:

349.09	308.88
238.45	196.20
385.59	318.99
330.00	257.63
388.63	299.04
348.43	321.47
339.85	297.10
348.20	218.23
361.45	286.23
357.33	316.69

Applying the transformation $Y = \ln X$, we find that $\bar{Y}_{20} = 5.732$ and $S_{20} = 0.184$. To obtain a lower tolerance limit for 90% of the log-compressive strengths with 95% confidence, we use the factor $k(0.05, 0.10, 20) = 1.901$. Thus, the lower tolerance limit for the transformed data is

$$\bar{Y}_{20} - kS_{20} = 5.732 - 1.901 \times 0.184 = 5.382$$

and the corresponding lower tolerance limit for the compressive strength is

$$e^{5.382} = 217.50 \ (kg/cm^2) \quad \blacksquare$$

If the tolerance limits are within the specification range we have a satisfactory production.

6.5.2 Distribution-Free Tolerance Intervals

The tolerance limits just described are based on the model of normal distribution. *Distribution-free tolerance limits* for a $1 - \beta$ proportion of the population, at confidence level $1 - \alpha$, can be obtained for any model of continuous c.d.f. $F(x)$. As we will show, if the sample size n is large enough, so that the following inequality is satisfied—that is,

$$\left(1 - \frac{\beta}{2}\right)^n - \frac{1}{2}(1 - \beta)^n \leq \frac{\alpha}{2} \tag{6.5.6}$$

then the order statistics $X_{(1)}$ and $X_{(n)}$ are lower and upper tolerance limits. This is based on the following important property:

> If X is a random variable having a continuous c.d.f. $F(x)$,
> then $U = F(x)$ has a uniform distribution on $(0, 1)$.

Indeed

$$\Pr\{F(X) \leq \eta\} = \Pr\{X \leq F^{-1}(\eta)\}$$
$$= F(F^{-1}(\eta)) = \eta \qquad 0 < \eta < 1$$

If $X_{(i)}$ is the ith order statistic of a sample of n i.i.d. random variables having a common c.d.f. $F(x)$, then $U_{(i)} = F(X_{(i)})$ is the ith order statistic of n i.i.d. random variables having a uniform distribution.

Now, the interval $(X_{(1)}, X_{(n)})$ contains at least a proportion $1 - \beta$ of the population if $X_{(1)} \leq \xi_{\beta/2}$ and $X_{(n)} \geq \xi_{1-\beta/2}$, where $\xi_{\beta/2}$ and $\xi_{1-\beta/2}$ are the $\beta/2$th and $(1 - \beta/2)$th quantiles of $F(x)$.

Equivalently, $(X_{(1)}, X_{(n)})$ contains at least a proportion $1 - \beta$ if

$$U_{(1)} \leq F(\xi_{\beta/2}) = \frac{\beta}{2}$$
$$U_{(n)} \geq F(\xi_{1-\beta/2}) = 1 - \beta/2$$

By using the joint p.d.f. of $(U_{(1)}, U_{(n)})$, we show that

$$\Pr\left\{U_{(1)} \leq \frac{\beta}{2}, U_{(n)} \geq 1 - \frac{\beta}{2}\right\} = 1 - 2\left(1 - \frac{\beta}{2}\right)^n + (1 - \beta)^n \tag{6.5.7}$$

This probability is the confidence that the interval $(X_{(1)}, X_{(n)})$ covers the interval $(\xi_{\beta/2}, \xi_{1-\beta/2})$. By finding n, which satisfies

$$1 - 2\left(1 - \frac{\beta}{2}\right)^n + (1 - \beta)^n \geq 1 - \alpha \tag{6.5.8}$$

we can ensure that the confidence level is at least $1 - \alpha$.

In Table 6.7, we give the values of n for some α and β values.

Table 6.7 can also be used to obtain the confidence level associated with fixed values of β and n. We see, for example, that with a sample of size 104, $(X_{(1)}, X_{(n)})$

TABLE   6.7

Sample size required for $(X_{(1)}, X_{(n)})$ to be a $(1 - \alpha, 1 - \beta)$-level tolerance interval

β	α	n
0.10	0.10	58
	0.05	72
	0.01	104
0.05	0.10	118
	0.05	146
	0.01	210
0.01	0.10	593
	0.05	734
	0.01	1057

is a tolerance interval for at least 90% of the population, with approximately 99% confidence level.

Other order statistics can be used to construct distribution-free tolerance intervals. That is, we can choose any integers j and k, where $1 \le j, k \le n/2$, and form the interval $(X_{(j)}, X_{(n-k+1)})$. When $j > 1$ and $k > 1$, the interval will be shorter than the interval $(X_{(1)}, X_{(n)})$, but its confidence level will be reduced.

6.6
Testing for Normality with Probability Plots

It is often assumed that a sample is drawn from a population that has a normal distribution. It is, however, important to test the assumption of normality. We present here a simple test based on the **normal scores** (NSCORES) of the sample values. The normal scores corresponding to a sample $x_1, x_2, \ldots, x_n$ are obtained in the following manner. First we let

$$r_i = \text{rank of } x_i \qquad i = 1, \ldots, n \qquad \text{(6.6.1)}$$

Here the rank of x_i is the position of x_i in a listing of the sample arranged in increasing order. Thus, the rank of the smallest value is 1, that of the second smallest is 2, and so on. We then let

$$p_i = (r_i - 3/8)/(n + 1/4) \qquad i = 1, \ldots, n \qquad \text{(6.6.2)}$$

Then the normal score of x_i is

$$z_i = \Phi^{-1}(p_i)$$

that is, the p_ith fractile of the standard normal distribution. If the sample is drawn at random from a normal distribution $N(\mu, \sigma^2)$, the relationship between the normal scores, NSCORES, and x_i should be approximately linear. Accordingly, the correlation between $x_1, \ldots, x_n$ and their NSCORES should be close to 1 in large samples.

The graphical display of the sample values versus their NSCORES is called a *normal QQ plot*.

In the following example, we provide a **normal probability plot** of $n = 50$ values simulated from $N(10, 1)$. If the simulation is good, and the sample is indeed generated from $N(10, 1)$, the X vs. NSCORES should be scattered randomly around the line $X = 10 + \text{NSCORES}$. We see in Figure 6.6 that this is indeed the case. Also, the correlation between the x values and their NSCORES is .976.

The linear regression of the x values on the NSCORES, is:

$$X = 10.043 + 0.953 * \text{NSCORES}$$

We see that both the intercept and slope of the regression equation are close to the nominal values of μ and σ.

Table 6.8 lists some critical values for testing whether the correlation between the sample values and their NSCORES is sufficiently close to 1. If the correlation is smaller than the critical value, an indication of nonnormality has been established. In the example of Figure 6.6, the correlation is $R = .976$. This value is almost equal to the critical value for $\alpha = .05$ given in Table 6.8. The hypothesis of normality is not rejected.

MINITAB provides a graph of the normal probability plot. In this graph, the probabilities corresponding to the NSCORES are plotted against x. The normal probability plot of the preceding example is given in Figure 6.7. A normal probability plot is obtained in MINITAB by the command

MTB> % NormPlot C1.

F I G U R E 6.6

Normal *QQ* plot of simulated values from $N(10, 1)$

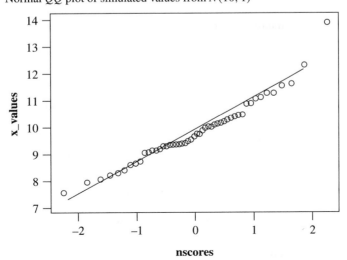

nscores

TABLE   6.8

Critical values for the correlation between sample values and their NSCORES

	α		
n	**0.10**	**0.05**	**0.01**
10	0.9347	0.9180	0.8804
15	0.9506	0.9383	0.9110
20	0.9600	0.9503	0.9290
30	0.9707	0.9639	0.9490
50	0.9807	0.9764	0.9664

Source: From Ryan, Joiner, and Ryan (——,p. 49).

FIGURE   6.7

Normal probability plot of 50 simulated $N(10, 1)$ values

Normal Probability Plot

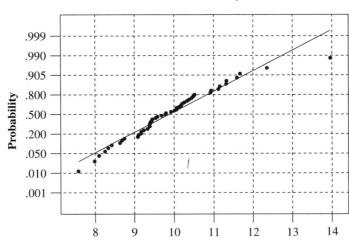

Average: 9.85749
Std Dev: 1.18390
N of data: 50

x_values

Anderson-Darling Normality Test
A-Squared = 0.471
p value: 0.235

To demonstrate the relationship between the sample values and their normal scores, when the sample is drawn from a nonnormal distribution, consider the following two examples.

EXAMPLE 6.10   Consider a sample of $n = 100$ observations from a log-normal distribution. The normal *QQ* plot of this sample is shown in Figure 6.8. The correlation here is only .788. It is apparent that the relation between the NSCORES and the sample values is not linear. We reject the hypothesis that the sample has been generated from a normal distribution.   ∎

FIGURE 6.8

Normal probability plot $n = 100$ random numbers generated from a log-normal distribution

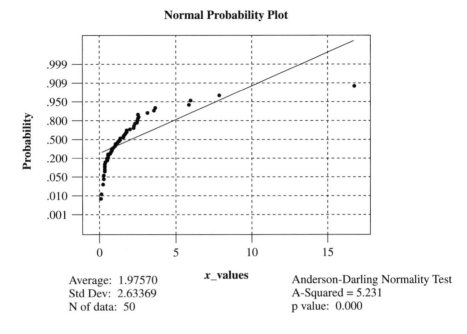

Normal Probability Plot

Average: 1.97570
Std Dev: 2.63369
N of data: 50

Anderson-Darling Normality Test
A-Squared = 5.231
p value: 0.000

EXAMPLE **6.11**   We consider here a sample of $n = 100$ values, with 50 of the values generated from $N(10, 1)$ and 50 from $N(15, 1)$. Thus, the sample represents a mixture of two normal distributions. The histogram is given in Figure 6.9 and a normal probability plot

FIGURE 6.9

Histogram of 100 random numbers, 50 generated from a $N(10, 1)$ and 50 from $N(15, 1)$

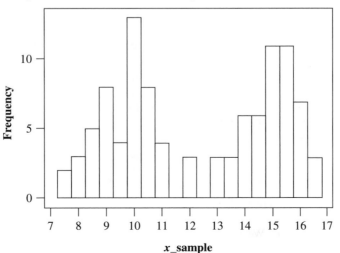

in Figure 6.10. The normal probability plot is definitely not linear. Although the correlation is .962 the hypothesis of a normal distribution is rejected.   ▪

FIGURE   6.10
Normal probability plot of 100 random numbers generated from a mixture of two normal distributions

Normal Probability Plot

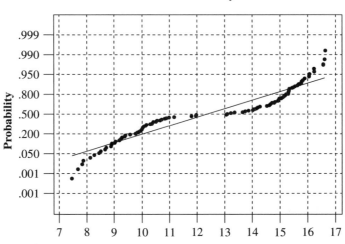

Average: 12.3249
Std Dev: 2.82919
N of data: 100

Anderson-Darling Normality Test
A-Squared = 4.230
p value: 0.000

6.7
Tests of Goodness of Fit
6.7.1   The Chi-Square Test (Large Samples)

The chi-square test is applied by comparing the observed frequency distribution of the sample to the expected one under the assumption of the model. More specifically, consider a (large) sample of size N. Let $\xi_0 < \xi_1 < \ldots < \xi_k$ be the limit points of k subintervals of the frequency distribution, and let f_i be the observed frequency in the ith subinterval. If, according to the model, the c.d.f. is specified by the distribution function $F(x)$, then the expected frequency E_i in the ith subinterval is

$$E_i = N(F(\xi_i) - F(\xi_{i-1})) \qquad i = 1, \ldots, k$$

The chi-square statistic is defined as

$$\chi^2 = \sum_{i=1}^{k} \frac{(f_i - E_i)^2}{E_i}$$

We notice that

$$\sum_{i=1}^{k} f_i = \sum_{i=1}^{k} E_i = N$$

and hence

$$\chi^2 = \sum_{i=1}^{k} \frac{f_i^2}{E_i} - N$$

The value of χ^2 is distributed approximately like $\chi^2[k-1]$. Thus, if $\chi^2 \geq \chi^2_{1-\alpha}[k-1]$, then the distribution $F(x)$ does not fit the observed data.

Often, the c.d.f. $F(x)$ is specified by its family—for example, normal or Poisson—but the values of the parameters have to be estimated from the sample. In this case, we reduce the number of degrees of freedom of χ^2 by the number of estimated parameters. For example, if $F(x)$ is $N(\mu, \sigma^2)$, where both μ and σ^2 are unknown, we use $N(\overline{X}, S^2)$ and compare χ^2 to $\chi^2_{1-\alpha}[k-3]$.

EXAMPLE 6.12 In Section 6.1, we considered the sampling distribution of sample means from the uniform distribution over the integers $\{1, \ldots, 100\}$. The frequency distribution of the means of samples of size $n = 10$ was given in Figure 6.1. We test here whether the model $N(50.5, 833.25)$ fits these data.

The observed and expected frequencies (for $N = 100$) are summarized in Table 6.9.

TABLE 6.9

Observed and expected frequencies of 100 sample means

Interval	f_i	E_i
27.5–32.5	3	1.84
32.5–37.5	11	5.28
37.5–42.5	12	11.32
42.5–47.5	11	18.08
47.4–52.5	19	21.55
52.5–57.5	24	19.7
57.5–62.5	14	12.73
62.5–67.5	4	6.30
67.5–72.5	2	2.33
TOTAL	100	99.13

Due to truncation of the tails of the normal distribution the sum of e_i here is 99.13. The value of χ^2 is 12.86. The value of $\chi^2_{.95}[8]$ is 15.5. Thus, deviation of the observed frequency distribution from the expected one is not significant at the $\alpha = .05$ level. ∎

EXAMPLE **6.13**   We consider here a sample of 100 cycle times of a piston, which is described in detail in Chapter 10. We make a chi-square test to determine whether the distribution of cycle times is normal. The estimated values of μ and σ are

$$\hat{\mu} = 0.1219 \quad \text{and} \quad \hat{\sigma} = 0.0109$$

Table 6.10 gives the observed and expected frequencies over $k = 8$ intervals.

The calculated value of χ^2 is 5.4036. We should consider the distribution of χ^2 with $k - 3 = 5$ degrees of freedom. The *P*-value of the test is .37. The hypothesis of normality is not rejected.   ■

T A B L E   6.10
Observed and expected frequencies of 100 cycle times

Lower limit	Upper limit	Observed frequency	Expected frequency
at or below	.1050	7	6.1
.1050	.1100	9	7.7
.1100	.1150	17	12.6
.1150	.1200	12	16.8
.1200	.1250	18	18.1
.1250	.1300	11	15.9
.1300	.1350	12	11.4
above .1350		14	11.4

6.7.2 The Kolmogorov-Smirnov Test

The **Kolmogorov-Smirnov (KS) test** is a more accurate test of goodness of fit than the chi-square test of the previous section.

Suppose the hypothesis is that the sample comes from a specified distribution with c.d.f. $F_0(x)$. The test statistic compares the empirical distribution of the sample, $\hat{F}_n(x)$, to $F_0(x)$ and considers the maximal value over all x values that the distance $|\hat{F}_n(x) - F_0(x)|$ may assume. Let $x_{(1)} \leq x_{(2)} \leq \ldots \leq x_{(n)}$ be the ordered sample values. Notice that $\hat{F}_n(x_{(i)}) = i/n$. The KS test statistic can be computed according to the formula

$$D_n = \max_{1 \leq i \leq n} \left\{ \max \left[\frac{i}{n} - F_0(x_{(i)}), F_0(x_{(i)}) - \frac{i-1}{n} \right] \right\} \tag{6.7.1}$$

We showed earlier that $U = F(X)$ has a uniform distribution on $(0, 1)$.

Accordingly, if the null hypothesis is correct, $F_0(X_{(i)})$ is distributed like the *i*th order statistic $U_{(i)}$ from a uniform distribution on $(0, 1)$, irrespective of the particular functional form of $F_0(x)$. The distribution of the KS test statistic, D_n, is therefore independent of $F_0(x)$, if the hypothesis H is correct. Tables of the critical values k_α for D_n are available. One can also estimate the value of k_α by simulation.

If $F_0(x)$ is a normal distribution—that is, if $F_0(x) = \Phi\left[\dfrac{x-\mu}{\sigma}\right]$—and if the mean μ and the standard deviation σ are unknown, we can consider the test statistic

$$D_n^* = \max_{1 \le i \le n} \left\{ \max \left\{ \frac{i}{n} - \Phi\left(\frac{X_{(i)} - \overline{X}_n}{S_n}\right), \ \Phi\left(\frac{X_{(i)} - \overline{X}_n}{S_n}\right) - \frac{i-1}{n} \right\} \right\} \quad \textbf{(6.7.2)}$$

where $\overline{X}_n$ and S_n are substituted for the unknown μ and σ. The critical values k_α^* for D_n^* are given approximately by

$$k_\alpha^* = \delta_\alpha^* / \left(\sqrt{n} - 0.01 + \frac{0.85}{\sqrt{n}} \right) \quad \textbf{(6.7.3)}$$

and δ_α^* is given in Table 6.11.

The following is a MINITAB program that computes the value of D_n^* for a sample of size $n = 100$, which is stored in C1. The sample data are the piston cycle times, which were discussed in Example 6.13 (file name OTURB.DAT).

```
MTB > Read 'C:\ISTAT\DATA\OTURB.DAT' C1.
MTB > sort C1 C2
MTB > LET C10 = C1/C1
MTB > LET k1=mean(C1)
MTB > LET k2=stan(C1)
MTB > LET C3=(C2-k1*C10)/k2*C10
MTB > CDF C3 C4;
SUBC>  Normal 0.0 1.0.
MTB > SET C5
DATA>  1( .01 : 1.00 / 0.01 )1
DATA> End.
MTB > LET C6=C5-.01
MTB > LET C7=ABSO(C4-C5)
MTB > LET C8=ABSO(C4-C6)
MTB > RMAX C7 C8 C9
MTB > Let k3=MAXI(C9)
MTB > PRINT k3
```

We obtain $D_n^* = 0.1107$. According to Table 6.11, the critical value for $\alpha = .05$ is $k_{.05}^* = 0.895/(10 - 0.01 + 0.085) = 0.089$. Thus, the hypothesis of normality for the piston cycle time data is rejected at $\alpha = .05$.

T A B L E   6.11
Some critical values δ_α^*

α	.15	.10	.05	.025	.01
δ_α^*	0.775	0.819	0.895	0.995	1.035

6.8

Bayesian Decision Procedures

It is often the case that the optimal decision depends on unknown parameters of statistical distributions. The **Bayesian decision framework** provides us with the tools to integrate information we may have on the unknown parameters with information we obtain from the observed sample in such a way that the expected loss due to erroneous decisions is minimized. The following example illustrates an industrial decision problem of this nature.

EXAMPLE **6.14**   The simplest inventory problem faced daily by organizations of all sizes worldwide concerns supply and demand. One such organization is Starbread Express, which supplies bread to a large community in the Midwest. Every night, the shift manager has to decide how many loaves of bread, s, to bake for the next day's consumption. Let X (a random variable) be the number of units demanded during the day. If a manufactured unit is left at the end of the day, we lose \$ c_1 on that unit. However, if a unit is demanded and is not available due to shortage, the loss is \$ c_2. How many units, s, should be manufactured so the total expected loss due to overproduction or to shortages will be minimized?

The loss at the end of the day is

$$L(s, X) = c_1(s - X)^+ + c_2(X - s)^+ \tag{6.8.1}$$

where $a^+ = \max(a, 0)$. The loss function $L(s, X)$ is a random variable. If the p.d.f. of X is $f(x), x = 0, 1, \ldots$, then the expected loss, a function of the quantity s, is

$$
R(s) = c_1 \sum_{x=0}^{s} f(x)(s - x) + c_2 \sum_{x=s+1}^{\infty} f(x)(x - s)
$$

$$
= c_2 E\{X\} - (c_1 + c_2) \sum_{x=0}^{s} xf(x) \tag{6.8.2}
$$

$$
+ s(c_1 + c_2)F(s) - c_2 s
$$

where $F(s)$ is the c.d.f. of X, at $X = s$, and $E\{X\}$ is the expected demand.

The optimal value of s, s^0, is the smallest integer s for which $R(s + 1) - R(s) \geq 0$. Because, for $s = 0, 1, \ldots$

$$R(s + 1) - R(s) = (c_1 + c_2)F(s) - c_2$$

we find that

$$s^0 = \text{Smallest nonnegative integer } s, \quad \text{such that} \quad F(s) \geq \frac{c_2}{c_1 + c_2} \tag{6.8.3}$$

In other words, s^0 is the $c_2/(c_1 + c_2)$th quantile of $F(x)$. We have seen that the optimal decision is a function of $F(x)$. If this distribution is unknown, or only partially known, we cannot determine the optimal value, s^0.

After observing a large number, N, of X values, we can consider the empirical distribution, $F_N(x)$, of the demand and determine the level:

$$S^0(F_N) = \text{smallest } s \text{ value} \quad \text{such that} \quad F_N(s) \geq \frac{c_2}{c_1 + c_2}$$

The question is what to do when N is small.   ∎

6.8.1 Prior and Posterior Distributions

In this section, we focus on parametric models. Let $f(x; \boldsymbol{\theta})$ denote the p.d.f. of some random variable X, which depends on a parameter $\boldsymbol{\theta}$. $\boldsymbol{\theta}$ could be a vector of several real parameters, as in the case of a normal distribution. Let Θ denote the set of all possible parameters $\boldsymbol{\theta}$. Recall that Θ is called the *parameter space*. For example, the parameter space Θ of the family of normal distribution is the set $\Theta = \{(\mu, \sigma); -\infty < \mu < \infty, 0 < \sigma < \infty\}$. In the case of Poisson distributions,

$$\Theta = \{\lambda; 0 < \lambda < \infty\}$$

In a Bayesian framework, we express our prior belief (based on prior information) concerning which $\boldsymbol{\theta}$ values are plausible by a p.d.f. on Θ, which is called the **prior probability density function**. Let $h(\boldsymbol{\theta})$ denote the prior p.d.f. of $\boldsymbol{\theta}$. To illustrate, suppose that X is a discrete random variable having a binomial distribution $B(n, \theta)$ and that n is known but θ is unknown. The parameter space is $\Theta = \{\theta; 0 < \theta < 1\}$. Suppose we believe that θ is close to 0.8, with small dispersion around this value. Figure 6.11 illustrates the p.d.f. of a beta distribution beta(80, 20), whose functional

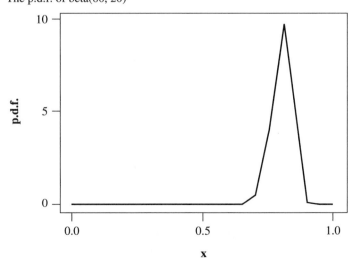

FIGURE 6.11
The p.d.f. of beta(80, 20)

form is

$$h(\theta; 80, 20) = \frac{99!}{79!19!}\theta^{79}(1 - \theta)^{19} \qquad 0 < \theta < 1$$

If we wish, however, to give more weight to small values of θ, we can choose beta(8, 2) as a prior density:

$$h(\theta; 8, 2) = 72\theta^{7}(1 - \theta) \qquad 0 < \theta < 1$$

(Figure 6.12).

FIGURE   6.12
The p.d.f. of beta(8, 2)

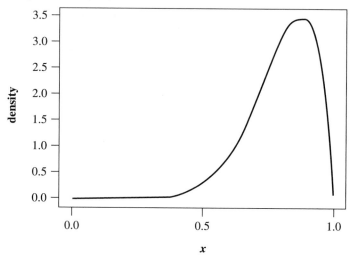

The average p.d.f. of X with respect to the prior p.d.f., $h(\theta)$, is called the **predictive** *p.d.f.* of X. This is given by

$$f_h(x) = \int_{\Theta} f(x; \theta)h(\theta)d\theta \qquad\qquad\qquad \textbf{(6.8.4)}$$

For the preceding example, the predictive p.d.f. is

$$f_h(x) = 72\binom{n}{x}\int_0^1 \theta^{7+x}(1 - \theta)^{n-x+1}d\theta$$

$$= 72\binom{n}{x}\frac{(7 + x)!(n + 1 - x)!}{(n + 9)!} \qquad x = 0, 1, \ldots, n$$

Before taking observations on X, we use the predictive p.d.f., $f_h(x)$, to predict the possible outcomes of observations on X. After observing the outcome of X—say, x—we convert the prior p.d.f. to a **posterior** p.d.f. by employing *Bayes' formula*. If

$h(\theta \mid x)$ denotes the posterior p.d.f. of θ, given that $\{X = x\}$, Bayes' formula yields

$$h(\theta \mid x) = \frac{f(x \mid \theta)h(\theta)}{f_h(x)} \qquad\qquad (6.8.5)$$

In the preceding example,

$$f(x \mid \theta) = \binom{n}{x}\theta^x(1-\theta)^{n-x} \qquad x = 0, 1, \ldots, n$$

$$h(\theta) = 72\theta^7(1-\theta), \quad 0 < \theta < 1$$

and hence

$$h(\theta \mid x) = \frac{(n+9)!}{(7+x)!(n+1-x)!}\theta^{7+x}(1-\theta)^{n+1-x} \qquad 0 < \theta < 1$$

This is again the p.d.f. of a beta distribution beta$(8 + x, n - x + 2)$.

Figure 6.13 shows some of these posterior p.d.f.'s for the case of $n = 10$, $x = 6$, 7, 8. Notice that the posterior p.d.f. $h(\theta \mid x)$ is the conditional p.d.f. of θ, given $\{X = x\}$. If we observe a random sample of n independent and identically distributed (i.i.d.) random variables, and the observed values of $X_1, \ldots, X_n$ are $x_1, \ldots, x_n$, then the posterior p.d.f. of θ is

$$h(\theta \mid x_1, \ldots, x_n) = \frac{\displaystyle\prod_{i=1}^{n}f(x_i \mid \theta)h(\theta)}{f_h(x_1, \ldots, x_n)} \qquad\qquad (6.8.6)$$

where

$$f_h(x_1, \ldots, x_n) = \int_{\Theta}\prod_{i=1}^{n}f(x_i \mid \theta)h(\theta)d\theta \qquad\qquad (6.8.7)$$

FIGURE   6.13
The posterior p.d.f. of θ, $n = 10$, $X = 6, 7, 8$

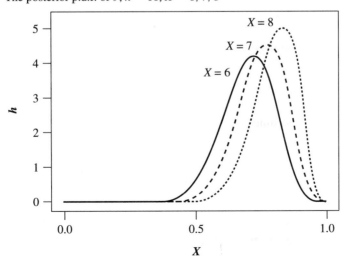

is the *joint predictive p.d.f.* of $X_1, \ldots, X_n$. If the i.i.d. random variables $X_1, X_2, \ldots$ are observed sequentially (timewise), then the posterior p.d.f. of θ, given $x_1, \ldots, x_n$, $n \geq 2$ can be determined recursively, by the formula

$$H(\theta \mid x_1, \ldots, x_n) = \frac{f(x_n \mid \theta) h(\theta \mid x_1, \ldots, x_{n-1})}{\int_{\Theta} f(x_n; \theta') h(\theta' \mid x_1, \ldots, x_{n-1}) d\theta'}$$

The function

$$f_h(x_n \mid x_1, \ldots, x_{n-1}) = \int_{\Theta} f(x_n \mid \theta) h(\theta \mid x_1, \ldots, x_{n-1}) d\theta$$

is called the *conditional predictive p.d.f.* of X_n, given $X_1 = x_1, \ldots, X_{n-1} = x_{n-1}$. Notice that

$$f_h(x_n \mid x_1, \ldots, x_{n-1}) = \frac{f_h(x_1, \ldots, x_n)}{f_h(x_1, \ldots, x_{n-1})} \tag{6.8.8}$$

6.8.2 Bayesian Testing and Estimation

6.8.2.1 Bayesian Testing

We discuss here the problem of testing hypotheses as a Bayesian decision problem. Suppose we consider a null hypothesis H_0 concerning a parameter θ of the p.d.f. of X. Suppose also that the parameter space Θ is partitioned into two sets: Θ_0 and Θ_1. Θ_0 is the set of θ values corresponding to H_0, and Θ_1 is the complementary set of elements of Θ not in Θ_0. If $h(\theta)$ is a prior p.d.f. of θ, then the prior probability that H_0 is true is $\pi = \int_{\Theta_0} h(\theta) d\theta$. The prior probability that H_1 is true is $\overline{\pi} = 1 - \pi$.

The statistician has to make a decision whether H_0 is true or H_1 is true. Let $d(\pi)$ be a decision function, assuming the values 0 and 1—that is,

$$d(\pi) = \begin{cases} 0 & \text{decision to accept } H_0 \ (H_0 \text{ is true}) \\ 1 & \text{decision to reject } H_0 \ (H_1 \text{ is true}) \end{cases}$$

Let w be an indicator of the true situation—that is,

$$w = \begin{cases} 0 & \text{if } H_0 \text{ is true} \\ 1 & \text{if } H_1 \text{ is true} \end{cases}$$

We also impose a loss function for erroneous decision

$$L(d(\pi), w) = \begin{cases} 0 & \text{if } d(\pi) = w \\ r_0 & \text{if } d(\pi) = 0, w = 1 \\ r_1 & \text{if } d(\pi) = 1, w = 0 \end{cases} \tag{6.8.9}$$

where r_0 and r_1 are finite positive constants. The prior risk associated with the decision function $d(\pi)$ is

$$R(d(\pi), \pi) = d(\pi) r_1 \pi + (1 - d(\pi)) r_0 (1 - \pi)$$
$$= r_0 (1 - \pi) + d(\pi)[\pi (r_0 + r_1) - r_0] \tag{6.8.10}$$

We wish to choose a decision function that minimizes the prior risk $R(d(\pi), \pi)$. Such a decision function is called the *Bayes' decision function*, and the prior risk associated with the Bayes' decision function is called the **Bayes' risk**. According to the formula for $R(d(\pi), \pi)$, we should choose $d(\pi)$ to be 1 if and only if $\pi(r_0 + r_1) - r_0 < 0$. Accordingly, the Bayes' decision function is

$$d^0(\pi) = \begin{cases} 0 & \text{if } \pi \geq \dfrac{r_0}{r_0 + r_1} \\[2ex] 1 & \text{if } \pi < \dfrac{r_0}{r_0 + r_1} \end{cases} \tag{6.8.11}$$

Let $\pi^* = r_0/(r_0 + r_1)$, and define the indicator function

$$I(\pi; \pi^*) = \begin{cases} 1 & \text{if } \pi \geq \pi^* \\ 0 & \text{if } \pi < \pi^* \end{cases}$$

Then, the Bayes' risk is

$$R^0(\pi) = r_0(1 - \pi)I(\pi; \pi^*) + \pi r_1(1 - I(\pi; \pi^*)) \tag{6.8.12}$$

Figure 6.14 shows the graph of the Bayes' risk function $R^0(\pi)$, for $r_0 = 1$ and $r_1 = 5$. We see that the function $R^0(\pi)$ attains its maximum at $\pi = \pi^*$. The maximal Bayes' risk is $R^0(\pi^*) = r_0 r_1/(r_0 + r_1) = 5/6$. If the value of π is close to π^*, the Bayes' risk is close to $R^0(\pi^*)$.

FIGURE   6.14
The Bayes' risk function

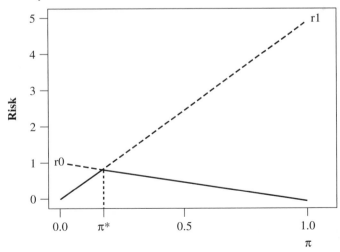

The preceding analysis can be performed even before observations begin. If π is close to 0 or to 1, the Bayes' risk, $R^0(\pi)$, is small, and we may be able to reach a decision concerning the hypotheses without even making observations. Recall that observations cost money, and making them may not be justifiable. However, if the cost of observations is negligible compared to a loss due to an erroneous decision, it might be prudent to take as many observations as required to reduce the Bayes' risk.

After observing a random sample, $x_1, \ldots, x_n$, we convert the prior p.d.f. of θ to posterior, and determine the posterior probability of H_0—namely,

$$\pi_n = \int_\Theta h(\theta \mid x_1, \ldots, x_n) d\theta$$

The analysis then proceeds as before, with the posterior probability π_n replacing π.

Accordingly, the Bayes' decision function is

$$d^0(x_1, \ldots, x_n) = \begin{cases} 0 & \text{if } \pi_n \geq \pi^* \\ 1 & \text{if } \pi_n < \pi^* \end{cases}$$

and the Bayes' posterior risk is

$$R^0(\pi_n) = r_0(1 - \pi_n) I(\pi_n; \pi^*) + \pi_n r_1(1 - I(\pi_n; \pi^*))$$

Under certain regularity conditions, $\lim_{n\to\infty} \pi_n = 1$ or 0, according to whether H_0 is true or false. We illustrate this with a simple example.

EXAMPLE **6.15**   Suppose that X has a normal distribution, with known $\sigma^2 = 1$. The mean μ is unknown. We wish to test $H_0 : \mu \leq \mu_0$ against $H_1 : \mu > \mu_0$. Suppose that the prior distribution of μ is also normal, $N(\mu^*, \tau^2)$. The posterior distribution of μ, given $X_1, \ldots, X_n$, is normal with mean

$$E\{\mu \mid X_1, \ldots, X_n\} = \mu^* \frac{1}{(1 + n\tau^2)} + \frac{n\tau^2}{1 + n\tau^2} \bar{X}_n$$

and posterior variance

$$V\{\mu \mid X_1, \ldots, X_n\} = \frac{\tau^2}{1 + n\tau^2}$$

Accordingly, the posterior probability of H_0 is

$$\pi_n = \Phi\left(\frac{\mu_0 - \dfrac{\mu^*}{1 + n\tau^2} - \dfrac{n\tau^2}{1 + n\tau^2}\bar{X}_n}{\sqrt{\dfrac{\tau^2}{1 + n\tau^2}}} \right)$$

According to the law of large numbers, $\bar{X}_n \to \mu$ (the true mean) as $n \to \infty$, with probability 1. Hence,

$$\lim_{n\to\infty} \pi_n = \begin{cases} 1 & \text{if } \mu < \mu_0 \\ \dfrac{1}{2} & \text{if } \mu = \mu_0 \\ 0 & \text{if } \mu > \mu_0 \end{cases}$$

Notice that the prior probability that $\mu = \mu_0$ is zero. Thus, if $\mu < \mu_0$ or $\mu > \mu_0$, $\lim_{n \to \infty} R^0(\pi_n) = 0$, with probability 1. That is, if n is sufficiently large, the Bayes' risk is, with probability close to 1, smaller than some threshold r^*. This suggests that we need to continue, stepwise or sequentially, collecting observations until the Bayes' risk $R^0(\pi_n)$ is, for the first time, smaller than r^*. When we stop, $\pi_n \geq 1 - r^*/r_0$ or $\pi_n \leq r^*/r_1$. We obviously choose $r^* < r_0 r_1/(r_0 + r_1)$.

6.8.2.2 Bayesian Estimation

In an estimation problem, the decision function is an estimator $\hat{\theta}(x_1, \ldots, x_n)$ that yields a point in the parameter space Θ. Let $L(\hat{\theta}(x_1, \ldots, x_n), \theta)$ be a loss function that is nonnegative, and $L(\theta, \theta) = 0$. The posterior risk of an estimator $\hat{\theta}(x_1, \ldots, x_n)$ is the expected loss with respect to the posterior distribution of θ, given $(x_1, \ldots, x_n)$—that is,

$$R_h(\hat{\theta}, \mathbf{x}_n) = \int_\Theta L(\hat{\theta}(\mathbf{x}_n), \theta) h(\theta \mid \mathbf{x}_n) d\theta \tag{6.8.13}$$

where $\mathbf{x}_n = (x_1, \ldots, x_n)$. We choose an estimator that minimizes the posterior risk. Such an estimator is called a **Bayes' estimator** and is designated by $\hat{\theta}_B(\mathbf{x}_n)$. Following are a few important cases.

Case A: θ real, $L(\hat{\theta}, \theta) = (\hat{\theta} - \theta)^2$
In this case, the Bayes' estimator of θ is the posterior expectation of θ:

$$\hat{\theta}_B(\mathbf{x}_n) = E_h\{\theta \mid \mathbf{x}_n\} \tag{6.8.14}$$

The Bayes' risk is the expected posterior variance:

$$R_h^0 = \int V_h\{\theta \mid \mathbf{x}_n\} f_h(x_1, \ldots, x_n) dx_1, \ldots, dx_n$$

Case B: θ real, $L(\hat{\theta}, \theta) = c_1(\hat{\theta} - \theta)^+ + c_2(\theta - \hat{\theta})^+$, with $c_1, c_2 > 0$ and $(a)^+ = \max(a, 0)$

As shown in the inventory example at the beginning of this section, the Bayes' estimator is

$$\hat{\theta}_B(\mathbf{x}_n) = \frac{c_2}{c_1 + c_2} \text{-th quantile of the posterior distribution of } \theta, \text{ given } \mathbf{x}_n$$

When $c_1 = c_2$, we obtain the posterior median. ∎

6.8.3 Credibility Intervals for Real Parameters

In this section, we focus on the case of a real parameter, θ. Given the values $x_1, \ldots, x_n$ of a random sample, let $h(\theta \mid \mathbf{x}_n)$ be the posterior p.d.f. of θ. An interval $C_{1-\alpha}(\mathbf{x}_n)$ such that

$$\int_{C_{1-\alpha}(\mathbf{x}_n)} h(\theta \mid \mathbf{x}_n)d\theta \geq 1 - \alpha \tag{6.8.15}$$

is called a **credibility interval** for θ. A credibility interval $C_{1-\alpha}(\mathbf{x}_n)$ is called a *highest posterior density (HPD) interval* if, for any $\theta \in C_{1-\alpha}(\mathbf{x}_n)$ and $\theta' \notin C_{1-\alpha}(\mathbf{x}_n)$, $h(\theta \mid \mathbf{x}_n) > h(\theta' \mid \mathbf{x}_n)$.

EXAMPLE **6.16**   Let $x_1, \ldots, x_n$ be the values of a random sample from a Poisson distribution $P(\lambda)$, $0 < \lambda < \infty$. We assign λ a gamma distribution, $G(\nu, \tau)$. The posterior p.d.f. of λ, given $\mathbf{x}_n = (x_1, \ldots, x_n)$ is

$$h(\lambda \mid \mathbf{x}_n) = \frac{(1 + n\tau)^{\nu + \Sigma x_i}}{\Gamma(\nu + \Sigma x_i)\tau^{\nu + \Sigma x_i}} \cdot \lambda^{\nu + \Sigma x_i - 1} e^{-\lambda \frac{1 + n\tau}{\tau}}$$

In other words, the posterior distribution is a gamma distribution $G\left[\nu + \sum_{i=1}^{n} x_i, \tau/(1 + n\tau)\right]$. From the relationship between the gamma and the χ^2 distributions, we can express the limits of a credibility interval for λ, at level $1 - \alpha$, as

$$\frac{\tau}{2(1 + n\tau)}\chi^2_{\alpha/2}[\phi] \quad \text{and} \quad \frac{\tau}{2(1 + n\tau)}\chi^2_{1-\alpha/2}[\phi]$$

where $\phi = 2\nu + 2\sum_{i=1}^{n} x_i$. This interval is called an *equal-tail credibility interval*. However, it is not an HPD credibility interval. Figure 6.15 shows the posterior

FIGURE   **6.15**

The posterior p.d.f. and credibility intervals

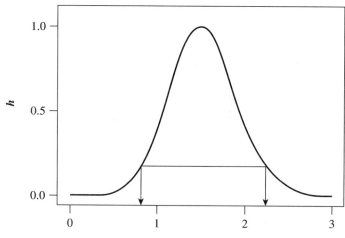

density for the special case of $n = 10$, $\nu = 2$, $\tau = 1$, and $\sum_{i=1}^{10} x_i = 15$. For these values, the limits of the credibility interval for λ, at level .95, are 0.9 and 2.364. As we see in the figure, $h(0.9 \mid \mathbf{x}_n) > h(2.364 \mid \mathbf{x}_n)$. Thus, the equal-tail credibility interval is not an HPD interval. The limits of the HPD interval can be determined by trial and error. In this case, they are approximately 0.86 and 2.29, as shown in Figure 6.15. ∎

6.9
Chapter Highlights

This chapter provides theoretical foundations for statistical inference. Inference on parameters of infinite populations is discussed using classical point estimation, confidence intervals, tolerance intervals, and hypothesis testing. Properties of point estimators such as moment equation estimators and maximum likelihood estimators are discussed in detail. Formulas for parametric confidence intervals and distribution-free tolerance intervals are provided. Statistical tests of hypothesis are presented with examples, including tests for normality with probability plots and the chi-square and Kolmogorov-Smirnov tests of goodness of fit. The chapter concludes with a section on Bayesian testing and estimation.

The main concepts and definitions introduced in this chapter include:

- Statistical inference
- Sampling distribution
- Unbiased estimator
- Consistent estimator
- Standard error
- Parameter space
- Statistic

- Point estimator
- Least-squares estimator
- Likelihood function
- Maximum likelihood estimator
- Null hypothesis
- Alternate hypothesis
- Testing statistical hypotheses
- Rejection region
- Acceptance region
- Type I error
- Type II error
- Significance level
- OC curve
- Power function
- *P*-value
- Confidence interval
- Tolerance interval
- Normal scores
- Normal probability plot
- Kolmogorov-Smirnov test
- Bayesian decision framework

6.10
Exercises

6.1.1 The consistency of the sample mean, $\overline{X}_n$, in RSWR, is guaranteed by the weak law of large numbers (WLLN) whenever the mean exists. Let $M_l = (1/n) \sum_{i=1}^{n} X_i^l$ be the sample estimate of the *l*th moment, which is assumed to exist $(l = 1, 2, \ldots)$. Show that M_r is a consistent estimator of μ_r.

6.1.2   Consider a population with mean μ and standard deviation $\sigma = 10.5$. Use the CLT to find approximately how large the sample size n should be so that $\Pr\{|\bar{X}_n - \mu| < 1\} = .95$.

6.2.1   Let $X_1, \ldots, X_n$ be a random sample from a normal distribution $N(\mu, \sigma)$. What is the moment equation estimator of the pth quantile $\xi_p = \mu + z_p\sigma$?

6.2.2   Let $(X_1, Y_1), \ldots, (X_n, Y_n)$ be a random sample from a bivariate normal distribution. What is the moment equation estimator of the correlation ρ?

6.2.3   Let $X_1, X_2, \ldots, X_n$ be a sample from a beta distribution beta(ν_1, ν_2); $0 < \nu_1, \nu_2 < \infty$. Find the moment equation estimators of ν_1 and ν_2.

6.2.4   Let $\bar{Y}_1, \ldots, \bar{Y}_k$ be the means of k independent RSWR from normal distributions, $N(\mu, \sigma_i)$, $i = 1, \ldots, k$, with common means and variances σ_i^2 known. Let $n_1, \ldots, n_k$ be the sizes of these samples. Consider a weighted average

$$\bar{Y}_w = \frac{\sum_{i=1}^{k} w_i \bar{Y}_i}{\sum_{i=1}^{k} w_i}$$

with $w_i > 0$. Show that for the estimator $\bar{Y}_w$ having smallest variance, the required weights are $w_i = n_i/\sigma_i^2$.

6.2.5   Using the formula

$$\hat{\beta}_1 = \sum_{i=1}^{n} w_i Y_i$$

with $w_i = (x_i - \bar{x}_n)/SS_x$, $i = 1, \ldots, n$, for the LSE of the slope β in a simple linear regression, derive the formula for $V\{\hat{\beta}_1\}$. We assume that $V\{Y_i\} = \sigma^2$ for all $i = 1, \ldots, n$.

6.2.6   Continuing with Exercise 6.2.5, derive the formula for the variance of the LSE of the intercept β_0 and cov$(\hat{\beta}_0, \hat{\beta}_1)$.

6.2.7   Show that the correlation between the LSEs, $\hat{\beta}_0$ and $\hat{\beta}_1$ in the simple linear regression, is

$$\rho = -\frac{\bar{x}_n}{\left(\frac{1}{n}\sum x_i^2\right)^{1/2}}$$

6.2.8   Let $X_1, \ldots, X_n$ be i.i.d. random variables having a Poisson distribution $P(\lambda)$, $0 < \lambda < \infty$. Show that the MLE of λ is the sample mean $\bar{X}_n$.

6.2.9   Let $X_1, \ldots, X_n$ be i.i.d. random variables from a gamma distribution, $G(\nu, \beta)$, with known ν. Show

that the MLE of β is $\hat{\beta}_n = (1/\nu)\bar{X}_n$, where $\bar{X}_n$ is the sample mean. What is the variance of $\hat{\beta}_n$?

6.2.10   Consider Example 6.4. Let $X_1, \ldots, X_n$ be a random sample from a negative-binomial distribution, NB$(2, p)$. Show that the MLE of p is

$$\hat{p}_n = \frac{2}{\bar{X}_n + 2}$$

where $\bar{X}_n$ is the sample mean.

a   On the basis of the WLLN, show that $\hat{p}_n$ is a consistent estimator of p. [*Hint:* $\bar{X}_n \rightarrow E\{X_1\} = 2[(1 - p)/p]$ in probability as $n \rightarrow \infty$.]

b   Using the fact that if $X_1, \ldots, X_n$ are i.i.d. like NB(k, p), then $T_n = \sum_{i=1}^{n} X_i$ is distributed like NB(nk, p), and the results of Example 6.4, show that, for large values of n,

$$\text{bias}(\hat{p}_n) \cong \frac{3p(1 - p)}{4n} \quad \text{and} \quad V\{\hat{p}_n\} \cong \frac{p^2(1 - p)}{2n}$$

6.2.11   Let $X_1, \ldots, X_n$ be a random sample from a shifted exponential distribution

$$f(x; \mu, \beta) = \frac{1}{\beta} \exp\left\{-\frac{x - \mu}{\beta}\right\} \quad x \geq \mu$$

where $0 < \mu, \beta < \infty$.

a   Show that the sample minimum $X_{(1)}$ is an MLE of μ.

b   Find the MLE of β.

c   What are the variances of these MLEs?

6.3.1   We wish to test that the proportion of defective items in a given lot is smaller than $p_0 = .03$. The alternative is that $p > p_0$. A random sample of $n = 20$ is drawn from the lot with replacement (RSWR). The number of observed defective items in the sample is $X = 2$. Is there sufficient evidence to reject the null hypothesis that $p \leq P_0$?

6.3.2   Compute and plot the operating characteristic curve, OC(p), for binomial testing of $H_0 : p \leq p_0$ versus $H_1 : p > p_0$, when the hypothesis is accepted if 2 or fewer defective items are found in a RSWR of size $n = 30$.

6.3.3   For testing the hypothesis $H_0 : p = .01$ versus $H_1 : p = .03$ concerning the parameter p of a binomial distribution, how large should the sample n be, and what should be the critical value k, if we want error probabilities to be $\alpha = .05$ and $\beta = .05$? [Use the normal approximation to the binomial.]

6.3.4 As will be discussed in Chapter 10, the Shewhart 3-σ control charts for statistical process control provide repeated tests of the hypothesis that the process mean is equal to the nominal mean, μ_0. If a sample mean $\overline{X}_n$ falls outside the limits $\mu_0 \pm 3(\sigma/\sqrt{n})$, the hypothesis is rejected.

a What is the probability that $\overline{X}_n$ will fall outside the control limits when $\mu = \mu_0$?

b What is the probability that when the process is in control, $\mu = \mu_0$, all sample means of 20 consecutive independent samples will be within the control limits?

c What is the probability that a sample mean will fall outside the control limits when μ changes from μ_0 to $\mu_1 = \mu_0 + 2(\sigma/\sqrt{n})$?

d What is the probability that a change from μ_0 to $\mu_1 = \mu_0 + 2(\sigma/\sqrt{n})$, will not be detected by the next 10 sample means?

6.3.5 Consider the data in file SOCELL.DAT. Use MINITAB to test whether the mean ISC at time t_1 is significantly smaller than 4 (amp). [Use a one-sample t-test.]

6.3.6 Is the mean of ISC for time t_2 significantly larger than 4 (amp)?

6.3.7 Consider a one-sided t-test based on a sample of size $n = 30$, with $\alpha = .01$. Compute the OC(δ) as a function of $\delta = (\mu - \mu_0)/\sigma$, $\mu > \mu_0$.

6.3.8 Compute the OC function for testing the hypothesis $H_0 : \sigma^2 \leq \sigma_0^2$ versus $H_1 : \sigma^2 > \sigma_0^2$, when $n = 31$ and $\alpha = .10$.

6.3.9 Compute the OC function in testing $H_0 : p \leq p_0$ versus $H_1 : p > p_0$ in the binomial case, when $n = 100$ and $\alpha = .05$.

6.3.10 Let $X_1, \ldots, X_n$ be a random sample from a normal distribution $N(\mu, \sigma)$. For testing $H_0 : \sigma^2 \leq \sigma_0^2$ against $H_1 : \sigma^2 > \sigma_0^2$, we use the test that rejects H_0 if $S_n^2 \geq (\sigma_0^2/n - 1)\chi_{1-\alpha}^2[n-1]$, where S_n^2 is the sample variance. What is the power function of this test?

6.3.11 Let $S_{n_1}^2$ and $S_{n_2}^2$ be the variances of two independent samples from normal distributions $N(\mu_i, \sigma_i)$, $i = 1, 2$. For testing $H_0 : (\sigma_1^2/\sigma_2^2) \leq 1$ against $H_1 : (\sigma_1^2/\sigma_2^2) > 1$, we use the F-test, which rejects H_0 when $F = \left(\dfrac{S_{n_1}^2}{S_{n_2}^2}\right) > F_{1-\alpha}[n_1 - 1, n_2 - 1]$.

What is the power of this test as a function of $\rho = \sigma_1^2/\sigma_2^2$?

6.4.1 A random sample of size $n = 20$ from a normal distribution gave the following values: 20.74, 20.85, 20.54, 20.05, 20.08, 22.55, 19.61, 19.72, 20.34, 20.37, 22.69, 20.79, 21.76, 21.94, 20.31, 21.38, 20.42, 20.86, 18.80, 21.41. Compute:

a The confidence interval for the mean μ, at level of confidence $1 - \alpha = .99$

b The confidence interval for the variance σ^2, at confidence level $1 - \alpha = .99$

c A confidence interval for σ, at level of confidence $1 - \alpha = .99$

6.4.2 Let C_1 be the event that a confidence interval for the mean μ covers it. Let C_2 be the event that a confidence interval for the standard deviation σ covers it. The probability that both μ and σ are simultaneously covered is

$$\begin{aligned} \Pr\{C_1 \cap C_2\} &= 1 - \Pr\{\overline{C_1 \cap C_2}\} \\ &= 1 - \Pr\{\overline{C}_1 \cup \overline{C}_2\} \\ &\geq 1 - \Pr\{\overline{C}_1\} - \Pr\{\overline{C}_2\} \end{aligned}$$

This inequality is called the *Bonferroni inequality*. Apply this inequality and the results of the previous exercise to determine the confidence interval for $\mu + 2\sigma$, at a level of confidence not smaller than .98.

6.4.3 Twenty independent trials yielded $X = 17$ successes. Assuming that the probability for success in each trial is the same, θ, determine the confidence interval for θ at level of confidence .95.

6.4.4 Let $X_1, \ldots, X_n$ be a random sample from a Poisson distribution with mean λ. Let $T_n = \displaystyle\sum_{i=1}^{n} X_i$. Using the relationship between the Poisson and the gamma c.d.f., we can show that a confidence interval for the mean λ, at level $1 - \alpha$, has lower and upper limits λ_L and λ_U, where

$$\lambda_L = \frac{1}{2n}\chi_{\alpha/2}^2[2T_n + 2]$$

and

$$\lambda_U = \frac{1}{2n}\chi_{1-\alpha/2}^2[2T_n + 2]$$

The following is a random sample of size $n = 10$ from a Poisson distribution 14, 16, 11, 19, 11, 9, 12, 15, 14, 13. Determine a confidence interval for λ at level of confidence .95. [*Hint:* For a large number of degrees of freedom, $\chi_p^2[\nu] \cong \nu + z_p\sqrt{2\nu}$, where z_p is the pth quantile of the standard normal distribution.]

6.5.1 The mean of a random sample of size $n = 20$, from a normal distribution with $\sigma = 5$, is $\overline{Y}_{20} = 13.75$.

Determine a $1 - \beta = .90$ content tolerance interval with confidence level $1 - \alpha = .95$.

6.5.2   Use the YARNSTRG.DAT data file to determine a $(.95, .95)$ tolerance interval for log yarn-strength. [*Hint*: Notice that the interval is $\overline{Y}_{100} \pm k S_{100}$, where $k = t(.025, .025, 100)$.]

6.5.3   Use the minimum and maximum of the log yarn-strength (see Exercise 6.5.2) to determine a distribution-free tolerance interval. What are the values of α and β for your interval? How does this interval compare with the interval in Exercise 6.5.2?

6.6.1   Make a normal QQ plot to test graphically whether the ISC t_1 of data file SOCELL.DAT is normally distributed.

6.6.2   Use MINITAB and data file CAR.DAT.

a   Test graphically whether the turn diameter is normally distributed.

b   Test graphically whether the log (horsepower) is normally distributed.

6.7.1   Use the CAR.DAT file. Make a frequency distribution of turn diameter, with $k = 11$ intervals. Fit a normal distribution to the data and make a chi-square test of the goodness of fit.

6.7.2   Use MINITAB and the CAR.DAT data file. Compute the KS test statistic D_n^* for the turn diameter variable, testing for normality. Compute k_α^* for $\alpha = .05$. Is D_n^* significant?

6.8.1   The daily demand (in number of loaves) for whole wheat bread at a certain bakery has a Poisson distribution with mean $\lambda = 100$. The loss to the bakery for undemanded units at the end of the day is $C_1 = \$0.10$. However, the penalty for a shortage of a unit is $C_2 = \$0.20$. How many loaves of whole wheat bread should be baked every day?

6.8.2   A random variable X has the binomial distribution $B(10, p)$. The parameter p has a beta prior distribution beta$(3, 7)$. What is the posterior distribution of p, given $X = 6$?

6.8.3   Continuing with Exercise 6.8.2, find the posterior expectation and posterior standard deviation of p.

6.8.4   A random variable X has a Poisson distribution with mean λ. The parameter λ has a gamma, $G(2, 50)$, prior distribution.

a   Find the posterior distribution of λ given $X = 82$.

b   Find the .025th and .975th quantiles of this posterior distribution.

6.8.5   A random variable X has a Poisson distribution with mean that is either $\lambda_0 = 70$ or $\lambda_1 = 90$. The prior probability of λ_0 is $1/3$. The losses due to wrong actions are $r_1 = \$100$ and $r_2 = \$150$. Observing $X = 72$, which decision would you take?

6.8.6   A random variable X is normally distributed, with mean μ and standard deviation $\sigma = 10$. The mean μ is assigned a prior normal distribution with mean $\mu_0 = 50$ and standard deviation $\tau = 5$. Determine a credibility interval for μ, at level .95, given $X = 60$. Is this credibility interval also an HPD interval?

7

Distribution Free-Inference: Computer-Intensive Techniques

In Chapter 6, we studied the theory and methods of statistical inference for parametric models. These models specify the distribution functions up to the values of their parameters. In this chapter, we develop methods for statistical inference that do not rely on specifying the functional form, or the parametric family, of the distribution. Such methods are called *distribution-free* or *nonparametric*. There is a highly developed statistical theory of nonparametric procedures (see Conover, 1980; Lehmann, 1975; Hettmansperger, 1984). Here, we present computer-intensive techniques for inference, as well as some classical nonparametric tests.

Random Sampling from Reference Distributions

In Section 2.4.1 we saw an example of blemishes on ceramic plates. In that example, (Table 2.2), the proportion of plates having *more* than one blemish was .23. Suppose that we decide to improve the manufacturing process and reduce this proportion. How can we test whether an alternative production process with new operating procedures and machine settings is indeed better so that the proportion of plates with more than one blemish is significantly smaller? The objective is to operate a process with a proportion of defective units (those with more than one blemish) that is smaller than .10. After various technological modifications, we are ready to test whether the modified process conforms with the new requirement. Suppose that a random sample of ceramic plates is drawn from the modified manufacturing process. One has to test whether the proportion of defective plates in the sample is not significantly larger than .10. In the parametric model (see Chapter 6), we assumed that the number of plates having more than one defect, in a random sample of n plates, has a binomial distribution $B(n, p)$. For testing $H_0 : p \leq .1$, a test was constructed based on the reference distribution $B(n, .1)$.

We can artificially create on a computer a population of 90 zeros and 10 ones. In this population, the proportion of 1s is $p_0 = .10$. From this population, we can draw a large number, M, of random samples with replacement (RSWR) of a given size n. In each sample, the sample mean, $\overline{X}_n$, is the proportion of 1s in the sample. The sampling distribution of the M sample means is our **empirical reference distribution** for the hypothesis that the proportion of defective plates is $p \leq p_0$. We pick a value α close to zero, and determine the $(1 - \alpha)$th quantile of the empirical reference distribution. If the observed proportion in the real sample is greater than this quantile, the hypothesis $H_0 : p \leq p_0$ is rejected.

EXAMPLE **7.1**   To illustrate, using MINITAB we created an empirical reference distribution of $M = 1000$ proportions of 1s in RSWRs of size $n = 50$. We did this by executing the following MACRO 1000 times:

```
Sample 50 C1 C2;
Replace.
Let k1 = mean (C2)
stack C3 k1 C3
end
```

We assumed that column $C1$ contained 90 zeros and 10 ones. The frequency distribution of column $C3$ represents the reference distribution. This is given in Table 7.1.

TABLE   **7.1**
Frequency distribution of $M = 1000$ means of RSWR from a set with 90 zeros and 10 ones

$\overline{x}$	f	$\overline{x}$	f
0.03	10	0.11	110
0.04	17	0.12	100
0.05	32	0.13	71
0.06	61	0.14	50
0.07	93	0.15	28
0.08	128	0.16	24
0.09	124	0.17	9
0.10	133	>0.17	10

For $\alpha = .05$, the .95 quantile of the empirical reference distribution is .15. Thus, if, in a real sample of size $n = 50$, the proportion of defectives is greater than .15, the null hypothesis is rejected.   ∎

EXAMPLE **7.2**   Consider a hypothesis on the length of aluminum pins (with cap), $H_0 : \mu \geq 60.1$ (mm), (See Example 3.4.) We now create an empirical reference distribution for this hypothesis. In the data set ALMPIN.DAT, we have the actual sample values. As we saw in Example 3.4, the mean of the variable Lengthwcp is $\overline{X}_{70} = 60.028$. Because the hypothesis states that the process mean is $\mu \geq 60.1$, we transform the sample values to $Y = X - 60.028 + 60.1$. This transformed sample has a mean of 60.1. We now create a reference distribution of sample means by drawing M RSWR of size $n = 70$ from the transformed sample. We can do this by using the following macro:

```
Sample 70 C7 C8;
Replace.
let k1 = mean(C8)
stack C9 k1 C9
end
```

Executing this macro $M = 1000$ times, we obtain an empirical reference distribution whose frequency distribution is given in Table 7.2.

TABLE   7.2

Frequency distribution of $\overline{X}_{70}$ from 1000 RSWR from Lengthwcp

Midpoint	Count
60.080	3
60.085	9
60.090	91
60.095	268
60.100	320
60.105	217
60.110	80
60.115	11
60.120	1

In the data set ALMPIN.DAT, there are six variables measuring various dimensions of aluminum pins. The data are stored in columns $C1$–$C6$. Column $C7$ contains the values of the transformed variable Y.

Because $\overline{X}_{70} = 60.028$ is smaller than 60.1, we consider as a test criterion the α-quantile of the reference distribution. If $\overline{X}_{70}$ is smaller than this quantile, we reject the hypothesis. For $\alpha = .01$, the .01 quantile in this reference distribution is 60.0869. Accordingly, we reject the hypothesis because it is very implausible (less than 1 chance in 100) that $\mu \geq 60.1$. The estimated P-value is less than 10^{-3} because the smallest value in the reference distribution is 60.0803. ■

7.2
Bootstrap Sampling
7.2.1   The Bootstrap Method

The **bootstrap method** was introduced in 1979 by B. Efron as an elegant method of performing statistical inference by computer and doing so without the need for extensive assumptions and intricate theory. Some of the ideas on how to make statistical inferences by computer sampling were presented in the previous sections. In this section, we present the bootstrap method in more detail.

Given a sample of size n, $S_n = \{x_1, \ldots, x_n\}$, let t_n denote the value of some specified sample statistic T. The bootstrap method draws M random samples with replacement (RSWRs) of size n from S_n. For each such sample, the statistic T is computed. Let $\{t_1^*, t_2^*, \ldots, t_M^*\}$ be the collection of these sample statistics. The distribution of these M values of T is called the *empirical bootstrap distribution (EBD)*. It provides an approximation—if M is large—to the *bootstrap distribution* of all possible values of the statistic T, that can be generated by repeatedly sampling from S_n. General properties of the EBD include:

1   The EBD is centered at the sample statistic t_n.

2   The mean of the EBD is an estimate of the mean of the sampling distribution of the statistic T over all possible samples.

3   The standard deviation of the EBD is the bootstrap estimate of the standard-error of T.

4   The $\alpha/2$-th and $(1 - \alpha/2)$th quantiles of the EBD are bootstrap confidence limits for the parameter that is estimated by t_n, at level of confidence $1 - \alpha$.

EXAMPLE **7.3**   We first illustrate the bootstrap method on an artificial example in which the population mean and variance are known.

Using MINITAB, we simulate sample values from a gamut of distributions by choosing from the menu the options: **CALC**; **RANDOM DATA**. There are 10 options for selecting a model of theoretical distribution. We select **NORMAL** and generate 100 rows of data into column $C1$ with mean $\mu = 20.5$ and standard deviation $\sigma = 12.5$. These two parameters should be estimated by the sample values. For estimating μ, we use the sample mean $\overline{X}_{100} = 18.709$, and for estimating σ, we employ the sample standard deviation $S_{100} = 12.040$. Using the following macro, we draw 100 RSWR samples from the original sample. For each such sample, we compute the bootstrap means and the bootstrap standard deviations using macro BOOTMEAN.

```
Sample 100 C1 C2;
Replace.
let k1 = mean(C2)
let k2 = stan(C2)
```

```
stack C3 k1 C3
stack C4 k2 C4
end
```

Column $C3$ contains the 100 bootstrap means and $C4$ contains the bootstrap standard deviations. The standard deviation of $C3$ is the bootstrap estimate of the standard error of $\overline{X}_{100}$. We denote it by $\text{SE}^*\{\overline{X}_{100}\}$. In this example, we obtained $\text{SE}^*\{\overline{X}_{100}\} = 1.2688$. This is close to the true standard error, $\text{SE}\{\overline{X}_{100}\} = \sigma/\sqrt{n} = 1.25$. The standard deviation of column $C4$ is $\text{SE}^*\{S_{100}\} = 0.8426$. This is close to the true standard error, which is $\frac{\sigma}{\sqrt{2n}} = 0.8839$. To obtain the bootstrap confidence limits, at confidence level $1 - \alpha = .95$, we sort the values in $C3$ and those in $C4$. The .025 quantile of $\overline{X}_{100}^*$ is 16.0498. The .975 quantile of $\overline{X}_{100}^*$ is 21.1142. The bootstrap interval $(16.05, 21.11)$ is called a *bootstrap confidence interval for* μ. We see that this interval indeed covers the true value, 20.5. Similarly, we find that the bootstrap confidence interval for σ, at level of confidence .95, is $(10.44, 13.59)$. The theory of parametric confidence intervals was discussed in Chapter 6.   ■

7.2.2 Examining the Bootstrap Method

In Section 7.2.1, we introduced the bootstrap method as a computer-intensive technique for making statistical inference. In this section, we examine some of the properties of the bootstrap methods in light of the theory of sampling from finite populations. Recall that the bootstrap method is based on repeatedly drawing M simple RSWRs from the original sample.

Let $\mathcal{S}_X = \{x_1, \ldots, x_n\}$ be the values of the n original observations on X. We can consider $\mathcal{S}_X$ as a finite population $\mathcal{P}$ of size n. Thus, the mean of this population, μ_n, is the sample mean $\overline{X}_n$, and the variance of this population is $\sigma_n^2 = (n-1)/nS_n^2$, where S_n^2 is the sample variance, $S_n^2 = 1/(n-1)\sum_{i=1}^{n}(x_i - \overline{X}_n)^2$. Let $\mathcal{S}_X^* = \{X_1^*, \ldots, X_n^*\}$ denote a simple RSWR from $\mathcal{S}_X$. $\mathcal{S}_X^*$ is the bootstrap sample. Let $\overline{X}_n^*$ denote the mean of the bootstrap sample.

We showed in Chapter 5 that the mean of a simple RSWR is an unbiased estimator of the corresponding population mean. Thus,

$$E^*\{\overline{X}_n^*\} = \overline{X}_n \tag{7.2.1}$$

where $E^*\{\cdot\}$ is the expected value with respect to the bootstrap sampling. Moreover, the bootstrap variance of $\overline{X}_n^*$ is

$$V^*\{\overline{X}_n^*\} = \frac{\frac{n-1}{n}S_n^2}{n} \tag{7.2.2}$$

$$= \frac{S_n^2}{n}\left(1 - \frac{1}{n}\right)$$

Thus, in large samples,

$$V^*\{\overline{X}_n^*\} \cong \frac{S_n^2}{n} \tag{7.2.3}$$

If the original sample, $\mathcal{S}_X$, is a realization of n i.i.d. random variables having a c.d.f. $F(x)$ with finite expected value μ_F and a finite variance σ_F^2, then, as shown in Section 4.8, the variance of $\overline{X}_n$ is σ_F^2/n. The sample variance S_n^2 is an unbiased estimator of σ_F^2. Thus, S_n^2/n is an unbiased estimator of σ_F^2/n. Finally, the variance of the EBD of $\overline{X}_1^*, \ldots, \overline{X}_M^*$, obtained by repeating the bootstrap sampling M times independently, is an unbiased estimator of $(S_n^2/n)(1-1/n)$. Thus, the variance of the EBD is an approximation to the variance of $\overline{X}_n$.

This estimation problem is a simple one, and there is no need for bootstrapping in order to estimate the variance or standard error of the estimator $\overline{X}_n$.

7.2.3   Harnessing the Bootstrap Method

The effectiveness of the bootstrap method manifests itself when a formula for the variance of an estimator is hard to obtain. In Section 5.2.1, we provided a formula for the variance of the estimator S_n^2 in simple RSWRs. By bootstrapping from the sample $\mathcal{S}_X$, we obtain an EBD of S_n^{*2}. The variance of this EBD is an approximation to the true variance of S_n^2. Thus, for example, when $\mathcal{P} = \{1, 2, \ldots, 100\}$ and $n = 20$, the true variance of S_n^2 is according to (5.2.6) 31,131.2, whereas the bootstrap approximation for a particular sample is 33,642.9. Another sample will yield a different approximation. The approximation obtained by the bootstrap method becomes more precise as the sample size grows. For the preceding problem, if $n = 100$, $V\{S_n\} = 5690.81$, and the bootstrap approximation is distributed around this value.

The following are values of four approximations of $V\{S_n\}$ for $n = 100$ when $M = 100$. Each approximation is based on different random samples from $\mathcal{P}$:

$$6293.28, \quad 5592.07, \quad 5511.71, \quad 5965.89$$

Each bootstrap approximation is an estimate of the true value of $V\{S_n\}$.

7.3

Bootstrap Testing of Hypotheses

In this section, we present some of the theory and the methods of testing hypotheses by bootstrapping. Given a test statistic $\mathbf{T} = T(X_1, \ldots, X_n)$, the critical level for the test, k_α, is determined according to the distribution of $\mathbf{T}$ under the null hypothesis, which is the **reference distribution**.

The bootstrapping method, as explained in Section 7.2, is a randomization method that resamples the sample values, and thus constructs a reference distribution for $\mathbf{T}$ independently of the unknown distribution F of $\mathbf{X}$. For each bootstrap sample, we compute the value of the test statistic $\mathbf{T}^* = T(x_1^*, \ldots, x_n^*)$. Let $\mathbf{T}_1^*, \ldots, \mathbf{T}_M^*$ be the M values of the test statistic obtained from the M samples from the **bootstrap**

population (BP). Let $F_M^*(t)$ denote the empirical c.d.f. of these values. $F_M^*(t)$ is an estimator of the bootstrap distribution $F^*(t)$, from which we can estimate the critical value k^*. Specific procedures are given in the following subsections.

7.3.1 Bootstrap Testing and Confidence Intervals for the Mean

Suppose that $\{x_1, \ldots, x_n\}$ is a random sample from a parent population having an unknown distribution, F, with mean μ and a finite variance σ^2.

We wish to test the hypothesis

$$H_0 : \mu \leq \mu_0 \quad \text{against} \quad H_1 : \mu > \mu_0$$

Let $\overline{X}_n$ and S_n be the sample mean and sample standard deviation, respectively. Suppose that we draw from the original sample M bootstrap samples. Let $\overline{X}_1^*, \ldots, \overline{X}_M^*$ be the means of the bootstrap samples. Recall that, because the bootstrap samples are RSWR, $E^*\{\overline{X}_j^*\} = \overline{X}_n$ for $j = 1, \ldots, M$, where $E^*\{\cdot\}$ designates the expected value with respect to the bootstrap sampling. Moreover, for large n,

$$\text{SE}^*\{\overline{X}_j^*\} \cong \frac{S_n}{\sqrt{n}} \qquad j = 1, \ldots, M$$

Thus, if n is not too small, the central limit theorem implies that $F_M^*(\overline{X}^*)$ is approximately $\Phi\left[(\overline{X}^* - \overline{X}_n)/(S_n/\sqrt{n})\right]$—that is, the bootstrap means $\overline{X}_1^*, \ldots, \overline{X}_M^*$ are distributed approximately normally around $\mu^* = \overline{X}_n$. We wish to reject H_0 if $\overline{X}_n$ is significantly larger than μ_0. According to this normal approximation to $F_M^*(\overline{X}^*)$, we should reject H_0 at level of significance α if $(\mu_0 - \overline{X}_n)/(S_n/\sqrt{n}) \leq z_\alpha$ or $\overline{X}_n \geq \mu_0 + z_{1-\alpha}(S_n/\sqrt{n})$. This is approximately the t-test of Section 6.5.2.B.

Notice that the reference distribution can be obtained from the EBD by subtracting $\Delta = \overline{X}_n - \mu_0$ from $\overline{X}_j^*$ ($j = 1, \ldots, M$). The reference distribution is centered at μ_0. The $(1 - \alpha/2)$th quantile of the reference distribution is $\mu_0 + z_{1-\alpha/2}(S_n/\sqrt{n})$. Thus, if $\overline{X}_n \geq \mu_0 + z_{1-\alpha/2}(S_n/\sqrt{n})$, we reject the null hypothesis $H_0 : \mu \leq \mu_0$.

If the sample size n is not large, we may not be justified in using the normal approximation. Instead, we use the bootstrap procedures of the following section.

7.3.2 Studentized Test for the Mean

A **studentized test statistic** for testing the hypothesis $H_0 : \mu \leq \mu_0$ is

$$t_n = \frac{\overline{X}_n - \mu_0}{S_n/\sqrt{n}} \tag{7.3.1}$$

H_0 is rejected if t_n is significantly greater than zero. To determine the rejection criterion, we construct an EBD by following procedure:

1   Draw a RSWR of size n from the original sample.

2   Compute $\overline{X}_n^*$ and S_n^* of the bootstrap sample.

3 Compute the studentized statistic

$$t_n^* = \frac{\overline{X}_n^* - \overline{X}_n}{S_n^*/\sqrt{n}} \qquad\qquad (7.3.2)$$

4 Repeat this procedure M times.

Let t_p^* denote the pth quantile of the EBD.

Case I. $H_0 : \mu \leq \mu_0$
The hypothesis H_0 is rejected if

$$t_n \geq t_{1-\alpha}^*$$

Case II. $H_0 : \mu \geq \mu_0$
We reject H_0 if

$$t_n \leq t_\alpha^*$$

Case III. $H_0 : \mu = \mu_0$
We reject H_0 if

$$|t_n| \geq t_{1-\alpha/2}^*$$

The corresponding P^*-levels are:

For Case I: The proportions of t_n^* values greater than t_n
For Case II: The proportions of t_n^* values smaller than t_n
For Case III: The proportion of t_n^* values greater than $|t_n|$ or smaller than $-|t_n|$.

H is rejected if P^* is small.

Notice the difference in definition between t_n and t_n^*: t_n is centered around μ_0 whereas t_n^* is centered around $\overline{X}_n$.

EXAMPLE **7.4**   In data file HYBRID1.DAT, we find the resistance (in ohms) of Res 3 in a hybrid microcircuit labeled hybrid 1 on $n = 32$ boards. The mean of Res 3 in hybrid 1 is $\overline{X}_{32} = 2143.4$. The question is whether Res 3 in hybrid 1 is significantly different from $\mu_0 = 2150$. We consider the hypothesis

$$H_0 : \mu = 2150 \qquad \text{(case III)}$$

Using program BOOT1SMP.EXE with $M = 500$, we obtain the following .95 confidence level bootstrap interval $(2109.50, 2179.91)$. We see that $\mu_0 = 2150$ is covered by this interval. We therefore infer that $\overline{X}_{32}$ is not significantly different from μ_0. The hypothesis H_0 is *not* rejected. To apply the bootstrap studentized test, we can use program STUDTEST.EXE, whose output is stored in C:\ISTAT\DATA\STUDTEST.DAT. We see that the studentized difference between the sample mean $\overline{X}_{32}$ and μ_0 is $t_n = -0.374$. $M = 500$ bootstrap replicas yield the value $P^* = .708$. The hypothesis is not rejected.   ∎

7.3.3   Studentized Test for the Difference of Two Means

The problem is whether two population means, μ_1 and μ_2, are the same. This problem is important in many branches of science and engineering when two "treatments" are compared.

Suppose we observe a random sample $X_1, \ldots, X_n$ from population 1 and another random sample $Y_1, \ldots, Y_{n_2}$ from population 2. Let $\overline{X}_{n_1}$, $\overline{Y}_{n_2}$, S_{n_1}, and S_{n_2} be the means and standard deviations of these two samples, respectively. Compute the studentized difference of the two sample means as

$$t = \frac{\overline{X}_{n_1} - \overline{Y}_{n_2} - \delta_0}{\left(\dfrac{S_{n_1}^2}{n_1} + \dfrac{S_{n_2}^2}{n_2}\right)^{1/2}} \qquad (7.3.3)$$

where $\delta = \mu_1 - \mu_2$. The question is whether this value is significantly different from zero. The hypothesis under consideration is

$$H_0 : \mu_1 = \mu_2 \qquad \text{or} \qquad \delta_0 = 0$$

By the bootstrap method, we draw RSWR of size n_1 from the x sample, and a RSWR of size n_2 from the y sample. Let $X_1^*, \ldots, X_{n_1}^*$ and $Y_1^*, \ldots, Y_{n_2}^*$ be these two bootstrap samples, with means and standard deviations $\overline{X}_{n_1}^*$, $\overline{Y}_{n_2}^*$ and $S_{n_1}^*$, $S_{n_2}^*$. We then compute the studentized difference

$$t^* = \frac{\overline{X}_{n_1}^* - \overline{Y}_{n_2}^* - (\overline{X}_{n_1} - \overline{Y}_{n_2})}{\left(\dfrac{S_{n_1}^{*2}}{n_1} + \dfrac{S_{n_2}^{*2}}{n_2}\right)^{1/2}} \qquad (7.3.4)$$

This procedure is repeated independently M times to generate an EBD of $t_1^*, \ldots, t_M^*$.

Let $(D_{\alpha/2}^*, D_{1-\alpha/2}^*)$ be a $(1 - \alpha)$-level confidence interval for δ, based on the EBD. If $t_{\alpha/2}^*$ is the $\alpha/2$-th quantile of t^* and $t_{1-\alpha/2}^*$ is its $(1 - \alpha/2)$th quantile, then

$$D_{\alpha/2}^* = (\overline{X}_{n_1} - \overline{Y}_{n_2}) + t_{\alpha/2}^* \left(\frac{S_{n_1}^2}{n_1} + \frac{S_{n_2}^2}{n_2}\right)^{1/2}$$

$$ \qquad (7.3.5)$$

$$D_{1-\alpha/2}^* = (\overline{X}_{n_1} - \overline{Y}_{n_2}) + t_{1-\alpha/2}^* \left(\frac{S_{n_1}^2}{n_1} + \frac{S_{n_2}^2}{n_2}\right)^{1/2}$$

If this interval does not cover the value $\delta_0 = 0$, we reject the hypothesis $H_0 : \mu_1 = \mu_2$. The P^*-value of the test is the proportion of t_i^* values that are either smaller than $-|t|$ or greater than $|t|$.

Program BOOT2SMP.EXE performs this bootstrapping test and stores the output in BOOT2SMP.DAT. For a bootstrap test comparing sample variances, see Section 7.3.5.1.

EXAMPLE 7.5   Here we compare the resistance coverage of Res 3 in hybrid 1 and in hybrid 2. The data file HYBRID2.DAT includes two columns. The first represents the sample of $n_1 = 32$ observations on hybrid 1, and the second column consists of $n_2 = 32$ observations on hybrid 2. The output file consists of $M = 500$ values of t_i^* ($i = 1, \ldots, M$).

We see that $\overline{X}_{n_1} = 2143.41$, $\overline{Y}_{n_2} = 1902.81$, $S_{n_1} = 99.647$, and $S_{n_2} = 129.028$. The studentized difference between the means is $t = 8.348$. The bootstrap $(1 - \alpha)$-level confidence interval for

$$\delta / \left(\frac{S_{n_1}^2}{n_1} + \frac{S_{n_2}^2}{n_2} \right)^{1/2}$$

is (6.326, 10.297). The hypothesis that $\mu_1 = \mu_2$ or $\delta = 0$ is rejected with $P^* \cong 0$. Figure 7.1 shows the histogram of the EBD of t^*.   ∎

FIGURE 7.1

Histogram of the EBD of $M = 500$ studentized differences

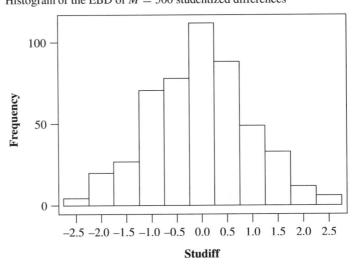

7.3.4 Bootstrap Tests and Confidence Intervals for the Variance

Let $S_1^{*2}, \ldots, S_M^{*2}$ be the variances of M bootstrap samples. These statistics are distributed around the sample variance S_n^2. Consider the problem of testing the hypotheses $H_0 : \sigma^2 \leq \sigma_0^2$ against $H_1 : \sigma^2 > \sigma_0^2$, where σ^2 is the variance of the parent population. As in Section 6.3.2.C, H_0 is rejected if S_n^2 / σ_0^2 is sufficiently large.

Let $G_M^*(x)$ be the bootstrap empirical c.d.f. of $S_1^{*2}, \ldots, S_M^{*2}$. The bootstrap P^*-value for testing H_0 is

$$P^* = Pr^* \left\{ \frac{S_i^{*2}}{S_n^2} > \frac{S_n^2}{\sigma_0^2} \right\} = 1 - G_M^* \left(\frac{S_n^4}{\sigma_0^2} \right) \tag{7.3.6}$$

If P^* is sufficiently small we reject H_0. For example, in a random sample of size $n = 20$ the sample standard deviation is $S_{20} = 24.812$. Suppose we wish to test whether it is significantly larger than $\sigma_0 = 20$. We can run program BOOT1SMP.EXE $M = 500$ times to find the bootstrap P^*-value of the test. The P^*-value for testing the hypothesis H_0 is the proportion of bootstrap standard deviations greater than $S_{20}/\sigma_0 = 30.781$. Running the program, we obtain $P^* = .028$. The hypothesis H_0 is rejected. S_{20} is significantly greater than $\sigma_0 = 20$. In a similar manner, we test the hypotheses $H_0 : \sigma^2 \geq \sigma_0^2$ against $H_1 : \sigma^2 < \sigma_0^2$, or the two-sided hypothesis $H_0 : \sigma^2 = \sigma_0^2$ against $H_0 : \sigma^2 \neq \sigma_0^2$. Percentile bootstrap confidence limits for σ^2, at level $1 - \alpha$, are given by the $\alpha/2$-th and $(1 - \alpha/2)$th quantiles of $G_M^*(x)$, or

$$S_{(j_{\alpha/2})}^{*2} \quad \text{and} \quad S_{(1+j_{1-\alpha/2})}^{*2}, \text{ where } j_{\alpha/2} = [M\alpha/2] \text{ and } j_{1-\alpha/2} = [M(1 - \tfrac{\alpha}{2})]$$

These bootstrap confidence limits for σ^2 at level .95, are 210.914 and 1024.315. The corresponding chi-square confidence limits (see Section 6.4.3) are 355.53 and 1312.80. Another type of **bootstrap confidence interval** is given by the limits

$$\frac{S_n^4}{S_{(j_{1-\alpha/2})}^{*2}}, \quad \frac{S_n^4}{S_{(j_{\alpha/2})}^{*2}}$$

These limits are similar to the chi-square confidence interval limits, but they use the quantiles of S_n^{*2}/S_n instead of those of $\chi^2[n-1]$. For the sample of size $n = 20$ with $S_{20} = 24.812$, the above confidence interval for σ^2 is (370.01, 1066.033).

7.3.5　Comparing Statistics of Several Samples

We often have to test whether the means or the variances of three or more populations are equal. In Chapters 12 and 13, we discuss the design and analysis of experiments where we study the effect of changing levels of different factors. Typically, we collect observations at different experimental conditions. The question is whether the observed differences between the means and variances of the samples observed under different factor levels are significant. The test statistic we will introduce to test differences between means might be affected also by differences between variances. It is therefore prudent to test first whether the population variances are the same. If this hypothesis is rejected, we should not use the test for means, discussed in the next section, but rather refer to a different type of analysis.

7.3.5.1　**Comparing Variances of Several Samples**

Suppose we have k samples, $k \geq 2$. Let $S_{n_1}^2, S_{n_2}^2, \ldots, S_{n_k}^2$ denote the variances of these samples. Let $S_{\max}^2 = \max\{S_{n_1}^2, \ldots, S_{n_k}^2\}$ and $S_{\min}^2 = \min\{S_{n_1}^2, \ldots, S_{n_k}^2\}$. The test statistic we want to consider is the ratio of the maximal to the minimal variances:

$$\tilde{F} = S_{\max}^2 / S_{\min}^2 \tag{7.3.7}$$

The hypothesis under consideration is

$$H_0 : \sigma_1^2 = \sigma_2^2 = \cdots = \sigma_k^2$$

To test this hypothesis, we construct the following EBD.

Step 1. Sample independently RSWRs of sizes $n_1, \ldots, n_k$, respectively, from the given samples. Let $S_{n_1}^{*2}, \ldots, S_{n_k}^{*2}$ be the sample variances of these bootstrap samples.

Step 2. Compute $W_i^{*2} = \dfrac{S_{n_i}^{*2}}{S_{n_i}^2}$, $i = 1, \ldots, k$.

Step 3. Compute $\tilde{F}^* = \max\limits_{1 \le i \le k} \{W_i^{*2}\} / \min\limits_{1 \le i \le k} \{W_i^{*2}\}$.

Repeat these steps M times to obtain the EBD of $\tilde{F}_1^*, \ldots, \tilde{F}_M^*$.

Let $\tilde{F}_{1-\alpha}^*$ denote the $(1 - \alpha)$th quantile of this EBD distribution. The hypothesis H is rejected with level of significance α, if $\tilde{F} > \tilde{F}_{1-\alpha}^*$. The corresponding P^* level is the proportion of $\tilde{F}^*$ values that are greater than $\tilde{F}$.

Program VARTEST.EXE performs this bootstrapping, storing the output in VARTEST.DAT.

EXAMPLE **7.6** We now compare the variances of the resistance Res 3 in three hybrids. The data file is HYBRID.DAT. In this example, $n_1 = n_2 = n_3 = 32$. We find that $S_{n_1}^2 = 9929.54$, $S_{n_2}^2 = 16648.35$, and $S_{n_3}^2 = 21001.01$. The ratio of the maximal to minimal variance is $\tilde{F} = 2.11$. Executing program VARTEST.EXE with $M = 500$, we find that in this particular application, $P^* = .582$. For $\alpha = .05$, we find that $\tilde{F}_{.95}^* = 2.515$. The sample $\tilde{F}$ is smaller than $\tilde{F}_{.95}^*$. The hypothesis of equal variances cannot be rejected at a level of significance of $\alpha = .05$. ■

7.3.5.2 Comparing Several Means: The One-Way Analysis of Variance

The one-way analysis of variance, ANOVA, is a procedure of testing the equality of means, assuming that the variances of the populations are all equal. The hypothesis under test is

$$H_0 : \mu_1 = \mu_2 \cdots = \mu_k$$

Let $\bar{X}_{n_1}, S_{n_1}^2, \ldots, \bar{X}_{n_k}, S_{n_k}^2$ be the means and variances of the k samples. We compute the test statistic

$$F = \frac{\sum\limits_{i=1}^{k} n_i(\bar{X}_{n_i} - \bar{\bar{X}})^2 / (k - 1)}{\sum\limits_{i=1}^{k} (n_i - 1)S_{n_i}^2 / (N - k)} \tag{7.3.8}$$

where

$$\bar{\bar{X}} = \frac{1}{N} \sum\limits_{i=1}^{k} n_i \bar{X}_{n_i} \tag{7.3.9}$$

is the weighted average of the sample means, called the *grand mean*, and $N = \sum\limits_{i=1}^{k} n_i$ is the total number of observations.

The following is a **bootstrap ANOVA**. According to the bootstrap method, we repeat the following procedure M times:

Step 1. Draw k RSWRs of sizes $n_1, \ldots, n_k$ from the k given samples.

Step 2. For each bootstrap sample, compute the mean and variance, $\bar{X}^*_{n_i}$ and $S^{*2}_{n_i}$, $i = 1, \ldots, k$.

Step 3. For each $i = 1, \ldots, k$ compute

$$\bar{Y}^*_i = \bar{X}^*_{n_i} - (\bar{X}_{n_i} - \bar{\bar{X}})$$

(Notice that $\bar{\bar{Y}}^* = (1/N)\sum_{i=1}^{k} n_i \bar{Y}^*_i = \bar{\bar{X}}^*$, which is the grand mean of the k bootstrap samples.)

Step 4. Compute

$$F^* = \frac{\sum_{i=1}^{k} n_i (\bar{Y}^*_i - \bar{\bar{Y}}^*)^2/(k-1)}{\sum_{i=1}^{k} (n_i - 1)S^{*2}_{n_i}/(N-k)}$$

$$= \frac{\left[\sum_{i=1}^{k} n_i (\bar{X}^*_{n_i} - \bar{X}_{n_i})^2 - N(\bar{\bar{X}} - \bar{\bar{X}}^*)^2\right]/(k-1)}{\sum_{i=1}^{k} (n_i - 1)S^{*2}_{n_i}/(N-k)}$$

(7.3.10)

After M repetitions we obtain the EBD of $F^*_1, \ldots, F^*_M$.

Let $F^*_{1-\alpha}$ be the $(1 - \alpha)$th quantile of this EBD. The hypothesis H_0 is rejected at level of significance α, if $F > F^*_{1-\alpha}$. Alternatively, H_0 is rejected if the P^*-level is small, where

$$P^* = \text{Proportion of } F^* \text{ values greater than } F$$

This bootstrapping procedure can be performed by executing program ANOV-TEST.EXE. The output is stored in C:\ISTAT\DATA\ANOVTEST.DAT.

EXAMPLE 7.7 Testing the equality of the means in the HYBRID.DAT file, we obtain, using program ANOVTEST.EXE $M = 500$ times, the following statistics:

Hybrid1: $\bar{X}_{32} = 2143.406$, $S^2_{32} = 9929.54$

Hybrid2: $\bar{X}_{32} = 1902.813$, $S^2_{32} = 16648.35$

Hybrid3: $\bar{X}_{32} = 1850.344$, $S^2_{32} = 21001.01$

The test statistic is $F = 49.274$. The P^* level for this F is 0.0000. Thus, the hypothesis H_0 is rejected. The histogram of the EBD of the F^* values is shown in Figure 7.2. ∎

FIGURE 7.2

Histogram of the EBD of $M = 500$ F^* values

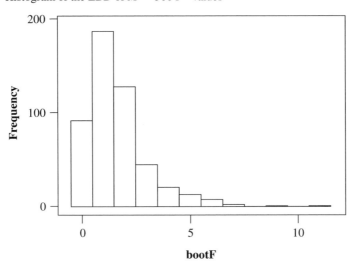

7.4
Bootstrap Tolerance Intervals
7.4.1 Bootstrap Tolerance Intervals for Bernoulli Samples

Trials (experiments) are called *Bernoulli trials* if the results of the trials are either 0 or 1 (head or tail; good or defective, and so on); the trials are independently performed, and the probability for 1 in a trial is a fixed constant p, $0 < p < 1$. A random sample (RSWR) of size n, from a population of 0s and 1s, whose mean is p (proportion of 1s) is called a *Bernoulli sample*. The number of 1s in such a sample has a binomial distribution. This is the sampling distribution of the number of 1s in all possible Bernoulli samples of size n and population mean p. The p is the probability that in a random drawing of an element from the population the outcome is 1.

Let X be the number of 1s in a RSWR of size n from such a population. If p is known, we can determine two integers $I_{\beta/2}(p)$ and $I_{1-\beta/2}(p)$ such that the proportion of Bernoulli samples for which $I_{\beta/2}(p) \leq X \leq I_{1-\beta/2}(p)$ is $1 - \beta$. Using MINITAB, we obtain these integers with the commands

```
MTB> let k1 = p
MTB> Inv_CDF β/2 k₂;
SUBC> Binomial n p
```

For example, if $n = 50$, $p = .1$, and $\beta = .05$, we obtain $I_{.025}(.1) = 1 (= k_2)$, and $I_{.975}(.1) = 9$.

If p is unknown and has to be estimated from a given Bernoulli sample of size n, we determine first the bootstrap $(1 - \alpha)$-level confidence interval for p. If the limits for this interval are $(p^*_{\alpha/2}, p^*_{1-\alpha/2})$, then the prediction interval $(I_{\beta/2}(p^*_{\alpha/2}), I_{1-\beta/2}(p^*_{1-\alpha/2}))$ is a **bootstrap tolerance interval** of confidence $1 - \alpha$ and content $1 - \beta$. The macro BINOPRED.MTB can be executed to obtain tolerance intervals. In this macro, $k1$ is the future Bernoulli sample size, n, $k3$ is $\beta/2$, and $k4$ is $1 - \beta/2$. Variable $k7$ is the size of the given Bernoulli sample, which is stored in column $C1$. Initiate $C3(1) = k1$, $C4(1) = k1$.

EXAMPLE **7.8**   Consider $n = 99$ electric voltage outputs of circuits, which are in data file OELECT.DAT. Suppose it is required that the output X be between 216 and 224 volts. We create a Bernoulli sample in which we give a circuit the value 1 if its electric output is in the interval $(216, 224)$ and the value 0 otherwise. This Bernoulli sample is stored in file ELECINDX.DAT.

The objective is to determine a $(.95, .95)$ tolerance interval for a future batch of $n = 100$ circuits from this production process. Using MINITAB, we import file ELECINDX.DAT to column $C1$. We set $k7 = 99$, $k1 = 100$, $k3 = 0.025$ and $k4 = 0.975$, and then apply MACRO BINOPRED.MTB $M = 500$ times. The next step is to order the columns $C3$ and $C4$, by the commands

```
MTB> SORT C3 C5
MTB> SORT C4 C6
```

Because $M\beta/2 = 500 \times .025 = 12.5$,

$$I_{.025}(p^*_{.025}) = (C5(12) + C5(13))/2 = 48$$

and

$$I_{.975}(p^*_{.975}) = (C6(487) + C6(488))/2 = 84$$

The bootstrap tolerance interval is $(48, 84)$. In other words, with confidence level of .95, we predict that 95% of future batches of $n = 100$ circuits will have between 48 and 84 circuits that comply to the standard. The exact tolerance interval limits are given by:

$$\text{Lower} = B^{-1}\left(\frac{\beta}{2}; n, \underline{p}_\alpha\right)$$

(7.4.1)

$$\text{Upper} = B^{-1}\left(1 - \frac{\beta}{2}; n, \bar{p}_\alpha\right)$$

where $(\underline{p}_\alpha, \bar{p}_\alpha)$ is a $1 - \alpha$ confidence interval for p. In these data, the .95 confidence interval for p is $(.585, .769)$. Thus, the $(.95, .95)$ tolerance interval is $(48, 84)$, which is equal to the bootstrap interval. ∎

7.4.2   Tolerance Intervals for Continuous Variables

In a RSWR of size n, the pth quantile—that is, $X_{(np)}$—is an estimator of the pth quantile of the distribution. Thus, we expect that the proportion of X values in the population falling in the interval $(X_{(n\beta/2)}, X_{(n(1-\beta/2))})$ is approximately $1 - \beta$ in large samples. As was explained in Chapter 2, $X_{(j)}$, $(j = 1, \ldots, n)$ is the jth order statistic of the sample, and for $0 < p < 1$, $X_{(j,p)} = X_{(j)} + p(X_{(j+1)} - X_{(j)})$. By the bootstrap method, we generate M replicas of the statistics $X^*_{(n\beta/2)}$ and $X^*_{(n(1-\beta/2))}$. The $(1 - \alpha, 1 - \beta)$ tolerance interval is given by $(Y^*_{(M\alpha/2)}, Y^{**}_{(M(1-\alpha/2))})$, where $Y^*_{(M\alpha/2)}$ is the $\alpha/2$-th quantile of the EBD of $X^*_{(n\beta/2)}$ and $Y^{**}_{(M(1-\alpha/2))}$ is the $(1 - \alpha/2)$-th quantile of the EBD of $X^*_{(n(1-\beta/2))}$. Macro CONTPRED.MTB provides M bootstrap copies of $X^*_{(n\beta/2)}$ and $X^*_{(n(1-\beta/2))}$, from which we determine the tolerance limits.

EXAMPLE **7.9**   Let us determine the $(.95, .95)$ tolerance interval for samples of size $n = 100$, of piston cycle times. Use the sample in the data file CYCLT.DAT. The original sample is of size $n_0 = 50$. Because future samples are of size $n = 100$, we draw from the original sample a RSWR of size $n = 100$. This bootstrap sample is put into column $C2$, and the ordered bootstrap sample is put into column $C3$. Because $n\beta/2 = 2.5$, $X^*_{(2.5)} = (C3(2) + C3(3))/2$. Similarly, $X^*_{(97.5)} = (C3(97) + C3(98))/2$. M replicas of $X^*_{(2.5)}$ and $X^*_{(97.5)}$ are put, respectively, into columns $C4$ and $C5$, $M = 500$. Finally, we sort column $C4$ and put it into $C6$, and sort $C5$ and put it into $C7$. $Y^*_{(M\alpha/2)} = (C6(12) + C6(13))/2$ and $Y^{**}_{(M(1-\alpha/2))} = (C7(487) + C7(488))/2$. A copy of this MINITAB session is given in the following box. Notice that the command **STORE** collects all commands written under STOR> into a macro whose name and address are given in the first line.

```
MTB> store 'C:\ISTAT\CHA7\CONTPRED.MTB'
Storing in file: C:\ISTAT\CHA7\CONTPRED.MTB
STOR> sample 100 C1 C2;
STOR> replace.
STOR> sort C2 C3
STOR> let k1 = (C3(2) + C3(3))/2
STOR> let k2 = (C3(97) + C3(98))/2
STOR> stack C4 k1 C4
STOR> stack C5 k2 C5
STOR> end
MTB> exec 'C:\ISTAT\CHA7\CONTPRED.MTB' 500
Executing from file: C:\ISTAT\CHA7\CONTPRED.MTB
MTB> sort C4 C6
MTB> sort C5 C7
MTB> let k3 = (C6(12) + C6(13))/2
MTB> let k4 = (C7(487) + C7(488))/2
```

To use macro CONTPRED.MTB on samples of size $n \neq 100$ and for $\beta \neq .05$, we need to edit it first. Change the sample size 100 in row 1 to the new n, and correspondingly modify $k1$ and $k2$. The bootstrap $(.95, .95)$ tolerance interval for $n = 100$ piston cycle times was estimated as $(0.175, 1.141)$.   ∎

7.5
Nonparametric Tests

Testing methods like the Z-tests, t-tests, and so on that were presented in Chapter 6 were designed for specific distributions. The Z- and t-tests are based on the assumption that the parent population is normally distributed. What would be the effect on the characteristics of the test if this basic assumption were wrong? This is an important question that deserves special investigation. If the population variance σ^2 is finite and the sample is large, then the t-test for the mean has approximately the required properties even if the parent population is not normal. In small samples, if it is doubtful whether the distribution of the parent population is normal, we should perform a distribution-free test, or compute the P-value of the test statistic by the bootstrapping method. In this section, we present three **nonparametric tests**, the so-called *sign test*, the *randomization test*, and the *Wilcoxon signed-rank test*.

7.5.1 The Sign Test

Suppose that $X_1, \ldots, X_n$ is a random sample from some continuous distribution F and has a positive p.d.f. throughout the range of X. Let ξ_p, for some $0 < p < 1$, be the pth quantile of F. We wish to test the hypothesis that ξ_p does not exceed a specified value ξ^*—that is,

$$H_0 : \xi_p \leq \xi^*$$

against the alternative

$$H_1 : \xi_p > \xi^*$$

If the null hypothesis H_0 is true, the probability of observing an X value smaller than ξ^* is greater or equal to p; and if H_1 is true, then this probability is smaller than p. The **sign test** of H_0 versus H_1 reduces the problem to a test for p in a binomial model. The test statistic is $K_n = \#\{X_i \leq \xi^*\}$—that is, the number of observed X values in the sample that do not exceed ξ^*. K_n has a binomial distribution $B(n, \theta)$, irrespective of the parent distribution F. According to H_0, $\theta \geq p$, and according to H_1, $\theta < p$. The test proceeds then as in Section 6.5.2.

EXAMPLE **7.10**   Continuing on with Example 7.8, we wish to test whether the median, $\xi_{.5}$, of the distribution of piston cycle times is greater than 0.50 (min). The sample data are in file CYCLT.DAT. The sample size is $n = 50$. Let $K_{50} = \#\{X_i \leq 50\}$. The null hypothesis is $H_0 : p \leq \frac{1}{2}$ vs. $H_1 : p > \frac{1}{2}$. From the sample values, we find $K_{50} = 24$. The P-value is $1 - B(23; 50, .5) = .664$. The null hypothesis H_0 is not rejected. The sample median is $M_e = 0.546$. This is, however, not significantly greater than 0.5. ∎

The sign test can be applied also to test whether tolerance specifications hold. Suppose that the standard specifications require that at least $1 - \beta$ proportion of products have an X value in the interval (ξ^*, ξ^{**}). If we wish to test this, with level of significance α, we can determine the $(1 - \alpha, 1 - \beta)$ tolerance interval for X based on the observed random sample, and not reject the hypothesis

$$H_0 : \xi^* \leq \xi_{\beta/2} \quad \text{and} \quad \xi_{1-\beta/2} \leq \xi^{**}$$

if the tolerance interval is included in (ξ^*, ξ^{**}).

We can also use the sign test. Given the random sample $X_1, \ldots, X_n$, we compute

$$K_n = \#\{\xi^* \leq X_i \leq \xi^{**}\}$$

The preceding null hypothesis H_0 is equivalent to the hypothesis

$$H_0^* : p \geq 1 - \beta$$

in the binomial test. H_0^* is rejected, with level of significance α, if

$$K_n < B^{-1}(\alpha; n, 1 - \beta)$$

where $B^{-1}(\alpha; n, 1 - \beta)$ is the α-quantile of the binomial distribution $B(n, 1 - \beta)$.

EXAMPLE **7.11** In Example 7.9, we found that the bootstrap $(.95, .95)$ tolerance interval for the CYCLT.DAT sample is $(0.175, 1.141)$. Suppose the specification requires that the piston cycle time in 95% of the cases be in the interval $(0.2, 1.1)$ (min). Can we reject the hypothesis

$$H_0^* : 0.2 \leq \xi_{.025} \quad \text{and} \quad \xi_{.975} \leq 1.1$$

with level of significance $\alpha = .05$? For the data CYCLT.DAT we find

$$K_{50} = \#\{.2 \leq X_i \leq 1.1\} = 41$$

Also $B^{-1}(.05, 50, .95) = 45$. Thus, because $K_{50} < 45$, H_0^* is rejected. This is in accord with the bootstrap tolerance interval because $(0.175, 1.141)$ contains the interval $(0.2, 1.1)$. ∎

7.5.2 The Randomization Test

The **randomization test** described here can be applied when testing whether two random samples come from the same distribution F without specifying the distribution F.

The null hypothesis, H_0, is that the two distributions from which the samples are generated are the same. The randomization test constructs a reference distribution for a specified test statistic by randomly assigning to the observations the labels of the samples. For example, let us consider two samples, which are denoted by A_1 and A_2. Each sample is of size $n = 3$. Suppose that we observed

$$
\begin{array}{cccccc}
A_2 & A_2 & A_2 & A_1 & A_1 & A_1 \\
1.5 & 1.1 & 1.8 & 0.75 & 0.60 & 0.80
\end{array}
$$

The sum of the values in A_2 is $T_2 = 4.4$ and that of A_1 is $T_1 = 2.15$. Is there an indication that the two samples are generated from different distributions? Let us consider the test statistic $D = (T_2 - T_1)/3$ and reject H_0 if D is sufficiently large. For the given samples $D = 0.75$. We now construct the reference distribution for D under H_0.

There are $\binom{6}{3} = 20$ possible assignments of the letters A_1 and A_2 to the six values. Each such assignment yields a value for D. The reference distribution assigns each such value of D an equal probability of $1/20$. The 20 assignments of letters and the corresponding D values are given in Table 7.3.

Under the reference distribution, each one of these values of D is equally probable, and the P-value of the observed value of the observed D is $P = \frac{1}{20} = .05$. The null hypothesis is rejected at the $\alpha = .05$ level. If n is large, it becomes impractical to construct the reference distribution in this manner. For example, if the number of samples is $t = 2$ and $n_1 = n_2 = 10$, we have $\binom{20}{10} = 184,756$ assignments.

We can, however, estimate the P-value from this reference distribution by sampling without replacement. This can be attained by MINITAB, using the macro RANDTES2.MTB. Before executing RANDTES2.MTB, we make the following preparation. We import the data file containing the two samples into column $C1$. In this column, sample A occupies the first n_1 rows, and sample B the last n_2 rows. After this, we perform the following MINITAB commands:

```
MTB> LET K1 = (n1).
MTB> LET K2 = (n2)
MTB> LET K3 = K1 + 1
MTB> LET K4 = K1 + K2
MTB> COPY C1 C3;
SUBC> USE 1 : K1.
MTB> COPY C1 C4;
SUBC> USE K3 : K4.
MTB> LET K5 = MEAN(C3) − MEAN(C4)
MTB> LET C5(1) = K5
```

TABLE 7.3

Assignments for the randomization test

Y_{ij}	Assignments									
0.75	1	1	1	1	1	1	1	1	1	1
0.60	1	1	1	1	2	2	2	2	2	2
0.80	1	2	2	2	1	1	1	2	2	2
1.5	2	1	2	2	1	2	2	1	1	2
1.1	2	2	1	2	2	1	2	1	2	1
1.8	2	2	2	1	2	2	1	2	1	1
D	0.750	0.283	0.550	0.083	0.150	0.417	−0.05	−0.050	−0.517	−0.250

Y_{ij}	Assignments									
0.75	2	2	2	2	2	2	2	2	2	2
0.60	1	1	1	1	1	1	2	2	2	2
0.80	1	1	1	2	2	2	1	1	1	2
1.5	1	2	2	1	1	2	1	1	2	1
1.1	2	1	2	1	2	1	1	2	1	1
1.8	2	2	1	2	1	1	2	1	1	1
D	0.250	0.517	0.050	0.050	−0.417	−0.150	−0.083	−0.550	−0.283	−0.750

Next we can execute macro RANDTES2.MTB. The first value in column $C5$ is the actual observed value of the test statistic. If we wish to compare the means of three or more samples, we should use the bootstrap ANOVA by ANOVTEST.EXE.

EXAMPLE 7.12   File OELECT.DAT contains $n_1 = 99$ random values of the output in volts of a rectifying circuit. File OELECT1.DAT contains $n_2 = 25$ values of outputs of another rectifying circuit. The question is whether the differences between the means of these two samples is significant. Let $\overline{X}$ be the mean of OELECT and $\overline{Y}$ be that of OELECT1. We find that $D = \overline{X} - \overline{Y} = -10.7219$. Executing macro RANDTES2.MTB 500 times yields results that, together with the original D, are described by the following sample statistics:

Descriptive Statistics						
Variable	N	Mean	Median	TrMEAN	StDev	SEMean
C5	501	−0.816	−0.0192	−0.0450	1.6734	0.0748
Variable	Min	Max	$Q1$	$Q3$		
C5	−10.7219	4.3893	−1.1883	1.0578		

Thus, the original mean -10.7219 is the minimum, and the test rejects the hypothesis of equal means with a P-value $P = \frac{1}{501}$. ■

7.5.3   The Wilcoxon Signed-Rank Test

In Section 7.5.1, we discussed the sign test. The **Wilcoxon signed-rank (WSR) test** is a modification of the sign test, which brings into consideration not only the signs of the sample values, but also their magnitudes. We construct the test statistic in two steps. First we rank the magnitudes (absolute values) of the sample values, giving the rank 1 to the value with smallest magnitude and the rank n to that with the maximal magnitude. In the second step, we sum the ranks of the positive values. For example, suppose that a sample of $n = 5$ is -1.22, -0.53, 0.27, 2.25, 0.89. The ranks of the magnitudes of these values are, respectively, 4, 2, 1, 5, 3. The signed-rank statistic is

$$W_5 = 0 \times 4 + 0 \times 2 + 1 + 5 + 3 = 9$$

Here we assigned each negative value the weight 0 and each positive value the weight 1.

The WSR test can be used for a variety of testing problems. If we wish to test whether the distribution median, $\xi_{.5}$, is smaller or greater than some specified value ξ^*, we can use the statistics

$$W_n = \sum_{i=1}^{n} I\{X_i > \xi^*\}R_i \qquad (7.5.1)$$

where

$$I\{X_i > \xi^*\} = \begin{cases} 1 & \text{if } X_i > \xi^* \\ 0 & \text{otherwise} \end{cases}$$

$R_i = \text{rank}(|X_i|)$.

The WSR test can also be applied to test whether two random samples are generated from the same distribution against the alternative that one comes from a distribution having a larger location parameter (median) than the other. In this case, we give the weight 1 to elements of sample 1 and the weight 0 to the elements of sample 2. The ranks of the values are determined by combining the two samples. For example, consider two random samples $X_1, \ldots, X_5$ and $Y_1, \ldots, Y_5$ generated from $N(0, 1)$ and $N(2, 1)$. These are

X	0.188	0.353	-0.257	0.220	0.168
Y	1.240	1.821	2.500	2.319	2.190

The ranks of the magnitudes of these values are

X	2	5	4	3	1
Y	6	7	10	9	8

The value of the WSR statistic is

$$W_{10} = 6 + 7 + 10 + 9 + 8 = 40$$

Notice that all ranks of the Y values are greater than those of the X values. This yields a relatively large value of W_{10}. Under the null hypothesis that the two samples are from the same distribution, the probability that the sign of a given rank is 1 is 1/2. Thus, the reference distribution for testing the significance of W_n is like that of

$$W_n^0 = \sum_{j=1}^{n} j B_j \left(1, \frac{1}{2}\right) \tag{7.5.2}$$

where $B_1 \left(1, \frac{1}{2}\right), \ldots, B_n \left(1, \frac{1}{2}\right)$ are mutually independent $B \left(1, \frac{1}{2}\right)$ random variables. The distribution of W_n^0 can be determined exactly. W_n^0 can assume the values $0, 1, \ldots, n(n+1)/2$ with probabilities that are the coefficients of the polynomial in t:

$$P(t) = \frac{1}{2^n} \prod_{j=1}^{n} \left(1 + t^j\right)$$

These probabilities can be computed exactly. For large values of n, W_n^0 is approximately normal with mean

$$E\{W_n^0\} = \frac{1}{2} \sum_{j=1}^{n} j = \frac{n(n+1)}{4} \tag{7.5.3}$$

and variance

$$V\{W_n^0\} = \frac{1}{4} \sum_{j=1}^{n} j^2 = \frac{n(n+1)(2n+1)}{24} \tag{7.5.4}$$

This can yield a large-sample approximation to the P-value of the test. The WSR test, to test whether the median of a symmetric continuous distribution F is equal to ξ^* or not, can be performed on MINITAB by using the command

```
MTB> WTest k1 C1
```

where $k1$ is the value of ξ^* and $C1$ is the column in which the sample resides.

7.6
MINITAB Macros and Executable Files Used in Chapter 7

ANOVTEST.EXE: Computes the difference in means in bootstrap distributions generated for comparing samples from several distributions in order to determine P-levels in testing statistical hypotheses on the difference in means between several distributions. The output is stored in ANOVTEST.DAT.

BINOPRED.MTB: Computes bootstrapped tolerance intervals from Bernoulli samples.

BOOT1SMP.EXE: Computes the mean and standard deviation of bootstrap distributions from a single sample.

BOOT1SMP.MTB: Computes the mean and standard deviation of bootstrap distributions from a single sample. (*Note*: Columns where stacking occurs have to be initiated.)

BOOT2SMP.EXE: Computes the studentized statistic of bootstrap distributions for comparing samples from two distributions in order to determine P-levels in testing statistical hypotheses on the difference in means and variances between the two distributions.

BOOTCORR.EXE: Computes the sample correlation and confidence intervals of bootstrap distributions from a single sample with two variables measured simultaneously.

BOOTPERC.MTB: Computes the 1st quartile, the mean, and the 3rd quartile of bootstrap distributions from a single sample.

BOOTREGR.MTB: Computes the least-square coefficients of a simple linear regression and their standard errors from bootstrap distributions from a single sample with two variables measured simultaneously.

CONFINT.MTB: Computes two sigma confidence intervals for a sample of size $k1$ to demonstrate the coverage probability of a confidence interval.

CONTPRED.MTB: Computes bootstrapped 95% tolerance intervals for means from one sample.

RANDTES2.MTB: Computes randomization distribution for comparing of two sample means.

STUDTEST.EXE: Computes the studentized statistic of bootstrap distributions from a single sample in order to determine P-levels in testing a statistical hypothesis.

VARTEST.EXE: Computes the maximum-to-minimum ratio of sample variances in bootstrap distributions generated for comparing samples from several distributions in order to determine P-levels in testing statistical hypotheses on the difference in variances between several distributions. The output is stored in VARTEST.DAT.

Chapter Highlights

Statistical inference is introduced by exploiting the power of the personal computer. Reference distributions are constructed through bootstrapping methods. Testing for statistical significance and the significance of least-square methods in simple linear regression using bootstrapping is demonstrated. Industrial applications are used throughout with specially written software simulations. Through this analysis, confidence intervals and reference distributions are derived and used to test statistical hypotheses. Bootstrap analysis of variance is developed for testing the equality of several population means. Tolerance distribution with bootstrapping is also presented. Three nonparametric procedures for testing are given: the sign test, randomization test, and Wilcoxon signed-rank test.

The main concepts and definitions introduced in this chapter include:

- Empirical reference distribution
- Bootstrap method
- Reference distribution

- Bootstrap population
- Studentized test statistic
- Bootstrap confidence interval
- Bootstrap ANOVA
- Bootstrap tolerance interval

- Nonparametric tests:
    Sign test
    Randomization test
    Wilcoxon signed-rank test

7.8
Exercises

7.1.1   Read file CAR.DAT into MINITAB. Five variables are stored in columns $C1$–$C5$. Write a macro that samples 64 values from column $C5$ (Mpg/City) with replacement and puts the sample in column $C6$. Let $k1$ be the mean of $C6$, and stack $k1$ in column $C7$. Execute this macro $M = 200$ times to obtain a sampling distribution of the sample means. Check graphically whether this sampling distribution is approximately normal. Also check whether the standard deviation of the sampling distribution is approximately $S/8$, where S is the standard deviation of $C5$.

7.1.2   Read file YARNSTRG.DAT into column $C1$ of MINITAB. Execute macro CONFINT.MTB (subdirectory ISTAT\CHA7) $M = 500$ times to obtain confidence intervals for the mean of $C1$. Use samples of size $n = 30$. Check in what proportion of samples the confidence intervals cover the mean of $C1$.

7.1.3   The average turn diameter of 58 U.S.-made cars in data file CAR.DAT is $\overline{X} = 37.203$ (m). Is this mean significantly larger than 37 (m)? To check this, use MINITAB with the following commands:

```
MTB> READ 'C: ISTAT\DATA\CAR.DAT' C1-C5
MTB> SORT C2 C3 C6 C7;
SUBC> BY C2.
MTB> COPY C7 C8;
SUBC> USE 1:58.
```

Column $C8$ contains the turn diameter of the 58 U.S.-made cars. Write a macro that samples 58 values from $C8$ with replacement, and puts them in $C9$. Stack the means of $C9$ in $C10$. Execute this macro 100 times. An estimate of the P-value is the proportion of means in $C10$ greater than $2 \times 37.203 - 37 = 37.406$. What is your estimate of the P-value?

7.2.1   You have to test whether the proportion of nonconforming units in a sample of size $n = 50$ from a production process is significantly greater than $p = .03$. Use MINITAB to determine when we should reject the hypothesis that $p \leq .03$ with $\alpha = .05$.

7.3.1   Use BOOT1SMP.EXE to generate 1000 bootstrap samples of the sample mean and sample standard deviation of the data in CYCLT.DAT on 50 piston cycle times.

a   Compute 95% confidence intervals for the sample mean and sample standard deviation.

b   Draw histograms of the EBD of the sample mean and sample standard deviation.

7.3.2   Use BOOTPERC.MTB to generate 1000 bootstrapped quartiles of the data in CYCLT.DAT.

a   Compute 95% confidence intervals for the 1st quartile, the median, and the 3rd quartile.

b   Draw histograms of the bootstrap quartiles.

7.3.3   Use program BOOTCORR.EXE to generate the EBD of size $M = 1000$ for the sample correlation ρ_{XY} between ISC1 and ISC2 in data file SOCELL.DAT. [For running the program, you have to prepare a temporary data file containing only the first two columns of SOCELL.DAT.] Compute the bootstrap confidence interval for ρ_{XY} at confidence level .95.

7.3.4   Generate the EBD of the regression coefficients (a, b) of miles per gallon/city, Y, versus horsepower, X, in data file CAR.DAT. See that X is in column $C1$, Y is in column $C2$. Run a simple regression with the command

```
MTB> Regr C2 1 C1;
SUBC> residuals C3.
```

The residuals are stored in $C3$. Let $k9$ be the sample size (109). Let $k1$ be the intercept a and $k2$ the slope b. Use the commands

```
MTB> Let k9 = 109
MTB> Let k1 = 30.7
MTB> Let k2 = −0.0736
MTB> Let C7(1) = k1
MTB> Let C8(1) = k2
```

Execute macro BOOTREGR.MTB $M = 100$ times.

a Determine a bootstrap confidence interval for the intercept a, at level .95.

b Determine a bootstrap confidence interval for the slope b, at level .95.

c Compare the bootstrap standard errors of a and b to those obtained from the formulas of Section 3.4.2.1.

7.3.5 Use STUDTEST.EXE to test the hypothesis that the data in CYCLT.DAT come from a distribution with mean $\mu_0 = 0.55$ sec.

a What is the P-value?

b Does the confidence interval derived in Exercise 7.3.1 include $\mu_0 = 0.55$?

c Could we have guessed the answer of part (b) after completing part (a)?

7.3.6 Use the program VARTEST.EXE to compare the variances of the two measurements recorded in data file ALMPIN2.DAT. First revise the file with the commands:

```
MTB> read 'c:\istat\data\almpin.dat' C1-C7
MTB> write 'c:\istat\data\almpin2.dat' C2 C3
```

a What is the P-value?

b Draw boxplots of the two measurements.

7.3.7 Use program BOOT2SMP.EXE to compare the means of the two measurements on the two variables Diameter 1 and Diameter 2 in ALMPIN2.DAT. What is the bootstrap estimate of the P-values for the means and variances?

7.3.8 Use program VARTEST.EXE to compare the variances of the gasoline consumption (mpg/city) of cars by origin. The data are in file MPG.DAT. There are $k = 3$ samples of sizes $n_1 = 58$, $n_2 = 14$, and $n_3 = 37$. Do you accept the null hypothesis of equal variances?

7.3.9 Use program ANOVTEST.EXE to test the equality of mean gas consumption (mpg/city) of cars by origin. The data file to use is MPG.DAT. The sample sizes are $n_1 = 58$, $n_2 = 14$, and $n_3 = 37$. The number of samples is $k = 3$. Do you accept the null hypothesis of equal means?

7.4.1 Use MINITAB to generate 50 random Bernoulli numbers with $p = .2$ and put them into $C1$. Use macro BINOPRED.MTB to obtain tolerance limits with $\alpha = .05$ and $\beta = .05$, for the number of nonconforming items in future batches of 50 items, when the process proportion of defectives is $p = .2$. Repeat this for $p = .1$ and $p = .05$.

7.4.2 Use macro CONTPRED.MTB to construct a $(.95, .95)$ tolerance interval for the piston cycle time from the data in OTURB.DAT.

7.4.3 Using macro CONTPRED.MTB and data file OTURB.DAT, determine a $(.95, .95)$ tolerance interval for the piston cycle times.

7.5.1 Using the sign test, test the hypothesis that the median, $\xi_{.5}$, of the distribution of cycle times of the piston is not exceeding $\xi^* = .7$ (min). The sample data are in file CYCLT.DAT. Use $\alpha = .10$ for level of significance.

7.5.2 Use the WSR test on the data of file OELECT.DAT to test whether the median of the distribution $\xi_{.5} = 220$ [Volts].

7.5.3. Apply the randomization test on the CAR.DAT file to test whether the turn diameter of foreign cars having four cylinders is different from that of U.S.-made cars with four cylinders.

8

Multiple Linear Regression and Analysis of Variance

In this chapter, we generalize the methodology of Section 3.4 to cases where the variability of a variable Y of interest can be explained to a large extent by the linear relationship between Y and k predicting or explaining variables $X_1, \ldots, X_k$. The number of explaining variables is $k \geq 2$. All the k variables $X_1, \ldots, X_k$ are continuous. The regression analysis of Y on several predictors is called **multiple regression**, in contrast to the simple regression discussed in Chapter 3. Multiple regression analysis is an important statistical tool for exploring the relationship between (the dependence of) one variable Y on a set of other variables. Applications of multiple regression analysis can be found in all areas of science and engineering. This method plays an important role in the statistical planning and control of industrial processes.

The statistical **linear model** for multiple regression is

$$y_i = \beta_0 + \sum_{j=1}^{k} \beta_j x_{ij} + e_i, \quad i = 1, \ldots, n$$

where $\beta_0, \beta_1, \ldots, \beta_k$ are the linear regression coefficients and e_i are random components. The commonly used method of estimating the regression coefficients and testing their significance is called *multiple regression analysis*. The method is based on the **principle of least squares**, according to which the regression coefficients are estimated by choosing $b_0, b_1, \ldots, b_k$ to minimize the sum of the squared residuals

$$\text{SSE} = \sum_{i=1}^{n} (y_i - (b_0 + b_1 x_{i1} + \cdots + b_k x_{ik}))^2$$

Sections 8.1–8.6 are devoted to the methods of regression analysis when both the response Y and the regressors $x_1, \ldots, x_k$ are quantitative variables. Section 8.7 examines quantal response regression, in which the response is a qualitative (binary) variable and the regressors $x_1, \ldots, x_k$ are quantitative. In particular the section looks at the logistic model and the logistic regression. Sections 8.8 and 8.9 discuss the analysis of variance for the comparison of sample means when the response is quantitative but the regressors are categorical variables. Finally, Section 8.10 examines comparison of proportions and testing hypotheses for contingency tables when both the response and the regressors are categorical.

8.1
Regression on Two Variables

The multiple regression linear model in the case of two predictors, assumes the form

$$y_i = \beta_0 + \beta_1 x_{i1} + \beta_2 x_{i2} + e_i \qquad i = 1, \ldots, n \tag{8.1.1}$$

The variables $e_1, \ldots, e_n$ are independent random variables, with $E\{e_i\} = 0$ and $V\{e_i\} = \sigma^2$, $i = 1, \ldots, n$. The principle of least squares calls for the minimization of the SSE. One can differentiate the SSE with respect to the unknown parameters. This yields the least-squares estimators, b_0, b_1, and b_2, of the regression coefficients, β_0, β_1, β_2. The formula for these estimators are:

$$b_0 = \bar{Y} - b_1 \bar{X}_1 - b_2 \bar{X}_2 \tag{8.1.2}$$

and b_1 and b_2 are obtained by solving the set of linear equations

$$\left. \begin{array}{l} S_{x_1}^2 b_1 + S_{x_1 x_2} b_2 = S_{x_1 y} \\ S_{x_1 x_2} b_1 + S_{x_2}^2 b_2 = S_{x_2 y} \end{array} \right\} \tag{8.1.3}$$

As before, $S_{x_1}^2$, $S_{x_1 x_2}$, $S_{x_2}^2$, $S_{x_1 y}$, and $S_{x_2 y}$ denote the sample variances and covariances of x_1, x_2, and y.

By simple substitution, we obtain the explicit formulas for b_1 and b_2:

$$b_1 = \frac{S_{x_2}^2 S_{x_1 y} - S_{x_1 x_2} S_{x_2 y}}{S_{x_1}^2 S_{x_2}^2 - S_{x_1 x_2}^2} \tag{8.1.4}$$

and

$$b_2 = \frac{S_{x_1}^2 S_{x_2 y} - S_{x_1 x_2} S_{x_1 y}}{S_{x_1}^2 S_{x_2}^2 - S_{x_1 x_2}^2} \tag{8.1.5}$$

The values $\hat{y}_i = b_0 + b_1 x_{1i} + b_2 x_{2i}$ $(i = 1, \ldots, n)$ are called the **predicted values**, or **FITs**, of the regression, and the residuals around the regression plane are

$$\hat{e}_i = y_i - \hat{y}_i$$
$$= y_i - (b_0 + b_1 x_{1i} + b_2 x_{2i}) \qquad i = 1, \ldots, n$$

The mean square of the residuals around the regression plane is

$$S_{y|(x_1, x_2)}^2 = S_y^2 (1 - R_{y|(x_1, x_2)}^2) \tag{8.1.6}$$

where

$$R_{y|(x_1, x_2)}^2 = \frac{1}{S_y^2} (b_1 S_{x_1 y} + b_2 S_{x_2 y}) \tag{8.1.7}$$

is the **multiple squared correlation** (multiple-R^2), and S_y^2 is the sample variance of y. The interpretation of the multiple-R^2 is as before—that is, the proportion of the variability of y that is explainable by the predictors (regressors) x_1 and x_2.

EXAMPLE **8.1**   We illustrate the fitting of a multiple regression on the following data, labeled GASOL.DAT. The data set consists of 32 measurements of distillation properties of crude oils (see Daniel and Wood, 1971, p. 165). There are five variables: $x_1, \ldots, x_4$ and y. These are

x_1 : crude oil gravity, °API

x_2 : crude oil vapor pressure, psi

x_3 : crude oil ASTM 10% point, °F

x_4 : gasoline ASTM endpoint, °F

y : yield of gasoline (in percentage of crude oil)

The measurements of crude oil and gasoline volatility ASTM in x_3 and x_4 measure the temperatures at which a given amount of liquid has been vaporized.

The sample correlations among these five variables are

	x_2	x_3	x_4	y
x_1	0.621	−0.700	−0.322	0.246
x_2		−0.906	−0.298	0.384
x_3			0.412	−0.315
x_4				0.712

We see that the yield y is highly correlated with x_4 and moderately corrected with x_2 (or x_3). The following is a MINITAB output of the regression of y on x_3 and x_4:

```
The regression equation is
Yield = 18.5 - 0.209 ASTM + 0.156 End_pt

Predictor        Coef       Stdev      t-ratio
Constant        18.468      3.009        6.14
ASTM           -0.20933     0.01274    -16.43
End_pt          0.155813    0.006855    22.73
s = 2.426    R-sq = 95.2%
```

We now compute these estimates of the regression coefficients using the preceding formulas. The variances and covariances of x_3, x_4, and y are (as derived from MINITAB):

	ASTM	End_pt	Yield
ASTM	1409.355		
End_pt	1079.565	4865.894	
Yield	−126.808	532.188	114.970

The means of these variables are $\overline{X}_3 = 241.500$, $\overline{X}_4 = 332.094$, and $\overline{Y} = 19.6594$. Thus, the least-squares estimators of b_1 and b_2 are obtained by solving the equations

$$1409.355b_1 + 1079.565b_2 = -126.808$$

$$1079.565b_1 + 4865.894b_2 = 532.188$$

The solution is

$$b_1 = -0.20933$$

and

$$b_2 = 0.15581$$

Finally, the estimate of β_0 is

$$b_0 = 19.6594 + 0.20933 \times 241.5 - 0.15581 \times 332.094$$

$$= 18.469$$

These are the same results as in the MINITAB output. Moreover, the multiple-R^2 is

$$R^2_{y|(x_3, x_4)} = \frac{1}{114.970}[0.20932 \times 126.808 + 0.15581 \times 532.88]$$

$$= .9521$$

In addition,

$$S^2_{y|(x_1, x_2)} = S^2_y(1 - R^2_{y|(x_1, x_2)})$$

$$= 114.97(1 - .9521)$$

$$= 5.50707$$

Figure 8.1 shows a scatterplot of the residuals $\hat{e}_i$ $(i = 1, \ldots, n)$ against the predicted values $\hat{y}_i$ $(i = 1, \ldots, n)$. This scatterplot does not reveal any pattern other than random. We can conclude that the regression of y on x_3 and x_4 accounts for all the systematic variability in the yield, y. Indeed, $R^2 = .952$, and no more than 4.8% of the variability in y is unaccounted by the regression. ∎

The following are formulas for the variances of the least-squares coefficients. First, we convert $S^2_{y|(x_1, x_2)}$ to S^2_e:

$$S^2_e = \frac{n-1}{n-3}S^2_{y|x} \tag{8.1.8}$$

FIGURE 8.1
Scatterplot of $\hat{e}$ vs. $\hat{Y}$

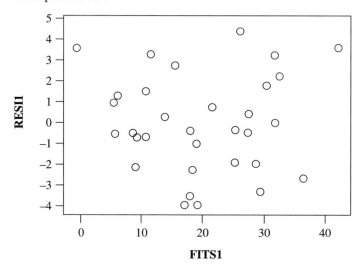

S_e^2 is an unbiased estimator of σ^2. The variance formulas are:

$$S_{b_0}^2 = \frac{S_e^2}{n} + \bar{x}_1^2 S_{b_1}^2 + \bar{x}_2^2 S_{b_2}^2 + 2\bar{x}_1 \bar{x}_2 S_{b_1 b_2}$$

$$S_{b_1}^2 = \frac{S_e^2}{n-1} \cdot \frac{S_{x_2}^2}{D}$$

$$S_{b_2}^2 = \frac{S_e^2}{n-1} \cdot \frac{S_{x_1}^2}{D}$$

$$S_{b_1 b_2} = -\frac{S_e^2}{n-1} \cdot \frac{S_{x_1 x_2}}{D}$$

(8.1.9)

where

$$D = S_{x_1}^2 S_{x_2}^2 - (S_{x_1 x_2})^2$$

EXAMPLE 8.2   Using the numerical results in Example 8.1 on the GASOL.DAT data, we find that

$$S_e^2 = 5.8869$$
$$D = 5,692,311.4$$
$$S_{b_1}^2 = 0.0001624$$
$$S_{b_2}^2 = 0.0000470$$
$$S_{b_1, b_2} = -0.0000332$$

and

$$S_{b_0}^2 = 9.056295$$

The squared roots of these variance estimates are the "Stdev" values printed in the MINITAB output, and $s = S_e$. ∎

8.2
Partial Regression and Correlation

In performing a multiple least-squares regression, we can study, in stages, the effect of the predictors on the response. This more pedestrian approach does not simultaneously provide all regression coefficients, but it does study the effect of predictors in more detail.

In stage I, we perform a simple linear regression of the yield y on one of the predictors—x_1, say. Let $a_0^{(1)}$ and $a_1^{(1)}$ be the intercept and slope coefficients of this simple linear regression. Let $\hat{\mathbf{e}}^{(1)}$ be the vector of residuals

$$\hat{e}_i^{(1)} = y_i - (a_0^{(1)} + a_1^{(1)} x_{1i}) \qquad i = 1, \ldots, n \qquad \text{(8.2.1)}$$

In stage II, we perform a simple linear regression of the second predictor, x_2, on the first predictor, x_1. Let $c_0^{(2)}$ and $c_1^{(2)}$ be the intercept and slope coefficients of this regression. Let $\hat{\mathbf{e}}^{(2)}$ be the vector of residuals,

$$\hat{e}_i^{(2)} = x_{2i} - (c_0^{(2)} + c_1^{(2)} x_{1i}) \qquad i = 1, \ldots, n \qquad \text{(8.2.2)}$$

In stage III, we perform a simple linear regression of $\hat{\mathbf{e}}^{(1)}$ on $\hat{\mathbf{e}}^{(2)}$. It can be shown that this linear regression must pass through the origin—that is, it has a zero intercept. Let $d^{(3)}$ be the slope coefficient.

The simple linear regression of $\hat{\mathbf{e}}^{(1)}$ on $\hat{\mathbf{e}}^{(2)}$ is called the **partial regression**. The correlation between $\hat{\mathbf{e}}^{(1)}$ and $\hat{\mathbf{e}}^{(2)}$ is called the **partial correlation** of y and x_2, given x_1, and is denoted by $r_{yx_2 \cdot x_1}$.

From the regression coefficients obtained in these three stages, we can determine the multiple regression coefficients of y on x_1 and x_2, according to the formulas:

$$
\begin{aligned}
b_0 &= a_0^{(1)} - d^{(3)} c_0^{(2)} \\
b_1 &= a_1^{(1)} - d^{(3)} c_1^{(2)} \\
b_2 &= d^{(3)}
\end{aligned}
\qquad \text{(8.2.3)}
$$

EXAMPLE 8.3 For the GASOL data, let us determine the multiple regression of the yield (y) on the ASTM (x_3) and the End_pt (x_4) in stages.

In stage I, the simple linear regression of y on ASTM is

$$\hat{y} = 41.4 - 0.08998 \cdot x_3$$

The residuals of this regression are $\hat{\mathbf{e}}^{(1)}$. Also, $R^2_{yx_3} = .099$. In stage II, the simple linear regression of x_4 on x_3 is

$$\hat{x}_4 = 147 + 0.766 \cdot x_3$$

The residuals of this regression are $\hat{\mathbf{e}}^{(2)}$. Figure 8.2 shows the scatterplot of $\hat{\mathbf{e}}^{(1)}$ versus $\hat{\mathbf{e}}^{(2)}$. The partial correlation is $r_{yx_4 \cdot x_3} = .973$. This high partial correlation means that, after adjusting the variability of y for the variability of x_3, and the variability of x_4 for that of x_3, the adjusted x_4 values—namely, $\hat{e}^{(2)}_i$ $(i = 1, \ldots, n)$— are still good predictors for the adjusted y values—namely, $\hat{e}^{(1)}_i$ $(i = 1, \ldots, n)$. The regression of $\hat{\mathbf{e}}^{(1)}$ on $\hat{\mathbf{e}}^{(2)}$, determined in stage III, is

$$\hat{\hat{e}}^{(1)} = 0.156 \cdot \hat{e}^{(2)}$$

We have found the following estimates:

$$a^{(1)}_0 = 41.4 \qquad a^{(1)}_1 = -0.090$$

$$c^{(2)} = 147.0 \qquad c^{(2)}_1 = 0.766$$

$$d^{(3)} = 0.156$$

From these formulas, we get

$$b_0 = 41.4 - 0.156 \times 147.0 = 18.468$$

$$b_1 = -0.0900 - 0.156 \times 0.766 = -0.2095$$

$$b_2 = 0.156$$

FIGURE 8.2

Scatterplot of $\hat{e}_1$ vs. $\hat{e}_2$

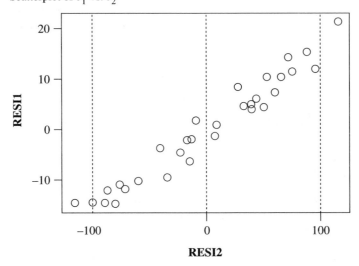

These values coincide with the previously determined coefficients. Finally, the relationship between the multiple and the partial correlations is

$$R^2_{y|(x_1,x_2)} = 1 - (1 - R^2_{yx_1})(1 - r^2_{yx_2 \cdot x_1})$$ (8.2.4)

In this example,

$$R^2_{y|(x_3,x_4)} = 1 - (1 - .099)(1 - .94673) = .9520$$ ∎

8.3

Multiple Linear Regression

In the general case, we have k predictors $(k \geq 1)$. Let (X) denote an array of n rows and $(k + 1)$ columns, in which the first column consists of the value 1 in all entries and the second to $(k + 1)$st columns consist of the values of the predictors $x_1, \ldots, x_k$. (X) is called the predictors' matrix. Let **y** be an array of n rows and one column consisting of the values of the response. The linear regression model can be written in matrix notation as

$$\mathbf{y} = (X)\boldsymbol{\beta} + \mathbf{e}$$ (8.3.1)

where $\boldsymbol{\beta}' = (\beta_0, \beta_1, \ldots, \beta_k)$ is the vector of regression coefficients, and **e** is a vector of random residuals. The ( )′ denotes the transpose of the vector. For the essentials of matrix algebra, see Appendix I.

The sum of squares of residuals can be written as

$$\text{SSE} = (\mathbf{Y} - (X)\boldsymbol{\beta})'(\mathbf{Y} - (X)\boldsymbol{\beta})$$ (8.3.2)

Notice that SSE defined at the beginning of the chapter is the same. Differentiating SSE partially with respect to the components of $\boldsymbol{\beta}$, and equating the partial derivatives to zero, we obtain a set of linear equations in **b**—namely,

$$(X)'(X)\mathbf{b} = (X)'Y$$ (8.3.3)

$(X)'$ is the transpose of the matrix (X). These linear equations are called the **normal equations**. **b** is the least-squares estimator LSE of $\boldsymbol{\beta}$.

If we define the matrix

$$B = [(X)'(X)]^{-1}(X)'$$ (8.3.4)

where $[\ ]^{-1}$ is the inverse of $[\ ]$, then the general formula of the least-squares regression coefficients vector $\mathbf{b}' = (b_0, \ldots, b_k)$ is given in matrix notation as

$$\mathbf{b} = (B)\mathbf{y}$$ (8.3.5)

The vector of predicted y values, or FITS, is given by $\hat{\mathbf{y}} = (H)\mathbf{y}$, where $(H) = (X)(B)$. The vector of residuals $\hat{\mathbf{e}} = \mathbf{y} - \hat{\mathbf{y}}$ is given by

$$\hat{\mathbf{e}} = (I - H)\mathbf{y}$$ (8.3.6)

where (I) is the $n \times n$ identity matrix. The variance of $\hat{\mathbf{e}}$ around the regression surface is

$$S_e^2 = \frac{1}{n-k-1} \sum_{i=1}^{n} \hat{e}_i^2$$

$$= \frac{1}{n-k-1} \mathbf{y}'(I-H)\mathbf{y}$$

The sum of squares of $\hat{e}_i$ $(i = 1, \ldots, n)$ is divided by $(n-k-1)$ to attain an unbiased estimator of σ^2. The multiple-R^2 is given by

$$R_{y|(x)}^2 = \frac{1}{(n-1)S_y^2} (\mathbf{b}'(X)'\mathbf{y} - n\bar{y}^2) \tag{8.3.7}$$

where $x_{i0} = 1$ for all $i = 1, \ldots, n$, and S_y^2 is the sample variance of y. Finally, an estimate of the variance–covariance matrix of the regression coefficients $b_0, \ldots, b_k$ is

$$(V_b) = S_e^2 [(X)'(X)]^{-1} \tag{8.3.8}$$

EXAMPLE **8.4**   The following is the MINITAB output of a multiple regression. We again use the ALMPIN data set, and regress the Cap Diameter (y) on Diameter 1 (x_1), Diameter 2 (x_2), and Diameter 3 (x_3). The Stdev of the regression coefficients are the square-roots of the diagonal elements of the (V_b) matrix. To see this, we present first the inverse of the $(X)'(X)$ matrix, which is given by the following symmetric matrix:

$$[(X)'(X)]^{-1} = \begin{bmatrix} 5907.11 & -658.57 & 557.99 & -490.80 \\ \cdot & 695.56 & -448.14 & -181.94 \\ \cdot & \cdot & 739.76 & -347.37 \\ \cdot & \cdot & \cdot & 578.75 \end{bmatrix}$$

The value of S_e^2 is the square of the printed s value—that is, $S_e^2 = 0.0000457$. Thus, the variances of the regression coefficients are:

$$S_{b_0}^2 = 0.0000457 \times 5907.11 = 0.2700206$$

$$S_{b_1}^2 = 0.0000457 \times 695.56 = 0.0317871$$

$$S_{b_2}^2 = 0.0000457 \times 739.76 = 0.033807$$

$$S_{b_3}^2 = 0.0000457 \times 578.75 = 0.0264489$$

Thus, S_{b_i} $(i = 0, \ldots, 3)$ are the Stdev in the printout. The t-ratios are given by

$$t_i = \frac{b_i}{S_{b_i}} \qquad i = 0, \ldots, 3$$

The t-ratios should be large to be considered significant. The significance criterion is given by the P-value. A large value for P indicates that the regression coefficient is not significantly different from zero. In the preceding list, we see that b_2 is not significant. Notice, however, that Diameter 2 by itself, as the sole predictor of Cap

Diameter, is very significant. This can be verified by running a simple regression of y on x_2. However, in the presence of x_1 and x_3, x_2 loses its significance.

The regression equation is
CapDiam = 4.04 + 0.755 Diam1 + 0.017 Diam2 + 0.323 Diam3

Predictor	Coef	Stdev	t-ratio	P
Constant	4.0411	0.5196	7.78	.000
Diam1	0.7555	0.1783	4.24	.000
Diam2	0.0172	0.1839	0.09	.926
Diam3	0.3227	0.1626	1.98	.051

s = 0.006761   R-sq = 87.9%   R-sq(adj) = 87.4%

Analysis of Variance

SOURCE	DF	SS	MS	F	P
Regression	3	0.0219204	0.0073068	159.86	.000
Error	66	0.0030167	0.0000457		
Total	69	0.0249371			

The analysis of variance table provides a global summary of the contribution of the various factors to the variability of y. The total sum of squares (SST) of y around its mean is

$$\text{SST} = (n-1)S_y^2 = \sum_{i=1}^{n}(y_i - \bar{y})^2 = .0249371$$

This value of SST is partitioned into the sum of the variability explainable by the regression (SSR) and that due to the residuals around the regression (error, SSE). These are given by

$$\text{SSE} = (n - k - 1)S_e^2$$

$$\text{SSR} = \text{SST} - \text{SSE}$$

$$= \mathbf{b}'(X)'\mathbf{y} - n\bar{Y}_n^2$$

$$= \sum_{j=0}^{k} b_j \cdot \sum_{i=1}^{n} X_{ij} y_i - \left(\sum_{i=1}^{n} y(i)\right)^2 / n$$

Figure 8.3 shows the scatterplot of the residuals, $\hat{e}_i$, versus the predicted values (FITS), $\hat{y}_i$. There is one point in this figure, corresponding to element 66 in the data set, whose x values have a strong influence on the regression. ∎

Multiple regression can be used to test whether two or more simple linear regressions are parallel (have the same slopes) or have the same intercepts. We will show this by comparing two simple linear regressions.

FIGURE 8.3

Scatterplot of the residual versus the predicted values of CapDiam

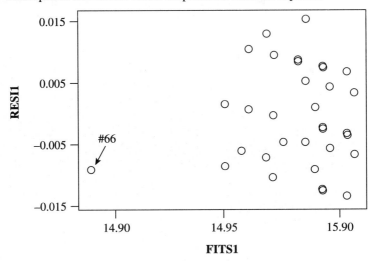

Let $(x_i^{(1)}, Y_i^{(1)})$, $i = 1, \ldots, n$, be the data set of the simple linear regression of Y on x, and let $(x_j^{(2)}, Y_j^{(2)})$, $j = 1, \ldots, n_2$ be that of the second regression. By combining the data on the regression x from the two sets, we get the $\mathbf{x}$ vector

$$\mathbf{x} = (x_1^{(1)}, \ldots, x_{n_1}^{(1)}, x_1^{(2)}, \ldots, x_{n_2}^{(2)})'$$

In a similar fashion, we combine the Y values and set

$$\mathbf{Y} = (Y_1^{(1)}, \ldots, Y_{n_1}^{(1)}, Y_1^{(2)}, \ldots, Y_{n_2}^{(2)})'$$

Introduce a dummy variable z. The vector $\mathbf{z}$ has n_1 0s at the beginning followed by n_2 1s. Consider now the multiple regression

$$\mathbf{Y} = b_0 \mathbf{1} + b_1 \mathbf{x} + b_2 \mathbf{z} + b_3 \mathbf{w} + \mathbf{e} \tag{8.3.9}$$

where $\mathbf{1}$ is a vector of $(n_1 + n_2)$ 1s, and $\mathbf{w}$ is a vector of length $(n_1 + n_2)$ whose ith component is the product of the corresponding components of $\mathbf{x}$ and $\mathbf{z}$—that is, $w_i = x_i z_i$ $(i = 1, \ldots, n_1 + n_2)$. Perform the regression analysis of $\mathbf{Y}$ on $(\mathbf{x}, \mathbf{z}, \mathbf{w})$. If b_2 is significantly different from 0, we conclude that the two simple regression lines have different intercepts. If b_3 is significantly different from 0, we conclude that the two lines have different slopes.

EXAMPLE 8.5 In this example, we compare the simple linear regressions of Turn Diameter (Y) on MPG/City (x) of U.S.-made cars and of Japanese cars. The data are in the file CAR.DAT. The simple linear regression for U.S. cars is

$$\hat{Y} = 49.0769 - 0.7565x$$

with $R^2 = .432$, $S_e = 2.735$ (56 degrees of freedom). The simple linear regression for Japanese cars is

$$\hat{Y} = 42.0860 - 0.5743x$$

with $R^2 = .0854$, $S_e = 3.268$ (35 degrees of freedom). The combined multiple regression of $\mathbf{Y}$ on $\mathbf{x}$, $\mathbf{z}$, $\mathbf{w}$ yields the following table of P-values of the coefficients:

Coefficients:

	Value	Std.Error	t value	$\Pr(> |t|)$
(Intercept)	49.0769	5.3023	9.2557	.0000
mpgc	−0.7565	0.1420	−5.3266	.0000
z	−6.9909	10.0122	−0.6982	.4868
w	0.1823	0.2932	0.6217	.5357

We see in this table that the P-values corresponding to $\mathbf{z}$ and $\mathbf{w}$ are .4868 and .5357, respectively. Accordingly, both b_2 and b_3 are not significantly different from zero. We conclude that the two regression lines are not significantly different. We can combine the data and have one regression line for both U.S. and Japanese cars— namely,

$$\hat{Y} = 44.8152 - 0.6474x$$

with $R^2 = .3115$, $S_e = 3.337$ (93 degrees of freedom). ∎

8.4

Partial-F Tests and the Sequential SS

In a MINITAB output for multiple regression, a column labeled "SEQ SS" (for **sequential SS**) partitions the regression sum of squares, SSR, into additive components of variance, each one with 1 degree of freedom. We have seen that the multiple-R^2, $R^2_{y|(x_1,...,x_k)} = \text{SSR}/\text{SST}$, is the proportion of total variability explainable by the linear dependence of Y on all the k regressors. A simple linear regression on the first variable, x_1, yields a smaller $R^2_{y|x_1}$. The first component of the SEQ SS is $\text{SSR}_{y|(x_1)} = \text{SST} \cdot R^2_{y|(x_1)}$. If we determine the multiple regression of Y on x_1 and x_2, then $\text{SSR}_{y|(x_1,x_2)} = \text{SST} \cdot R^2_{y|(x_1,x_2)}$ is the amount of variability explained by the linear relationship with the two variables. The difference

$$\text{DSS}_{x_2|x_1} = \text{SST}(R^2_{y|(x_1,x_2)} - R^2_{y|(x_1)}) \tag{8.4.1}$$

is the additional amount of variability explainable by x_2, after accounting for x_1. Generally, for $i = 2, \ldots, k$

$$\text{DSS}_{x_i|x_1...,x_{i-1}} = \text{SST}(R^2_{y|(x_1,...,x_i)} - R^2_{y|(x_1,...,x_{i-1})}) \tag{8.4.2}$$

is the additional contribution of the ith variable after controlling for the first $(i - 1)$ variables.

Let

$$S^2_{e(i)} = \frac{\text{SST}}{n-i-1}(1 - R^2_{y|(x_1,\ldots,x_i)}) \qquad i = 1, \ldots, k \qquad \text{(8.4.3)}$$

Then

$$F^{(i)} = \frac{\text{DSS}_{x_i|x_1,\ldots,x_{i-1}}}{S^2_{e(i)}} \qquad i = 1, \ldots, k \qquad \text{(8.4.4)}$$

is called the **partial-F** for testing the significance of the contribution of the variable x_i, after controlling for $x_1, \ldots, x_{i-1}$. If $F^{(i)}$ is greater than the $(1 - \alpha)$th quantile, $F_{1-\alpha}[1, n-i-1]$ of the F distribution, then the additional contribution of X_i is significant. The partial-F test is used to assess whether the addition of the ith regressor significantly improves the prediction of Y, given that the first $i - 1$ regressors have already been included.

EXAMPLE 8.6   In Example 8.4, we examined the multiple regression of CapDiam, y, on Diam1, Diam2, and Diam3 in the ALMPIN.DAT file. Here we compute partial-F statistics corresponding to the SEQ SS values.

Variable	SEQ SS	SSE	d.f.	Partial-F	P-value
Diam1	.0216567	.003280	68	448.98	0
Diam2	.0000837	.003197	67	1.75	.190
Diam3	.0001799	.003167	66	3.75	.057

We see from these partial-F values and their corresponding P-values that, after using Diam1 as a predictor, the additional contribution of Diam2 is insignificant. Diam3, however, when added to the regressor Diam1, significantly decreases the variability that is left unexplained.   ∎

The partial-F test is sometimes called the *sequential-F* test (Draper and Smith, 1981, p. 612). We use the terminology *partial-F statistic* because of the following relationship between the partial-F and the partial correlation. In Section 8.2, we defined the partial correlation $r_{yx_2 \cdot x_1}$ as the correlation between $\hat{\mathbf{e}}^{(1)}$, which is the vector of residuals around the regression of Y on x_1, and $\hat{\mathbf{e}}^{(2)}$, the vector of residuals of x_2 around its regression on x_1. Generally, suppose that we have determined the multiple regression of Y on $(x_1, \ldots, x_{i-1})$. Let $\hat{\mathbf{e}}(y \mid x_1, \ldots, x_{i-1})$ be the vector of residuals around this regression. Let $\hat{\mathbf{e}}(x_i \mid x_1, \ldots, x_{i-1})$ be the vector of residuals around the multiple regression of x_i on $x_1, \ldots, x_{i-1}$ $(i \geq 2)$. The correlation between $\hat{\mathbf{e}}(y \mid x_1, \ldots, x_{i-1})$ and $\hat{\mathbf{e}}(x_i \mid x_1, \ldots, x_{i-1})$ is the partial correlation between Y and x_i, given $x_1, \ldots, x_{i-1}$. We denote this partial correlation by $r_{yx_i \cdot x_1, \ldots, x_{i-1}}$. The following relationship holds between the partial-F, $F^{(i)}$, and the partial correlation:

$$F^{(i)} = (n - i - 1)\frac{r^2_{yx_i \cdot x_1,\ldots,x_{i-1}}}{1 - r^2_{yx_i \cdot x_1,\ldots,x_{i-1}}} \qquad i \geq 2 \qquad \text{(8.4.5)}$$

This relationship is used to test whether $r_{yx_i \cdot x_1,\ldots,x_{i-1}}$ is significantly different from zero. $F^{(i)}$ should be larger than $F_{1-\alpha}[1, n - i - 1]$.

Model Construction: Stepwise Regression

Often, data can be collected on a large number of regressors, which might help us predict the outcomes of a certain variable, Y. However, different regressors vary generally with respect to the amount of variability in Y that they can explain. Moreover, different regressors or predictors are sometimes highly correlated and therefore they might not all be needed to explain the variability in Y, or to be used as predictors.

The following example is given by Draper and Smith (1981, p. 615). The amount of steam (in lb) used monthly, Y, in a plant may depend on nine regressors:

$x_1 =$ Pounds of real fatty acid in storage per month

$x_2 =$ Pounds of crude glycerine made in a month

$x_3 =$ Monthly average wind velocity (miles/hour)

$x_4 =$ Plant operating days per month

$x_5 =$ Number of days per month with temperature below $32°$F

$x_6 =$ Monthly average atmospheric temperature ($°$F)

and three additional regressors that are not listed. Are all regressors required in order for us to be able to predict Y? If not, which variables should be used? This is the problem of model construction.

There are several techniques for constructing a regression model. MINITAB, S-PLUS, and other computer packages use a stepwise method (**stepwise regression**) that is based on forward selection, backward elimination, and user intervention, which can force certain variables to be included. Here we discuss only the forward selection procedure.

In the first step, we select the variable x_j $(j = 1, \ldots, k)$ whose correlation with Y has maximal magnitude, provided it is significantly different from zero.

At each step, the procedure computes a partial-F, or partial correlation, for each variable, x_l, that was not selected in the previous steps. The variable having the largest significant partial-F is selected. The procedure stops when no additional variables can be selected. We illustrate the forward stepwise regression in the following example.

EXAMPLE 8.7 In Example 8.1, we introduced the data file GASOL.DAT and performed a multiple regression of y on x_3 and x_4. In this example, we apply the MINITAB stepwise regression procedure to arrive at a linear model that includes all the variables that contribute significantly to the prediction. Following is the MINITAB output:

Stepwise Regression
F-to-Enter: 4.00   F-to-Remove: 4.00
Response is gasy on 4 predictors, with N = 32

Step	1	2	3
Constant	−16.662	18.468	4.032
gasx4	0.1094	0.1558	0.1565
T-Ratio	5.55	22.73	24.22
gasx3		−0.209	−0.187
T-Ratio		−16.43	−11.72
gasx1			0.22
T-Ratio			2.17
S	7.66	2.43	2.28
R-sq(%)	50.63	95.21	95.90

More? (Yes, No, Subcommand, or Help)
SUBC> No.

At each stage, the MINITAB procedure includes the variable whose partial-F value is maximal but greater than "F-to-enter," which is 4.00. Because each partial-F statistic has 1 degree of freedom in the denominator, and because

$$(F_{1-\alpha}[1, v])^{1/2} = t_{1-\alpha/2}[v]$$

the output prints the corresponding values of t. Thus, in step 1, variable x_4 is selected (gasx4). The fitted regression equation is

$$\hat{Y} = -16.662 + 0.1094x_4$$

with $R^2_{y|(x_4)} = .5063$. The partial-F for x_4 is $F = (5.55)^2 = 30.8025$. Because this value is greater than "F-to-remove," which is 4.00, x_4 remains in the model. In step 2, the maximal partial correlation of Y and x_1, x_2, x_3 given x_4, is that of x_3, with partial-$F = 269.9449$. Variable x_3 is selected, and the new regression equation is

$$\hat{Y} = 18.468 + 0.1558x_4 - 0.2090x_3$$

with $R^2_{y|(x_4, x_3)} = .9521$. Because the partial-$F$ of x_4 is $(22.73)^2 = 516.6529$, the two variables remain in the model. In step 3, the variable x_1 is chosen. Because its partial-F is $(2.17)^2 = 4.7089$, it is included as well. The final regression equation is

$$\hat{Y} = 4.032 + 0.1565x_4 - 0.1870x_3 + 0.2200x_1$$

with $R^2_{y|(x_4, x_3, x_1)} = .959$. Only 4.1% of the variability in Y is left unexplained.

To conclude the example, we provide the three regression analyses from MINITAB. One can compute the partial-F values from these tables.

MTB> Regress 'gasy' 1 'gasx4';
SUBC> Constant.

Regression Analysis

The regression equation is gasy = −16.7 + 0.109 gasx4

Predictor	Coef	Stdev	t-ratio	p
Constant	−16.662	6.687	−2.49	.018
gasx4	0.10937	0.01972	5.55	.000

s = 7.659   R-sq = 50.6%   R-sq(adj) = 49.0%

Analysis of Variance

SOURCE	DF	SS	MS	F	p
Regression	1	1804.4	1804.4	30.76	.000
Error	30	1759.7	58.7		
Total	31	3564.1			

Unusual observations

Obs.	gasx4	gasy	Fit	Stdev.Fit	Residual	St.Resid
4	407	45.70	27.85	2.00	17.85	2.41R

R denotes an obs. with a large st. resid.

MTB> Regress 'gasy' 2 'gasx4' 'gasx3';
SUBC> Constant.

Regression Analysis

The regression equation is gasy = 18.5 + 0.156 gasx4 − 0.209 gasx3

Predictor	Coef	Stdev	t-ratio	p
Constant	18.468	3.009	6.14	.000
gasx4	0.155813	0.006855	22.73	.000
gasx3	−0.20933	0.01274	−16.43	.000

s = 2.426   R-sq = 95.2%   R - sq(adj) = 94.9%

Analysis of Variance

SOURCE	DF	SS	MS	F	p
Regression	2	3393.5	1696.7	288.41	.000
Error	29	170.6	5.9		
Total	31	3564.1			

SOURCE	DF	SEQ SS
gasx4	1	1804.4
gasx3	1	1589.1

MTB> Regress 'gasy' 3 'gasx4' 'gasx3' 'gasx1';
SUBC> Constant.

Regression Analysis

The regression equation is gasy = 4.03 + 0.157 gasx4 − 0.187 gasx3 + 0.222 gasx1

Predictor	Coef	Stdev	t-ratio	p
Constant	4.032	7.223	0.56	.581
gasx4	0.156527	0.006462	24.22	.000
gasx3	−0.18657	0.01592	−11.72	.000
gasx1	0.2217	0.1021	2.17	.038

$s = 2.283$ R-sq $= 95.9\%$ R-sq(adj) $= 95.5\%$

Analysis of Variance

SOURCE	DF	SS	MS	F	p
Regression	3	3418.1	1139.4	218.51	.000
Error	28	146.0	5.2		
Total	31	3564.1			

SOURCE	DF	SEQ SS
gasx4	1	1804.4
gasx3	1	1589.1
gasx1	1	24.6

Unusual Observations

Obs.	gasx4	gasy	Fit	Stdev.Fit	Residual	St.Resid
17	340	30.400	25.634	0.544	4.766	2.15R

R denotes an obs. with a large st. resid.

MTB>

8.6
Regression Diagnostics

As mentioned in Chapter 3, the least-squares regression line is sensitive to extreme x or y values of the sample elements. Sometimes even one point may change the characteristics of the regression line substantially. We illustrate this in the following example.

EXAMPLE 8.8   Consider again the SOCELL data from Chapter 3. We saw earlier that the regression line (L1) of ISC at time $t2$ on ISC at time $t1$ is $\hat{y} = 0.536 + 0.929x$, with $R^2 = .959$.

The point having the largest x value has a y value of 5.37. If the y value of this point is changed to 4.37, we obtain a different regression line (L2), given by $\hat{y} = 2.04 + 0.532x$, with $R^2 = .668$. MINITAB diagnostics singles out this point as unusual, as seen in the following box:

Unusual Observations

Obs.	ISC1	ISC2	Fit	Stdev.Fit	Residual	St.Resid
9	5.11	4.3700	4.7609	0.1232	−0.3909	−3.30RX

In this section, we discuss **regression diagnostics** and present diagnostic tools that are commonly used. The objective is to measure the degree of influence points have on the regression line.

We start with the notion of the **x-leverage of a point**. Consider the matrix (H) defined in Section 8.3. The vector of predicted values, $\hat{\mathbf{y}}$, is obtained as $(H)\mathbf{y}$. The x-leverage of the ith point is measured by the ith diagonal element of (H), which is

$$h_i = \mathbf{x}_i'((X)'(X))^{-1}\mathbf{x}_i \qquad i = 1, \ldots, n \tag{8.6.1}$$

Here $\mathbf{x}_i'$ denotes the ith row of the predictors' matrix (X):

$$\mathbf{x}_i' = (1, x_{i1}, \ldots, x_{ik})$$

In the special case of simple linear regression $(k = 1)$, we obtain the formula

$$h_i = \frac{1}{n} + \frac{(x_i - \bar{x})^2}{\displaystyle\sum_{j=1}^{n}(x_j - \bar{x})^2} \qquad i = 1, \ldots, n \tag{8.6.2}$$

Notice that $S_e\sqrt{h_i}$ is the **standard error** (square root of variance) **of the predicted value** $\hat{y}_i$. This interpretation holds also in the multiple regression case $(k > 1)$. Figure 8.4 shows the x-leverage values of the various points in the SOCELL example.

From formula 8.6.2 we deduce that, when $k = 1$, $\sum_{i=1}^{n} h_i = 2$. Generally, for any k, $\sum_{i=1}^{n} h_i = k + 1$. Thus, the average x-leverage is $\bar{h} = (k+1)/n$. In the solar cells example, the average x-leverage of the 16 points is $\frac{2}{16} = 0.125$. Point 9 has a leverage value of $h_9 = 0.521$. This is indeed a high x-leverage.

FIGURE 8.4

x-Leverage of ISC values

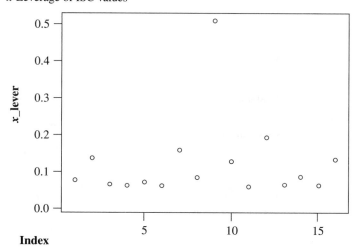

The standard error of the ith residual, $\hat{e}_i$, is given by

$$S\{\hat{e}_i\} = S_e\sqrt{1 - h_i} \tag{8.6.3}$$

The **standardized residuals** are therefore given by

$$\hat{e}_i^* = \frac{\hat{e}_i}{S\{\hat{e}_i\}} = \frac{\hat{e}_i}{S_e\sqrt{1 - h_i}} \qquad i = 1, \ldots, n \tag{8.6.4}$$

Several additional indices measure the effects of the points on the regression. Two such measures are the **Cook's distance** and the **FITs distance**.

If we delete the ith point from the data set and recompute the regression, we obtain a vector of regression coefficients $\mathbf{b}^{(i)}$ and standard deviation of residuals $S_e^{(i)}$. The standardized squared distance of $\mathbf{b}^{(i)}$ from $\mathbf{b}$.

$$D_i^2 = \frac{(\mathbf{b}^{(i)} - \mathbf{b})'((X)'(X))(\mathbf{b}^{(i)} - \mathbf{b})}{(k + 1)S_e^2} \tag{8.6.5}$$

is the so-called Cook's distance.

The influence of the predicted values, denoted by DFIT, is defined as

$$\text{DFIT}_i = \frac{\hat{Y}_i - \hat{Y}_i^{(i)}}{S_e^{(i)}\sqrt{h_i}} \qquad i = 1, \ldots, n \tag{8.6.6}$$

where $\hat{Y}_i^{(i)} = b_0^{(i)} + \sum_{j=1}^{k} b_j^{(i)} x_{ij}$ are the predicted values of Y at $(1, x_{i1}, \ldots, x_{ik})$, when the regression coefficients are $\mathbf{b}^{(i)}$.

Figure 8.5 shows the Cook's distance for the ALMPIN data set.

FIGURE   8.5

Cook's distance for aluminum pins data

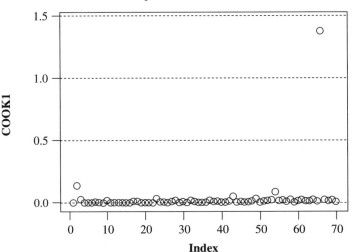

8.7

Quantal Response Analysis: Logistic Regression

In this section, we consider the case where the regressant Y is a binary random variable and the regressors are quantitative. The distribution of Y at a given combination of x values $\mathbf{x} = (x_1, \ldots, x_k)$ is binomial $B(n, p(\mathbf{x}))$, where n is the number of identical and independent repetitions of the experiment at $\mathbf{x}$ and $p(\mathbf{x}) = P\{Y = 1 \mid \mathbf{x}\}$. The question is how to model the function $p(\mathbf{x})$. An important class of models is the so-called **quantal response models**, according to which

$$p(\mathbf{x}) = F(\boldsymbol{\beta}'\mathbf{x}) \tag{8.7.1}$$

where $F(\cdot)$ is a c.d.f., and

$$\boldsymbol{\beta}'\mathbf{x} = \beta_0 + \beta_1 x_1 + \cdots + \beta_k x_k \tag{8.7.2}$$

Logistic regression is a method of estimating the regression coefficients $\boldsymbol{\beta}$, in which

$$F(z) = e^z/(1 + e^z) \qquad -\infty < z < \infty \tag{8.7.3}$$

is the logistic c.d.f.

The experiment is conducted at m different and linearly independent combinations of $\mathbf{x}$ values. Thus, let

$$(X) = (1, \mathbf{x}_1, \mathbf{x}_2, \ldots, \mathbf{x}_k)$$

be the predictors' matrix of m rows and $(k + 1)$ columns. We assumed that $m > (k + 1)$ and the rank of (X) is $(k + 1)$. Let $\mathbf{x}^{(i)}$ $(i = 1, \ldots, m)$ denote the ith row vector of (X).

As mentioned before, we replicate the experiment at each $\mathbf{x}^{(i)}$ n times. Let $\hat{p}_i$ $(i = 1, \ldots, m)$ be the proportion of 1s observed at $\mathbf{x}^{(i)}$,—that is, $\hat{p}_{i,n} = \dfrac{1}{n}\sum_{j=1}^{n} Y_{ij}$, $i = 1, \ldots, m$, where $Y_{ij} = 0, 1$ is the observed value of the regressant at the jth replication $(j = 1, \ldots, n)$.

We have proven before that $E\{\hat{p}_{i,n}\} = p(\mathbf{x}^{(i)})$ and $V\{\hat{p}_{i,n}\} = (1/n)p(\mathbf{x}^{(i)})(1 - p(\mathbf{x}^{(i)}))$, $i = 1, \ldots, m$. Also, the estimators $\hat{p}_i$ $(i = 1, \ldots, m)$ are independent. According to the logistic model,

$$p(\mathbf{x}^{(i)}) = \frac{e^{\boldsymbol{\beta}'\mathbf{x}^{(i)}}}{1 + e^{\boldsymbol{\beta}'\mathbf{x}^{(i)}}} \qquad i = 1, \ldots, m \tag{8.7.4}$$

The problem is to estimate the regresson coefficients $\boldsymbol{\beta}$. Notice that the log-odds at $\mathbf{x}^{(i)}$ is

$$\log \frac{p(\mathbf{x}^{(i)})}{1 - p(\mathbf{x}^{(i)})} = \boldsymbol{\beta}'\mathbf{x}^{(i)} \qquad i = 1, \ldots, m \tag{8.7.5}$$

Define $Y_{i,n} = \log[\hat{p}_{i,n}/(1 - \hat{p}_{i,n})]$, $i = 1, \ldots, m$. $Y_{i,n}$ is finite if n is sufficiently large. Because $\hat{p}_{i,n} \to p(\mathbf{x}^{(i)})$ in probability as $n \to \infty$ (WLLN), and because $\log[x/(1 - x)]$ is a continuous function of x on $(0, 1)$, $Y_{i,n}$ is a consistent estimator of $\boldsymbol{\beta}'\mathbf{x}^{(i)}$. For large values of n, we can write the regression model

$$Y_{i,n} = \boldsymbol{\beta}'\mathbf{x}^{(i)} + e_{i,n} + e_{i,n}^* \qquad i = 1, \ldots, m \qquad\qquad (8.7.6)$$

where

$$e_{i,n} = (\hat{p}_{i,n} - p(\mathbf{x}^{(i)}))/[p(\mathbf{x}^{(i)})(1 - p(\mathbf{x}^{(i)}))] \qquad\qquad (8.7.7)$$

Variable $e_{i,n}^*$ is a negligible remainder term if n is large. $e_{i,n}^* \to 0$ in probability at the rate of $1/n$. If we omit the remainder term $e_{i,n}^*$, we have the approximate regression model

$$Y_{i,n} \cong \boldsymbol{\beta}'\mathbf{x}^{(i)} + e_{i,n} \qquad i = 1, \ldots, m \qquad\qquad (8.7.8)$$

where

$$E\{e_{i,n}\} = 0$$

$$V\{e_{i,n}\} = \frac{1}{n} \cdot \frac{1}{p(\mathbf{x}^{(i)})(1 - p(\mathbf{x}^{(i)}))}$$

$$= \frac{(1 + e^{\boldsymbol{\beta}'\mathbf{x}^{(i)}})^2}{n \cdot e^{\boldsymbol{\beta}'\mathbf{x}^{(i)}}} \qquad\qquad (8.7.9)$$

$i = 1, \ldots, m$. The problem here is that $V\{e_{i,n}\}$ depends on the unknown $\boldsymbol{\beta}$ and varies from one $\mathbf{x}^{(i)}$ to another. An ordinary LSE of $\boldsymbol{\beta}$ is given by $\hat{\boldsymbol{\beta}} = [(X)'(X)]^{-1}(X)'Y$, where $\mathbf{Y}' = (Y_{1,n}, \ldots, Y_{m,n})$. Because the variances of $e_{i,n}$ are different, an estimator having smaller variances is the weighted LSE:

$$\hat{\boldsymbol{\beta}}_w = [(X)'W(\boldsymbol{\beta})(X)]^{-1}(X)'W(\boldsymbol{\beta})Y \qquad\qquad (8.7.10)$$

where $W(\boldsymbol{\beta})$ is a diagonal matrix whose ith term is

$$W_i(\boldsymbol{\beta}) = \frac{n e^{\boldsymbol{\beta}'\mathbf{x}^{(i)}}}{(1 + e^{\boldsymbol{\beta}'\mathbf{x}^{(i)}})^2} \qquad i = 1, \ldots, m \qquad\qquad (8.7.11)$$

The problem is that the weights $W_i(\boldsymbol{\beta})$ depend on the unknown vector $\boldsymbol{\beta}$. An iterative approach to obtain $\hat{\boldsymbol{\beta}}_w$ is to substitute on the right-hand side the value of $\hat{\boldsymbol{\beta}}$ obtained in the previous iteration, starting with the ordinary LSE, $\hat{\boldsymbol{\beta}}$. Other methods of estimating the coefficients $\boldsymbol{\beta}$ of the logistic regression are based on the maximum likelihood method. For additional information, see Kotz and Johnson (1985, p. 128).

8.8

The Analysis of Variance: Comparison of Means

8.8.1   The Statistical Model

When the regressors $x_1, x_2, \ldots, x_k$ are qualitative (categorical) variables and the variable of interest Y is quantitative, the previously discussed methods of multiple regression are invalid. The different values that the regressors obtain are different categories of the variables. For example, suppose that we study the relationship between film speed (Y) and the type of gelatine x used in the preparation of the chemical emulsion for coating the film. Here the regressor is a categorical variable. The values it obtains are the various types of gelatine, as classified according to manufacturers.

When we have k, $k \geq 1$, such qualitative variables—the combination of categorical levels of the k variables—are called **treatment combinations** (a term introduced by experimentalists). Several observations, n_i, can be performed at the ith treatment combination. These observations are considered a random sample from the (infinite) population of all possible observations under the specified treatment combination. The statistical model for the jth observation is

$$Y_{ij} = \mu_i + e_{ij} \qquad i = 1, \ldots, t; \ \ j = 1, \ldots, n_i$$

where μ_i is the population mean for the ith treatment combination, t is the number of treatment combinations, and e_{ij} ($i = 1, \ldots, t; j = 1, \ldots, n_i$) are assumed to be independent random variables (experimental errors) with $E\{e_{ij}\} = 0$ for all (i,j) and $v\{e_{ij}\} = \sigma^2$ for all (i,j). The comparison of the means μ_i ($i = 1, \ldots, t$) provides information on the various effects of the different treatment combinations. The method used to do this analysis is called *analysis of variance* (ANOVA).

8.8.2 The One-Way Analysis of Variance (ANOVA)

In Section 7.3.5.2, we introduced the ANOVA F-test statistics and presented the algorithm for bootstrap ANOVA for comparing the means of k populations. In this section, we develop the rationale for the ANOVA. We assume here that the errors e_{ij} are independent and normally distributed. For the ith treatment combination (sample), let

$$\bar{Y}_i = \frac{1}{n_i} \sum_{j=1}^{n_i} Y_{ij} \qquad i = 1, \ldots, t \tag{8.8.1}$$

and

$$\text{SSD}_i = \sum_{j=1}^{n_i} (Y_{ij} - \bar{Y}_i)^2 \qquad i = 1, \ldots, t \tag{8.8.2}$$

Let $\bar{\bar{Y}} = (1/N) \sum_{i=1}^{t} n_i \bar{Y}_i$ be the grand mean of all the observations.

The one-way ANOVA is based on the following partition of the total sum of squares of deviations around $\bar{\bar{Y}}$,

$$\sum_{i=1}^{t} \sum_{j=1}^{n_i} (Y_{ij} - \bar{\bar{Y}})^2 = \sum_{i=1}^{t} \text{SSD}_i + \sum_{i=1}^{t} n_i (\bar{Y}_i - \bar{\bar{Y}})^2 \tag{8.8.3}$$

We denote the left-hand side by SST and the right-hand side by SSW and SSB— that is

$$\text{SST} = \text{SSW} + \text{SSB} \tag{8.8.4}$$

SST, SSW, and SSB are symmetric quadratic forms in deviations like $Y_{ij} - \overline{\overline{Y}}$, $\overline{Y}_{ij} - \overline{Y}_i$, and $\overline{Y}_i - \overline{\overline{Y}}$. Because $\sum_i \sum_j (Y_{ij} - \overline{\overline{Y}}) = 0$, only $N - 1$ linear functions $Y_{ij} - \overline{\overline{Y}} = \sum_{i'} \sum_{j'} c_{ij'} Y_{i'j'}$, with

$$c_{ij'} = \begin{cases} 1 - \dfrac{1}{N} & i' = i, j' = j \\ -\dfrac{1}{N} & \text{otherwise} \end{cases}$$

are linearly independent, where $N = \sum_{i=1}^{t} n_i$. For this reason, we say that the quadratic form SST has $(N - 1)$ degrees of freedom (d.f.). Similarly, SSW has $(N - t)$ degrees of freedom because SSW $= \sum_{i=1}^{t} SSD_i$, and the number of degrees of freedom of SSD_i is $(n_i - 1)$. Finally, SSB has $(t - 1)$ degrees of freedom. Notice that SSW is the total sum of squares of deviations *within* the t samples, and *SSB* is the sum of squares of deviations *between* the t sample means.

Dividing a quadratic form by its number of degrees of freedom, we obtain the mean-squared statistic. We summarize all these statistics in a table called the *ANOVA table*. The ANOVA table for comparing t treatments is given in Table 8.1.

Generally, in an ANOVA table, "d.f." designates degrees of freedom, "SS" designates the sum of squares of deviations, and "MS" designates the mean-square. In all tables,

$$MS = \frac{SS}{d.f.} \tag{8.8.5}$$

We show now that

$$E\{MSW\} = \sigma^2 \tag{8.8.6}$$

Indeed, according to the model, and because $\{Y_{ij}, j = 1, \ldots, n_i\}$ is a RSWR from the population corresponding to the ith treatment,

$$E\left\{ \frac{SSD_i}{n_i - 1} \right\} = \sigma^2 \qquad i = 1, \ldots, t$$

T A B L E   8.1
ANOVA table for one-way layout

Source of variation	d.f.	SS	MS
Between treatments	$t - 1$	SSB	MSB
Within treatments	$N - t$	SSW	MSW
Total (Adjusted for mean)	$N - 1$	SST	—

Because MSW $= \sum\limits_{i=1}^{t} v_i \left[\text{SSD}_i/(n_i - 1) \right]$, where $v_i = (n_i - 1)/(N - t)$, $i = 1, \ldots, t$,

$$E\{\text{MSW}\} = \sum_{i=1}^{t} v_i E \left\{ \frac{\text{SSD}_i}{n_i - 1} \right\} = \sigma^2 \sum_{i=1}^{t} v_i$$

$$= \sigma^2$$

Another important result is

$$E\{\text{MSB}\} = \sigma^2 + \frac{1}{t - 1} \sum_{i=1}^{t} n_i \tau_i^2 \tag{8.8.7}$$

where $\tau_i = \mu_i - \overline{\mu}$ $(i = 1, \ldots, t)$ and $\overline{\mu} = (1/N) \sum\limits_{i=1}^{t} n_i \mu_i$. Thus, under the null hypothesis $H_0 : \mu_1 = \cdots = \mu_t$, $E\{\text{MSB}\} = \sigma^2$. This motivates us to use the F statistic for testing H_0:

$$F = \frac{\text{MSB}}{\text{MSW}} \tag{8.8.8}$$

H_0 is rejected at level of significance α if

$$F > F_{1-\alpha}[t - 1, N - t]$$

EXAMPLE **8.9**   Three different vendors are being considered to supply cases for floppy disk drives. The question is whether the latch mechanism that opens and closes the disk loading slot is sufficiently reliable. In order to test the reliability of this latch, three independent samples of cases, each of size $n = 10$, were randomly selected from the production lots of these vendors. The testing was performed on a special apparatus that opens and closes a latch until it breaks. The number of cycles required until latch failure was recorded. To avoid uncontrollable environmental factors that might bias the results, the order in which the cases of the different vendors were tested was completely randomized. We can find the results of this experiment in file VENDOR.DAT arranged in three columns. Column 1 represents the sample from vendor A_1, column 2 that of vendor A_2, and column 3 that of vendor A_3. An ANOVA was performed using MINITAB. The analysis was done on $Y = \text{Number of Cycles}^{1/2}$ in order to have data that were approximately normally distributed. The original data are expected to have a positively skewed distribution because it reflects the life length of the latch. The following table shows the one-way ANOVA as performed by MINITAB at the command

```
MTB> Oneway C6 C4
```

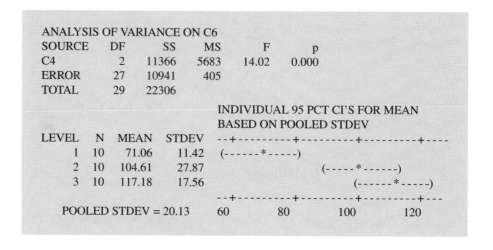

ANALYSIS OF VARIANCE ON C6

SOURCE	DF	SS	MS	F	p
C4	2	11366	5683	14.02	0.000
ERROR	27	10941	405		
TOTAL	29	22306			

INDIVIDUAL 95 PCT CI'S FOR MEAN
BASED ON POOLED STDEV

LEVEL	N	MEAN	STDEV
1	10	71.06	11.42
2	10	104.61	27.87
3	10	117.18	17.56

POOLED STDEV = 20.13

```
                   --+---------+---------+---------+----
     (------*-----)
                                   (-----*------)
                                        (------*-----)
                   --+---------+---------+---------+---
                   60        80       100       120
```

Figure 8.6 shows boxplots of the three samples, by vendor.

F I G U R E   8.6

Boxplots of Y by vendor

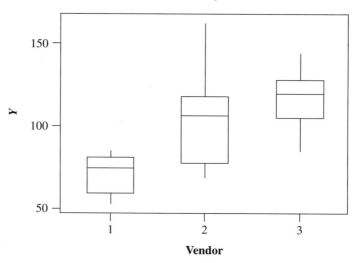

In columns $C1$–$C3$, we stored the cycles data (importing file VENDOR.DAT). Column $C4$ has indices of the samples. Column $C5$ contains the data of $C1$–$C3$ in a stacked form. Column $C6 = \mathrm{sqrt}(C5)$.

The ANOVA table shows that the F statistic is significantly large, having a P-value close to 0. The null hypothesis H_0 is rejected. The reliability of the latches from the three vendors is not the same. The .95 confidence intervals for the means show that vendors A_2 and A_3 manufacture latches with similar reliability. That for vendor A_1 is significantly lower.   ∎

8.9
Simultaneous Confidence Intervals: Multiple Comparisons

Whenever the hypothesis of no difference between the treatment means is rejected, the question arises: Which of the treatments have similar effects, and which ones differ significantly? In Example 8.9, we analyzed data on the strength of latches supplied by three different vendors. It was shown that the differences are very significant. We also saw that the latches from vendor A_1 were weaker than those of vendors A_2 and A_3, which were of similar strength. Generally, if there are t treatments, and the ANOVA shows that the differences between the treatment means are significant, we may have to perform up to $\binom{t}{2}$ comparisons to rank the different treatments in term of their effects.

If we compare the means of all pairs of treatments, we wish to determine $\binom{t}{2} = t(t-1)/2$ confidence intervals to the true differences between the treatment means. If each confidence interval has confidence level $1 - \alpha$, the probability that all $\binom{t}{2}$ confidence intervals cover the true differences simultaneously is smaller than $1 - \alpha$. The simultaneous confidence level might be as low as $1 - t\alpha$.

There are different types of **simultaneous confidence intervals**. In this section, we present **Scheffé's method** for simultaneous confidence intervals for any number of contrasts (Scheffé, 1959, p. 66). A **contrast** between t means $\overline{Y}_1, \ldots, \overline{Y}_t$ is a linear combination $\sum_{i=1}^{t} c_i \overline{Y}_i$, such that $\sum_{i=1}^{t} c_i = 0$. Thus, any difference between two means is a contrast—for example, $\overline{Y}_2 - \overline{Y}_1$. Any second-order difference—for example,

$$(\overline{Y}_3 - \overline{Y}_2) - (\overline{Y}_2 - \overline{Y}_1) = \overline{Y}_3 - 2\overline{Y}_2 + \overline{Y}_1$$

is a contrast. In Chapter 12, we will see various estimators of effects, which are all linear contrasts. The space of all possible linear contrasts has dimension $(t - 1)$. For this reason, according to Scheffé's method, the coefficient we use to obtain simultaneous confidence intervals of level $1 - \alpha$ is

$$S_\alpha = ((t - 1)F_{1-\alpha}[t - 1, t(n - 1)])^{1/2} \tag{8.9.1}$$

where $F_{1-\alpha}[t - 1, t(n - 1)]$ is the $(1 - \alpha)$th quantile of the F distribution. It is assumed that all the t samples are of equal size n. Let $\hat{\sigma}_p^2$ denote the pooled estimator of σ^2:

$$\hat{\sigma}_p^2 = \frac{1}{t(n - 1)} \sum_{i=1}^{t} \text{SSD}_i \tag{8.9.2}$$

Then the simultaneous confidence intervals for all contrasts of the form $\sum_{i=1}^{t} c_i \mu_i$ have limits

$$\sum_{i=1}^{t} c_i \overline{Y}_i \pm S_\alpha \frac{\hat{\sigma}_p}{\sqrt{n}} \left(\sum_{i=1}^{t} c_i^2 \right)^{1/2} \tag{8.9.3}$$

EXAMPLE **8.10**   In data file HADPAS.DAT, we have the resistance values (in ohms) of several resistors on six different hybrids on 32 cards. Here we analyze the differences between the means of the $n = 32$ resistance values of resistor Res 3, where the treatments are the $t = 6$ hybrids. The boxplots of the samples corresponding to the six hybrids are presented in Figure 8.7.

FIGURE **8.7**

Boxplots of six hybrids

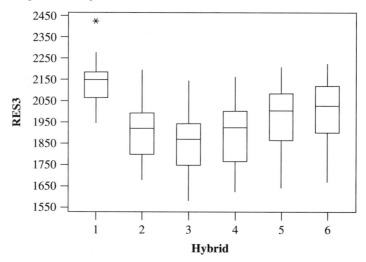

In Table 8.2, we present the means and standard deviations of these six samples (treatments).

The pooled estimator of σ is

$$\hat{\sigma}_p = 133.74$$

The Scheffé coefficient, for $\alpha = .05$, is

$$S_{.05} = (5F_{.95}[5, 186])^{1/2} = 3.332$$

TABLE **8.2**

Means and standard deviations of resistance Res 3 by hybrid

Hybrid	$\overline{Y}$	S_y
1	2143.41	99.647
2	1902.81	129.028
3	1850.34	144.917
4	1900.41	136.490
5	1980.56	146.839
6	2013.91	139.816

Upper and lower simultaneous confidence limits, with .95 level of significance, are obtained by adding $\pm S_\alpha(\hat{\sigma}_p/\sqrt{16}) = \pm 111.405$ to the differences between means.

Differences that are smaller in magnitude than 111.405 are considered insignificant. Thus, if we order the sample means, we obtain:

Hybrid	Mean	Group mean
1	2143.41	2143.41
6	2013.91	1997.24
5	1980.56	
2	1902.81	1884.52
4	1900.41	
3	1850.34	

The difference between the means of hybrid 1 and all the others is significant. The mean of hybrid 6 is significantly different from those of 2, 4, and 3. The mean of hybrid 5 is significantly larger than that of hybrid 3. We suggest, therefore, the following *homogeneous grouping* of treatments (all treatments within the same homogeneous group have means that are not significantly different):

Homogeneous group	Means of groups
{1}	2143.41
{5, 6}	1997.24
{2, 3, 4}	1884.52

The difference between the means of {5, 6} and {2, 3, 4} is the contrast

$$-\frac{1}{3}\overline{Y}_2 - \frac{1}{3}\overline{Y}_3 - \frac{1}{3}\overline{Y}_4 + \frac{1}{2}\overline{Y}_5 + \frac{1}{2}\overline{Y}_6$$

This contrast is significant if it is greater than

$$S_\alpha \frac{\hat{\sigma}_p}{\sqrt{32}}\sqrt{\left(\frac{1}{2}\right)^2 + \left(\frac{1}{2}\right)^2 + \left(\frac{1}{3}\right)^2 + \left(\frac{1}{3}\right)^2 + \left(\frac{1}{3}\right)^2} = 71.912$$

Thus the difference is significant. ∎

8.10
Categorical Data Analysis

If all variables $x_1, \ldots, x_k$ and Y are categorical, we cannot perform the ANOVA without special modifications. In this section, we discuss the analysis appropriate for such cases.

8.10.1   Comparison of Binomial Experiments

Suppose we have performed t independent binomial experiments, each one corresponding to a treatment combination. In the ith experiment, we ran n_i independent trials. The yield variable, J_i, is the number of successes among the n_i trials $(i = 1, \ldots, t)$. We further assume that in each experiment, the n_i trials are independent and have the same, unknown probability for success, θ_i—that is, J_i has a binomial distribution $B(n_i, \theta_i)$, $i = 1, \ldots, t$. We wish to compare the probabilities of success, θ_i $(i = 1, \ldots, k)$. Accordingly, the null hypothesis is of equal success probabilities:

$$H_0 : \theta_1 = \theta_2 = \cdots = \theta_k$$

The following test is good for large samples. Because, by the CLT, $\hat{p}_i = J_i/n_i$ has a distribution that is approximately normal for large n_i, with mean θ_i and variance $\theta_i(1 - \theta_i)/n_i$, we can show that the large-sample distribution of

$$Y_i = 2\arcsin\left(\sqrt{\frac{J_i + 3/8}{n_i + 3/4}}\right) \tag{8.10.1}$$

(in radians) is approximately normal, with mean $\eta_i = 2\arcsin(\sqrt{\theta_i})$ and variance $V\{Y_i\} = 1/n_i$, $i = 1, \ldots, t$.

Using this result, we find that, under the assumption of H_0, the sampling distribution of the test statistic

$$Q = \sum_{i=1}^{k} n_i(Y_i - \overline{Y})^2 \tag{8.10.2}$$

where

$$\overline{Y} = \frac{\displaystyle\sum_{i=1}^{k} n_i Y_i}{\displaystyle\sum_{i=1}^{k} n_i} \tag{8.10.3}$$

is approximately chi-square with $k - 1$ d.f., $\chi^2[k - 1]$. In this test, we reject H_0 at level of significance α if $Q > \chi^2_{1-\alpha}[k - 1]$.

Another test statistic for general use in contingency tables will be given in Section 8.10.2.

EXAMPLE **8.11**   In Table 3.14, we presented the frequency of failure of nine different components when a large number of components is inserted automatically. In this example, we test the hypothesis that the failure probabilities, θ_i, are the same for all components. Table 8.3 gives the values of J_i (number of failures), n_i, and $Y_i = 2\arcsin\left[\sqrt{(J_i + 3/8)/(n_i + 3/4)}\right]$ for each component. Using MINITAB, if the values of J_i are stored in C1 and those of n_i in C2, we compute Y_i (stored in C3) with the command

TABLE 8.3

The arcsin transformation

i	J_i	n_i	Y_i
1	61	108119	0.0476556
2	34	136640	0.0317234
3	10	107338	0.0196631
4	23	105065	0.0298326
5	25	108854	0.0305370
6	9	96873	0.0196752
7	12	107391	0.0214697
8	3	105854	0.0112931
9	13	180630	0.0172102

```
MTB> let C3 = 2 * asin(sqrt((C1 + .375)/(C2 + .75)))
```

The test statistic Q can be computed by the MINITAB command (the constant $k1$ stands for Q):

```
MTB> let k1 = sum(C2 * C3 ** 2) − sum(C2 * C3)**2/sum(C2)
```

The value of Q is 105.43. The P-value of this statistic is 0. The null hypothesis is rejected. To determine this P-value using MINITAB, because the distribution of Q under H_0 is $\chi^2[8]$, we use the command

```
MTB> CDF 105.43;
SUBC> Chisquare 8.
```

We find that $\Pr\{\chi^2[8] \leq 105.43\} \doteq 1$. This implies that $P = 0$. ∎

8.10.2 The Chi-Square Test for Contingency Tables

In Section 3.5, we discussed several indices of association for contingency tables based on the statistic X^2. In the present context, if we have a 2-factor contingency table with k rows and m columns, let O_{ij} denote the *observed frequency* in the (i, j)

cell. We then compute the *expected frequency*, E_{ij}, for cell (i, j) according to the null hypothesis. The observed and expected frequencies are compared by the statistic

$$X^2 = \sum_{i=1}^{k} \sum_{j=1}^{m} \frac{(O_{ij} - E_{ij})^2}{E_{ij}} \tag{8.10.4}$$

Suppose the null hypothesis is that the rows factor and the columns factor are independent. In other words, the conditional probabilities of the row categories, given different column categories, should be the same. The expected frequencies, when we have to estimate the conditional probabilities from the observed frequencies, are computed in the following manner.

Let

$$N_{i.} = \sum_{j=1}^{m} O_{ij} \qquad i = 1, \ldots, k$$

and

$$N_{.j} = \sum_{i=1}^{k} O_{ij} \qquad j = 1, \ldots, m$$

be the rows and column sums, respectively. Let $N = \sum_{i=1}^{k} N_{i.}$ be the table grand sum. Then, under the hypothesis of independence,

$$E_{ij} = \frac{N_{i.} N_{.j}}{N} \qquad i = 1, \ldots, k; \; j = 1, \ldots, m$$

Notice that $\sum_{i} \sum_{j} E_{ij} = \sum_{i} \sum_{j} O_{ij} = N$.

Accordingly,

$$X^2 = \sum_{i=1}^{k} \sum_{j=1}^{m} \frac{O_{ij}^2}{E_{ij}} - N \tag{8.10.5}$$

In large samples (each $O_{ij} > 6$), X^2 has, under H_0, approximately a χ^2 distribution with $(k-1)(m-1)$ degrees of freedom. We therefore reject H_0 at level of significance α if

$$X^2 > \chi_{1-\alpha}^2[(k-1)(m-1)]$$

EXAMPLE **8.12**   Here we compute the X^2 statistic for the contingency table given in Table 3.15. The hypothesis of independence implies that the probabilities of failure (and of success) are the same for all components. This is the same null hypothesis tested in the previous example. In Table 8.4, we present the observed and expected frequencies in the contingency table. The values of O_{ij} are at the top and those of E_{ij} at the bottom of each cell. The values in this table yield the statistic

$$X^2 = 131.50$$

T A B L E   8.4

Observed and expected frequencies for insertion result by component

Component	Failure	Success	Total
C1	61	108058	108119
	19.44	108099.56	
C2	34	136606	136640
	24.57	136615.43	
C3	10	107328	107338
	19.30	107318.70	
C4	23	105042	105065
	18.89	105046.11	
C5	25	108829	108854
	19.57	108834.43	
C6	9	96864	96873
	17.42	96855.58	
C7	12	107379	107391
	19.31	107371.69	
C8	3	105851	105854
	19.03	105834.97	
C9	13	180617	180630
	32.47	180597.53	
Total	190	1,056,574	1,056,764

Also, here the *P*-value of the test is zero. As in the previous example, the null hypothesis is rejected. ■

Chapter Highlights

The chapter provides an introduction to multiple regression methods, in which the relationship of k explanatory (predicting) quantitative variables to a variable of interest is explored. In particular, the least-squares estimation procedure is presented in detail for regression on two variables. Partial regression and correlation are discussed. The least-squares estimation of the regression coefficients for multiple regressions ($k > 2$) is presented with matrix formulas. The contributions of the individual regressors are tested by the partial-*F* test. The sequential SS partition of the total sum of squares due to the departure on the regressors is defined and explained. The partial correlation, given a set of predictors, is defined and its relationship to the partial-F statistic is given.

The analysis of variance (ANOVA) for testing the significance of differences between several sample means is introduced, as is the method of multiple comparisons, which protects the overall level of significance. Comparisons of proportions for categorical data (binomial or multinomial) are also discussed. The chapter contains a section on regression diagnostics as well, in which the influence of individual points on the regression is studied. In particular, the section covers measuring the effects of points that seem to deviate considerably from the rest of the sample.

The main concepts and definitions include:

- Multiple regression
- Linear model
- Principle of least squares
- Predicted values, or FITs
- Multiple squared correlation

- Partial regression
- Partial correlation
- Sequential SS
- Partial-F test
- Stepwise regression
- Regression diagnostics
- x-leverage of a point
- Standard error of the predicted value
- Standardized residual
- Cook's distance
- FITs distance
- Quantal response models
- Logistic regression
- Treatment combinations
- Simultaneous confidence intervals
- Scheffé's method
- Contrast

8.12

Exercises

8.1.1

a   Differentiate partially the quadratic function

$$\text{SSE} = \sum_{i=1}^{n} (Y_i - \beta_0 - \beta_1 X_{i1} - \beta_2 X_{i2})^2$$

with respect to β_0, β_1, and β_2 to obtain the linear equations in the least-squares estimates b_0, b_1, b_2. These linear equations are called the *normal equations*.

b   Obtain the formulas for b_0, b_1, and b_2 from the normal equations.

8.1.2   Consider the variables Miles per Gallon, Horsepower, and Turn Diameter in the data set CAR.DAT. Find the least-squares regression line of MPG (y) on Horsepower (x_1) and Turn Diameter (x_2). For this purpose, first use the equations in Section 8.1 and then verify your computations by using the MINITAB command **REGRESS**.

8.2.1   Compute the partial correlation between Miles per Gallon and Horsepower, given Number of Cylinders, in data file CAR.DAT.

8.2.2   Compute the partial regression of Miles per Gallon and Turn Diameter, given Horsepower, in data file CAR.DAT.

8.2.3   Use the 3-stage algorithm of Section 8.2 to obtain the multiple regression of Exercise 8.1.2 from the results of Exercise 8.2.2.

8.3.1   Consider Example 8.4. From the MINITAB output, we see that, when CapDiam is regressed on Diam1, Diam2, and Diam3, the regression coefficient of Diam2 is not significant (P-value = .926), and this variable can be omitted. Perform a regression of CapDiam on Diam2 and Diam3. Is the regression coefficient for Diam2 significant? How can you explain the difference between the results of the two regressions?

8.3.2 Regress the yield in GASOL.DAT on the four variables x_1, x_2, x_3, x_4.

a What is the regression equation?

b What is the value of R^2?

c Which regression coefficient(s) is(are) nonsignificant?

d Which factors are important to control the yield?

e Are the residuals from the regression distributed normally? Make a graphical test.

8.3.3

a Show that the matrix $(H) = (X)(B)$ is idempotent—that is, $(H)^2 = (H)$.

b Show that the matrix $(Q) = (I - H)$ is idempotent, and therefore $s_e^2 = \mathbf{y}'(Q)\mathbf{y}/(n - k - 1)$.

8.3.4 Show that the vectors of fitted values, $\hat{\mathbf{y}}$, and of the residuals, $\hat{\mathbf{e}}$, are orthogonal—that is, $\hat{\mathbf{y}}'\hat{\mathbf{e}} = 0$.

8.3.5 Show that the $1 - R_{y|(x)}^2$ is proportional to $||\hat{\mathbf{e}}||^2$, which is the squared Euclidean norm of $\hat{\mathbf{e}}$.

8.3.6 In Section 4.5, we presented properties of the $\text{cov}(X, Y)$ operator. Prove the following generalization of property (iv). Let $\mathbf{X}' = (X_1, \ldots, X_n)$ be a vector of n random variables. Let $(\mathbf{\Sigma})$ be an $n \times n$ matrix whose (i, j)th element is $\mathbf{\Sigma}_{ij} = \text{cov}(X_i, X_j)$, $i, j = 1, \ldots, n$. Notice that the diagonal elements of $(\mathbf{\Sigma})$ are the variances of the components of $\mathbf{X}$. Let $\boldsymbol{\beta}$ and $\boldsymbol{\gamma}$ be two n-dimensional vectors. Prove that $\text{cov}(\boldsymbol{\beta}'\mathbf{X}, \boldsymbol{\gamma}'\mathbf{X}) = \boldsymbol{\beta}'(\mathbf{\Sigma})\boldsymbol{\gamma}$. [The matrix $(\mathbf{\Sigma})$ is called the variance–covariance matrix of $\mathbf{X}$.]

8.3.7 Let $\mathbf{X}$ be an n-dimensional random vector having a variance–covariance matrix $(\mathbf{\Sigma})$. Let $\mathbf{W} = (\mathbf{B})\mathbf{X}$, where $(\mathbf{B})$ is an $m \times n$ matrix. Show that the variance–covariance matrix of $\mathbf{W}$ is $(\mathbf{B})(\mathbf{\Sigma})(\mathbf{B})'$.

8.3.8 Consider the linear regression model $\mathbf{y} = (X)\boldsymbol{\beta} + \mathbf{e}$. The $\mathbf{e}$ is a vector of random variables such that $E\{e_i\} = 0$ for all $i = 1, \ldots, n$ and

$$\text{cov}(e_i, e_j) = \begin{cases} \sigma^2 & \text{if } i = j \\ 0 & \text{if } i \neq j \end{cases}$$

$i, j = 1, \ldots, n$. Show that the variance–covariance matrix of the LSE $\mathbf{b} = (\mathbf{B})\mathbf{y}$ is $\sigma^2[(\mathbf{X})'(\mathbf{X})]^{-1}$.

8.3.9 Consider CAR.DAT data file. Employ the method of Section 8.3 to compare the slopes and intercepts of the two simple regressions of MPG on horsepower for U.S. and Japanese cars.

8.4.1 The following data (see Draper and Smith, 1981, p. 629) give the amount of heat evolved in the hardening of a cement cube (in calories per gram of cement), and the percentage of four different chemicals in the cement (relative to the total weight of the mixture from which the cement was made). The four regressors are

x_1 :amount of tricalcium aluminate

x_2 :amount of tricalcium silicate

x_3 :amount of tetracalcium alumino ferrite

x_4 :amount of dicalcium silicate

The response Y is the amount of heat evolved. The data are given in the following table:

X_1	X_2	X_3	X_4	Y
7	29	6	60	78
1	29	15	52	74
11	56	8	20	104
11	31	8	47	87
7	52	6	33	95
11	55	9	22	109
3	71	17	6	102
1	31	22	44	72
2	54	18	22	93
21	47	4	26	115
1	40	23	34	83
11	66	9	12	113
10	68	8	12	109

Compute in a sequence the regressions of Y on x_1; of Y on x_1, x_2; of Y on x_1, x_2, x_3; of Y on x_1, x_2, x_3, x_4. For each regression, compute the partial-F of the new regression added, the corresponding partial correlation with Y, and the sequential SS.

8.5.1 For the data of Exercise 8.4.1, construct a linear model of the relationship between Y and $x_1, \ldots, x_4$ by the forward stepwise regression method.

8.6.1 Consider the linear regression of Miles per Gallon on Horsepower for the cars in data file CAR.DAT, with Origin = 3. Compute for each car the residuals, ResI; the standardized residuals, SRes; the leverage HI; and the Cook's distance, D^2.

8.8.1 In Chapter 10, we describe a gas piston and the various factors that affect the cycle time of the piston. Program TURB1.EXE simulates the cycle time of a piston. In order to test whether changing the piston weight from 30 to 60 (kg) affects the cycle time significantly, run program TURB1.EXE four times at weight 30, 40, 50, 60 (kg), keeping all other factors at their low level. In each run, make $n = 5$ observations. Perform a one-way ANOVA of the results, and state your conclusions. [You can use MINITAB.]

8.8.2   In experiments performed to study the effects of some factors on the integrated circuits fabrication process, the following results were obtained on the pre-etch line width (μ_m):

Exp. 1	Exp. 2	Exp.3
2.58	2.62	2.22
2.48	2.77	1.73
2.52	2.69	2.00
2.50	2.80	1.86
2.53	2.87	2.04
2.46	2.67	2.15
2.52	2.71	2.18
2.49	2.77	1.86
2.58	2.87	1.84
2.51	2.97	1.86

Perform an ANOVA to find whether the results of the three experiments are significantly different, (a) by using MINITAB and (b) by using the bootstrap program ANOVTEST.EXE. Do the two test procedures yield similar results?

8.8.3   In manufacturing film for industrial use, samples from two different batches gave the following film speeds:

Batch A: 103, 107, 104, 102, 95, 91, 107, 99, 105, 105

Batch B: 104, 103, 106, 103, 107, 108, 104, 105, 105, 97

Test whether the differences between the two batches are significant by using (a) a randomization test; (b) an ANOVA.

8.8.4   Use the MINITAB macro RANDTES3.MTB to test the significance of the differences between the results of the three experiments in Exercise 8.8.2.

8.9.1   In data file PLACE.DAT we have 26 samples, each one of size $n = 16$, of x, y, θ deviations of component placements. Make an ANOVA to test the significance of the sample means in the x deviation. Classify the samples into homogeneous groups such that the differences between sample means in the same group are not significant, and those in different groups are significant. Use the Scheffé coefficient S_α for $\alpha = .05$.

8.10.1   The frequency distribution of cars by origin and number of cylinders is given in the following table:

Number of cylinders	U.S.	Europe	Asia	Total
4	33	7	26	66
6 or more	25	7	11	43
Total	58	14	37	109

Perform a chi-square test of the dependence of number of cylinders and the origin of car.

8.10.2   Make a chi-square test of the association between turn diameter and miles/gallon based on Table 3.19.

II

Methods of Industrial Statistics

The objective of Part II is to provide the reader with a working knowledge of industrial statistics. Topics covered include sampling techniques, control charts, basic improvement and planning tools, design of experiments, quality by design tools, and reliability methods. Different problems facing industry are discussed throughout Part II, thus creating the context for the application of industrial statistics methods.

The organization of Part II follows the quality ladder sequence discussed in Chapter 1. Chapter 9 deals with sampling plans for product inspection. The chapter covers single-sampling plans, double-sampling plans, and sequential sampling. The chapter takes the reader an extra step and discusses skip-lot sampling, where consistent good performance leads to skipping some lots that are accepted without inspection.

Chapter 10 covers basic issues in statistical process control. After establishing the motivation for statistical process control, the chapter introduces the reader to process capability studies, process capability indices, the seven tools for process improvement, and the basic Shewhart charts. Chapter 11 covers more advanced topics such as the economic design of Shewhart control charts, CUSUM procedures, and multivariate statistical process control techniques. The chapter concludes with special sections on Bayesian detection, process tracking, and automatic process control.

Chapter 12 presents the classical approach to the design and analysis of experiments. Chapter 13 builds on Chapter 12 to introduce the concepts of quality by design as they were developed by Genichi Taguchi of Japan.

Finally, Chapter 14 deals with reliability methods, including reliability demonstration tests and accelerated life testing.

9

Sampling Plans for
Product Inspection

9.1
General Discussion

Sampling plans for product inspection are quality assurance schemes designed to test whether the quality level of a product conforms with the required standards. These methods of quality inspection are especially important when products are received from suppliers or vendors on whom we have no other assessment of the quality level of their production processes. Generally, if a supplier has in his or her plant established procedures of statistical process control (see Chapters 10 and 11) that assure the required quality standards, then sampling inspection of shipments may not be necessary. However, periodic auditing of the quality level of certified suppliers might be prudent to ensure that this level does not drop below the acceptable standards. Quality auditing or inspection by sampling techniques can also be applied within the plant at various stages of the production process—for example, when lots are transferred from one department to another.

In this chapter, we discuss various sampling and testing procedures designed to maintain quality standards. In particular, we examine single-, double-, and sequential-sampling plans for attributes and single-sampling plans for continuous measurements. We also discuss testing via tolerance limits. The chapter concludes with a section describing some of the established standards, and in particular the skip-lot procedure, which appears in modern standards.

Following are some general concepts associated with sampling inspection schemes. In recent years, some of the terminology that was used for almost 50 years has changed. A product unit that did not meet quality specifications or requirements, for example, used to be called *defective*. This term was recently changed to *nonconforming*. Thus, in early standards—MIL-STD-105D and others—we find the terms "defective items" and "number of defects." In modern standards, such as ANSI/ASQC Z1.4 and the international standard ISO 2859, the term used is "nonconforming." We will use the two terms interchangeably. Similarly, the terms

LTPD and LQL, which will be explained later, will be used interchangeably. We also furnish a program (see subdirectory \CHA9\) that determines the sampling plans and their characteristics for single-stage, double-stage, and sequential-sampling by attributes. The name of this program is ACCEPT.

A **lot** is a collection of N elements that are subject to quality inspection. Accordingly, a lot is a finite real population of products. *Acceptance* of a lot means the quality level has been approved, thus giving a "green light" to subsequent use of the elements of the lot. Generally, when we refer to lots, we are referring to lots of raw material, of semi-finished or finished products, and so on that are purchased from vendors or produced by subcontractors. Before acceptance, a lot is typically subjected to quality inspection unless the vendor has been certified and its products are delivered directly, without inspection, to the production line. Purchase contracts typically specify the acceptable quality level and the method of inspection.

In general, it is expected that a lot contains no more than a certain percentage of nonconforming (defective) items, where the test conditions that classify an item as defective are usually well specified. A decision needs to be made as to whether a lot has to be subjected to a complete inspection, item by item or whether acceptance can be determined using a sample from the lot. If we decide to inspect a sample, we must determine how large it will be and what the criterion will be for accepting or rejecting the lot. Furthermore, the performance characteristics of the procedures in use should be understood.

The proportion of nonconforming items in a lot is the ratio $p = M/N$, where M is the number of defective items in the whole lot and N is the size of the lot. If we choose to accept only lots with zero defectives, we have to inspect each lot completely, item by item. This approach is called 100% inspection. This is the approach used, for example, when items are used in a critical or very expensive system, such as a communication satellite. In such cases, the cost of inspection is negligible compared to the cost of failure. In many situations, though, complete inspection is impossible (for example, because of destructive testing) or impractical (because of the large expense involved). In such cases, the two parties involved (the customer and the supplier) specify an **acceptable quality level (AQL)** and a **limiting quality level (LQL)**. When the proportion of defectives, p, in the lot is not larger than the AQL, the lot is considered good and should be accepted with high probability. If, however, the proportion of defectives in the lot is greater than the LQL, the lot should be rejected with high probability. If p is between the AQL and the LQL, then the lot can either be accepted or rejected, with various probability levels. In the past, the LQL was called *lot tolerance percent defectives (LTPD)*. In program ACCEPT, for CHA9, the LTPD appears instead of LQL.

How should the parties specify the AQL and LQL levels? Usually, the AQL is determined by the quality requirements of the customer, who is going to use the product. The producer of the product, which is the supplier, tries generally to demonstrate to the customer that his production processes maintain a capability level (see Section 10.4) in accordance with the customer's or consumer's requirements. Both the AQL and LQL are specified in terms of proportions, p_0 and p_t, of nonconforming in the process.

The risk of rejecting a good lot—that is, a lot with $p \leq$ AQL—is called the **producer's risk**, whereas the risk of accepting a bad lot—that is, a lot for which $p \geq$

LQL—is called the **consumer's risk**. Thus, the problem of designing an acceptance sampling plan is that of choosing:

a   the method of sampling,

b   the sample size,

and

c   the appropriate test of the hypothesis

$$H_0 : p \le \text{AQL}$$

against the alternative

$$H_1 : p \ge \text{LQL}$$

so that the probability of rejecting a good lot will not exceed a value α (the level of significance) and the probability of accepting a bad lot will not exceed β. In this context, α and β are called the **producer's risk** and the **consumer's risk**, respectively.

Single-Stage Sampling Plans for Attributes

A **single-stage sampling** plan for an attribute is an acceptance/rejection procedure for a lot of size N, according to which a random sample of size n is drawn from the lot, without replacement. Let M be the number of defective items (elements) in the lot, and let X be the number of defective items in the sample. Obviously, X is a random variable whose range is $\{0, 1, 2, \ldots, n^*\}$, where $n^* = \min(n, M)$. The distribution function of X is the hypergeometric distribution $H(N, M, n)$ (see Section 4.3.2) with the probability distribution function (p.d.f.)

$$h(x; N, M, n) = \frac{\binom{M}{x}\binom{N-M}{n-x}}{\binom{N}{n}} \qquad x = 0, \ldots, n^* \tag{9.2.1}$$

and the cumulative distribution function (c.d.f.)

$$H(x; N, M, n) = \sum_{j=0}^{x} h(j; N, M, n) \tag{9.2.2}$$

Suppose we consider a lot of $N = 100$ items to be acceptable if it has no more than $M = 5$ nonconforming items, and nonacceptable if it has more than $M = 10$ nonconforming items. For a sample of size $n = 10$, we derive the hypergeometric distribution $H(100, 5, 10)$ and $H(100, 10, 10)$.

From Tables 9.1a and 9.1b, we see that, if such a lot is accepted whenever $X = 0$, the consumer's risk of accepting a lot that should be rejected is

$$\beta = H(0; 100, 10, 10) = .3305$$

■ **TABLE 9.1**

a. The p.d.f. and c.d.f. of $H(100, 5, 10)$

j	$h(j; 100, 5, 10)$	$H(j; 100, 5, 10)$
0	.5838	.5838
1	.3394	.9231
2	.0702	.9934
3	.0064	.9997
4	.0003	1.0000

b. The p.d.f. and c.d.f. of $H(100, 10, 10)$

j	$h(j; 100, 10, 10)$	$H(j; 100, 10, 10)$
0	.3305	.3305
1	.4080	.7385
2	.2015	.9400
3	.0518	.9918
4	.0076	.9993
5	.0006	1.0000

The producer's risk of rejecting an acceptable lot is

$$\alpha = 1 - H(0; 100, 5, 10) = .4162$$

As before, let p_0 denote the AQL and p_t the LQL. Obviously, $0 < p_0 < p_t < 1$. Suppose that the decision is to accept a lot whenever the number of nonconforming Xs is not greater than c—that is, $X \leq c$. Variable c is called the **acceptance number**. For specified values of p_0, p_t, α, and β, we can determine n and c so that

$$\Pr\{X \leq c \mid p_0\} \geq 1 - \alpha \qquad \text{(9.2.3)}$$

and

$$\Pr\{X \leq c \mid p_t\} \leq \beta \qquad \text{(9.2.4)}$$

Notice that n and c should satisfy the inequalities

$$H(c; N, M_0, n) \geq 1 - \alpha \qquad \text{(9.2.5)}$$
$$H(c; N, M_t, n) \leq \beta \qquad \text{(9.2.6)}$$

where $M_0 = [Np_0]$ and $M_t = [Np_t]$ and $[a]$ is the integer part of a. Table 9.2 gives numerical results showing how n and c depend on p_0 and p_t when the lot is of size $N = 100$ and $\alpha = \beta = .05$. The values of n and c in this table were computed by program ACCEPT, Single Sampling for Attributes (Option A).

We see that even if the requirements are not very stringent—for example, when $p_0 = .01$ and $p_t = .05$—the required sample size is $n = 65$. If in such a sample there is more than one defective item, then the entire lot is rejected. Similarly, if $p_0 = .03$ and $p_t = .05$, then the required sample size is $n = 92$, which is almost the entire lot.

TABLE 9.2

Sample size n and critical level c for single-stage acceptance sampling, $N = 100$ and $\alpha = \beta = .05$

p_0	p_t	n	c	p_0	p_t	n	c
.01	.05	65	1	.03	.05	92	3
.01	.08	46	1	.03	.08	71	3
.01	.11	36	1	.03	.11	56	3
.01	.14	29	1	.03	.14	37	3
.01	.17	24	1	.03	.17	31	2
.01	.20	20	1	.03	.20	27	2
.01	.23	18	1	.03	.23	24	2
.01	.26	16	1	.03	.26	21	2
.01	.29	14	1	.03	.29	19	2
.01	.32	13	1	.03	.32	13	1

However, if $p_0 = .01$ and p_t is greater than .20, we need no more than 20 items in the sample. If we relax the requirement concerning α and β and allow higher producer's and consumer's risks, the required sample size will be smaller, as shown in Table 9.3.

An important characterization of an acceptance sampling plan is given by its **operating characteristic (OC)** function. This function, denoted by OC(p), yields the probability of accepting a lot having proportion p of defective items. If we let $M_p = [Np]$, then we can calculate the OC function by

$$OC(p) = H(c; N, M_p, n) \tag{9.2.7}$$

In Table 9.4, we present a few values of the OC function for single stage acceptance sampling when the lot is of size $N = 100$, based on sample size $n = 50$ and acceptance number $c = 1$.

Figure 9.1 shows the graph of the OC function that corresponds to Table 9.4.

TABLE 9.3

Sample size n and critical level c for single-stage acceptance sampling, $N = 100$, $\alpha = .10$, $\beta = .20$

p_0	p_t	n	c	p_0	p_t	n	c
.01	.05	49	1	.03	.05	83	3
.01	.08	33	1	.03	.08	58	3
.01	.11	25	1	.03	.11	35	2
.01	.14	20	1	.03	.14	20	1
.01	.17	9	0	.03	.17	16	1
.01	.20	7	0	.03	.20	14	1
.01	.23	6	0	.03	.23	12	1
.01	.26	6	0	.03	.26	11	1
.01	.29	5	0	.03	.29	10	1
.01	.32	5	0	.03	.32	9	1

TABLE 9.4

The OC function of a single-stage acceptance sampling plan, $N = 100$, $n = 50$, $c = 1$

p	$OC(p)$	p	$OC(p)$
.000	1.0000	.100	.0078
.010	1.0000	.110	.0038
.020	.7525	.120	.0019
.030	.5000	.130	.0009
.040	.3087	.140	.0004
.050	.1811	.150	.0002
.060	.1022	.160	.0001
.070	.0559	.170	.0000
.080	.0297	.180	.0000
.090	.0154	.190	.0000

FIGURE 9.1

Operating characteristics curve for a single-stage acceptance sampling plan, $N = 100$, $n = 50$, $c = 1$

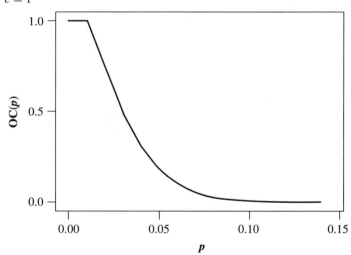

9.3
Approximate Determination of the Sampling Plan

If the sample size, n, is not too small, the c.d.f. of the hypergeometric distribution can be approximated by the normal distribution. More specifically, for large values of n we have the following approximation:

$$H(x; N, M, n) \cong \Phi \left(\frac{x + .5 - nP}{\left(nPQ \left(1 - \dfrac{n}{N}\right) \right)^{1/2}} \right) \tag{9.3.1}$$

where $P = M/N$ and $Q = 1 - P$.

The first question is: How large should n be? The answer to this question depends on how close we wish the approximation to be. Generally, if $.2 < P < .8, n = 20$ is large enough to yield a good approximation, as illustrated in Table 9.5.

TABLE **9.5**

Hypergeometric c.d.f.'s and their normal approximations

	$N = 100, M = 30, n = 20$		$N = 100, M = 50, n = 20$		$N = 100, M = 80, n = 20$	
a	**Hypergeometric**	**Normal**	**Hypergeometric**	**Normal**	**Hypergeometric**	**Normal**
0	0.00030	0.00140	0.00000	0.00000	0.00000	0.00000
1	0.00390	0.00730	0.00000	0.00000	0.00000	0.00000
2	0.02270	0.02870	0.00000	0.00000	0.00000	0.00000
3	0.08240	0.08740	0.00040	0.00060	0.00000	0.00000
4	0.20920	0.20780	0.00250	0.00310	0.00000	0.00000
5	0.40100	0.39300	0.01140	0.01260	0.00000	0.00000
6	0.61510	0.60700	0.03920	0.04080	0.00000	0.00000
7	0.79540	0.79220	0.10540	0.10680	0.00000	0.00000
8	0.91150	0.91260	0.22700	0.22780	0.00000	0.00000
9	0.96930	0.97130	0.40160	0.40180	0.00000	0.00000
10	0.99150	0.99270	0.59840	0.59820	0.00060	0.00030
11	0.99820	0.99860	0.77300	0.77220	0.00390	0.00260
12	0.99970	0.99980	0.89460	0.89320	0.01810	0.01480
13	1.00000	1.00000	0.96080	0.95920	0.06370	0.06000
14			0.98860	0.98740	0.17270	0.17550
15			0.99750	0.99690	0.36470	0.37790
16			0.99960	0.99940	0.60840	0.62210
17			1.00000	0.99990	0.82420	0.82450
18					0.95020	0.94000
19					0.99340	0.98520

If $P < .2$ or $P > .8$, we usually need larger sample sizes to attain a good approximation. We now show how the constants (n, c) can be determined. The two requirements to satisfy are $\text{OC}(p_0) = 1 - \alpha$ and $\text{OC}(p_t) = \beta$. These requirements are expressed approximately by the following two equations:

$$c + \frac{1}{2} - np_0 = z_{1-\alpha} \left(np_0 q_0 \left(1 - \frac{n}{N} \right) \right)^{1/2}$$

$$c + \frac{1}{2} - np_t = -z_{1-\beta} \left(np_t q_t \left(1 - \frac{n}{N} \right) \right)^{1/2}$$

(9.3.2)

Approximate solutions to n and c, n^* and c^*, respectively, are:

$$n^* \cong \frac{n_0}{1 + n_0/N}$$

(9.3.3)

where

$$n_0 = \frac{(z_{1-\alpha}\sqrt{p_0 q_0} + z_{1-\beta}\sqrt{p_t q_t})^2}{(p_t - p_0)^2} \qquad (9.3.4)$$

and

$$c^* \cong n^* p_0 - \frac{1}{2} + z_{1-\alpha}\sqrt{n^* p_0 q_0 (1 - n^*/N)} \qquad (9.3.5)$$

Table 9.6 gives several single-stage sampling plans, (n, c), as obtained from program ACCEPT, and their approximations (n^*, c^*). The table also provides the corresponding attained risk levels, $\hat{\alpha}$ and $\hat{\beta}$. We see that the approximation provided for n and c yields risk levels that are generally close to the nominal ones. Program ACCEPT starts with the approximate solution (n^*, c^*) and then searches for the values of (n, c) so that $\hat{\alpha} \leq \alpha$ and $\hat{\beta} \leq \beta$.

TABLE   9.6

Exact and approximate single-stage sampling plans for $\alpha = \beta = .05$, $N = 500, 1000, 2000$, $p_0 = .01, p_t = .03, .05$

N	Method	$p_0 = .01, p_t = .03$				$p_0 = .01, p_t = .05$			
		n	c	$\hat{\alpha}$	$\hat{\beta}$	n	c	$\hat{\alpha}$	$\hat{\beta}$
500	Exact	254	4	.033	.050	139	3	.023	.050
	Approximate	248	4	.029	.060	127	2	.107	.026
1000	Exact	355	6	.028	.050	146	3	.045	.049
	Approximate	330	5	.072	.036	146	3	.045	.049
2000	Exact	453	8	.022	.050	176	4	.026	.050
	Approximate	396	6	.082	.032	157	3	.066	.037

9.4
Double-Sampling Plans for Attributes

A **double-sampling plan** for attributes is a two-stage procedure. In the first stage, a random sample of size n_1 is drawn without replacement from the lot. Let X_1 denote the number of defective items in this first-stage sample. The rules for the second stage are the following:

If $X_1 \leq c_1$, sampling terminates and the lot is accepted.

If $X_1 \geq c_2$, sampling terminates and the lot is rejected.

If X_1 is between c_1 and c_2, a second-stage random sample of size n_2 is drawn without replacement from the remaining items in the lot.

Let X_2 be the number of defective items in this second-stage sample. Then, if $X_1 + X_2 \leq c_3$, the lot is accepted, and if $X_1 + X_2 > c_3$, the lot is rejected.

Generally, if there are very few (or very many) defective items in the lot, the decision to accept or reject the lot can be reached after the first stage of sampling. Because the first-stage samples are smaller than those needed in a single-stage sampling, a considerable saving in inspection cost may be attained.

In this type of sampling plan, there are five parameters to select—namely, n_1, n_2, c_1, c_2, and c_3. Variations in the values of these parameters affect the operating characteristics of the procedure, as well as the expected number of observations required (that is, the total sample size). Theoretically, we could determine the optimal values of these five parameters by imposing five independent requirements on the OC function and the function of expected total sample size, called the **average sample number (ASN) function**, at various values of p. However, to simplify this procedure, it is common practice to set $n_2 = 2n_1$ and $c_2 = c_3 = 3c_1$. This reduces the problem to that of selecting just n_1 and c_1. Every such selection will specify a particular double-sampling plan. For example, if the lot consists of $N = 150$ items, and we choose a plan with $n_1 = 20$, $n_2 = 40$, $c_1 = 2$, and $c_2 = c_3 = 6$, we will achieve certain properties. However, if we set $n_1 = 20$, $n_2 = 40$, $c_1 = 1$, and $c_2 = c_3 = 3$, the plan will have different properties.

The formula of the OC function associated with a double-sampling plan $(n_1, n_2, c_1, c_2, c_3)$ is

$$OC(p) = H(c_1; N, M_p, n_1) + $$

$$\sum_{j=c_1+1}^{c_2-1} h(j; N, M_p, n_1)H(c_3 - j; N - n_1, M_p - j, n_2) \qquad \text{(9.4.1)}$$

where $M_p = [Np]$. Obviously, we must have $c_2 \geq c_1 + 2$; otherwise, the plan is a single-stage plan. We can compute the OC function using the computer program ACCEPT. The probability $\Pi(p)$ of stopping after the first stage of sampling is

$$\Pi(p) = H(c_1; N, M_p, n_1) + 1 - H(c_2 - 1; N, M_p, n_1)$$
$$= 1 - [H(c_2 - 1; N, M_p, n_1) - H(c_1; N, M_p, n_1)] \qquad \text{(9.4.2)}$$

The expected total sample size, ASN, is given by the formula

$$ASN(p) = n_1 \Pi(p) + (n_1 + n_2)(1 - \Pi(p))$$
$$= n_1 + n_2[H(c_2 - 1; N, M_p, n_1) - H(c_1; N, M_p, n_1)] \qquad \text{(9.4.3)}$$

In Table 9.7, we present the OC function and the ASN function for the double-sampling plan (20, 40, 2, 6, 6,) for a lot of size $N = 150$.

We see from the table that the double-sampling plan illustrated here is not stringent. The probability of accepting a lot with 10% defectives is .785 and the probability of accepting a lot with 15% defectives is .395. If we consider the plan (20, 40, 1, 3, 3), a more stringent procedure is obtained, as shown in Table 9.8. The probability of accepting a lot having 10% defectives has dropped to .39, and that of accepting a lot with 15% defectives has dropped to .15. Table 9.8 shows that the ASN is 23.1 when $p = .025$ (the sampling usually terminates after the first stage), and the ASN is 29.1 when $p = .15$. The maximum ASN occurs around $p = .10$.

TABLE   9.7

The OC and ASN of a double-sampling plan (20,40,2,6,6), $N = 150$

p	OC(p)	ASN(p)	P	OC(p)	ASN(p)
.000	1.0000	20.0	.250	.0714	41.4
.025	1.0000	20.3	.275	.0477	38.9
.050	.9946	22.9	.300	.0268	35.2
.075	.9472	26.6	.325	.0145	31.6
.100	.7849	32.6	.350	.0075	28.4
.125	.5759	38.3	.375	.0044	26.4
.150	.3950	42.7	.400	.0021	24.3
.175	.2908	44.6	.425	.0009	22.7
.200	.1885	45.3	.450	.0004	21.6
.225	.1183	44.1	.475	.0002	21.1
			.500	.0001	20.6

TABLE   9.8

The OC and ASN for the double-sampling plan (20,40,1,3,3), $N = 150$

p	OC(p)	ASN(p)
.000	1.0000	20.0
.025	.9851	23.1
.050	.7969	28.6
.075	.6018	31.2
.100	.3881	32.2
.125	.2422	31.2
.150	.1468	29.1
.175	.0987	27.3
.200	.0563	25.2
.225	.0310	23.5

Computer programs such as ACCEPT can be used to determine an acceptable double-sampling plan for attributes. Suppose the population size is $N = 1000$. Suppose that AQL = .01 and LQL = .03. If $n_1 = 200$, $n_2 = 400$, $c_1 = 3$, $c_2 = 9$, and $c_3 = 9$, then OC(.01) = .9892, and OC(.03) = .1191. Thus, $\alpha = .011$ and $\beta = .119$. The double-sampling plan with $n_1 = 120$, $n_2 = 240$, $c_1 = 0$, and $c_2 = c_3 = 7$ yields $\alpha = .044$ and $\beta = .084$. For the last plan, the expected sample sizes are ASN(.01) = 294 and ASN(.03) = 341. These expected sample sizes are smaller than the required sample size of $n = 355$ in a single-stage plan (see Table 9.6). Moreover, if $p \leq p_0$ or $p \geq p_t$, the sampling will terminate with high probability after the first stage with only $n_1 = 120$ observations. This is a threefold decrease in the sample size over the single-sampling plan. Other double-sampling plans can do even better.

If the lot is very large, and we use large samples in stage 1 and stage 2, the formulas for the OC and ASN function can be approximated by

$$
\begin{aligned}
OC(p) \cong\ & \Phi\left(\frac{c_1 + 1/2 - n_1 p}{(n_1 pq(1 - n_1/N))^{1/2}}\right) \\
& + \sum_{j=c_1+1}^{c_2-1}\left[\Phi\left(\frac{j + 1/2 - n_1 p}{(n_1 pq(1 - n_1/N))^{1/2}}\right)\right. \\
& \left. - \Phi\left(\frac{j - 1/2 - n_1 p}{(n_1 pq(1 - n_1/N))^{1/2}}\right)\right] \cdot \Phi\left(\frac{c_3 - j + 1/2 - n_2 p}{\left(n_2 pq\left(1 - \dfrac{n_2}{N - n_1}\right)\right)^{1/2}}\right)
\end{aligned}
$$

(9.4.4)

and

$$
\begin{aligned}
ASN(p) = n_1 + n_2 \Bigg[& \Phi\left(\frac{c_2 - 1/2 - n_1 p}{n_1 pq(1 - n_1/N))^{1/2}}\right) \\
& - \Phi\left(\frac{c_1 + 1/2 - n_1 p}{(n_1 pq(1 - n_1/N))^{1/2}}\right)\Bigg]
\end{aligned}
$$

(9.4.5)

Table 9.9 gives the OC and the ASN functions for double sampling from a population of size $N = 1000$, when the parameters of the plan are $(100, 200, 3, 9, 9)$. The exact values thus obtained are compared to the values obtained from the large-sample approximation formulas. In the next section, we generalize the idea of double sampling to try to reach acceptance decisions more quickly and therefore at reduced cost.

TABLE   9.9

The exact and approximate OC and ASN functions for the double-sampling plan $(100,200,3,9,9)$, $N = 1000$

p	OC Exact	OC Approximate	ASN Exact	ASN Approximate
.01	.999	.999	102.5	100.8
.02	.962	.968	126.2	125.9
.03	.746	.737	170.1	175.7
.04	.457	.445	213.4	219.7
.05	.251	.248	240.7	244.1
.06	.131	.137	247.1	246.6
.07	.064	.075	235.5	231.6
.08	.030	.040	212.6	207.4
.09	.014	.021	185.8	181.1

9.5
Sequential Sampling

Sometimes it is possible to subject a selected item to immediate testing before selecting the next item. Such an on-line testing environment allows for considerable savings in acceptance sampling procedures. In this situation, we can decide after each observation if sufficient information is available for accepting or rejecting the lot, or if an additional item (or items) should be randomly selected and tested. This approach may lead to substantial reductions in the number of required observations, especially when the proportion of nonconforming items in the lot is very small or very high. For example, from a lot of size $N = 5000$, a single-sampling plan—for $\alpha = \beta = .05$, AQL $= .01$, and LQL $= .05$—is $n = 179$, $c = 4$. Under this plan, if we find at least five defective items in the sample, the lot is rejected. Suppose that the lot under inspection contains more than 10% defective items. In such a case, five defective items are expected in 50 observations. We can reject such a lot as soon as we observe five defective items and avoid observing all other items. Sequential sampling procedures allow us to decide after each observation whether to continue or terminate sampling. When stopping occurs, we decide whether to accept or reject the lot.

In this section, we discuss **sequential sampling** by attributes, also known as the Wald **sequential probability ratio test (SPRT)**, which can guarantee that the sum of the producer's and consumer's risks does not exceed the sum of the preassigned values, α and β. Furthermore, this procedure has the property that, compared to all other procedures with producer's and consumer's risks not greater than α and β, the SPRT has the smallest ASN values at p_0 and p_t. This is known as the *optimality* of the SPRT. For simplicity, we limit the discussion to the SPRT method for binomial distributions and not for hypergeometric ones. This is valid when the lot is very large compared to the sample size.

Given n observations, with X_n defectives, the *likelihood ratio* is defined as the ratio of the binomial p.d.f. $b(X_n; n, p_t)$ to $b(X_n; n, p_0)$—that is,

$$\Lambda(X_n; n, p_0, p_t) = \left(\frac{p_t}{p_0}\right)^{X_n} \left(\frac{1 - p_t}{1 - p_0}\right)^{n - X_n} \tag{9.5.1}$$

Two critical values, A and B, satisfying $0 < A < B < \infty$ are selected. The SPRT, with limits (A, B), is a procedure that terminates sampling at the first $n \geq 1$ for which either $\Lambda(X_n; n, p_0, p_t) \geq B$ or $\Lambda(X_n; n, p_0, p_t) \leq A$. In the first case, the lot is rejected; in the second, the lot is accepted. If $A < \Lambda(X_n; n, p_0, p_t) < B$, an additional item is randomly sampled from the lot. Any specified values of A and B yield risk values $\alpha^*(A, B)$ and $\beta^*(A, B)$, which are the actual producer's and consumer's risks. We can show that if $A = \beta/(1 - \alpha)$ and $B = (1 - \beta)/\alpha$ (with both α and β smaller than $1/2$), then $\alpha^*(A, B) + \beta^*(A, B) \leq \alpha + \beta$. Therefore, it is customary to use these particular critical values. Thus, the SPRT, with $A = \beta/(1 - \alpha)$ and $B = (1 - \beta)/\alpha$, reduces to the following set of rules:

1 Stop sampling and reject the lot if $X_n \geq h_2 + sn$.

2 Stop sampling and accept the lot if $X_n \leq -h_1 + sn$.

3 Continue sampling if $-h_1 + sn < X_n < h_2 + sn$.

Here

$$h_1 = \log\left(\frac{1-\alpha}{\beta}\right) / \log\left(\frac{p_t(1-p_0)}{p_0(1-p_t)}\right)$$

$$h_2 = \log\left(\frac{1-\beta}{\alpha}\right) / \log\left(\frac{p_t(1-p_0)}{p_0(1-p_t)}\right)$$

(9.5.2)

and

$$s = \frac{\log\left(\dfrac{1-p_0}{1-p_t}\right)}{\log\left(\dfrac{p_t}{1-p_t} \cdot \dfrac{1-p_0}{p_0}\right)}$$

(9.5.3)

For example, if $p_0 = .01$, $p_t = .05$, and $\alpha = \beta = .05$, we obtain

$$h_1 = 1.78377$$

$h_2 = h_1$, and

$$s = 0.02499$$

The SPRT can be performed graphically by plotting the two boundary lines,

$$a_n = -1.78 + .025n$$

and

$$r_n = 1.78 + .025n$$

After each observation, the point (n, X_n) is plotted. As long as the points lie between the parallel lines, a_n and r_n, sampling continues. As soon as $X_n \geq r_n$ or $X_n \leq a_n$, sampling stops and the proper action is taken. Notice that the smallest number of observations required to reach the acceptance region is $n_0 = [h_1/s] + 1$, and the smallest number needed to reach the rejection region is

$$n_1 = [h_2/(1-s)] + 1$$

(9.5.4)

In this example, $n_0 = 72$ and $n_1 = 2$. This shows that if the first two observations are defectives, the lot is immediately rejected. However, the lot cannot be accepted until 72 observations are made, and these must all be observations of conforming items.

To illustrate the use of the SPRT, we first determine the acceptance numbers a_n and the rejection numbers r_n (see Table 9.10).

Then we use simulation to show the sequential decision process. Using MINITAB, we generate a random sequence of Bernoulli trials (0s and 1s) with a specified probability of 1 (which corresponds to a defective item). The total number of defects among the first n items in the sample, X_n, is then computed. Table 9.11 shows this simulation of 100 binomial outcomes with $P = .01$, and lists the partial sums, X_n, from left to right.

We see that there are no 1s (defective items) among the first 100 numbers generated at random with $P = .01$. According to Table 9.10, the SPRT would have stopped sampling after the 72nd observation and accepted the lot.

TABLE 9.10

Acceptance and rejection boundaries for a SPRT, AQL = .01, LQL = .05, $\alpha = \beta = .05$

sample size (n)	accept (a_n)	reject(r_n)
1–8	–	2
9–48	–	3
49–71	–	4
72–88	0	4
88–111	0	5
112–128	1	5
129–151	1	6
152–168	2	6
169–191	2	7

TABLE 9.11

Simulated results of 100 binomial experiments, $n = 1, p = .01$

0	0	0	0	0	0	0	0	0	0	0	0	0	0	0	0	0	0	0	0
0	0	0	0	0	0	0	0	0	0	0	0	0	0	0	0	0	0	0	0
0	0	0	0	0	0	0	0	0	0	0	0	0	0	0	0	0	0	0	0
0	0	0	0	0	0	0	0	0	0	0	0	0	0	0	0	0	0	0	0
0	0	0	0	0	0	0	0	0	0	0	0	0	0	0	0	0	0	0	0

Now we simulate binomial variables with $P = .05$. The result is given in Table 9.12.

Here we see that $X_n = r_n$ for the first time at $n = 38$. Accordingly, the SPRT would have stopped sampling and rejected the lot after 38 observations. It can be proven theoretically that every SPRT stops sooner or later and that a decision is reached at that point.

A SPRT can be characterized by computing its OC and ASN functions. The OC function provides the points: $(p, \mathrm{OC}(p))$. We can express these points as functions of a parameter τ, according to the following two formulas:

$$p(\tau) = \frac{1 - \left(\dfrac{1 - p_t}{1 - p_0}\right)^\tau}{\left(\dfrac{p_t}{p_0}\right)^\tau - \left(\dfrac{1 - p_t}{1 - p_0}\right)^\tau} \tag{9.5.5}$$

and

$$\mathrm{OC}(p(\tau)) = \frac{\left(\dfrac{1 - \beta}{\alpha}\right)^\tau - 1}{\left(\dfrac{1 - \beta}{\alpha}\right)^\tau - \left(\dfrac{\beta}{1 - \alpha}\right)^\tau} \tag{9.5.6}$$

For $\tau = 1$, we obtain $p(1) = p_0$, and for $\tau = -1$, we obtain $p(-1) = p_t$. We can also show that for $\tau = 0$, $p(0) = s$. Table 9.13 gives the OC function, computed according to these two formulas for the SPRT, with $p_0 = .01$, $p_t = .05$, and $\alpha = \beta = .05$.

TABLE   9.12

Simulated results of 100 binomial experiments, $n = 1, p = .05$

Partial Sums, X_n					(Count)
0	0	0	0	0	(5)
0	0	0	0	0	(10)
0	0	0	0	0	(15)
0	1	1	1	1	(20)
1	1	1	1	1	(25)
1	1	1	1	1	(30)
1	2	2	2	2	(35)
2	2	3	3	3	(40)
3	3	3	3	3	(45)
3	4	4	4	4	(50)
4	4	4	4	4	(55)
4	4	4	4	4	(60)
4	5	5	5	5	(65)
5	5	5	5	5	(70)
5	5	5	5	5	(75)
5	5	5	5	5	(80)
5	5	5	5	5	(85)
5	5	5	5	5	(90)
5	5	5	5	5	(95)
5	5	5	5	5	(100)

TABLE   9.13

The OC and ASN function of the SPRT, $p_0 = .01, p_t = .05, \alpha = \beta = .05$

τ	$p(\tau)$	$OC(p(\tau))$	ASN
1.0000	.0100	.950	107
0.5000	.0160	.813	128
0.2500	.0200	.676	134
0.1250	.0230	.591	133
0.0625	.0237	.546	132
0.0000	.0250	.500	130
−0.1250	.0280	.409	125
−0.2500	.0300	.324	117
−0.5000	.0360	.187	101
−0.7500	.0430	.099	79
−1.0000	.0500	.050	64

The ASN function of a SPRT is given by the formula

$$ASN(p) = \frac{OC(p) \log\left(\dfrac{\beta}{1-\alpha}\right) + (1 - OC(p)) \log\left(\dfrac{1-\beta}{\alpha}\right)}{p \log(p_t/p_0) + (1 - p) \log((1 - p_t)/(1 - p_0))} \qquad (9.5.7)$$

for $p \neq s$, and

$$\text{ASN}(s) = \frac{h_1 h_2}{s(1-s)} \tag{9.5.8}$$

The values of the ASN function corresponding to the SPRT with $p_0 = .01$, $p_t = .05$, and $\alpha = \beta = .05$ are given in Table 9.13 and plotted in Figure 9.2. In Table 9.6, we presented the sample size n and the acceptance number c required for a single-sampling plan with $p_0 = .01$, $p_t = .05$, and $\alpha = \beta = .05$, for various lots. We found that if $N = 2000$, we need a single sample for $n = 176$ observations. In the SPRT, we stop, on the average, after 107 observations if $p = .01$ and after 64 if $p = .05$. If $p > .05$, the ASN is considerably smaller. This illustrates the potential savings associated with the SPRT.

FIGURE 9.2

The OC curve and the ASN curve of the SPRT, $p_0 = .01$, $p_t = .05$, and $\alpha = \beta = .05$

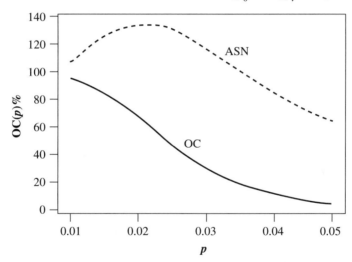

9.6
Acceptance Sampling Plans for Variables

It is sometimes possible to determine whether an item is defective or not by first performing a measurement on it that gives a value of a continuous random variable X and then comparing that value to specification limits. For example, in Chapter 2, we discussed measuring the strength of yarn. In this case, a piece of yarn is deemed defective if its strength X is less than ξ, where ξ is the required minimum strength—that is, its lower specification limit. The proportion of defective yarn pieces in the population (or very large lot) is the probability that $X \leq \xi$.

Suppose now that X has a normal distribution with mean μ and variance σ^2. (If the distribution is not normal, we can often reduce it to a normal one by a proper transformation.) Accordingly, the proportion of defectives in the population is

$$p = \Phi\left(\frac{\xi - \mu}{\sigma}\right) \tag{9.6.1}$$

We have to decide whether $p \leq p_0$ (= AQL) or $p \geq p_t$ (= LQL) in order to accept or reject the lot.

Let x_p represent the pth quantile of a normal distribution with mean μ and standard deviation σ. Then

$$x_p = \mu + z_p \sigma \tag{9.6.2}$$

If $x_{p_0} \geq \xi$, we should accept the lot because the proportion of defectives is less than p_0. Because we do not know μ and σ, we must make our decision on the basis of estimates from a sample of n measurements. We decide to reject the lot if

$$\overline{X} - kS < \xi$$

and accept the lot if

$$\overline{X} - kS \geq \xi$$

Here, $\overline{X}$ and S are the usual sample mean and standard deviation, respectively. The factor k is chosen so that the producer's risk (the risk of rejecting a good lot) does not exceed α. Values of the factor k are given approximately by the formula (given also in Section 6.4.1)

$$k \cong t_{1-\alpha, p_0, n}$$

where

$$t_{1-a, p_0, n} = \frac{z_{1-p_0}}{1 - z_{1-a}^2/2n} + \frac{z_{1-a}\left(1 + \dfrac{z_{p_0}^2}{2} - \dfrac{z_{1-a}^2}{2n}\right)^{1/2}}{\sqrt{n}(1 - z_{1-a}^2/2n)} \tag{9.6.3}$$

The OC function of such a test is given approximately (for large samples) by

$$OC(p) \cong 1 - \Phi\left(\frac{(z_p + k)\sqrt{n}}{(1 + k^2/2)^{1/2}}\right) \tag{9.6.4}$$

where $k = t_{1-\alpha, p_0, n}$. We can thus determine n and k so that

$$OC(p_0) = 1 - \alpha$$

and

$$OC(p_t) = \beta$$

These two conditions yield the equations

$$(z_{p_t} + k)\sqrt{n} = z_{1-\beta}(1 + k^2/2)^{1/2}$$

and $\tag{9.6.5}$

$$(z_{p_0} + k)\sqrt{n} = z_\alpha(1 + k^2/2)^{1/2}$$

The solution for n and k yields:

$$n = \frac{(z_{1-\alpha} + z_{1-\beta})^2(1 + k^2/2)}{(z_{p_t} - z_{p_0})^2} \tag{9.6.6}$$

and

$$k = (z_{p_t}z_\alpha + z_{p_0}z_\beta)/(z_{1-\alpha} + z_{1-\beta}) \tag{9.6.7}$$

In other words, if the sample size n is given by formula 9.6.6, we can replace $t_{1-\alpha,p_0,n}$ by the simpler term k and accept the lot if

$$\overline{X} - kS \geq \xi$$

The statistic $\overline{X} - kS$ is called a *lower tolerance limit*.

EXAMPLE 9.1 Consider the example from Chapter 2 in which we tested the compressive strength of concrete cubes. The compressive strength must be greater than 240 kg/cm². We found that $Y = \ln X$ had an approximately normal distribution. Suppose we need to decide whether to accept or reject this lot with the following specifications: $p_0 = .01$, $p_t = .05$, and $\alpha = \beta = .05$. According to the normal distribution,

$$z_{p_0} = -2.326 \qquad z_{p_t} = -1.645$$

and

$$z_{1-\alpha} = z_{1-\beta} = 1.645$$

Thus, according to these formulas, we find $k = 1.9855$ and $n = 70$. Hence, with a sample size of 70, we can accept the lot if $\overline{Y} - 1.9855S \geq \xi$, where $\xi = \ln(240) = 5.48$. ∎

The sample size required in this single-stage sampling plan for variables is substantially smaller than the one we determined for the single-stage sampling plan for attributes (which was $n = 176$). However, the sampling plan for attributes was free of any assumption about the distribution of X; in Example 9.1, we had to assume that $Y = \ln X$ is normally distributed. Thus, there is a certain trade-off between the two approaches. In particular, if our assumptions concerning the distribution of X are erroneous, we may not have the desired producer's and consumer's risks.

The procedure of acceptance sampling for variables can be generalized to upper and lower tolerance limits, double sampling, and sequential sampling. The interested reader can find more information on the subject in Duncan (1986, ch. 12–15). Fuchs and Kenett (1997) applied tolerance limits to the appraisal of ceramic substrates in the multivariate case.

Rectifying Inspection of Lots

A **rectifying inspection** is a complete inspection of a rejected lot made for the purpose of replacing the defective items by nondefective items. (Lots that are accepted are not subjected to rectification.) We shall assume that the tests are nondestructive, that all the defective items in the sample are replaced by good ones, and that the sample is replaced in the lot.

If a lot contains N items and has a proportion p of defectives before the inspection, the proportion of defectives in the lot after inspection is

$$p' = \begin{cases} 0 & \text{if lot is rejected} \\ p(N - X)/N & \text{if lot is accepted} \end{cases} \tag{9.7.1}$$

where X is the number of defectives in the sample. If the probability of accepting a lot by a given sampling plan is $\mathrm{OC}(p)$, then the expected proportion of outgoing defectives, when sampling is single stage by attribute, is

$$E\{p'\} = p\, \mathrm{OC}(p) \left(1 - \frac{n}{N} R_s^*\right) \tag{9.7.2}$$

where

$$R_s^* = \frac{H(c - 1; N - 1, [Np] - 1, n - 1)}{H(c; N, [Np], n)} \tag{9.7.3}$$

If n/N is small, then

$$E\{p'\} \cong p\, \mathrm{OC}(p) \tag{9.7.4}$$

The expected value of p' is called the **average outgoing quality (AOQ)**.

The formula for R_s^* depends on the method of sampling inspection. If the inspection is by double sampling, the formula is considerably more complicated.

Table 9.14 gives the AOQ values corresponding to a rectifying plan where $N = 1000$, $n = 250$, and $c = 5$.

TABLE   9.14
AOQ values for rectifying plan, $N = 1000$, $n = 250$, $c = 5$

p	$\mathrm{OC}(p)$	R_s^*	AOQ
.005	1.000	1.0000	.004
.010	.981	.9710	.007
.015	.853	.8730	.010
.020	.618	.7568	.010
.025	.376	.6546	.008
.030	.199	.5715	.005
.035	.094	.5053	.003

The average outgoing quality limit (AOQL) of a rectifying plan is defined as the maximal value of AOQ. Thus, the AOQL corresponding to the plan of Table 9.14 is approximately .01. The AOQ given is presented graphically in Figure 9.3.

FIGURE   9.3
AOQ curve for single-sampling plan, $N = 1000$, $n = 250$, and $c = 5$

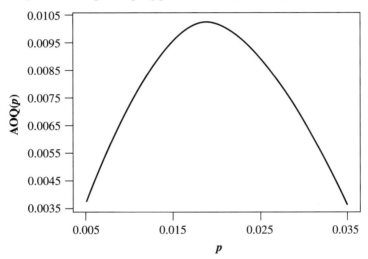

We also characterize a rectifying plan by the **average total inspection (ATI)** associated with a given value of p. If a lot is accepted, only n items (the sample size) have been inspected, whereas if it is rejected, the number of items inspected is N. Thus,

$$\text{ATI}(p) = n\,\text{OC}(p) + N(1 - \text{OC}(p))$$
$$= n + (N - n)(1 - \text{OC}(p)) \qquad \text{(9.7.5)}$$

This function is increasing from n (when $p = 0$) to N (when $p = 1$).

In our example, the lot contains $N = 1000$ items and the sample size is $n = 250$. The graph of the ATI function is given in Figure 9.4.

Dodge and Romig (1959) published tables for the design of single- and double-sampling plans for attributes for which the AOQL is specified and the ATI is minimized at a specified value of p. In Table 9.15, we provide a few values of n and c for such a single-sampling plan, for which the AOQL $= .01$.

According to this table, for a lot of size 2000, to guarantee an AOQL of 1% and minimal ATI at $p = .01$, one needs a sample of size $n = 180$, with $c = 3$. For another method of determining n and c, see Duncan (1986, ch. 16).

Rectifying sampling plans with less than 100% inspection of rejected lots have been developed and are available in the literature.

FIGURE   **9.4**

ATI curve for single-sampling plan, $N = 1000$, $n = 250$, and $c = 5$

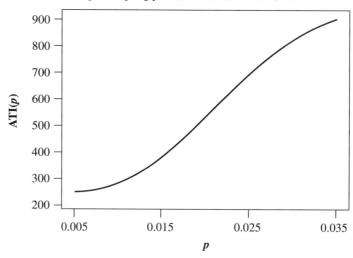

TABLE   **9.15**

Selected values of (n, c) for a single-sampling plan, AOQL $= .01$ and ATI minimum at p

p	.004–.006		.006–.008		.008–.01	
N	n	c	n	c	n	c
101–200	32	0	32	0	32	0
201–300	33	0	33	0	65	1
501–600	75	1	75	1	75	1
1001–2000	130	2	130	2	180	3

9.8

National and International Standards

During World War II, the U.S. Army developed standards for sampling acceptance schemes by attributes. Army ordinance tables were prepared in 1942, and the Navy issued its own tables in 1945. Joint Army and Navy standards were issued in 1949. These standards were superseded in 1950 by the common standards named MIL-STD-105A. The MIL-STD-105D was issued by the U.S. government in 1963 and slightly revised as MIL-STD-105E in 1989. Now these standards are being gradually phased out by the Department of Defense and replaced by a certification program, which is described later. The American National Standards Institute, ANSI, adopted the military standards with some minor modifications as ANSI Z1.4 standards. These were adopted in 1974 by the International Organization for Standardization as ISO 2859. In 1981, ANSI Z1.4 standards were adopted by the American Society for Quality Control with some additions, and the standards issued were named ANSI/ASQC Z1.4.

The military standards were designed to inspect incoming lots from a variety of suppliers. All suppliers are required to satisfy specified quality levels for the products. These quality levels are indexed by the AQL. It is assumed that a supplier will send lots (shipments) continuously. All these lots are subjected to quality inspection. At the beginning, an AQL value is specified for the product, and the type of sampling plan is decided (single, double, sequential, and so on). For a given lot size and type of sampling, the parameters of the sampling procedure are then determined. For example, if the sampling is single stage by attribute, the parameters (n, c) are read from the tables. The special feature of the MIL-STD-105E is that lots can be subjected to normal, tightened, or reduced inspection. Inspection starts at a **normal inspection level**. If 2 or more out of 5 consecutive lots have been rejected, the level switches to **tightened inspection level**. Normal inspection is reinstituted if 5 consecutive lots have been accepted. If 10 consecutive lots remain under tightened inspection, an action may take place to discontinue the contract with the supplier. However, if the last 10 lots have all been accepted at a normal inspection level, and the total number of defective units found in the samples from these ten lots is less than a specified value, then a switch from normal to **reduced inspection level** can take place.

In this text, we do not reproduce the MIL-STD-105 tables, but readers can find a detailed explanation and examples in Duncan (1986, ch. 10).

We conclude with the following example. Suppose that for a given product AQL $= .01$ (1%). The size of the lots is $N = 1000$. The military standard specifies that a single-stage sampling for attributes, under normal inspection, has the parameters $n = 80$ and $c = 2$. Program ACCEPT yields the following OC values for this plan:

p	.01	.02	.03	.04	.05
OC(p)	.961	.789	.564	.365	.219

Thus, if the proportion of nonconforming, p, items of the supplier is less than the AQL, the probability of accepting a lot is larger than .961. A supplier that continues to ship lots with $p = .01$ has a probability of $(.961)^{10} = .672$ that all 10 lots will be accepted, and the inspection level will be switched to a reduced one. Under the reduced level, the sample size from the lot is reduced to $n = 32$. The corresponding acceptance number is $c = 1$. Thus, despite the fact that the quality level of the supplier remains good, there is a probability of .33 that no switch will be made to reduced inspection level after the tenth lot. However, the probability that no switch will be made to a tightened level of inspection before the sixth lot is inspected is .9859. This is the probability that, after each inspection, the next lot will continue to be inspected under normal level. If there is no deterioration in the quality level of the supplier, and $p = .03$, the probability that the inspection level will be switched to tightened after five inspections is .722.

Skip-Lot Sampling Plans for Attributes

We saw in the previous section that, according to MIL-STD-105E, if a supplier keeps shipping high-quality lots, then after a while his lots are subjected to inspection under reduced level. All lots are inspected under a reduced-level inspection scheme as long as their quality level remains high. The **skip-lot sampling plan (SLSP)**, which was proposed by Liebesman and Saperstein (1983), introduces a new element of savings if lots continue to have a very low proportion of nonconforming items. As we will see, instead of just a reduced level of inspection for high-quality lots, under SLSP, such lots are not necessarily even inspected. If the lots coming in from a given supplier qualify for skipping, then their probability of inspection is only .5. This probability is later reduced to .33 and to .2 if the inspected lots continue to be almost free of nonconforming items. Thus, suppliers that continue to manufacture their product with a proportion of defectives p that is considerably smaller than the specified AQL stand a good chance of having only a small fraction of their lots inspected. The SLSP specified next was adopted as the ISO 2859/3 standard in 1986.

9.9.1   ISO 2859 Skip-Lot Sampling Procedures

A SLSP has to address three main issues:

1   What are the conditions for beginning or reinstating the skip-lot (SL) state?

2   What is the fraction of lots to be skipped?

3   Under what conditions should one stop skipping lots on a temporary or permanent basis?

The fraction of lots to be skipped is the probability that a given lot will not be inspected. If this probability, for example, is .8, we generate a random number, U, with uniform distribution on $(0, 1)$. If $U < .8$, inspection is skipped; otherwise, the lot is inspected.

We define three states:

State 1.   Every lot is inspected.

State 2.   Some lots are skipped and not inspected.

State 3.   All lots are inspected, pending a decision of disqualification (back to state 1) or resumption of SL (back to state 2).

Lot-by-lot inspection is performed during state 3, but the requirements to requalify for skip-lot inspection are less stringent than the initial qualification requirements.

Switching rules apply to four transitions between states: Qualification (state 1 to state 2), Interruption (state 2 to state 3), Resumption (state 3 to state 2), Disqualification (state 3 to state 1).

The switching rules for the SLSP procedure are as follows:.

Skip-Lot Switching Rules   The following rules are appropriate for single sampling by attributes. Other rules are available for other sampling schemes.

A. Qualification   (state 1 → state 2).

1   Ten consecutive lots are accepted.

2   The total number of defective items in the samples from the 10 lots is smaller than the critical level given in Table 9.16.

3   The number of defective items in each one of the last two lots is smaller than the values specified in Table 9.17.

4   The supplier has a stable manufacturing organization, continuous production, and other traits that qualify him/her to be a high-quality stable manufacturer.

B. Interruption   (state 2 → state 3)

1   In the sample from an inspected lot there are more defectives than specified in Table 9.17.

TABLE   9.16

Minimum cumulative sample size in 10 lots for skip-lot qualifications

Cumulative number of defectives	AQL(%)						
	0.65	**1.0**	**1.5**	**2.5**	**4.0**	**6.5**	**10.0**
0	400	260	174	104	65	40	26
1	654	425	284	170	107	65	43
2	883	574	383	230	144	88	57
3	1098	714	476	286	179	110	71
4	1306	849	566	340	212	131	85
5	1508	980	653	392	245	151	98
6	1706	1109	739	444	277	171	111
7	1902	1236	824	494	309	190	124
8	2094	1361	907	544	340	209	136
9	2285	1485	990	594	371	229	149
10	2474	1608	1072	643	402	247	161
11	2660	1729	1153	692	432	266	173
12	2846	1850	1233	740	463	285	185
13	3031	1970	1313	788	493	303	197
14	3214	2089	1393	836	522	321	209
15	3397	2208	1472	883	552	340	221
16	3578	2326	1550	930	582	358	233
17	3758	2443	1629	977	611	376	244
18	3938	2560	1707	1024	640	394	256
19	4117	2676	1784	1070	669	412	268
20	4297	2793	1862	1117	698	430	279

TABLE   9.17

Individual lot acceptance numbers for skip-lot qualification

Sample Size	AQL(%)						
	0.65	1.0	1.5	2.5	4.0	6.5	10.0
2	-	-	-	-	-	0	0
3	-	-	-	-	0	0	0
5	-	-	-	0	0	0	1
8	-	-	0	0	0	1	1
13	-	0	0	0	1	1	2
20	0	0	0	1	1	2	3
32	0	0	1	1	2	3	5
50	0	1	1	2	3	5	7
80	1	1	2	3	5	7	11
125	1	2	3	4	7	11	16
200	2	3	4	7	11	17	25
315	3	4	7	11	16	25	38
500	5	7	10	16	25	39	58
800	7	11	16	25	38	60	91
1250	11	16	23	38	58	92	138
2000	17	25	36	58	91	144	217

C. Resumption   (state 3 → state 2)

1   Four consecutive lots are accepted.

2   The last two lots satisfy the requirements of Table 9.17.

D. Disqualification   (state 3 → state 1)

1   Two lots are rejected within 10 consecutively inspected lots; or

2   The supplier violates the supplier qualification criteria (see item A4).

We saw in the previous section that, under normal inspection, MIL-STD-105E specifies that, for AQL = .01 and lots of size $N = 1000$, random samples of size $n = 80$ should be drawn. The critical level was $c = 2$. If 10 lots have been accepted consecutively, the total number of observed defectives is $S_{10} \leq 20$. The total sample size is 800 and, according to Table 9.16, S_{10} should not exceed 3 to qualify for a switch to state 2. Moreover, according to Table 9.17, the last two samples should each have fewer than 1 defective item. Thus, the probability of qualifying for state 2 on the basis of the last 10 samples, when $p = $ AQL $ = .01$ is

$$QP = b^2(0; 80, .01)B(3; 640, .01)$$
$$= (.4475)^2 \times .1177 = .0236$$

Thus, if the fraction of defectives level is exactly at the AQL value, the probability of qualifying is only .02. However, if the supplier maintains the production at a fraction of defectives of $p = .001$, then the qualification probability is

$$QP = b^2(0; 80, .001)B(3; 640, .001)$$
$$= (.9231)^2 \times .9958 = .849$$

Thus, a supplier who maintains a level of $p = .001$ when the AQL $= .01$ will probably be qualified after the first 10 inspections and will switch to state 2: skipping lots. Eventually, only 20% of his lots will be inspected under this SLSP standard, with great savings to both producer and consumer. This illustrates the importance of maintaining high-quality production processes. In Chapters 10 and 11, we will study how to control the production processes statistically in order to maintain stable processes of high quality. Generally, for the SLSP to be effective, the fraction of defectives level of the supplier should be smaller than half the AQL. For a p level close to the AQL, the SLSP and the MIL-STD-105D are very similar in performance characteristics.

Currently, the Department of Defense (DoD) requires contractors and suppliers to become certified quality providers. This certification demands ongoing statistical process control (SPC) at the supplier's plant. The DoD relies on the internal control system of a certified supplier to detect any quality problems of subcontractors.

Certification of suppliers leads to "ship to stock" agreements, where the supplier supplies products directly to the customer's stockroom without any incoming inspection. Skip-lot sampling is halfway between implementing MIL-STD-105 and supplier certification programs.

The Deming Inspection Criterion

Deming (1982) has derived a formula to express the expected cost to the firm of sampling lots of incoming material. We define the following:

N = Number of items in a lot

k_1 = Cost of inspecting one item at the beginning of the process

q = Probability of a conforming item

p = Probability of a nonconforming item

$Q = \mathrm{OC}(p)$ = Probability of accepting a lot

k_2 = Cost to the firm when one nonconforming item is moved downstream to a customer or to the next stage of the production process

p'' = Probability of nonconforming items being in an accepted lot

n = Sample size inspected from a lot of size N

Thus, the total expected cost per lot is

$$EC = (Nk_1/q)[1 + Qq\{(k_2/k_1)p'' - 1\}\{1 - n/N\}] \qquad (9.10.1)$$

If $(k_2/k_1)p'' > 1$, then any sampling plan increases the cost to the firm and $n = N$ (100% inspection) becomes the least costly alternative.

If $(k_2/k_1)p'' < 1$, then the value $n = 0$ yields the minimum value of EC so that no inspection is the alternative of choice.

Now p'' can be only somewhat smaller than p. For example, if $N = 50$, $n = 10$, $c = 0$, and $p = .04$, then $p'' = .0345$. Substituting p for p'' gives us the following rules:

If $(k_2/k_1)p > 1$, inspect every item in the lot.

If $(k_2/k_1)p < 1$, accept the lot without inspection.

The Deming assumption is that the process is under control and that p is known. Sampling plans such as MIL-STD-105D do not make such assumptions and, in fact, are designed for catching shifts in process levels.

To keep the process under control, Deming suggests the use of control charts and statistical process control (SPC) procedures, which are discussed in Chapters 10 and 11.

The assumption that a process is under control means that the firm has absorbed the cost of SPC as internal overhead or as a piece-cost. Deming's assertion then is that assuming up front the cost of SPC implementation is cheaper in the long run than doing business in an environment where a process may go out of control undetected until its output undergoes acceptance sampling.

In Chapter 10, we introduce the reader to the basic tools and principles of statistical process control.

Published Tables for Acceptance Sampling

This section provides information on published tables and schemes for sampling inspection by attribute and by variables. The material given here follows Chapters 24–25 of Juran (1979). We do not explain here how to use these tables. The interested practitioner can follow the instructions attached to the tables and/or read more about the tables in Juran (1979, ch. 24–25) or in Duncan (1986).

I. Sampling by Attributes

1   MIL-STD-105E

Type of sampling: Single, double, and multiple
Type of application: General
Key features: Maintains average quality at a specified level. Aims to minimize rejection of good lots. Provides single-sampling plans for specified AQL and producer's risk.
Reference: *MIL-STD-105E* (1989)

2   Dodge-Romig

Type of sampling: Single and double
Type of application: Where 100% rectifying of lots is applicable
Key features: One type of plan uses a consumer's risk of $\beta = .10$. Another type limits the AOQL. Protection is provided with minimum inspection per lot.
Reference: Dodge and Romig (1959)

3 H-107

Type of sampling: Continuous single stage

Type of application: When production is continuous and inspection is nondestructive

Key features: Plans are indexed by AQL, which generally start with 100% inspection until some consecutive number of defect-free units is found. Then inspection continues on a sampling basis until a specified number of defectives is found.

Reference: *H-107* (1959)

II. Sampling by Variables

1 MIL-STD-414

Assumed distribution: Normal

Criteria specified: AQL

Features: Lot evaluation by AQL. Includes tightened and reduced inspection.

Reference: *MIL-STD-414* (1957)

2 H-108

Assumed distribution: Exponential

Criteria specified: Mean life (MTBF)

Features: Life testing for reliability specifications

Reference: *H-108* (1959)

9.12
Chapter Highlights

Traditional supervision consists of keeping close control over operations and progress. The focus of attention is the product or process outputs. A direct implication of this approach is guaranteeing product quality through inspection screening. The chapter discusses sampling techniques and measures of inspection effectiveness. Performance characteristics of sampling plans are discussed and guidelines for choosing economic sampling plans are presented. The basic theory of single-stage acceptance sampling plans for attributes is first presented, including the concepts of acceptable quality level and limiting quality level. Formulas for determining sample size, acceptance levels, and operating characteristic functions are provided. Moving from single-stage sampling, the chapter covers double-sampling and sequential sampling using Wald's sequential probability ratio test. One section deals with acceptance sampling for variable data. Other topics covered include computations of average sample numbers and average total inspection for rectifying inspection plans. Modern skip-lot sampling procedures are introduced and compared to the standard application of sampling plans, where every lot is inspected. The Deming "all-or-nothing" inspection criteria are presented, and the connection between sampling inspection and statistical process control is made. Throughout the chapter, we refer to a specially designed software program called ACCEPT, which is used to perform various calculations and generate appropriate tables and graphs. The main concepts and definitions introduced in this chapter include:

- Lot
- Acceptable quality level
- Limiting quality level
- Producer's risk

- Consumer's risk
- Single-stage sampling
- Acceptance number
- Operating characteristic
- Double-sampling plan
- Average sample number function
- Sequential sampling

- Sequential probability ratio test
- Rectifying inspection
- Average outgoing quality
- Average total inspection
- Normal, tightened, or reduced inspection levels
- Skip-lot sampling plans

Exercises

9.2.1   Use program ACCEPT to determine single-sampling plans for attributes when the lot is $N = 2500$, $\alpha = \beta = .01$, and

i   AQL = .005, LQL = .01

ii   AQL = .01, LQL = .03

iii   AQL = .01, LQL = .05

9.2.2   Use program ACCEPT to investigate how the lot size, N, influences the single-sampling plans for attributes when $\alpha = \beta = .05$, AQL = .01, LQL = .03, by computing the plans for $N = 100$, $N = 500$, $N = 1000$, and $N = 2000$.

9.2.3   Compute the OC(p) function for the sampling plan computed in Exercise 9.2.1(iii). What is the probability of accepting a lot having 2.5% nonconforming items?

9.3.1   Compute the large-sample approximation to a single-sample plan for attributes (n^*, c^*), with $\alpha = \beta = .05$ and AQL = .025, LQL = .06. Compare these to the exact results. The lot size is $N = 2000$.

9.3.2   Repeat the previous exercise with $N = 3000$, $\alpha = \beta = .10$, AQL = .01 and LQL = .06.

9.4.1   Using program ACCEPT, obtain the OC and ASN functions of the double-sampling plan with $n_1 = 200$, $n_2 = 2n_1$ and $c_1 = 5$, $c_2 = c_3 = 15$ when $N = 2000$.

a   What are the attained α and β when AQL = .015 and LQL = .05?

b   What is the ASN when $p = $ AQL?

c   What is a single-sampling plan having the same α and β? How many observations do we expect to save if $p = $ AQL? Notice that if $p = $ LQL, the

present double-sampling plan is less efficient than the corresponding single-sampling plan.

9.4.2   Compute the OC and ASN values for a double-sampling plan with $n_1 = 150$, $n_2 = 200$, $c_1 = 5$, $c_2 = c_3 = 10$, when $N = 2000$. Notice how high β is when LQL = .05. This plan is reasonable if LQL = .06. Compare this plan to a single-sampling plan for $\alpha = .02$ and $\beta = .10$, AQL = .02, and LQL = .06.

9.5.1   Determine a sequential plan for the case of AQL = .02, LQL = .06, and $\alpha = \beta = .05$. Compute the OC and ASN functions of this plan. What are the ASN values when $p = $ AQL, $p = $ LQL, and $p = .035$?

9.5.2   Compare the single-sampling plan and the sequential plan when AQL = .01, LQL = .05, $\alpha = \beta = .01$ and $N = 10,000$. What are the expected savings in sampling cost if each observation costs \$1, and $p = $ AQL?

9.6.1   Determine n and k for a continuous variable-size sampling plan when $(p_0) = $ AQL = .01 and $(p_t) = $ LQL = .05, $\alpha = \beta = .05$.

9.6.2   Consider data file ALMPIN.DAT. An aluminum pin is considered defective if its cap diameter is smaller than 14.9 mm. For the parameters $p_0 = .01$ and $\alpha = .05$, compute k and decide whether to accept or reject the lot on the basis of a sample of $n = 70$ pins. What is the probability of accepting a lot with a proportion of defectives of $p = .03$?

9.6.3   Determine the sample size and k for a single-sampling plan by a normal variable, with the parameters AQL = .02, LQL = .04, and $\alpha = \beta = .10$.

9.7.1. A single-sampling plan for attributes, from a lot of size $N = 500$, is given by $n = 139$ and $c = 3$. Each lot that is not accepted is rectified. Compute the AOQ, when $p = .01$, $p = .02$, $p = .03$ and $p = .05$. What are the corresponding ATI values?

9.8.1 A single-sampling plan, under normal inspection, has probability $\alpha = .05$ of rejection, when $p = $ AQL. What is the probability, when $p = $ AQL in five consecutive lots, that there will be a switch to tightened inspection? What is the probability of switching to a tightened inspection if p increases so that $OC(p) = .7$?

9.9.1 Compute the probability for qualifying for state 2, in a SLSP, when $n = 100$, $c = 1$. What is the upper bound on S_{10} in order to qualify for state 2, when AQL $= .01$? Compute the probability for state 2 qualification.

Basic Tools and Principles of
Statistical Process Control

Basic Concepts of Statistical Process Control

In this chapter, we present the basics of **statistical process control (SPC)**. The general approach is prescriptive and descriptive rather than analytical. With SPC, we do not aim at modeling the distribution of data collected from a given process. Our goal is to control the process with the aid of decision rules that will signal significant discrepancies between the observed data and the standards of the process under control. We demonstrate the application of SPC to various processes by referring to the examples of piston cycle time and fiber strength, which were discussed in Chapter 2. Other examples used include power failures in a computer center and office procedures for scheduling the appointments of a university dean. The data on the piston cycle time are generated by the piston software simulator TURB1.EXE (see Appendix II). To study the causes of variability in the piston cycle time, we show a sketch of a piston in Figure 10.1, and, in Table 10.1, seven factors that can be controlled to change the cycle time of a piston.

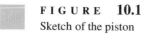

F I G U R E   10.1

Sketch of the piston

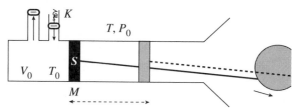

TABLE 10.1

Operating factors of the piston simulator and their operational levels

Factor	units	Minimum	Maximum
Piston weight	M (Kg)	30	60
Piston surface area	S (m^2)	0.005	0.020
Initial gas volume	V_0 (m^3)	0.002	0.010
Spring coefficient	K (N/m)	1000	5000
Atmospheric pressure	P_0 (N/m^2)	9×10^4	11×10^4
Ambient temperature	T (K)	290	296
Filling gas temperature	T_0 (K)	340	360

Figure 10.2 is a run chart (also called *connected time plot*) and Figure 10.3 is a histogram of 50 piston cycle times (in sec) measured under stable operating conditions. Throughout the measurement time frame, the piston-operating factors remained fixed at their maximum levels. The data can be found in file OTURB1.DAT. The average cycle time of the 50 cycles is 0.413 sec with a standard deviation of 0.121 sec. A word of caution: Rerunning TURB1.EXE at different operating conditions and/or a different sampling interval will change OTURB1.DAT.

Even though no changes occurred in the operating conditions of the piston, we observe variability in the cycle times. From Figure 10.2 we note that cycle times vary between 0.24 and 0.64 sec. The histogram in Figure 10.3 indicates some skewness in

FIGURE 10.2

Run chart (connected time plot) of 50 piston cycle times (sec)

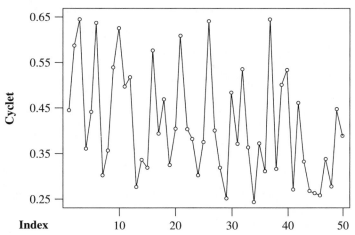

the data. The normal probability plot of the 50 cycle times (Figure 10.4) also leads to the conclusion that the cycle time distribution is not normal, but skewed.

Another example of variability is provided by the yarn-strength data presented in Chapter 2. The yarn-strength test results show variability in the properties of

F I G U R E   10.3
Histogram of 50 piston cycle times

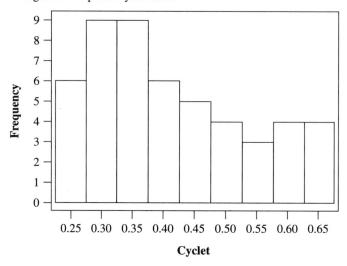

F I G U R E   10.4
Normal probability plot of 50 piston cycle times

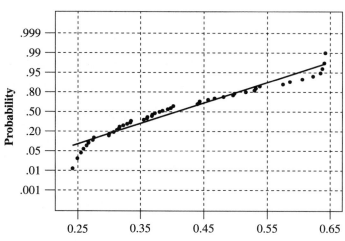

Average: 0.41292
Std Dev: 0.121126
N of data: 50

W-test for Normality
R: 0.9697
p value (approx): 0.0234

the product. High yarn strength indicates good spinning and weaving performance. Yarn strength is considered a function of the fiber length, fiber fineness, and fiber tensile strength. As a general rule, longer cottons are fine-fibered and shorter cottons coarse-fibered. Very fine fibers, however, tend to reduce the rate of processing, so the degree of fiber fineness depends on the specific end product use. Variability in fiber fineness is a major cause of variability in yarn strength and processing time.

In general, a production process has many sources or causes of variation. These can be further subdivided into process inputs and process operational characteristics, including equipment, procedures, and environmental conditions. Environmental conditions consist of factors such as temperature and humidity or work tools. Visual guides, for instance, might not allow operators to position parts on fixtures precisely. The complex interactions among material, tools, machine, work methods, operators, and the environment combine to create variability in the process. Factors that are permanent—that are a natural part of the process—cause **chronic problems** and are called **common causes** of variation. The combined effect of common causes can be described using probability distributions. Such distributions were introduced in Chapter 2 and their theoretical properties in Chapter 4. It is important to recognize that recurring causes of variability affect every work process and that even under a stable process there are differences in performance over time. Failure to recognize variation leads to wasteful actions, such as those described in Section 1.3. The only way to reduce the negative effects of chronic, common causes of variability is to modify the process. This modification can occur at the level of the process inputs, the process technology, the process controls, or the process design. Some of these changes are technical (for example, different process settings), some are strategic (for example, different product specifications), and some are related to human resources management (for example, training of operators). **Special causes**, **assignable causes**, or **sporadic spikes** arise from external temporary sources that are not inherent to the process. (The terms are used here interchangeably.) For example, an increase in temperature can potentially affect the piston's performance. It can have an impact in terms of changes in the average cycle times and/or the variability in cycle times.

To signal the occurrence of special causes, we need a control mechanism. Specifically, in the case of the piston, such a mechanism can consist of taking samples or subgroups of five consecutive piston cycle times. Within each subgroup, we compute the subgroup average and standard deviation. Program TURB2.EXE generates sample means and standard deviations of such grouped data. Figures 10.5 and 10.6 display charts of the average and standard deviations of 20 samples of five cycle time measurements, which are stored in file OTURB2.DAT. The chart of averages is called an X-bar chart; the chart of standard deviations is called an S-chart. All 100 measurements were taken under fixed operating conditions of the piston (all factors set at the maximum levels). The average of cycle time averages is 0.414 sec and the average of the standard deviations of the 20 subgroups is 0.11 sec. All these numbers were generated by the piston computer simulation model, which allows us to change the factors affecting the operating conditions of the piston. Again, we know that no changes were made to the control factors. The observed variability is due to common causes only—variability in atmospheric pressure, filling gas temperature, and the like.

We now rerun the piston simulator, introducing a forced change in the piston ambient temperature. At the beginning of the 8th sample, temperature begins to

F I G U R E   10.5

X-bar chart of cycle times under stable operating conditions

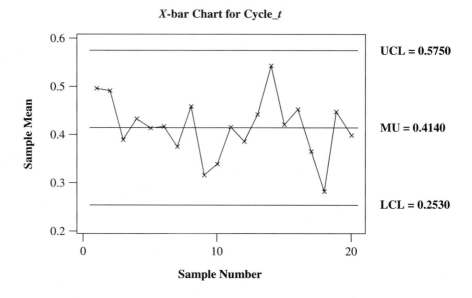

F I G U R E   10.6

S-chart of cycle times under stable operating conditions

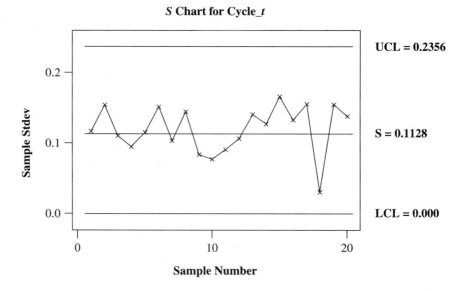

rise at a rate of 10% per cycle. Can we flag this special cause? The *X*-bar chart of this new simulated data is presented in Figure 10.7. Up to the 10th sample, the chart is similar to that of Figure 10.5. As of the 11th sample, the subgroup averages are consistently above 0.414 sec. This run persists until the 21st sample, when we

F I G U R E   10.7

X-bar chart of cycle times with a trend in ambient temperature

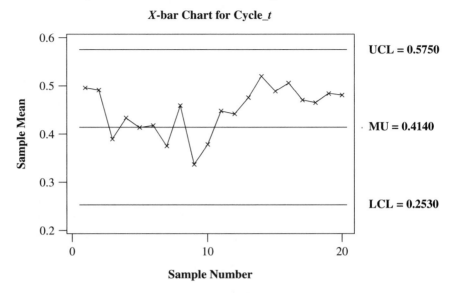

stopped the simulation. To have 10 points in a row above the average is unlikely to occur by chance alone. The probability of such an event is $(1/2)^{10} = .00098$. The implication of the 10-point run is that common causes are no longer the only causes of variation, and that a special factor has begun affecting the piston's performance. In this particular case, we know that it is an increase in ambient temperature. The *S*-chart of the same data (Figure 10.8) shows a run below the average of 0.11 beginning at the 9th sample. This indication occurs earlier than that in the *X*-bar chart. The information obtained from both charts indicates that a special cause has been in effect from the 9th sample onward. Its effect has been to increase cycle time and reduce variability. The new average cycle time appears to be around 0.49 sec.

   The piston simulator allows us to try other types of changes in the operational parameters of the piston. For example, we can change the spring that controls the intake valve in the piston gas chamber. In the next simulation, the standard deviation of the spring coefficient is increasing at a 5% rate past the 8th sample. Figures 10.9 and 10.10 are *X*-bar and *S*-charts corresponding to this scenario. Until the 8th sample, these charts are identical to those in Figures 10.5 and 10.6. After the 13th sample, changes appear in the chart. Points 14 and 18 fall outside the control limits, and it seems that variability has increased. This is seen also in Figure 10.10. We see that after the 9th sample, there is a run upward of 6 points, and points 18 and 19 are way above the upper control limit.

   Control charts have wide applicability throughout an organization. Top managers can use a control chart to study variation in sales and decide on new marketing strategies. Operators can use the same tool to determine if and when to adjust a manufacturing process. An example with universal applicability is the scheduling process of daily appointments in a university dean's office. At the end of each working day, the

FIGURE   10.8

S-chart of cycle times with a trend in ambient temperature

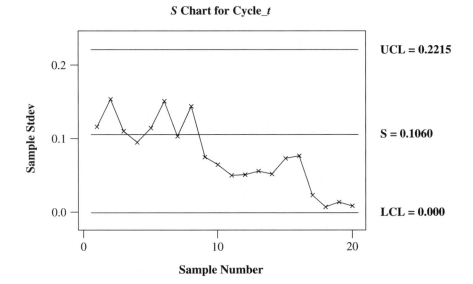

S Chart for Cycle_t

FIGURE   10.9

X-bar chart of cycle times with a trend in spring coefficient precision

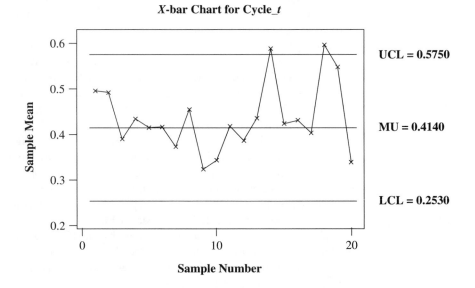

X-bar Chart for Cycle_t

various meetings and appointments coordinated by the office of the dean were classified as "on time" or as having a problem, such as "late beginning," "did not end on
time," "was interrupted," and so on. The ratio of problem appointments to the total

S-chart of cycle times with a trend in spring coefficient precision

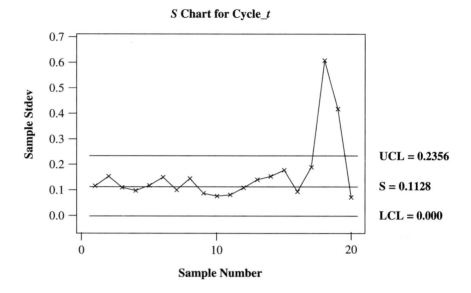

S Chart for Cycle_t

number of daily appointments was tracked and control limits computed. Figure 10.11 is the dean's control chart (see Kelly, Kenett, Newton, Roodman, and Wowk, 1991).

Another example of a special cause is the miscalibration of spinning equipment. Miscalibration can be identified by the ongoing monitoring of yarn strength. Process operators analyzing X-bar and S-charts can stop and adjust the process as trends

FIGURE **10.11**

Control chart for proportion of appointments with scheduling problems

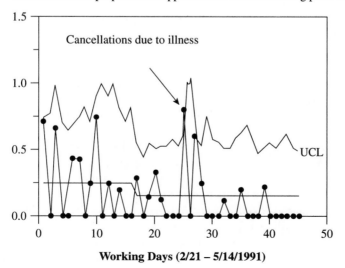

Working Days (2/21 – 5/14/1991)

develop or sporadic spikes appear. Timely indication of a sporadic spike is crucial to the effectiveness of process control mechanisms. *Ongoing chronic problems, however, cannot be resolved by using local operator adjustments.* The statistical approach to process control allows us to distinguish between chronic problems and sporadic spikes. This is crucial because these two different types of problems require different approaches. Process control ensures that a process performs at a level determined "doable" by a process capability study. In Section 10.3, we discuss how to conduct such studies and how to set control limits.

So far we have focused on the analysis of data for process control. Another essential component of process control is the generation and routing of relevant and timely data through proper feedback loops. We distinguish between two types of feedback loops: **external feedback loops** and **internal feedback loops**. An external feedback loop consists of information gathered at a subsequent downstream process or by direct inspection of the process outputs.

To illustrate these concepts and ideas, let us look at the process of driving to work. The time it takes you to get to work is a variable that depends on various factors, such as the number of other cars on the road, how you happen to catch the traffic lights, your mood that morning, and so on. These factors are part of the process, and you have little or no control over them. Such common causes create variation in the time it takes you to reach work. One day it may take you 15 minutes and the next day 12 minutes. If you are particularly unlucky and have to stop at all the red lights, it might take 18 minutes. Suppose, however, that on one particular day it takes 45 minutes to reach work. Such a long trip is outside the normal range of variation and is probably associated with a special cause such as a flat tire, a traffic jam, or road construction.

External feedback loops rely on measurements of the process outcome. They provide information akin to that obtained from looking at a rear-view mirror. The previous example consisted of monitoring time after you reached work. In most cases, identifying a special cause at that point is too late. Suppose a local radio station provided its listeners with live coverage of traffic conditions. If we monitor, on a daily basis, the volume of traffic reported by the radio, we can avoid traffic jams, road construction, and other unexpected delays. Such information will help us eliminate certain special causes of variation. Moreover, if we institute a preventive maintenance program for our car, we can eliminate many types of engine and other car problems, further reducing the impact of special causes. To eliminate the occasional flat tire would also involve improving road maintenance—a much larger task. The radio station is a source of internal feedback; it provides information that can be used to correct our route and thus allow us to arrive at work on time almost every day. This is equivalent to driving the process while looking ahead. Most drivers are able to avoid getting off the road, even when obstacles present themselves unexpectedly.

We now describe how control charts are used for "staying on course." Manufacturing examples we'll look at in our discussion include the physical dimensions of holes drilled by a computerized numerically controlled (CNC) machine, and piston cycle times. The finished part leaving a CNC machine can be inspected immediately after the drilling operation or later, when the part is assembled into another part. Piston cycle times can be recorded on-line or stored for off-line analysis. Or, electrical parameters can be tested during the final assembly of an electronic product. The test data reflect, among other things, the performance of the components' assembly process. Information on defects such as missing components and wrong or misaligned components can be fed back, through an external feedback loop, to the assembly operators. Data collected on process variables, measured internally to the process, are the basis of an internal feedback loop information flow. An example of such data is the air pressure in the hydraulic system of a CNC machine. Air pressure can be measured so that trends or deviations in pressure are detected early enough to allow for corrective action to take place. Another example involves tracking the temperature around a piston. Such information will directly point out the trend in temperature that was indirectly observed in Figures 10.7 and 10.8. Moreover, routine direct measurements of the precision of the piston's spring coefficient will exhibit the trend that went unnoticed in Figures 10.9 and 10.10. The relationship among a process, its suppliers, and its customers is presented in Figure 10.12. Internal and external feedback loops depend on a coherent structure of suppliers, processes, and customers. It is in this context that we can achieve effective statistical process control.

FIGURE 10.12

The supplier-process-customer structure and its feedback loops

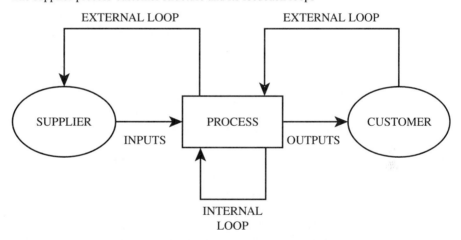

In this section, we have discussed the concepts of feedback loops, chronic problems (common causes), and sporadic spikes (special causes). Data funnelled through feedback loops are used to indicate the types of forces affecting the measured process. Statistical process control is "a rule of behavior that will strike a balance for the net economic loss from two sources of mistake: (1) looking for special causes too often, or overadjusting; (2) not looking often enough" (Deming, 1967). In the implementation of statistical process control, we distinguish between two phases: (1) achieving control and (2) maintaining control. To achieve control, we study the causes of variation and then try to eliminate the special causes while gaining a thorough understanding of the remaining permanent factor—the common causes—that affect the process. Tools such as graphic displays (Chapters 2 and 3), correlation and regression analysis (Section 3.3), control charts (Sections 10.3 and 10.6 and

Chapter 11) and designed experiments (Chapters 12 and 13) are typically used in a process capability study whose objective is to achieve control. Section 10.3 will discuss the major steps of a process capability study and the determination of control limits on the control charts. Once control is achieved, it has to be maintained. The next section describes how control is maintained with the help of control limits.

Driving a Process with Control Charts

Control charts allow us to determine when action should be taken to adjust a process that has been affected by a special cause. Control charts also tell us when to leave a process alone and not misinterpret variations due to common causes. Special causes need to be addressed by corrective action. Common causes are the focus of ongoing efforts aimed at improving the process.

We distinguish between control charts for attribute data and control charts for variable data. Attribute data require an operational definition of what constitutes a problem or defect. When the observation unit is classified in one of two alternate categories (for example, pass vs. fail or conforming vs. nonconforming), we can track the proportion of nonconforming units in the observation sample. Such a chart is called a *p-chart*. If the size of the observation sample is fixed, we can simply track the number of nonconforming units and derive an *np-chart*. When an observation consists of the number of nonconformities per unit of observation, we track either number of nonconformities (*c-charts*) or rates of nonconformities (*u-charts*). Rates are computed by dividing the number of nonconformities by the number of opportunities for errors or problems. For variable data, we distinguish between processes that can be repeatedly sampled under uniform conditions and processes where measurements were derived one at a time (for example, a monthly sales figure). In the latter case, we use control charts for individual data—also called *moving range charts*. When data can be grouped, we can use a variety of charts such as the *X*-bar chart or the median chart (discussed in detail in Chapter 11).

Using the piston cycle times discussed earlier, we now examine how *X*-bar control charts actually work. An *X*-bar control chart for a piston's cycle time is constructed by first grouping observations by time period and then summarizing the location and variability in these subgroups. An example of this was provided in Figures 10.5 and 10.6, where the average and standard deviations of five consecutive cycle times were tracked over 20 such subgroups. The three lines that are added to the simple run charts are the *center line*, positioned at the grand average, the **upper control limits (UCL)**, and the **lower control limits (LCL)**. The UCL and LCL indicate the range of variability we expect to observe around the center line, under stable operating conditions. Figure 10.5 shows averages of 20 subgroups of five consecutive cycle times each. The center line and control limits are computed from the average of the 20 subgroup averages and the estimated standard deviation for averages of samples of size 5. The center line is at 0.414 sec. When using the classic 3-sigma charts developed by Shewhart, the control limits are positioned at 3 standard deviations of $\overline{X}$—namely, $3\sigma/\sqrt{n}$, away from the center line. Using MINITAB, we find that UCL = 0.585 sec and LCL = 0.243 sec. Under stable operating conditions, with only common causes affecting performance, the chart will typically have all points within the control limits. Specifically, with 3-sigma control limits, we expect to have, on average, only 1 out of 370 points (1/0.0027) outside these limits—a rather rare event. Therefore, when a point falls beyond the control limits, we can safely question the stability of the process. The risk that such an alarm will turn out to be false is 0.0027. A false alarm occurs when the sample mean falls outside the control limits and we suspect an assignable cause, but only common causes are operating. Moreover, stable random variation does not exhibit patterns such as upward or downward trends, or consecutive runs of points above or below the center line. We saw earlier how a control chart was used to detect an increase in the ambient temperature of a piston from the cycle times. The *X*-bar chart (Figure 10.7) indicates a run of 6 or more points above the center line. Figure 10.13 shows several patterns that indicate nonrandomness. These are:

1 *Pattern 1*: A single point outside the control limits

2 *Pattern 2*: A run of seven or more points in a row above (or below) the centerline

3 *Pattern 3*: Six consecutive points increasing (trend up) or decreasing (trend down)

4 *Pattern 4*: Two out of three points in a region between $\mu \pm 2\sigma/\sqrt{n}$ and $\mu \pm 3\sigma/\sqrt{n}$

A comprehensive discussion of detection rules and properties of classic 3-sigma control charts and other modern control chart techniques is presented in Chapter 11.

As we saw earlier, there are many types of control charts. Selection of the control chart to use in a particular application primarily depends on the type of data that will flow through the feedback loops. The piston provided us with an example of variable data, and we used an *X*-bar and *S*-chart to monitor the piston's performance. Examples of attribute data are blemishes on a given surface, wave solder defects, friendly service at the bank, and missed shipping dates. Again, each type of data leads to a different type of control chart. All control charts have a center line and upper and lower control limits (UCL and LCL). In general, the rule for flagging special causes is the same in every type of control chart. Figure 10.14 presents a classification of the various control charts. Properties of the different types of charts, including the more advanced EWMA and CUSUM charts, are given in Chapter 11.

Earlier, we discussed several examples of control charts and introduced different types of control charts. The block diagram in Figure 10.14 organizes control charts by the type of data flowing through feedback loops. External feedback loops typically rely on properties of the process's products and lead to control charts based on counts or classification. If products are classified using "pass" versus "fail" criteria, we use *np*-charts or *p*-charts, depending on whether the products are tested in fixed or variable subgroups. The advantage of such charts is that several criteria can be combined to produce a definition of what constitutes a "fail," or defective product. When counting nonconformities or incidences of a certain event or phenomenon, we use *c*-charts or *u*-charts. These charts provide more information than *p*-charts or

FIGURE   10.13

Patterns used to detect special causes

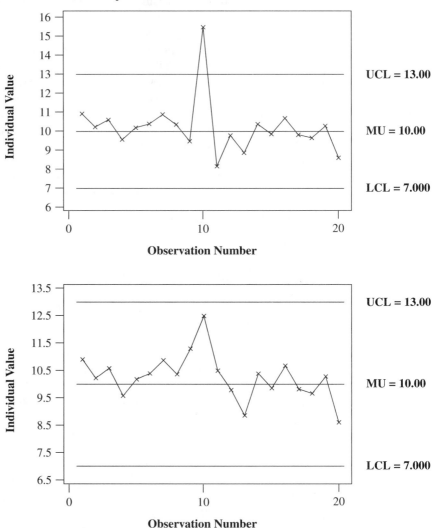

np-charts because the actual number of nonconformities in a product is accounted for. The drawback is that several criteria cannot be combined without weighing the different types of nonconformities. *C*-charts assume a fixed likelihood of incidence; *u*-charts are used in cases of varying likelihood levels. For large subgroups (subgroup sizes larger than 1000), the number of incidences, incidences per unit, number of defectives, or percentage of defectives can be considered as individual measurements, and an *X*-bar chart for subgroups of size 1 can be used. Internal feedback

FIGURE 10.13 (continued)
Patterns used to detect special causes

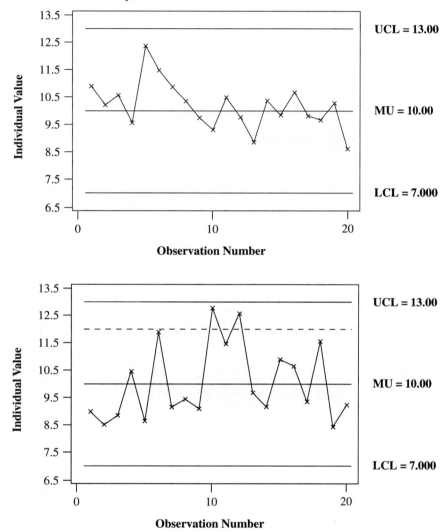

loops and, in some cases, external feedback loops rely on variable data derived from measuring product or process characteristics. If measurements are grouped in samples, we can combine *X*-bar charts with *R*-charts or *S*-charts. Such combinations provide a mechanism to control the stability of a process with respect to both location and variability. *X*-bar charts track the sample averages, *R*-charts track sample ranges (maximum − minimum), and *S*-charts, sample standard deviations. For samples larger than 10, *S*-charts are recommended over *R*-charts. For small samples and

FIGURE   10.14

Classification of control charts

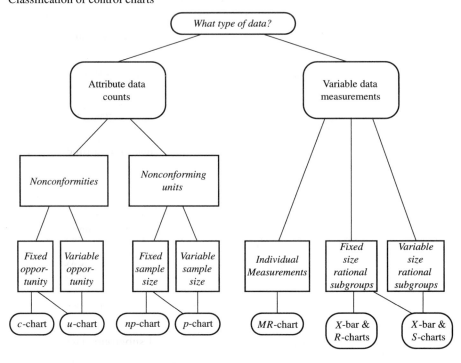

manual maintenance of control charts, *R*-charts are preferred. When sample sizes vary, only *S*-charts should be used to track variability.

10.3
Setting Up a Control Chart: Process Capability Studies

Setting up the limits of a control chart requires a detailed study of process variability and of the causes creating this variability. Control charts are used to detect the occurrence of special, sporadic causes while minimizing the risk of misinterpreting special causes as common causes. To achieve this objective, we need to assess the effect of chronic, common causes, and then set up control limits that reflect the variability resulting from such common causes. The study of process variability, which precedes the setting up of control charts, is called a **process capability study**. We distinguish between attribute process capability studies and variable process capability studies.

Attribute process capability studies determine process capability in terms of fraction of defective or nonconforming output. Such studies begin with data collected over several time periods. A rule of thumb is to collect data over three time periods with 20 to 25 samples of size 50 to 100 units each. For each sample, the control chart statistic is computed and a control chart is drawn. This will lead to a *p*-, *np*-, *c*-, or *u*-chart and to investigation patterns flagging special causes, such as those in Figure 10.14. Special causes are then investigated and possibly removed. This action requires that the process be changed to justify removal of the measurements taken during the time periods when those special causes were active. The new control chart, computed without these points, shows the capability of the process. Its center line is typically used as a measure of process capability. For example, in Figure 10.11, we can see that the process capability of the scheduling of appointments at the dean's office improved from 25% of appointments with problems to 15% after a change was introduced in the process. The change consisted of acknowledging appointments with a confirmation note spelling out time, date, and topic of appointment; a brief agenda; and a scheduled ending time. On the 25th working day, there was one sporadic spike caused by illness. The dean had to leave early that day and several appointments were cancelled. When sample sizes are large (over 1000 units), control charts for attribute data become ineffective because of their narrow control limits, and *X*-bar charts for individual measurements are used.

Variable process capability studies determine the capability of a process in terms of the distribution of measurements on product or process characteristics. Setting up control charts for variable data requires far fewer data than attribute data control charts. Data are collected in samples called **rational subgroups**, which are selected from a time frame such that relatively homogeneous conditions exist within each subgroup. The design strategy of rational subgroups is to measure variability due to common causes only. Control limits are then determined based on measures of location and variability in each rational subgroup. The control limits are set to account for variability due to these common causes. Any deviation from stable patterns relative to the control limits (see Figure 10.13) indicates a special cause. For example, in the piston case study, a rational subgroup consists of five consecutive cycle times. The statistics used are the average and standard deviation of the subgroups. The 3-sigma control limits are computed to be UCL = 0.585 and LCL = 0.243. From an analysis of Figure 10.5, we conclude that the *X*-bar chart, based on a connected time plot (or run chart) of 20 consecutive averages, exhibits a pattern that is consistent with a stable process. We can now determine the process capability of the piston movement within the cylinder.

Process capability for variable data is a characteristic that reflects the probability that individual outcomes of a process will be within the engineering specification limits. Assume that the piston engineering specifications stipulate a nominal value of 0.3 sec and maximum and minimum values of 0.5 and 0.1 sec respectively. Table 10.2 shows the output from the process capability analysis option included in the SPC window of MINITAB. The 100 measurements that were produced under stable conditions have a mean (average) of 0.414 sec and a standard deviation of 0.127 sec. The predicted proportion of cycle times beyond the specification limits is computed using the normal distribution as an approximation. The computations show that, under stable operating conditions, an estimated 25% of future cycle times will be above 0.5 sec, and that 0.6% will be below below 0.1 sec. We clearly see that the nominal value of 0.3 sec is slightly lower than the process average, having a *z*-score of -0.90 and that the upper limit, or maximum specification limit, is 0.67 standard deviation above the average. The probability that a standard normal random variable is larger than 0.67 is .251. This is an estimate of the future percentage of

TABLE   10.2

MINITAB output of process capability analysis of piston cycle time

Process Capability Analysis for OTRUB1.Cyclet			

Sample size = 100
Sample mean = 0.41433
Sample standard
deviation = 0.127329

Specification	Observed Beyond Spec.	z-score	Estimated Beyond Spec.
Upper limit: 0.5	25.000%	0.67	25.053%
Nominal: 0.3		−0.90	
Lower limit: 0.1	0.000%	−2.47	0.678%
	25.000%		25.731%

Capability Indices	Goodness-of-Fit Tests
CP: 0.52	Shapiro-Wilk's W: 0.908685
CPK: 0.22	P value: 0.000000
(upper): 0.22	Chi-square test: 40.160000
(lower): 0.82	P value: 0.007100

cycle times above the upper limit of 0.5 sec, provided stable conditions prevail. It is obvious from this analysis that the piston process is incapable of complying with the engineering specifications. The lower right-hand side of the table presents two tests for normality. Both tests reject the hypothesis of normality.

Process Capability Indices

In assessing the process capability for variable data, two indices have recently gained popularity: C_p and C_{pk}. The first **process capability index** is an indicator of the potential of a process to meet two-sided specifications with as few defects as possible. For symmetric specification limits, an upper specification limit (USL) and a lower one (LSL), the full potential is actually achieved when the process is centered between the specification limits. In order to compute C_p, one simply divides the process tolerance by 6 standard deviations:

$$C_p = (\text{USL} - \text{LSL})/(6 * \text{Standard deviation}) \tag{10.4.1}$$

The numerator indicates how wide the specifications are; the denominator measures the width of the process. Under normal assumptions, the denominator is a range of values that accounts for 99.73% of the observations from a centered process, operating under stable conditions, with variability due only to common causes. When $C_p = 1$, we expect that 0.27% of the observations will fall outside the specification limits. A target for many modern industries is, on every process, to reach a level of $C_p = 2$, which practically guarantees that under stable conditions, and for processes kept under control around the process nominal values, there will be no defective products ("zero defects"). With $C_p = 2$, the theoretical estimate under normal assumptions, allowing for a possible shift in the location of the process mean by as much as 1.5 standard deviations, is 3.4 cases per million observations outside specification limits.

Another measure of process capability is:

$$C_{pk} = \text{Minimum}(C_{pu}, C_{pl}) \tag{10.4.2}$$

where

$$C_{pu} = (\text{USL} - \text{Process mean})/(3 * \text{Standard deviation})$$

and $\qquad\qquad\qquad\qquad\qquad\qquad\qquad\qquad\qquad$ **(10.4.3)**

$$C_{pl} = (\text{Process mean} - \text{LSL})/(3 * \text{Standard deviation})$$

When the process mean is centered between the specification limits, C_{pk} is equal to C_p. Noncentered processes have their potential capability measured by C_p and their actual capability measured by C_{pk}. As shown in Table 10.2 for the piston data, estimates of C_p and C_{pk} are $\hat{C}_p = 0.52$ and $\hat{C}_{pk} = 0.22$. This indicates that something could be gained by centering the piston cycle times around 0.3 sec. Even if this is possible to achieve, there will still be observations outside the upper and lower limits because the standard deviation is too large. The validity of the C_p and C_{pk} indices is questionable in cases where the measurements on X are not normally distributed, but have skewed distributions. The proper form of a capability index under nonnormal conditions has yet to be developed (Kotz and Johnson, 1994).

It is common practice to estimate C_p or C_{pk} by substituting the sample mean, $\overline{X}$, and the sample standard deviation, S, for the process mean, μ, and the process standard deviation, σ—that is,

$$\hat{C}_{pu} = \frac{\xi_U - \overline{X}}{3S} \qquad \hat{C}_{pl} = \frac{\overline{X} - \xi_L}{3S} \tag{10.4.4}$$

and $\hat{C}_{pk} = \min(\hat{C}_{pu}, \hat{C}_{pl})$, where ξ_L and ξ_U are the lower and upper specification limits. The question is how close $\hat{C}_{pk}$ is to the true process capability value.

In the following discussion, we develop confidence intervals for C_{pk} that have confidence levels close to the nominal $1 - \alpha$ in large samples. The derivation of these intervals depends on the following results:

1 In a large-size random sample from a normal distribution, the sampling distribution of S is approximately normal, with mean σ and variance $\sigma^2/2n$.

2 In a random sample from a normal distribution, the sample mean, $\overline{X}$, and the sample standard deviation, S, are independent.

3 If A and B are events such that $\Pr\{A\} = 1 - \alpha/2$ and $\Pr\{B\} = 1 - \alpha/2$, then $\Pr\{A \cap B\} \geq 1 - \alpha$. (This inequality is called the *Bonferroni inequality*.)

In order to simplify notation, let

$$\rho_1 = C_{pl}, \qquad \rho_2 = C_{pu}, \qquad \text{and} \qquad \omega = C_{pk}$$

Notice that because $\overline{X}$ is distributed like $N(\mu, \sigma^2/n)$, $\overline{X} - \xi_L$ is distributed like $N(\mu - \xi_L, \sigma^2/n)$. Furthermore, by results 1 and 2, the distribution of $\overline{X} - \xi_L - 3S\rho_1$ in large samples is like that of $N[0, (\sigma^2/n)(1 + (9/2)\rho_1^2)]$. It follows that, in large samples

$$\frac{(\overline{X} - \xi_L - 3S\rho_1)^2}{\dfrac{S^2}{n}\left(1 + \dfrac{9}{2}\rho_1^2\right)}$$

is distributed like $F[1, n-1]$. Or,

$$\Pr\left\{\frac{(\overline{X} - \xi_L - 3S\rho_1)^2}{\dfrac{S^2}{n}\left(1 + \dfrac{9}{2}\rho_1^2\right)} \leq F_{1-\alpha/2}[1, n-1]\right\} = 1 - \alpha/2 \qquad \text{(10.4.5)}$$

Thus, let $\rho_{1,\alpha}^{(L)}$ and $\rho_{1,\alpha}^{(U)}$ be the two real roots (if they exist) of the quadratic equation in ρ_1

$$(\overline{X} - \xi_L)^2 - 6S\rho_1(\overline{X} - \xi_L) + 9S^2\rho_1^2$$
$$= F_{1-\alpha/2}[1, n-1]\frac{S^2}{n}\left(1 + \frac{9}{2}\rho_1^2\right) \qquad \text{(10.4.6)}$$

Equivalently, $\rho_{1,\alpha}^{(L)}$ and $\rho_{1,\alpha}^{(U)}$ are the two real roots $(\rho_{1,\alpha}^{(L)} \leq \rho_{1,\alpha}^{(U)})$ of the quadratic equation

$$9S^2\left(1 - \frac{F_{1-\alpha/2}[1, n-1]}{2n}\right)\rho_1^2 - 6S(\overline{X} - \xi_L)\rho_1$$
$$+ \left((\overline{X} - \xi_L)^2 - S^2\frac{F_{1-\alpha/2}[1, n-1]}{n}\right) = 0 \qquad \text{(10.4.7)}$$

Substituting in this equation $(\overline{X} - \xi_L) = 3S\hat{C}_{pl}$, we obtain the equation

$$\left(1 - \frac{F_{1-\alpha/2}[1, n-1]}{2n}\right)\rho_1^2 - 2\hat{C}_{pl}\rho_1 + \left(\hat{C}_{pl}^2 - \frac{F_{1-\alpha/2}[1, n-1]}{9n}\right)$$
$$= 0 \qquad \text{(10.4.8)}$$

We assume that n satisfies $n > (F_{1-\alpha}[1, n-1]/2)$. Under this condition, $1 - (F_{1-\alpha/2}[1, n-1]/2n) > 0$ and the two real roots of the quadratic equation are

$$\rho_{1,\alpha}^{(U,L)} = \qquad \text{(10.4.9)}$$

$$\frac{\hat{C}_{pl} \pm \sqrt{\dfrac{F_{1-\alpha/2}[1, n-1]}{n}\left(\dfrac{\hat{C}_{pl}^2}{2} + \dfrac{1}{9}\left(1 - \dfrac{F_{1-\alpha/2}^2[1, n-1]}{2n}\right)\right)^{1/2}}}{\left(1 - \dfrac{F_{1-\alpha/2}[1, n-1]}{2n}\right)}$$

From the inequalities it follows that $(\rho_{1,\alpha}^{(L)}, \rho_{1,\alpha}^{(U)})$ is a confidence interval for ρ_1 at confidence level $1 - \alpha/2$.

Similarly, $(\rho_{2,\alpha}^{(L)}, \rho_{2,\alpha}^{(U)})$ is a confidence interval for ρ_2 at confidence level $1 - \alpha/2$, where $\rho_{2,\alpha}^{(U,L)}$ are obtained by replacing $\hat{C}_{pl}$ by $\hat{C}_{pu}$ in formula 10.4.9. Finally, from the Bonferroni inequality and the fact that $C_{pk} = \min\{C_{pl}, C_{pu}\}$, we obtain the confidence limits for C_{pk} at level of confidence $1 - \alpha$ given in Table 10.3.

TABLE 10.3
Confidence limits for C_{pk} at level $1 - \alpha$

Lower limit	Upper limit	Condition
$\rho_{1,\alpha}^{(L)}$	$\rho_{1,\alpha}^{(U)}$	$\rho_{1,\alpha}^{(U)} < \rho_{2,\alpha}^{(L)}$
$\rho_{1,\alpha}^{(L)}$	$\rho_{1,\alpha}^{(U)}$	$\rho_{1,\alpha}^{(L)} < \rho_{2,\alpha}^{(L)} < \rho_{1,\alpha}^{(U)} < \rho_{2,\alpha}^{(U)}$
$\rho_{1,\alpha}^{(L)}$	$\rho_{2,\alpha}^{(U)}$	$\rho_{1,\alpha}^{(L)} < \rho_{2,\alpha}^{(L)} < \rho_{2,\alpha}^{(U)} < \rho_{1,\alpha}^{(U)}$
$\rho_{2,\alpha}^{(L)}$	$\rho_{1,\alpha}^{(U)}$	$\rho_{2,\alpha}^{(L)} < \rho_{1,\alpha}^{(L)} < \rho_{1,\alpha}^{(U)} < \rho_{2,\alpha}^{(U)}$
$\rho_{2,\alpha}^{(L)}$	$\rho_{2,\alpha}^{(U)}$	$\rho_{2,\alpha}^{(L)} < \rho_{1,\alpha}^{(L)} < \rho_{2,\alpha}^{(U)} < \rho_{1,\alpha}^{(U)}$
$\rho_{2,\alpha}^{(L)}$	$\rho_{2,\alpha}^{(U)}$	$\rho_{2,\alpha}^{(U)} < \rho_{1,\alpha}^{(L)}$

EXAMPLE 10.1 In this example, we illustrate the computation of the confidence interval for C_{pk}. Suppose that the specification limits are $\xi_L = -1$ and $\xi_U = 1$. Suppose that $\mu = 0$ and $\sigma = 1/3$. In this case, $C_{pk} = 1$. Now, using MINITAB, we simulate a sample of size $n = 20$ from a normal distribution with mean $\mu = 0$ and standard deviation $\sigma = 1/3$. We obtain a random sample with $\overline{X} = 0.01366$ and standard deviation $S = 0.3757$. For this sample, $\hat{C}_{pl} = 0.8994$ and $\hat{C}_{pu} = 0.8752$. Thus, the estimate of C_{pk} is $\hat{C}_{pk} = 0.8752$. For $\alpha = .05$, $F_{.975}[1, 19] = 5.9216$. Obviously $n = 20 > (F_{.975}[1, 19]/2) = 2.9608$. According to the formula,

$$\rho_{1,.05}^{(U,L)} = \frac{0.8994 \pm \sqrt{\dfrac{5.9216}{20}\left(\dfrac{(.8994)^2}{2} + \dfrac{1 - \dfrac{5.9216}{40}}{9}\right)^{1/2}}}{1 - \dfrac{5.9216}{40}}$$

Thus, $\rho_{1,.05}^{(L)} = 0.6845$ and $\rho_{1,.05}^{(U)} = 1.5060$. Similarly, $\rho_{2,.05}^{(L)} = 0.5859$ and $\rho_{2,.05}^{(U)} = 1.4687$. Therefore, the confidence interval, at level .95, for C_{pk} is $(0.5859, 1.4687)$. ■

10.5
Seven Tools for Process Control and Process Improvement

Seven tools have proven extremely effective in helping organizations control processes and implement process improvement projects. Some of these tools were presented in the preceding chapters. For completeness, all the tools are briefly reviewed here and references are given to the earlier chapters.

The preface to the English edition of the famous text by Kaoru Ishikawa (1986) on quality control states: "The book was written to introduce quality control practices in Japan which contributed tremendously to the country's economic and industrial development." The Japanese work force did indeed master an elementary set of tools that helped them improve processes. Seven of the tools were nicknamed the "magnificent seven": The flowchart, check sheet, run chart, histogram, Pareto chart, scatterplot, and cause-and-effect diagram.

Flowcharts

Flowcharts are used to describe a process being studied or to describe the desired sequence of a new, improved process. Often, creating a flowchart is the first step taken by a team looking for ways to improve a process. The differences between how a process could work and how it actually does work exposes redundancies, misunderstandings, and general inefficiencies.

Check Sheets

Check sheets are basic manual data collection mechanisms. They consist of forms designed to tally the total number of occurrences of certain events by category. Developing a check sheet is usually the starting point in data collection efforts. In setting up a check sheet, one needs to agree on category definitions, the data collection time frame, and the actual data collection method. An example of a check sheet is shown in Figure 10.15.

FIGURE 10.15

A typical check sheet

Check sheet			
Product: Receiver unit XYZ		DATE: 9/09/89	
		Name: Smith	
		Lot: 17	
Total examined: 200			
Defect type	Defect count		Subtotal
Chipped	### ### ###		15
Off color	### ### ### ### //		22
Bent	###		5
		Grand total:	42

Run Charts

Run charts are used to visually represent data collected over time. They are also called *connected time plots*. Trends and consistent patterns are easily identified on run charts. An example is given in Figure 10.2.

Histograms

The **histogram** was presented in Section 2.4 as a graphical display of the distribution of measurements collected as a sample. It shows the frequency or number of observations of a particular value or within a specified group. Histograms are used extensively in process capability studies to provide clues about the characteristics of the process generating the data. However, as we saw in Section 10.3, they ignore information on the order in which data are collected.

Pareto Charts

Pareto charts are used extensively in modern organizations. These charts help to focus on the few important causes of trouble. When observations are collected and classified into different categories using valid and clear criteria, one can construct a Pareto chart. The Pareto chart is a display, using bar graphs sorted in descending order, of the relative importance of events such as errors, by category. Superimposed on the bars is a cumulative curve that helps point out the few important categories that contain most of cases. Pareto charts are used to choose the starting point for problem solving, monitor changes, or identify the basic cause of a problem. Their usefulness stems from the Pareto principle, which states that in any group of factors contributing to a common effect, a relative few (20%) account for most of the effect (80%). A Pareto chart of software errors found in testing a PBX electronic switch is shown in Figure 10.16. Errors are labeled according to the software unit where they occurred. For example, the "EKT" (electronic key telephone) category makes up 6.5% of the errors. What can we learn from this about the software development process? The "GEN", "VHS," and "HI" categories account for over 80% of the

FIGURE 10.16

Pareto chart of software errors

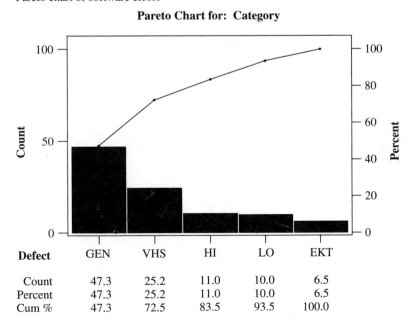

Pareto Chart for: Category

Defect	GEN	VHS	HI	LO	EKT
Count	47.3	25.2	11.0	10.0	6.5
Percent	47.3	25.2	11.0	10.0	6.5
Cum %	47.3	72.5	83.5	93.5	100.0

errors. These are the causes of problems on which major improvement efforts should initially concentrate. Section 10.6 discusses a statistical test for comparing Pareto charts. Such tests are necessary if we want to distinguish between differences that can be attributed to random noise and significant differences that should be investigated to identify an assignable cause.

Scatterplots

Scatterplots were introduced in Chapter 3. They are used to exhibit what happens to one variable when another variable changes. Such information is needed in order to test a theory or make forecasts. For example, we might want to verify the theory that the relative number of errors found in engineering drawings declines with increased drawing size.

Cause-and-Effect Diagrams

Cause-and-effect diagrams (also called *fishbone charts* or *Ishikawa diagrams*) are used to identify, explore, and display all possible causes of a problem or event. The diagram is usually completed in a meeting of individuals who have firsthand knowledge of the problem being investigated. Figure 10.17 shows a cause-and-effect diagram listing causes of problems in teaching and the effects that result. The ideal group

FIGURE   10.17

A cause-and-effect diagram of problems with teaching

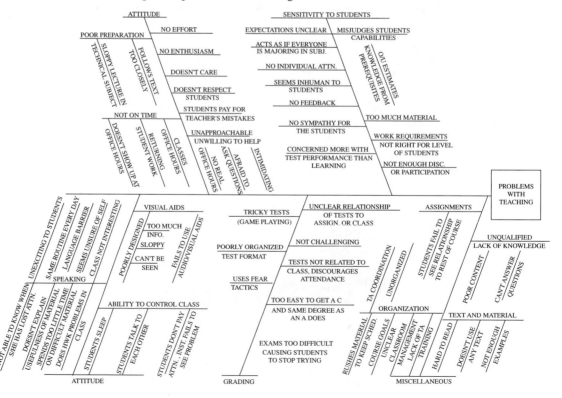

for completing such a diagram would consist of past and present students, instructors, teaching assistants, and administrators who handle classroom assignments and schedules. It is standard practice to weight the causes by impact on the problem being investigated and then initiate projects to reduce the harmful effects of the main causes.

In attempting to reduce levels of defects or output variability, a team will typically begin by collecting data and charting the process. Flowcharts and check sheets are used in these early stages. Run charts, histograms, and Pareto charts can then be prepared from the data collected on check sheets or by other methods. Next, the current process is diagnosed using addition scatterplots and cause-and-effect diagrams. Once solutions for improvement are implemented, their impact can be assessed using run charts, histograms, and Pareto charts. A statistical test for comparing Pareto charts is presented in the next section.

Statistical Analysis of Pareto Charts

Pareto charts are often compared over time or across processes. In such comparisons, we need to know whether differences between two Pareto charts should be attributed to random variation or to special, significant causes. In this section, we discuss a statistical test that is used to flag statistically significant differences between two Pareto charts (Kenett, 1991). Once the classification of observations into different categories is completed, we have the actual number of observations per category. The *reference Pareto chart* is one constructed in an earlier time period or on a different, but comparable, process. Other terms for the reference Pareto chart are the *benchmark* or *standard Pareto chart*. The proportion of observations in each category of the reference Pareto chart is the proportion we expect to find in Pareto charts of data collected under the same conditions. The expected number of observations in the different categories of the Pareto chart is computed by multiplying the total number of observations in the chart by the corresponding expected proportion. The standardized residuals assess the significance of the deviations between the new Pareto chart and the reference Pareto chart. The statistical test relies on computation of standardized residuals:

$$Z_i = (n_i - Np_i)/[Np_i(1 - p_i)]^{1/2} \qquad i = 1, \ldots, K \qquad \text{(10.6.1)}$$

where

N = Total number of observations in the Pareto chart

p_i = Proportion of observations in category i in the reference Pareto chart

Np_i = Expected number of observations in category i, given a total of N observations

n_i = Actual number of observations in category i

K = Number of categories

In performing this statistical test we assume that observations are independently classified into distinct categories. However, the actual classification into categories might depend on the data-gathering protocol. Typically, classification relies on the first error cause encountered, and therefore a different test procedure could produce different data. Also, this test is more powerful than the standard chi-square test. It recognizes differences between a reference Pareto chart and a current Pareto chart that a chi-square test would not deem significant.

To perform the statistical analysis, we first list the error categories in a fixed unsorted order. One natural order is listing the category names in alphabetic order. Organizing the data is necessary if we are to make meaningful comparisons. The test itself consists of seven steps, the last being an interpretation of the results. To demonstrate these steps, we use data concerning time card errors; these are given in Table 10.4.

TABLE   **10.4**

Time card error data in 15 departments

Department #	Reference Pareto	Current Pareto
1	23	14
2	42	7
3	37	85
4	36	19
5	17	23
6	50	13
7	60	48
8	74	59
9	30	2
10	25	0
11	10	12
12	54	14
13	23	30
14	24	20
15	11	0

The data come from a system that monitors time card entries in a medium-size company with 15 departments. During a management meeting, the human resources manager was asked to initiate a project aimed at reducing time card errors. The manager asked to see a Pareto chart (which now serves as a reference Pareto chart) of last month's time card errors by department and found that departments 6, 7, 8, and 12 were responsible for 46% of these errors. The manager then appointed a special team to look into the causes of the errors. The team recommended changing the format of the time card, and this change was implemented throughout the company. Three weeks later, a new Pareto chart was prepared from 346 newly reported time card errors and a statistical analysis of the chart was performed to determine what department had a significant change in its relative number of time card errors.

The steps in performing the statistical analysis are as follows:

1 For each department, compute the proportion of observations in the reference Pareto chart:

$$p_1 = 23/516 = .04457$$
$$\vdots$$
$$p_{15} = 11/516 = .0213$$

2 Compute the total number of observations in the new Pareto chart:

$$N = 14 + 7 + 85 + \cdots + 20 = 346$$

3 Compute the expected number of observations in department i, $E_i = Np_i$, $i = 1, \ldots, 15$:

$$E_1 = 346 \times 0.04457 = 15.42$$
$$\vdots$$
$$E_{15} = 346 \times 0.0213 = 7.38$$

4 Compute the standardized residuals: $Z_i = (n_i - Np_i)/(Np_i(1 - p_i))^{1/2}$, $i = 1, \ldots, 15$:

$$Z_1 = (14 - 15.42)/[15.42(1 - 0.04457)]^{1/2} = -0.37$$
$$\vdots$$
$$Z_{15} = (0 - 7.38)/[7.38(1 - 0.0213)]^{1/2} = -2.75$$

5 Look up in Table 10.5 the critical value for $K = 15$. Interpolate between $K = 10$ and $K = 20$. For the $\alpha = .01$ significance level, the critical value is approximately $(3.10 + 3.30)/2 = 3.20$.

6 Identify categories with standardized residuals larger in absolute value than 3.20. Table 10.6 indicates with a star the departments where the proportion of errors was significantly different from that in the reference Pareto.

T A B L E   10.5

Critical values for standardized residuals

K	Significance level		
	10%	5%	1%
4	1.95	2.24	2.81
5	2.05	2.32	2.88
6	2.12	2.39	2.93
7	2.18	2.44	2.99
8	2.23	2.49	3.04
9	2.28	2.53	3.07
10	2.32	2.57	3.10
20	2.67	2.81	3.30
30	2.71	2.94	3.46

TABLE 10.6

Table of standardized residuals for the time cards error data

Department Number	Pareto	E_i	Z_i
1	14	15.42	−0.37
2	7	28.16	−4.16*
3	85	24.81	12.54*
4	19	24.14	−1.08
5	23	11.40	3.49*
6	13	33.53	−3.73*
7	48	40.23	1.30
8	59	49.62	1.44
9	2	20.12	−4.16*
10	0	16.76	−4.20*
11	12	6.71	2.06
12	14	36.21	−3.90*
13	30	15.42	3.80*
14	20	16.09	1.00
15	0	7.38	−2.75

7 Data from departments 2, 3, 5, 6, 9, 10, 12, and 13 vary significantly from the reference Pareto chart to the new Pareto chart. In department 3, for example, we expected 24.81 occurrences, a much smaller number than the actual 85.

The statistical test enables us to compare systematically two Pareto charts with the same categories. Focusing on the differences between Pareto charts complements the analysis of trends and changes in overall process error levels. Increases or decreases in such error levels may result from changes across all error categories. Or, there may be no changes in error levels but, instead, significant changes in the mix of errors across categories. The statistical analysis reveals such changes. Another advantage of the statistical procedure is that it can apply to different time frames. For example, the reference Pareto can cover a period of one year and the current Pareto can span a period of three weeks.

Critical values are computed on the basis of the Bonferroni inequality approximation. This inequality states that, because we are examining simultaneously K standardized residuals, the overall significance level is not more than K times the significance level of an individual comparison. Dividing the overall significance level of choice by K, and using the normal approximation, produces the critical values in Table 10.5. For more details on this procedure see Kenett (1991).

10.7

The Shewhart Control Charts

The Shewhart control charts are devices used to check statistical control in which a sample of size n is drawn from the process every h units of time. Let θ denote a parameter of the distribution of the observed random sample $X_1, \ldots, X_n$. Let $\hat{\theta}_n$

denote an appropriate estimate of θ. If θ_0 is a desired operation level for the process, we construct around θ_0 two limits: UCL and LCL. As long as LCL $\leq \hat{\theta}_n \leq$ UCL, we say the process is under statistical control.

More specifically, suppose that $X_1, X_2, \ldots$ are normally distributed and independent. Every h hours (time units) a sample of n observations is taken.

Suppose that when the process is under control, the distribution mean is θ_0 and its variance σ^2 is known. We set $\hat{\theta}_n \equiv \overline{X}_n = (1/n)\sum_{j=1}^{n} X_j$. The control limits are

$$\mathrm{UCL} = \theta_0 + 3\frac{\sigma}{\sqrt{n}}$$

$$\mathrm{LCL} = \theta_0 - 3\frac{\sigma}{\sqrt{n}}$$

(10.7.1)

In addition to these control limits, we set warning limits. A point $\overline{X}_n$ outside the warning limits requires us to be watchful. The warning limits are

Upper warning limit (UWL) $= \theta_0 + 2\dfrac{\sigma}{\sqrt{n}}$

Lower warning limit (LWL) $= \theta_0 - 2\dfrac{\sigma}{\sqrt{n}}$

(10.7.2)

Notice that:

1 The samples are independent.

2 All $\overline{X}_n$ are identically $N[\theta_0, (\sigma^2/n)]$ as long as the process is under control.

3 If α is the probability of observing $\overline{X}_n$ outside the control limits, when $\theta = \theta_0$, then $\alpha = .0027$. We expect one of every $N = 370$ samples to yield a value of $\overline{X}_n$ outside the control limits.

4 We expect about 5% of the $\overline{X}_n$ points to lie outside the warning limits when the process is under control. Thus, although for testing the null hypothesis $H_0 : \theta = \theta_0$ against $H_1 : \theta \neq \theta_0$, we may choose a level of significance $\alpha = .05$ and use the limits UWL, LWL as rejection limits, in the control case the situation is equivalent to that of simultaneously (or repeatedly) testing many hypotheses. For this reason, we have to consider a much smaller α, such as $\alpha = .0027$, of the 3-sigma limits.

5 In most practical applications of the Shewhart 3-sigma control charts, the samples taken are of small size—$n = 4$ or $n = 5$, for example—and the frequency of samples is high (h small). Shewhart recommended small samples to reduce the possibility that a shift in θ will occur during sampling. However, if the samples are picked frequently, there is a greater chance of detecting a shift early. The questions of how frequently to sample and what should the sample size be are related to the concept of rational subgroups, which was discussed earlier in Section 10.3. An economic approach to the determination of rational subgroups will be presented in Section 11.5.

Figure 10.18 shows a Shewhart control chart for means. The variable is the cycle time of a gas turbine. We see on the chart the warning and the control limits. The term SL in the chart is $\sigma/\sqrt{n}$.

FIGURE   10.18
Shewhart control chart

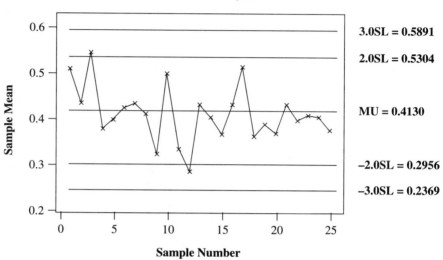

10.7.1   Control Charts for Attributes

In this section, we look at control charts when the control statistic is the sample fraction of defectives $\hat{p}_i = (x_i/n_i)$, $i = 1, \ldots, N$. Here n_i is the size of the ith sample and x_i is the number of defective items in the ith sample. It is preferable for the sample size, n_i, to be the same for all samples.

Given N samples, we estimate the common parameter θ by $\hat{\theta} = \sum_{i=1}^{N} x_i / \sum_{i=1}^{N} n_i$. The upper and lower control limits may change with the sample point and are given by

$$\text{UCL} = \hat{\theta} + 3\sqrt{\frac{\hat{\theta}(1 - \hat{\theta})}{n_i}}$$

$$i = 1, \ldots, N \qquad \text{(10.7.3)}$$

$$\text{LCL} = \hat{\theta} - 3\sqrt{\frac{\hat{\theta}(1 - \hat{\theta})}{n_i}}$$

Table 10.7 gives the number of defective items found in random samples of size $n = 100$, drawn daily from a production line.

Figure 10.19 shows the control chart for the data of Table 10.7. Note that the fraction of defectives on two days was significantly high but that the process on the whole remained under control during the month. We can revise the control chart by deleting these two days and computing a modified estimate of θ. We obtain a new value of $\hat{\theta} = 139/2900 = .048$. This new estimator yields a revised upper control limit

$$\text{UCL}' = 0.112$$

TABLE 10.7

Number of defects in daily samples ($n = 100$)

Sample/Day	Number of defects	Sample/Day	Number of defects
i	x_i	i	x_i
1	6	16	6
2	8	17	4
3	8	18	6
4	13	19	8
5	6	20	2
6	6	21	7
7	9	22	4
8	7	23	4
9	1	24	2
10	8	25	1
11	5	26	5
12	2	27	15
13	4	28	1
14	5	29	4
15	4	30	1
		31	5

FIGURE 10.19

p-chart for January data

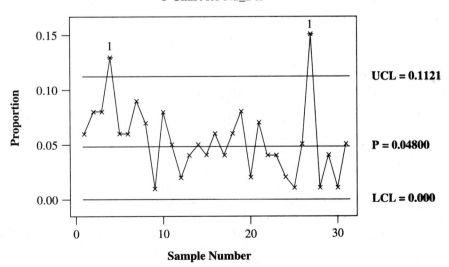

P Chart for Nu_Def

10.7.2   Control Charts for Variables

10.7.2.1 $\overline{X}$-Charts

After the process has been observed for k sampling periods, we can compute estimates of the process mean and standard deviation. The estimate for the process mean is

$$\overline{\overline{X}} = \frac{1}{k} \sum_{i=1}^{k} \overline{X}_i$$

This is the center line of the control chart. The process standard deviation can be estimated using either the average sample standard deviation

$$\hat{\sigma} = \overline{S}/c(n) \tag{10.7.4}$$

where

$$\overline{S} = \frac{1}{k} \sum_{i=1}^{k} S_i$$

or the average sample range,

$$\hat{\sigma} = \overline{R}/d(n) \tag{10.7.5}$$

where

$$\overline{R} = \frac{1}{k} \sum_{i=1}^{k} R_i$$

The factors $c(n)$ and $d(n)$ guarantee that we obtain unbiased estimates of σ. We can show, for example, that $E(\overline{S}) = \sigma c(n)$, where

$$c(n) = \left[\Gamma(n/2) / \Gamma\left(\frac{n-1}{2}\right) \right] \sqrt{2/(n-1)} \tag{10.7.6}$$

Moreover, $E\{R_n\} = \sigma d(n)$, where, from the theory of order statistics (see Section 4.7), we obtain that in the mormal case

$$d(n) = \tag{10.7.7}$$

$$\frac{n(n-1)}{2\pi} \int_0^\infty y \int_{-\infty}^\infty \exp\left\{ -\frac{x^2 + (y+x)^2}{2} \right\} [\Phi(x+y) - \Phi(x)]^{n-2} dx dy$$

Table 10.8 lists the factors $c(n)$ and $d(n)$ for $n = 2, 3, \ldots, 10$.

The control limits are now computed as

$$\text{UCL} = \overline{\overline{X}} + 3\hat{\sigma}/\sqrt{n}$$

and $\tag{10.7.8}$

$$\text{LCL} = \overline{\overline{X}} - 3\hat{\sigma}/\sqrt{n}$$

Despite the wide use of the sample ranges to estimate the process standard deviation, the method is neither very efficient nor robust. It is popular only because the sample range is easier to compute than the sample standard deviation. However, because many hand calculators now have built-in programs for computing the sample standard deviation, the computational advantage of the range should not be considered. In any case, the sample ranges should not be used when the sample size is greater than 10. We now illustrate the construction of an $\overline{X}$-chart by using the data in Table 10.9.

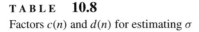

TABLE **10.8**

Factors $c(n)$ and $d(n)$ for estimating σ

n	$c(n)$	$d(n)$
2	0.7979	1.1283
3	0.8862	1.6926
4	0.9213	2.0587
5	0.9400	2.3259
6	0.9515	2.5343
7	0.9594	2.7044
8	0.9650	2.8471
9	0.9693	2.9699
10	0.9727	3.0774

TABLE **10.9**

20 samples of five electric contact lengths

Hour i	x_1	x_2	x_3	x_4	x_5	$\overline{X}$	S	R
1	1.9890	2.1080	2.0590	2.0110	2.0070	2.0348	0.04843	0.11900
2	1.8410	1.8900	2.0590	1.9160	1.9800	1.9372	0.08456	0.21800
3	2.0070	2.0970	2.0440	2.0810	2.0510	2.0560	0.03491	0.09000
4	2.0940	2.2690	2.0910	2.0970	1.9670	2.1036	0.10760	0.30200
5	1.9970	1.8140	1.9780	1.9960	1.9830	1.9536	0.07847	0.18300
6	2.0540	1.9700	2.1780	2.1010	1.9150	2.0436	0.10419	0.26300
7	2.0920	2.0300	1.8560	1.9060	1.9750	1.9718	0.09432	0.23600
8	2.0330	1.8500	2.1680	2.0850	2.0230	2.0318	0.11674	0.31800
9	2.0960	2.0960	1.8840	1.7800	2.0050	1.9722	0.13825	0.31600
10	2.0510	2.0380	1.7390	1.9530	1.9170	1.9396	0.12552	0.31200
11	1.9520	1.7930	1.8780	2.2310	1.9850	1.9678	0.16465	0.43800
12	2.0060	2.1410	1.9000	1.9430	1.8410	1.9662	0.11482	0.30000
13	2.1480	2.0130	2.0660	2.0050	2.0100	2.0484	0.06091	0.14300
14	1.8910	2.0890	2.0920	2.0230	1.9750	2.0140	0.08432	0.20100
15	2.0930	1.9230	1.9750	2.0140	2.0020	2.0014	0.06203	0.17000
16	2.2300	2.0580	2.0660	2.1990	2.1720	2.1450	0.07855	0.17200
17	1.8620	2.1710	1.9210	1.9800	1.7900	1.9448	0.14473	0.38100
18	2.0560	2.1250	1.9210	1.9200	1.9340	1.9912	0.09404	0.20500
19	1.8980	2.0000	2.0890	1.9020	2.0820	1.9942	0.09285	0.19100
20	2.0490	1.8790	2.0540	1.9260	2.0080	1.9832	0.07760	0.17500
					Average:	2.0050 $\overline{\overline{X}}$	0.09537 $\overline{S}$	0.23665 $\overline{R}$

These measurements represent the length (in cm) of the electrical contacts of relays in samples of size 5, taken hourly from the operating process. For purposes of illustration, both the sample standard deviation and the sample range are computed for each sample. The center line for the control chart is $\overline{\overline{X}} = 2.005$. From Table 10.8 we find for $n = 5$, $c(5) = 0.9400$. Let

$$A_1 = 3/(c(5)\sqrt{5}) = 1.427$$

The control limits are given by

$$\text{UCL} = \overline{\overline{X}} + A_1\overline{S} = 2.141$$

and

$$\text{LCL} = \overline{\overline{X}} - A_1\overline{S} = 1.869$$

The resulting control chart is shown in Figure 10.20. If we use the sample ranges to determine the control limits, we first find that $d(5) = 2.326$ and

$$A_2 = 3/(d(5)\sqrt{5}) = 0.577$$

This gives us control limits of

$$\text{UCL}' = \overline{\overline{X}} + A_2\overline{R} = 2.142$$
$$\text{LCL}' = \overline{\overline{X}} - A_2\overline{R} = 1.868$$

FIGURE   10.20
$\overline{X}$ control chart for contact data

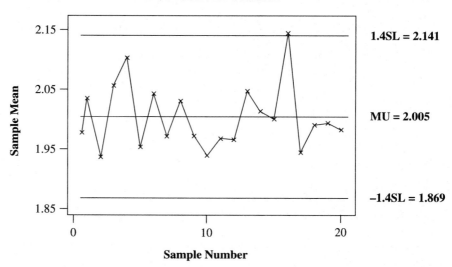

X–bar Chart for ContactL

10.7.2.2   *S*-Charts and *R*-Charts

As discussed earlier, control of process variability can be as important as control of the process mean. Two types of control charts are commonly used for this purpose: an *R*-chart, based on sample ranges, and an *S*-chart, based on sample standard deviations. Because ranges are easier to compute than standard deviations, *R*-charts are probably more commonly used in practice. The R-chart is not very efficient, however. In fact, its efficiency declines rapidly as sample size increases and sample range should not be used for samples greater than size 5. Nevertheless, we shall discuss both types of charts.

To construct control limits for the S-chart, we use a normal approximation to the sampling distribution of the sample standard deviation, S. This means we will use control limits

$$\text{UCL} = \bar{S} + 3\hat{\sigma}_s$$

and

$$\text{LCL} = \bar{S} - 3\hat{\sigma}_s$$

(10.7.9)

where $\hat{\sigma}_s$ represents an estimate of the standard deviation of S. This standard deviation is

$$\sigma_s = \sigma/\sqrt{2(n-1)}$$

(10.7.10)

Using the unbiased estimate $\hat{\sigma} = \bar{S}/c(n)$, we obtain

$$\hat{\sigma}_s = \bar{S}/(c(n)\sqrt{2(n-1)})$$

(10.7.11)

and hence the control limits

$$\text{UCL} = \bar{S} + 3\bar{S}/(c(n)\sqrt{2(n-1)}) = B_4\bar{S}$$

and

$$\text{LCL} = \bar{S} - 3\bar{S}/(c(n)\sqrt{2(n-1)}) = B_3\bar{S}$$

(10.7.12)

The factors B_3 and B_4 can be determined from Table 10.8. If $B_3 < 0$ set $B_3 = 0$.

Using the electrical contact data in Table 10.9, we find

$$\text{Center line} = \bar{S} = 0.095$$
$$\text{UCL} = B_4\bar{S} = 2.021(.095) = .192$$

and

$$\text{LCL} = B_3\bar{S} = 0$$

The S-chart is given in Figure 10.21.

An R-chart is constructed using similar techniques, with center line $= \bar{R}$ and control limits:

$$\text{UCL} = D_4\bar{R}$$

and

$$\text{LCL} = D_3\bar{R}$$

(10.7.13)

where

$$D_3 = \left(1 - \frac{3 \cdot e(n)}{d(n)}\right)$$

and

$$D_4 = \left(1 + \frac{3 \cdot e(n)}{d(n)}\right)$$

(10.7.14)

$(e(n)/d(n))\bar{R}$ is an estimate of the standard deviation of R, for a sample of size n. Values of $e(n)$ for $n = 2, \ldots, 6$ are given in Table 10.10. If $D_3 < 0$ set $D_3 = 0$.

F I G U R E   10.21

S-chart for contact data

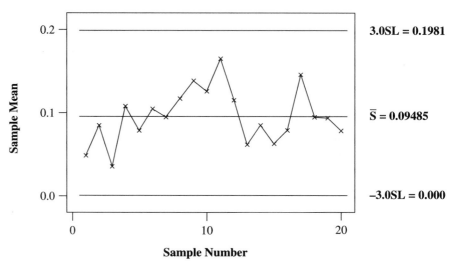

S Chart for LengCont

TA B L E   **10.10**

Factor $e(n)$ for estimating Stdev of R_n

n	2	3	4	5	6
$e(n)$	0.8526	0.8883	0.8802	0.8644	0.8487

Using the data of Table 10.9, we find

$$\text{Centerline} = \overline{R} = 0.237$$
$$\text{UCL} = D_4\overline{R} = (2.114)(.237) = 0.501$$

and

$$\text{LCL} = D_3\overline{R} = 0$$

The *R*-chart is shown in Figure 10.22.

The decision of whether to use an *R*-chart or *S*-chart to control variability ultimately depends on which method works best in any given situation. Both methods are based on several approximations. There is, however, one additional point that should be considered. The average value of the range of n variables depends to a great extent on the sample size n. As n increases, the range increases. An *R*-chart based on 5 observations per sample will look quite different from an *R*-chart based on 10 observations per sample. For this reason, it is difficult to visualize the variability characteristics of the process directly from the data. However, the sample standard deviation, *S*, used in the *S*-chart, is a good estimate of the process standard deviation, σ. As the sample size increases, *S* will tend to be even closer to the true value of σ. The process standard deviation is the key to understanding the variability of the process.

F I G U R E **10.22**
R-chart for contact data

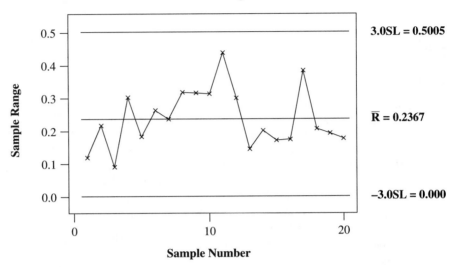

R Chart for LengCont

3.0SL = 0.5005

$\bar{R}$ = 0.2367

−3.0SL = 0.000

Sample Number

10.8
Chapter Highlights

Competitive pressures are forcing many management teams to focus on process control and process improvement as an alternative to screening and inspection. This chapter discusses techniques used effectively in industrial organizations that have adopted such ideas. Classic control charts, quality control, and quality planning tools are presented along with modern statistical process control procedures, including new statistical techniques for constructing confidence intervals of process capability indices and analyzing Pareto charts. Throughout the chapter, a software piston simulator is used to demonstrate how control charts are set up and used in real-life applications. The main concepts and definitions introduced in this chapter include:

- Statistical process control
- Chronic problems
- Common causes
- Special causes
- Assignable causes
- Sporadic spikes
- External feedback loops
- Internal feedback loops
- Control charts
- Upper control limit
- Lower control limit
- Process capability study
- Rational subgroups
- Process capability index
- Flowchart
- Check sheet
- Run chart
- Histogram
- Pareto chart

- Scatterplot
- Cause-and-effect diagram

- Upper warning limit
- Lower warning limit

Exercises

10.1.1 Use MINITAB and file OELECT.DAT to chart the individual electrical outputs of the 99 circuits. Do you observe any trend or nonrandom pattern in the data? [Use under Control charts the option of Individual chart. For Mu and Sigma use "historical" values, which are $\overline{X}$ and S.]

10.1.2 Chart the individual variability of the length of steel rods in the STEELROD.DAT file. Is there any perceived assignable cause of nonrandomness?

10.2.1 Examine the chart of Exercise 10.1.2 for possible patterns of nonrandomness.

10.2.2 Test the data in file OTURB2.DAT for lack of randomness. In this file we have three columns. In the first, we have the sample size. In the second and third, we have the sample means and standard deviation. If you use MINITAB, you can chart individual means. For the historical mean, use the mean of column $c2$. For historical standard deviation, use $(\hat{\sigma}^2/5)^{1/2}$, where $\hat{\sigma}^2$ is the pooled sample variance. Before solving run TURB.EXE with 20 samples of size 5.

10.2.3 A sudden change in a process lowers the process mean by 1 standard deviation. It has been determined that the quality characteristic being measured is approximately normally distributed and that the change had no effect on the process variance.

a What proportion of points is expected to fall outside the control limits on the $\overline{X}$-chart if the subgroup size is 4?

b Answer the same question for subgroups of size 6.

c Answer the same question for subgroups of size 9.

10.3.1 Make a capability analysis of the electric output (in volts) of the 99 circuits in data file OELECT.DAT, with target values of $\mu_0 = 220$ and LSL $= 210$, USL $= 230$.

10.4.1 Estimate the capability index C_{pk} for the output of the electronic circuits, based on data file OELECT.DAT when LSL $= 210$ and USL $= 230$. Determine the point estimate as well as its confidence interval, with confidence level .95.

10.4.2 Estimate the capability index for the steel rods given in data file STEELROD.DAT, when the length specifications are $\xi_L = 19$ and $\xi_U = 21$ (in cm) and the level of confidence is $1 - \alpha = .95$.

10.4.3 The specification limits of the piston cycle times generated by TURB1.EXE are 0.5 ± 0.2 sec. Generate 20 cycle times at the lower level of the 7 control parameters.

a Compute C_p and C_{pk}.

b Compute a 95% confidence interval for C_{pk}. Generate 20 cycle times at the upper level of the 7 control factors.

c Recompute C_p and C_{pk}.

d Recompute a 95% confidence interval for C_{pk}.

e Is there a significant difference in process capability between lower and upper operating levels in the piston simulator?

10.4.4 A fiber manufacturer has a large contract that stipulates that its fiber, among other properties, have tensile strength greater than 1.800 (grams/fiber) in 95% of the fiber used. The manufacturer states that the standard deviation of the process is 0.015 g.

a Assuming a process is under statistical control, what is the smallest nominal value of the mean that will ensure compliance with the contract?

b Given the nominal value in part a, what are the control limits of $\overline{X}$-chart for subgroups of size 6?

c What is the process capability if the process mean is $\mu = 1.82$?

10.4.5 The output voltage of a power supply is specified as 370 ± 5 volts DC. Subgroups of four units are drawn from every batch and submitted to special quality-control tests. The data from 30 subgroups on output voltage produced $\sum_{i=1}^{30} \overline{X} = 10{,}950.0$ and $\sum_{i=1}^{30} S_i = 34.6$.

a Compute the control limits for $\overline{X}$.

b Assuming statistical control and a normal distribution of output voltage, what is the proportion of defective products?

c If the power supplies are set to a nominal value of 370 volts, what is the proportion of defective products?

d Compute the new control limits for $\overline{X}$.

e If these new control limits are used but the adjustment to 370 volts is not carried out, what is the probability that this fact will not be detected on the first subgroup?

f What is the process capability before and after the adjustment of the nominal value to 370 volts? Compute both C_p and C_{pk}.

10.6.1 The following data were collected in a circuit pack production plant during October:

	Number of nonconformities
Missing component	293
Wrong component	431
Too much solder	120
Insufficient solder	132
Failed component	183

An improvement team recommended several changes that were implemented in the first week of November. The following data were collected in the second week of November:

	Number of nonconformities
Missing component	34
Wrong component	52
Too much solder	25
Insufficient solder	34
Failed component	18

a Construct Pareto charts of the nonconformities in October and the second week of November.

b Has the improvement team produced significant differences in the type of nonconformities?

10.7.1 Control charts for $\overline{X}$ and R are maintained on total soluble solids produced at $20°C$ in parts per million (ppm). Samples are drawn from production containers every hour and tested in a special test device. The test results are organized into subgroups of $n = 5$ measurements, corresponding to 5 hours of production. After 125 hours of production, we find that $\sum_{i=1}^{25} \overline{X}_i = 390.8$ and $\sum_{i=1}^{25} R_i = 84$. The specification on the process states that containers with more than 18 ppm of total soluble solids should be reprocessed.

a Compute an appropriate capability index.

b Assuming a normal distribution and statistical control, what proportion of the sample measurements is expected to be out of spec?

c Compute the control limits for $\overline{X}$ and R.

10.7.2 *Part I*: Run TURB1.EXE at the lower levels of the 7 piston parameters. Generate 100 cycle times and store them in OTURB1.DAT. Add 2.0 to the last 50 cycle times.

a Compute control limits of $\overline{X}$ and R by constructing subgroups of size 5, and analyze the control charts. Use 0.5 for the "historical" value of μ.

Part II: Assign a random number, R_i, from $U(0, 1)$ to each cycle time. Sort the 100 cycle times by R_i, $i = 1, \ldots, 100$.

b Recompute the control limits of $\overline{X}$ and R and reanalyze the control charts.

c Explain the differences between parts a and b.

10.7.3 *Part I*: Run TURB2.EXE by specifying the 7 piston parameters within their acceptable range. Record the 7 operating levels you used and generate 20 subgroups of size 5. The output is stored in OTURB2.DAT.

a Compute the control limits for $\overline{X}$ and S.

Part II: Rerun TURB2.EXE at the same operating conditions and generate 20 subgroups of size 10. The output overrides OTURB2.DAT.

b Recompute the control limits for $\overline{X}$ and S.

c Explain the differences between parts a and b.

11

Advanced Methods of Statistical Process Control

In this chapter, we discuss more advanced methods of statistical process control. We start with tests of whether data are randomly distributed around a mean level or whether there is a trend or a shift in the data. The tests we consider are nonparametric tests called **runs tests**. This is followed by a section on modified Shewhart-type control charts for the mean. Modifications of Shewhart charts were introduced into the collection of SPC tools in order to increase the power of the procedures to detect changes. Section 11.3 is devoted to the problem of determining the size and frequency of samples for the statistical control of processes by Shewhart control charts. In Section 11.4, we discuss control charts for the means of multivariate data. In Section 11.5, we introduce an alternative control tool based on cumulative sums: the famous CUSUM procedures, which come out of Page's control schemes. In Section 11.6, Bayesian detection procedures are presented. Section 11.7 is devoted to procedures of process control that track the process level. The last section introduces some notions from engineering control theory that are useful in automatically controlled processes.

Tests of Randomness

In performing process capability analysis (see Chapter 10) or retroactively analyzing data for constructing a control chart, the first thing we would like to test is whether these data are randomly distributed around their mean. If they are, it means the process is statistically stable and only common causes affect the variability. In this section, we discuss such tests of randomness.

Consider a sample $x_1, x_2, \ldots, x_n$, where the index of the values of x indicates some kind of ordering. For example, x_1 is the first observed value of X, x_2 is the second observed value, and so on, and x_n is the value observed last. If the sample is indeed random, there should be no significant relationship between the values of X and their place in the sample. Thus, tests of randomness usually test the hypothesis that all possible configurations of the x's are equally probable, against the alternative hypothesis that some significant clustering of members takes place. For example, suppose we have a sequence of five 0s and five 1s. The ordering 0, 1, 1, 0, 0, 0, 1, 1, 0, 1 seems to be random, whereas the ordering 0, 0, 0, 0, 0, 1, 1, 1, 1, 1 seems to be conspicuously not random.

11.1.1 Testing the Number of Runs

In a sequence of m_1 0s and m_2 1s, we distinguish between runs of 0s—that is, an uninterrupted string of 0s—and runs of 1s. Accordingly, in the sequence 0 1 1 1 0 0 1 0 1 1, there are four 0s and six 1s and there are three runs of 0s and three runs of 1s. We denote the total number of runs by R. The probability distribution of the total number of runs, R, is determined under the model of randomness. It can be shown that, if there are m_1 0s and m_2 1s, then

$$\Pr\{R = 2k\} = \frac{2\binom{m_1 - 1}{k - 1}\binom{m_2 - 1}{k - 1}}{\binom{n}{m_2}} \tag{11.1.1}$$

and

$$\Pr\{R = 2k + 1\} = \frac{\binom{m_1 - 1}{k - 1}\binom{m_2 - 1}{k} + \binom{m_1 - 1}{k}\binom{m_2 - 1}{k - 1}}{\binom{n}{m_2}} \tag{11.1.2}$$

Here, n is the sample size, $m_1 + m_2 = n$.

One alternative to the hypothesis of randomness is clustering of the 0s (or 1s). In such a case, we expect to observe longer runs of 0s (or 1s) and, consequently, a smaller number of total runs. In this case, the hypothesis of randomness is rejected if the total number of runs, R, is too small. However, there could be an alternative to randomness that is the reverse of clustering. This alternative is called *mixing*. For example, the following sequence of ten 0s and ten 1s is completely mixed and is obviously not random:

$$0, 1, 0, 1, 0, 1, 0, 1, 0, 1, 0, 1, 0, 1, 0, 1, 0, 1, 0, 1$$

The total number of runs here is $R = 20$. Thus, if there are too many runs, we should also reject the hypothesis of randomness. Consequently, if we consider the null hypothesis H_0 of randomness against the alternative H_1 of clustering, the lower (left) tail of the distribution should be used for the rejection region. If the alternative, H_1, is the hypothesis of mixing, then the upper (right) tail of the distribution should be used. If the alternative is either clustering or mixing, the test should be two-sided.

We test the hypothesis of randomness by using the test statistic R, the total number of runs. The critical region for the one-sided alternative that there is clustering is of the form:

$$R \leq R_\alpha$$

where R_p is the pth quantile of the null distribution of R. The p.d.f. and c.d.f. of R are computed by the program DISTRUNS.EXE. For the one-sided alternative of mixing, we reject H_0 if $R \geq R_{1-\alpha}$. In cases of large samples, we can use the normal approximations

$$R_\alpha = \mu_R - z_{1-\alpha}\sigma_R$$

and

$$R_{1-\alpha} = \mu_R + z_{1-\alpha}\sigma_R$$

where

$$\mu_R = 1 + 2m_1 m_2/n \qquad (11.1.3)$$

and

$$\sigma_R = [2m_1 m_2(2m_1 m_2 - n)/n^2(n-1)]^{1/2} \qquad (11.1.4)$$

are the mean and standard deviation, respectively, of R under the hypothesis of randomness. We can also use the normal distribution to approximate the P-value of the test. For one-sided tests, we have

$$\alpha_L = \Pr\{R \leq r\} \cong \Phi((r - \mu_R)/\sigma_R) \qquad (11.1.5)$$

and

$$\alpha_U = \Pr\{R \geq r\} \cong 1 - \Phi((r - \mu_R)/\sigma_R) \qquad (11.1.6)$$

where r is the observed number of runs. For the two-sided alternative, the P-value of the test is approximated by

$$\alpha' = \begin{cases} 2\alpha_L & \text{if } R < \mu_R \\ 2\alpha_U & \text{if } R > \mu_R \end{cases} \qquad (11.1.7)$$

11.1.2 Runs Above and Below a Specified Level

The runs test for the randomness of a sequence of 0s and 1s can be applied to test whether the values in a sequence, which are continuous in nature, are randomly distributed. We can also consider whether the values are above or below the sample average or the sample median. In such a case, every value above the specified level will be assigned the value 1, whereas all the others will be assigned the value 0. Once this is done, the previous runs test can be applied.

For example, suppose we are given a sequence of $n = 30$ observations and we wish to test for randomness using the number of runs, R, above and below the median, M_e. There are 15 observations below and 15 above the median. In this case, we take $m_1 = 15$, $m_2 = 15$, and $n = 30$. In Table 11.1, we present the p.d.f. and c.d.f. of the number of runs, R, below and above the median of a random sample of size $n = 30$.

For a level of significance of $\alpha = .05$ if $R \leq 10$ or $R \geq 21$, the two-sided test rejects the hypothesis of randomness. Critical values for a two-sided runs test above

T A B L E   11.1

Distribution of R in a random sample of size $n = 30$, $m_1 = 15$

R	p.d.f.	c.d.f.
2	0.00000	0.00000
3	0.00000	0.00000
4	0.00000	0.00000
5	0.00002	0.00002
6	0.00011	0.00013
7	0.00043	0.00055
8	0.00171	0.00226
9	0.00470	0.00696
10	0.01292	0.01988
11	0.02584	0.04572
12	0.05168	0.09739
13	0.07752	0.17491
14	0.11627	0.29118
15	0.13288	0.42407
16	0.15187	0.57593
17	0.13288	0.70882
18	0.11627	0.82509
19	0.07752	0.90261
20	0.05168	0.95428
21	0.02584	0.98012
22	0.01292	0.99304
23	0.00470	0.99774
24	0.00171	0.99945
25	0.00043	0.99987
26	0.00011	0.99998
27	0.00002	1.00000
28	0.00000	1.00000
29	0.00000	1.00000
30	0.00000	1.00000

and below the median can also be obtained by the large-sample approximation

$$R_{\alpha/2} = \mu_R - z_{1-\alpha/2}\sigma_R$$
$$R_{1-\alpha/2} = \mu_R + z_{1-\alpha/2}\sigma_R$$

(11.1.8)

Substituting $m = m_1 = m_2 = 15$ and $\alpha = .05$, we have $\mu_R = 16$, $\sigma_R = 2.69$, $z_{.975} = 1.96$. Hence, $R_{\alpha/2} = 10.7$ and $R_{1-\alpha/2} = 21.3$. Thus, according to the large-sample approximation if $R \leq 10$ or $R \geq 22$, the hypothesis of randomness is rejected.

This test of the total number of runs, R, above and below a given level (such as the mean or the median of a sequence) can be performed by using MINITAB.

EXAMPLE 11.1   In this example, we use MINITAB to perform a runs test on a simulated random sample of size $n = 28$ from the normal distribution $N(10, 1)$. The test is of runs above and below the distribution mean 10. We obtain a total of $R = 14$, with $m_1 = 13$ values below and $m_2 = 15$ values above the mean. Figure 11.1 shows this random sequence. The following MINITAB analysis shows that we can accept the hypothesis of randomness.

```
MTB > print C1
C1
                                                                    Count
   10.917   10.751    9.262   11.171   10.807    8.630   10.097   9.638     08
   10.785   10.256   10.493    8.712    8.765   10.613   10.943   8.727     16
   11.154    9.504    9.477   10.326   10.735    9.707    9.228   9.879     24
    9.976    9.699    8.266    9.139                                        28
MTB > Runs 10.0 c1.
C1
K = 10.0000
THE OBSERVED NO. OF RUNS = 14
THE EXPECTED NO. OF RUNS = 14.9286
13 OBSERVATIONS ABOVE K 15 BELOW
            THE TEST IS SIGNIFICANT AT 0.7193
            CANNOT REJECT AT ALPHA = 0.05
```

The MINITAB test is one-sided. H_0 is rejected if R is large. Program DIST-RUNS.EXE yields, for the same data, the P-level .7117. ■

F I G U R E   11.1

Random normal sequence and the runs above and below the level 10

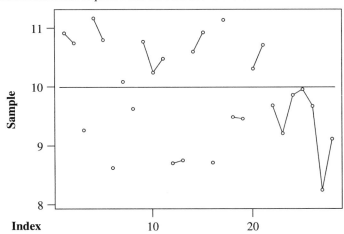

11.1.3   Runs Up and Down

Tests of the total number of runs above or below a specified level may not be sufficient in cases where the alternative hypothesis to randomness is possible cyclical fluctuations in the level of the process. For example, a sequence may show consistent fluctuations up and down, as in the following example:

$$-1, \ -.75, \ -.50, \ -.25, \ 0, \ .5, \ 1, \ .5, \ .25, \ -.75, \ \ldots$$

Here we see a steady increase from -1 to 1 and then a steady decrease. This sequence is obviously not random, and even the previous test of the total number of runs above and below 0 will reject the hypothesis of randomness. If the development of the sequence is not as conspicuous as this—say, for example, the sequence

$$-1, \ -.75, \ -.50, \ 1, \ .5, \ -.25, \ 0, \ .25, \ -.25, \ 1, \ .5, \ .25, \ -.75, \ \ldots$$

the runs test above and below 0 may not reject the hypothesis of randomness. Indeed, in this case, if we replace every negative number by 0 and every nonnegative number by $+1$, we find that $m_1 = 6$, $m_2 = 7$, and $R = 7$. In this case, the exact value of α_L is .5, and the hypothesis of randomness is not rejected.

To test for possible cyclical effects, we use a test of runs up and down. Let $x_1, x_2, \ldots, x_n$ be a given sequence, and let us define

$$y_i = \begin{cases} +1 & \text{if } x_i < x_{i+1} \\ -1 & \text{if } x_i \geq x_{i+1} \end{cases}$$

for $i = 1, \ldots, n-1$. We then count the total number of runs, R, up and down. A *run up* is a string of $+1$s, whereas a *run down* is a string of -1s. In the previous sequence, we have the following values of x_i and y_i:

x_i	-1.00	-0.75	-0.50	1.00	.050	-0.25	0.00	0.25	-0.25	1.00	0.50	0.25	-0.75
y_i		1	1	1	-1	-1	1	1	-1	1	-1	-1	-1

We thus have a total of $R^* = 6$ runs, 3 up and 3 down, with $n = 13$.

To test the hypothesis of randomness based on the number of runs up and down, we need the null distribution of R^*.

When the sample size is sufficiently large, we can use the normal approximation

$$R_\alpha^* = \mu_{R^*} - z_{1-\alpha}\sigma_{R^*}$$

and

$$R_{1-\alpha}^* = \mu_{R^*} + z_{1-\alpha}\sigma_{R^*}$$

where

$$\mu_{R^*} = (2n-1)/3 \tag{11.1.9}$$

and

$$\sigma_{R^*} = [(16n - 29)/90]^{1/2} \tag{11.1.10}$$

The attained significance levels are approximated by

$$\alpha_L^* = \Phi((r^* - \mu_{R^*})/\sigma_{R^*}) \tag{11.1.11}$$

and

$$\alpha_U^* = 1 - \Phi((r^* - \mu_{R^*})/\sigma_{R^*}) \tag{11.1.12}$$

EXAMPLE 11.2   The sample in data file YARNSTRG.DAT contains 100 values of log yarn-strength. In this sample, there are $R^* = 64$ runs up or down, 32 runs up, and 32 runs down. The expected value of R^* is

$$\mu_{R^*} = \frac{199}{3} = 66.33$$

and its standard deviation is

$$\sigma_{R^*} = \left(\frac{1600 - 29}{90}\right)^{1/2} = 4.178$$

The attained level of significance is

$$\alpha_L = \Phi\left(\frac{64 - 66.33}{4.178}\right) = 0.289$$

The hypothesis of randomness is not rejected.   ■

11.1.4 Testing the Length of Runs Up and Down

In the previous sections, we considered tests of randomness based on the total number of runs. If the number of runs is small relative to the size of the sequence, n, we obviously expect some of the runs to be rather long. We now consider the question of just how long the runs can be under a state of randomness.

Consider runs up and down, and let R_k $(k = 1, 2, \ldots)$ be the total number of runs, up or down, of length greater than or equal to k. Thus, $R_1 = R^*$, R_2 is the total number of runs, up or down, of length 2 or more, and so on. The formulas in Table 11.2 are for the expected values of each R_k—that is, $E\{R_k\}$—under the

TABLE 11.2
Expected values of R_k as a function of n

k	$E\{R_k\}$
1	$(2n - 1)/3$
2	$(3n - 5)/12$
3	$(4n - 11)/60$
4	$(5n - 19)/360$
5	$(6n - 29)/2520$
6	$(7n - 41)/20160$
7	$(8n - 55)/181440$

assumption of randomness. Each expected value is expressed as a function of the size n of the sequence.

In general, we have

$$E\{R_k\} = \frac{2[n(k + 1) - k^2 - k + 1]}{(k + 2)!} \qquad 1 \le k \le n - 1 \qquad \text{(11.1.13)}$$

If $k \ge 5$, we have $E\{R_k\} \cong V\{R_k\}$, and the Poisson approximation to the probability distribution of R_k is considered good, provided $n > 20$. Thus, if $k \ge 5$, according to the Poisson approximation, we find

$$\Pr\{R_k \ge 1\} \cong 1 - \exp(-E\{R_k\}) \qquad \text{(11.1.14)}$$

For example, if $n = 50$, then $E\{R_k\}$ and $\Pr\{R_k \ge 1\}$ are as follows:

k	$E\{R_k\}$	$\Pr\{R_k \ge 1\}$
5	0.1075	0.1020
6	0.0153	0.0152
7	0.0019	0.0019

The probability of observing even 1 run, up or down, of length 6 or more is quite small. This is the reason for the following rule of thumb: **Reject the hypothesis of randomness if a run is of length 6 or more**. This and other rules of thumb were presented in Chapter 10 for ongoing process control.

Modified Shewhart Control Charts for $\overline{X}$

The modified Shewhart control chart for $\overline{X}$, used to detect possible shifts in the means of the parent distributions, gives the signal to stop whenever the sample means $\overline{X}$ fall outside the control limits $\theta_0 \pm a(\sigma/\sqrt{n})$ or whenever a run of r sample means falls outside the warning limits (all on the same side) $\theta_0 \pm w(\sigma/\sqrt{n})$.

We denote the modified scheme by (a, w, r). For example, 3-σ (3-sigma) control charts, with warning lines at 2-σ and a run of $r = 4$, is denoted by $(3, 2, 4)$. If $r = \infty$, the scheme $(3, 0, \infty)$ is reduced to the common Shewhart 3-σ procedure. Similarly, the scheme $(a, 3, 1)$ for $a > 3$ is equivalent to the Shewhart 3-σ control charts. A control chart for a $(3, 1.5, 2)$ procedure is shown in Figure 11.2. The means are of samples of size 5. There is no run of length 2 or more between the warning and action limits.

The *run length (RL)* of a control chart is the number of samples taken until the "out of control" alarm is given. The **average run length (ARL)** of an (a, w, r) plan is smaller than that of the simple Shewhart a-σ procedure. We denote the average run length of an (a, w, r) procedure by $\text{ARL}(a, w, r)$. Obviously, if w and r are small, we will tend to stop too soon, even when the process is under control. For example, if $r = 1$ and $w = 2$, then any procedure $(a, 2, 1)$ is equivalent to a Shewhart 2-σ procedure, which stops on the average every 20 samples when the process is under

F I G U R E   11.2

A Modified Shewhart $\overline{X}$-chart

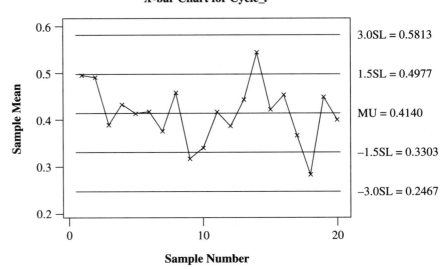

X-bar Chart for Cycle_*t*

control. Weindling (1967) and Page (1962) derived the formula for the average run length ARL(a, w, r). Page used the theory of runs; Weindling used another theory (Markov chains theory). An excellent expository paper discussing the results was published by Weindling, Littauer, and de Oliviera (1970).

The basic formula for the determination of the average run length is

$$\mathrm{ARL}_\theta(a, w, r) = \tag{11.2.1}$$

$$\left[P_\theta(a) + H_\theta^r(a, w)\frac{1 - H_\theta(a, w)}{1 - H_\theta^r(a, w)} + L_\theta^r(a, w)\frac{1 - L_\theta(a, w)}{1 - L_\theta^r(a, w)} \right]^{-1}$$

where:

$$P_\theta(a) = P_\theta\left\{ \overline{X} \le \theta_0 - a\frac{\sigma}{\sqrt{n}} \right\} + P_\theta\left\{ \overline{X} \ge \theta_0 + a\frac{\sigma}{\sqrt{n}} \right\}$$

$$H_\theta(a, w) = P_\theta\left\{ \theta_0 + w\frac{\sigma}{\sqrt{n}} \le \overline{X} \le \theta_0 + a\frac{\sigma}{\sqrt{n}} \right\} \tag{11.2.2}$$

$$L_\theta(a, w) = P_\theta\left\{ \theta_0 - a\frac{\sigma}{\sqrt{n}} \le \overline{X} \le \theta_0 - w\frac{\sigma}{\sqrt{n}} \right\}$$

Table 11.3 gives some values of ARL(a, w, r) for $a = 3$, $w = 1(.5)2.5$, $r = 2(1)7$, when the samples are of size $n = 5$ from a normal distribution, and the shift in the mean is of size $\delta\sigma$. We see in the table that the procedures (3, 1, 7), (3, 1.5, 5), (3, 2, 3), and (3, 2.5, 2) yield similar ARL functions. However, these modified procedures are more efficient than the Shewhart 3-σ procedure. They all have close

TABLE 11.3

Values of ARL(a, w, r), $a = 3.00$ against $\delta = (\mu_1 - \mu_0)/\sigma$, $n = 5$

| | | \multicolumn{6}{c}{r} | | | | | |
w	δ	2	3	4	5	6	7
1.0	0.00	22.0	107.7	267.9	349.4	366.9	369.8
	0.25	11.2	32.1	67.1	101.3	121.0	129.0
	0.50	4.8	9.3	14.9	20.6	25.4	28.8
	0.75	2.8	4.4	5.8	7.1	8.1	8.9
	1.00	2.0	2.7	3.3	3.7	3.9	4.1
	1.25	1.6	2.0	2.2	2.3	2.3	2.4
	1.50	1.4	1.5	1.5	1.6	1.6	1.6
	1.75	1.2	1.2	1.2	1.2	1.2	1.2
	2.00	1.1	1.1	1.1	1.1	1.1	1.1
	2.25	1.0	1.0	1.0	1.0	1.0	1.0
1.5	0.00	93.1	310.1	365.7	370.1	370.4	370.4
	0.25	31.7	88.1	122.8	130.3	132.8	133.1
	0.50	9.3	18.3	26.8	31.0	32.6	33.2
	0.75	4.0	6.4	8.2	9.4	10.0	10.4
	1.00	2.4	3.0	3.7	4.1	4.3	4.4
	1.25	1.7	2.2	2.2	2.3	2.3	2.4
	1.50	1.4	1.5	1.5	1.6	1.6	1.6
	1.75	1.2	1.2	1.2	1.2	1.2	1.2
	2.00	1.1	1.1	1.1	1.1	1.1	1.1
	2.25	1.0	1.0	1.0	1.0	1.0	1.0
2.0	0.00	278.0	367.8	370.3	370.4	370.4	370.4
	0.25	84.7	128.3	132.8	133.1	133.2	133.2
	0.50	19.3	30.0	32.8	33.3	33.4	33.4
	0.75	6.5	9.2	10.3	10.6	10.7	10.7
	1.00	3.1	3.9	4.3	4.4	4.5	4.5
	1.25	1.9	2.2	2.3	2.4	2.4	2.4
	1.50	1.4	1.5	1.6	1.6	1.6	1.6
	1.75	1.2	1.2	1.2	1.2	1.2	1.2
	2.00	1.1	1.1	1.1	1.1	1.1	1.1
	2.25	1.0	1.0	1.0	1.0	1.0	1.0
2.5	0.00	364.0	370.3	370.4	370.4	370.4	370.4
	0.25	127.3	133.0	133.2	133.2	133.2	133.2
	0.50	30.6	33.2	33.4	33.4	33.4	33.4
	0.75	9.6	10.6	10.7	10.8	10.8	10.8
	1.00	4.0	4.4	4.5	4.5	4.5	4.5
	1.25	2.2	2.3	2.4	2.4	2.4	2.4
	1.50	1.5	1.6	1.6	1.6	1.6	1.6
	1.75	1.2	1.2	1.2	1.2	1.2	1.2
	2.00	1.1	1.1	1.1	1.1	1.1	1.1
	2.25	1.0	1.0	1.0	1.0	1.0	1.0

ARL values when $\delta = 0$, but when $\delta > 0$, their ARL values are considerably smaller than the Shewhart procedure.

The Size and Frequency of Sampling for Shewhart Control Charts

In this section we discuss the design of the sampling procedure for Shewhart control charts. We start with the problem of the economic design of sampling for $\overline{X}$-charts.

11.3.1 The Economic Design for $\overline{X}$-Charts

Duncan (1956, 1971, 1978) studied the question of optimally designing the $\overline{X}$ control charts. We show here, in a somewhat simpler fashion, how this problem can be approached. More specifically, assume that we sample from a normal population and that σ^2 is known. A shift of size $\delta = (\theta_1 - \theta_0)/\sigma$ or larger should be detected with high probability.

Let c (dollars/hour) be the hourly cost of a shift in the mean of size δ. Let $d(\$)$ be the cost of sampling (and testing the items). Assuming that the time of shift from θ_0 to $\theta_1 = \theta_0 + \delta\sigma$ is exponentially distributed with mean $1/\lambda$ (hr) and that a penalty of $1(\$)$ is incurred for every unneeded inspection, the total expected cost is

$$K(h, n) \cong \frac{ch + dn}{1 - \Phi(3 - \delta\sqrt{n})} + \frac{1 + dn}{\lambda h} \tag{11.3.1}$$

This function can be minimized with respect to h and n to determine the optimal sample size and frequency of sampling. Differentiating partially with respect to h and equating to zero, we obtain the formula of the optimal h for a given n—namely,

$$h_{\text{opt}} = \left(\frac{1 + d \cdot n}{c\lambda}\right)^{1/2} (1 - \Phi(3 - \delta\sqrt{n}))^{1/2} \tag{11.3.2}$$

However, the function $K(h, n)$ is increasing with n due to the contribution of the second term on the RHS. Thus, for this expected cost function, we take a sample of size $n = 4$ every h_{opt} hours. Some values of h_{opt} are

δ	d	c	λ	h_{opt}
2	.5	3.0	.0027	16.20
1	.1	30.0	.0027	0.66

For additional reading on this subject, see Gibra (1971).

11.3.2 Increasing the Sensitivity of p-charts

The **operating characteristic (OC) function** for a Shewhart p-chart is the probability, as a function of p, that the statistic $\hat{p}_n$ falls between the upper and lower control limits. Thus, the operating characteristic of a p-chart, with control limits $p_0 \pm 3\sqrt{[p_0(1 - p_0)/n]}$ is

$$\mathrm{OC}(p) = \mathrm{Pr}_\theta \left\{ p_0 - 3\sqrt{\frac{p_0(1 - p_0)}{n}} < \hat{p}_n < p_0 + 3\sqrt{\frac{p_0(1 - p_0)}{n}} \right\} \quad \textbf{(11.3.3)}$$

where $\hat{p}_n$ is the proportion of defective items in the sample. $n \times \hat{p}_n$ has the binomial distribution, with c.d.f. $B(j; n, p)$. Accordingly,

$$\mathrm{OC}(p) = B(np_0 + 3\sqrt{np_0(1 - p_0)}; \ n, p)$$
$$- B(np_0 - 3\sqrt{np_0(1 - p_0)}; \ n, p) \quad \textbf{(11.3.4)}$$

For large samples, we can use the normal approximation to $B(j; n, p)$ and obtain

$$\mathrm{OC}(p) \cong \Phi\left(\frac{(\mathrm{UCL} - p)\sqrt{n}}{\sqrt{p(1 - p)}}\right) - \Phi\left(\frac{(\mathrm{LCL} - p)\sqrt{n}}{\sqrt{p(1 - p)}}\right) \quad \textbf{(11.3.5)}$$

The value of the $\mathrm{OC}(p)$ at $p = p_0$ is $2\Phi(3) - 1 = .997$. The values of $\mathrm{OC}(p)$ for $p \neq p_0$ are smaller. A typical $\mathrm{OC}(p)$ function looks like that in Figure 11.3.

When the process is in control, with a process fraction defective p_0, we have $\mathrm{OC}(p_0) = .997$; otherwise, $\mathrm{OC}(p) < .997$. The probability that we will detect a change in quality to level p_1, with a single point outside the control limits, is $1 - \mathrm{OC}(p_1)$. As an example, suppose we have estimated p_0 as $\bar{p} = .15$ from past data. With a sample of size $n = 100$, our control limits are

$$\mathrm{UCL} = 0.15 + 3((.15)(.85)/100)^{1/2} = 0.257$$

and

$$\mathrm{LCL} = 0.15 - 3((.15)(.85)/100)^{1/2} = 0.043$$

FIGURE   11.3

Typical OC curve for a p-chart

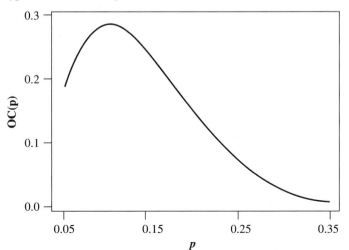

In Table 11.4, we see that it is almost certain that a single point will fall outside the control limits when $p = .40$, but it is unlikely that it will fall there when $p = .20$. However, if the process fraction defective remains at the $p = .20$ level for several measurement periods, the probability of detecting the shift increases. The probability that at least one point falls outside the control limits when $p = .20$ for five consecutive periods is

$$1 - [OC(.20)]^5 = .3279$$

The probability of detecting shifts in the fraction defective is even greater than .33 if we apply run tests on the data.

The OC curve can also be useful for determining the required sample size for detecting, with high probability, a change in the process fraction defective in a single measurement period. To see this, suppose that the system is in control at level p_0, and we wish to detect a shift to level p_t with specified probability $1 - \beta$. For example, to be 90% confident that the sample proportion will be outside the control limits immediately after the process fraction defective changes to p_t, we require that

$$1 - OC(p_t) = .90$$

We can solve this equation to find that the required sample size is

$$n \cong \frac{(3\sqrt{p_0(1 - p_0)} + z_{1-\beta}\sqrt{p_t(1 - p_t)})^2}{(p_t - p_0)^2} \tag{11.3.6}$$

If we want the sample proportion, with probability $1 - \beta$, to be outside the limits at least once within k sampling periods when the precise fraction defective is p_t, the sample size must be

$$n \cong \frac{(3\sqrt{p_0(1 - p_0)} + z_{1-b}\sqrt{(p_t(1 - p_t))})^2}{(p_t - p_0)^2} \tag{11.3.7}$$

where $b = \beta^{1/k}$.

These results are illustrated in Table 11.5 for a process with $p_0 = .15$. It is practical to take a small sample of $n = 5$ at each period. We see in the table that in this case a change from .15 to .40 would be detected within five periods with probability of .9. To detect smaller changes requires larger samples.

TABLE 11.4
Operating characteristic values for p-chart, $\bar{p} = .15$ and $n = 100$

p	$OC(p)$	$1 - OC(p)$	$1 - [OC(p)]^5$
.05	.6255	.3745	.9043
.10	.9713	.0287	.1355
.15	.9974	.0026	.0130
.20	.9236	.0764	.3280
.25	.5636	.4364	.9432
.30	.1736	.8264	.9998
.40	.0018	.9982	1.0000

░ **TABLE 11.5**

Sample size required for probability .9 of detecting a shift to level p_t from level $p_0 = .15$ (in one period and within five periods)

p_t	One period	Five periods
.05	183	69
.10	848	217
.20	1003	156
.25	265	35
.30	123	14
.40	47	5

11.4
Multivariate Control Charts

Univariate control charts track observations on one dimension. However, multivariate data are much more informative than a collection of one-dimensional variables. Simultaneously accounting for variation in several variables requires both an overall measure of departure of the observation from the targets and an assessment of the data covariance structure. **Multivariate control charts** were developed for that purpose. Here we discuss the construction of multivariate control charts, using the multivariate data on aluminum pins that were introduced in Chapter 3 in file ALMPIN.DAT.

We will construct a multivariate control chart using the first 30 cases of the data file as a base sample and the other 40 cases as observations on a production process we wish to control.

The observations in the base sample provide estimates of the means, variance, and covariances of the six variables being measured. Let $\overline{X}_i$ denote the mean of variable X_i $(i = 1, \ldots, k)$ in the base sample. The size of the base sample is n.

Let S_{ij} denote the covariance between X_i and X_j $(i, j = 1, \ldots, k)$—namely,

$$S_{ij} = \frac{1}{n-1} \sum_{l=1}^{n} (X_{il} - \overline{X}_{i.})(X_{jl} - \overline{X}_{j.}) \tag{11.4.1}$$

Notice that S_{ii} is the sample variance of X_i $(i = 1, \ldots, k)$. Let **S** denote the $k \times k$ covariance matrix:

$$\mathbf{S} = \begin{bmatrix} S_{11} & S_{12} & \cdots & S_{1k} \\ S_{21} & S_{22} & \cdots & S_{2k} \\ \vdots & & & \\ S_{k1} & S_{k2} & \cdots & S_{kk} \end{bmatrix} \tag{11.4.2}$$

Notice that $S_{ij} = S_{ji}$ for every i, j. Thus, **S** is a symmetric matrix.

Let **m** denote the $k \times 1$ vector of sample means whose transpose is

$$\mathbf{m}' = (\overline{X}_{1.}, \ldots, \overline{X}_{k.})$$

Finally, we compute the inverse of $\mathbf{S}$—namely, $\mathbf{S}^{-1}$. This inverse exists unless one (or some) of the variables is a linear combination of the others. Such variables should be excluded.

Suppose now that, in every time unit, we draw a sample of size m $(m \geq 1)$ from the production process and observe on each element the k variables of interest. To distinguish between the sample means from the production process and those of the base sample, we will denote by $\overline{Y}_{i.}(t)$ $(t = 1, 2, \ldots)$ the sample mean of variable X_i from the sample at time t. Let $\mathbf{Y}(t)$ be the vector of these k means—that is, $\mathbf{Y}'(t) = (\overline{Y}_{1.}(t), \ldots, \overline{Y}_{k.}(t))$. To monitor the means $\mathbf{Y}(t)$ of the samples from the production process in order to detect when a significant change from $\mathbf{m}$ occurs, we construct a ontrol chart called a T^2-*chart*. We assume that the covariances do not change in the production process. Thus, for every time period t $(t = 1, 2, \ldots)$, we compute the T^2 statistics:

$$T_t^2 = (\mathbf{Y}(t) - \mathbf{m})'\mathbf{S}^{-1}(\mathbf{Y}(t) - \mathbf{m}) \tag{11.4.3}$$

It can be shown that as long as the process mean and covariance matrix are the same as those of the base sample, the distribution of T^2 is like that of

$$\frac{n+m}{nm} \cdot \frac{(n-1)k}{n-k}F[k, n-k]$$

Accordingly, we set the (upper) control limit for T^2 at

$$\text{UCL} = \frac{n+m}{nm} \cdot \frac{(n-1)k}{n-k}F_{.997}[k, n-k] \tag{11.4.4}$$

If a point $T(t)$ falls above this control limit, there is an indication of a significant change in the mean vector.

EXAMPLE **11.3**   The base sample consists of the first 30 rows of data file ALMPIN.DAT. The mean vector of the base sample is

$$\mathbf{m}' = (9.99, 9.98, 9.97, 14.98, 49.91, 60.05)$$

The covariance matrix of the base sample is $\mathbf{S}$, where $10^3\mathbf{S}$ is

$$\begin{bmatrix} 0.1826 & 0.1708 & 0.1820 & 0.1826 & -0.0756 & -0.0054 \\ & 0.1844 & 0.1853 & 0.1846 & -0.1002 & -0.0277 \\ & & 0.2116 & 0.1957 & -0.0846 & 0.0001 \\ & & & 0.2309 & -0.0687 & -0.0054 \\ & & & & 1.3179 & 1.0039 \\ & & & & & 1.4047 \end{bmatrix}$$

(Because $\mathbf{S}$ is symmetric, we show only the upper matrix.) The inverse of $\mathbf{S}$ is

$$\mathbf{S}^{-1} = \begin{bmatrix} 5319.3 & -2279.10 & -17079.7 & -9343.4 & 145.0 & -545.3 \\ & 66324.1 & -28342.7 & -10877.9 & 182.0 & 1522.8 \\ & & 50553.9 & -6467.9 & 853.0 & -1465.1 \\ & & & 25745.6 & -527.5 & 148.6 \\ & & & & 1622.3 & -1156.0 \\ & & & & & 1577.6 \end{bmatrix}$$

We now compute the T_t^2 values for the last 40 rows of this data file, thinking of each row as a vector of a sample of size $m = 1$ taken every 10 minutes. Table 11.6 gives these 40 vectors and their corresponding T^2 values. For example, T_1^2 of the table is

TABLE   **11.6**

Dimensions of aluminum pins in a production process and their T^2 values

X_1	X_2	X_3	X_4	X_5	X_6	T^2
10.00	9.99	9.99	14.99	49.92	60.03	6.982
10.00	9.99	9.99	15.00	49.93	60.03	9.054
10.00	10.00	9.99	14.99	49.91	60.02	12.222
10.00	9.99	9.99	14.99	49.92	60.02	8.362
10.00	9.99	9.99	14.99	49.92	60.00	12.070
10.00	10.00	9.99	15.00	49.94	60.05	11.706
10.00	9.99	9.99	15.00	49.89	59.98	12.865
10.00	10.00	9.99	14.99	49.93	60.01	16.556
10.00	10.00	9.99	14.99	49.94	60.02	16.775
10.00	10.00	9.99	15.00	49.86	59.96	15.059
10.00	9.99	9.99	14.99	49.90	59.97	15.698
10.00	10.00	10.00	14.99	49.92	60.00	27.379
10.00	10.00	9.99	14.98	49.91	60.00	21.239
10.00	10.00	10.00	15.00	49.93	59.98	32.144
10.00	9.99	9.98	14.98	49.90	59.98	7.483
9.99	9.99	9.99	14.99	49.88	59.98	18.609
10.01	10.01	10.01	15.01	49.87	59.97	27.736
10.00	10.00	9.99	14.99	49.81	59.91	22.821
10.01	10.00	10.00	15.01	50.07	60.13	29.148
10.01	10.00	10.00	15.00	49.93	60.00	14.096
10.00	10.00	10.00	14.99	49.90	59.96	33.359
10.01	10.01	10.01	15.00	49.85	59.93	37.126
10.00	9.99	9.99	15.00	49.83	59.98	11.714
10.01	10.01	10.00	14.99	49.90	59.98	28.278
10.01	10.01	10.00	15.00	49.87	59.96	21.375
10.00	9.99	9.99	15.00	49.87	60.02	7.248
9.99	9.99	9.99	14.98	49.92	60.03	17.712
9.99	9.98	9.98	14.99	49.93	60.03	9.197
9.99	9.99	9.98	14.99	49.89	60.01	6.080
10.00	10.00	9.99	14.99	49.89	60.01	11.507
9.99	9.99	9.99	15.00	50.04	60.15	29.372
10.00	10.00	10.00	14.99	49.84	60.03	22.348
10.00	10.00	9.99	14.99	49.89	60.01	11.507
10.00	9.99	9.99	15.00	49.88	60.01	7.743
10.00	10.00	9.99	14.99	49.90	60.04	10.441
9.90	9.89	9.91	14.88	49.99	60.14	50.041
10.00	9.99	9.99	15.00	49.91	60.04	6.285
9.99	9.99	9.99	14.98	49.92	60.04	16.726
10.01	10.01	10.00	15.00	49.88	60.00	16.232
10.00	9.99	9.99	14.99	49.95	60.10	6.020

computed according to the formula

$$T_1^2 = (\mathbf{Y}(1) - \mathbf{m})'\mathbf{S}^{-1}(\mathbf{Y}(1) - \mathbf{m}) = 6.9817$$

The 40 T_t^2 values of Table 11.6 are plotted in Figure 11.4. The UCL in this chart is 34.559.

These computations can be performed by a version of MINITAB that performs matrix manipulations. For a comprehensive treatment of multivariate quality control with MINITAB application, see Fuchs and Kenett (1997).   ■

F I G U R E   11.4
T^2-chart for aluminum pins

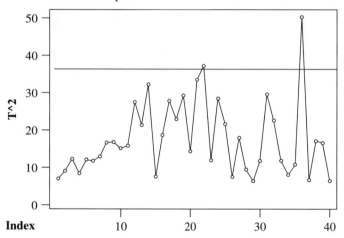

11.5
Cumulative Sum Control Charts
11.5.1 Upper Page's Scheme

When the process level changes from a past or specified level, we expect a control procedure to trigger an alarm. Depending on the size of the change and the size of the sample, it may take several sampling periods before the alarm occurs. A method that has a smaller ARL than the standard Shewhart control charts for detecting certain types of changes is the **cumulative sum** (or **CUSUM**) **control chart**, which was introduced by Barnard (1959) and Page (1954). See also Yashchin (1985a, 1991b).

CUSUM charts differ from the common Shewhart control charts in several respects. The main difference is that instead of plotting the individual value of the statistic of interest, such as X, $\overline{X}$, S, R, p, or c, a statistic based on the cumulative sums is computed and tracked. By summing deviations of the individual statistic from a target value, T, we get a consistent increase or decrease of the cumulative sum when the process is above or below the target. Figure 11.5 shows the behavior of the cumulative sums

$$S_t = \sum_{i=1}^{t}(X_i - 10) \tag{11.5.1}$$

of data simulated from a normal distribution with mean

$$\mu_t = \begin{cases} 10 & \text{if } t \le 20 \\ 13 & \text{if } t > 20 \end{cases}$$

and $\sigma_t = 1$ for all t.

As soon as the shift in the mean of the data occurs, a pronounced drift begins in S_t. Page (1954) suggested that, to detect an upward shift in the mean, we should consider the sequence

$$S_t^+ = \max\{S_{t-1}^+ + (X_k - K^+), 0\} \qquad t = 1, 2, \ldots \tag{11.5.2}$$

where $S_0^+ \equiv 0$, and that we could decide that a shift has occurred as soon as $S_t^+ > h^+$. The statistics X_t, $t = 1, 2, \ldots$, upon which the (truncated) cumulative sums are constructed could be means of samples of n observations, standard deviations, sample proportions, or individual observations.

In the next section we will see how the parameters K^+ and h^+ are determined. We will see that if X_t are means of samples of size n, with process variance σ^2, and if the desired process mean is θ_0, whereas the maximal tolerated process mean is θ_1, $\theta_1 - \theta_0 > 0$, then

$$K^+ = \frac{\theta_0 + \theta_1}{2} \qquad \text{and} \qquad h^+ = -\frac{\sigma^2 \log \alpha}{n(\theta_1 - \theta_0)} \qquad 0 < \alpha < 1 \tag{11.5.3}$$

FIGURE   11.5

A plot of cumulative sums with drift after $t = 20$

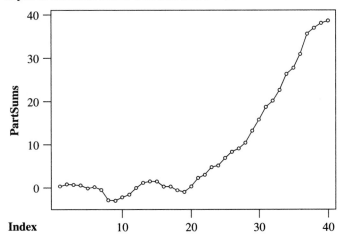

EXAMPLE **11.4**   In this example, we illustrate Page's procedure. The data in Table 11.7 represent the number of computer crashes per month due to power failures experienced at a computer center over a period of 28 months. After a crash, the computers are made operational with an "Initial Program Load." Thus, we refer to the data as the IPL data set.

TABLE **11.7**

Number of monthly computer crashes due to power failures

t	X_t	t	X_t	t	X_t
1	0	11	0	21	0
2	2	12	0	22	1
3	0	13	0	23	3
4	0	14	0	24	2
5	3	15	0	25	1
6	3	16	2	26	1
7	0	17	2	27	3
8	0	18	1	28	5
9	2	19	0		
10	1	20	0		

Power failures are potentially harmful. A computer center might be able to tolerate such failures if they are far enough apart. However, if they become too frequent, the center might decide to invest in an uninterruptable power supply. It seems intuitively clear from Table 11.7 that computer crashes due to power failures do become more frequent. Is the variability in failure rates due to chance alone (common causes) or can it be attributed to special causes that should be investigated? Suppose that the computer center can tolerate at most an average of one power failure every three weeks (21 days), or $30/21 = 1.43$ crashes per month. Even better would be fewer than one failure per six weeks, or 0.71 crash per month. In Table 11.8, we show the computation of Page's statistics S_t^+, with $K^+ = \frac{1}{2}(0.71 + 1.43) = 1.07$. For $\alpha = .05$, $\sigma = 1$, $n = 1$, we obtain the critical level $h^+ = 4.16$. Thus, we see that an alarm is first triggered after the 27th month. Figure 11.6 shows the graph of S_t^+ versus t, which is called a *CUSUM chart*. We see in the figure that although S_6^+ is close to 4, the graph falls back toward zero and no alarm is triggered until the 27th month. ∎

11.5.2   Some Theoretical Background

In Section 9.5, we discussed the Wald sequential probability ratio test (SPRT) for the special case of testing hypotheses about the parameter θ of a binomial distribution. Generally, if $X_1, X_2, \ldots$ is a sequence of i.i.d. random variables (continuous or discrete) having a p.d.f. $f(x; \theta)$, and we wish to test two simple hypotheses— $H_0 : \theta = \theta_0$ versus $H_1 : \theta = \theta_1$—with type I and type II error probabilities α and β,

TABLE 11.8

The S_t^+ statistics for the IPL data

t	X_t	$X_t - 1.07$	S_t^+
1	0	−1.07	0
2	2	0.93	0.93
3	0	−1.07	0
4	0	−1.07	0
5	3	1.93	1.93
6	3	1.93	3.86
7	0	−1.07	2.79
8	0	−1.07	1.72
9	2	0.93	2.65
10	1	−0.07	2.58
11	0	−1.07	1.51
12	0	−1.07	0.44
13	0	−1.07	0
14	0	−1.07	0
15	0	−1.07	0
16	2	0.93	0.93
17	2	0.93	1.86
18	1	−0.07	1.79
19	0	−1.07	0.72
20	0	−1.07	0
21	0	−1.07	0
22	1	−0.07	0
23	3	1.93	1.93
24	2	0.93	2.86
25	1	−0.07	2.79
26	1	−0.07	2.72
27	3	1.93	4.65
28	5	3.93	8.58

respectively, we use the Wald SPRT, which is a sequential procedure that, after t observations, $t \geq 1$, considers the likelihood ratio

$$\Lambda(X_1, \ldots, X_t) = \prod_{i=1}^{t} \frac{f(X_i; \theta_1)}{f(X_i; \theta_0)} \tag{11.5.4}$$

If $\beta/(1-\alpha) < \Lambda(X_1, \ldots, X_t) < (1-\beta)/\alpha$, then another observation is taken; otherwise, sampling terminates. If $\Lambda(X_1, \ldots, X_t) < \beta/(1-\alpha)$, then H_0 is accepted; and if $\Lambda(X_1, \ldots, X_t) > (1-\beta)/\alpha$, H_0 is rejected.

In an upper control scheme, we consider only the upper boundary, by setting $\beta = 0$. Thus, we can decide that the true hypothesis is H_1 as soon as

$$\sum_{i=1}^{t} \log \frac{f(X_i; \theta_1)}{f(X_i; \theta_0)} \geq -\log \alpha$$

Next, we examine the structure of this testing rule in a few special cases.

FIGURE 11.6

Page's CUSUM chart of IPL data

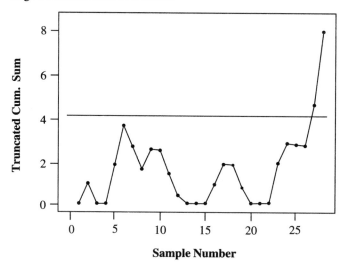

Normal Distribution　We consider X_i to be normally distributed with known variance σ^2 and mean θ_0 or θ_1. In this case

$$\log \frac{f(X_i; \theta_1)}{f(X_i; \theta_0)} = -\frac{1}{2\sigma^2}\{(X_i - \theta_1)^2 - (X_i - \theta_0)^2\}$$

$$= \frac{\theta_1 - \theta_0}{\sigma^2}\left(X_i - \frac{\theta_0 + \theta_1}{2}\right)$$

(11.5.5)

Thus, the criterion

$$\sum_{i=1}^{t} \log \frac{f(X_i; \theta_1)}{f(X_i; \theta_0)} \geq -\log \alpha$$

is equivalent to

$$\sum_{i=1}^{t} \left(X_i - \frac{\theta_0 + \theta_1}{2}\right) \geq -\frac{\sigma^2 \log \alpha}{\theta_1 - \theta_0}$$

For this reason, we use the upper Page control scheme with $K^+ = (\theta_0 + \theta_1)/2$ and $h^+ = -\sigma^2 \log(\alpha)/(\theta_1 - \theta_0)$. If X_t is an average of n independent observations, then we replace σ^2 by σ^2/n.

Binomial Distributions　Suppose that X_t has a binomial distribution $B(n, \theta)$. If $\theta \leq \theta_0$, the process level is under control. If $\theta \geq \theta_1$, the process level is out of control

$(\theta_1 > \theta_0)$. Because

$$f(x; \theta) = \binom{n}{x} \left(\frac{\theta}{1-\theta}\right)^x (1-\theta)^n$$

then

$$\sum_{i=1}^{t} \log \frac{f(X_i; \theta_1)}{f(X_i; \theta_0)} \geq -\log \alpha \tag{11.5.6}$$

if

$$\sum_{i=1}^{t} \left(X_i - \frac{n \log\left(\dfrac{1-\theta_0}{1-\theta_1}\right)}{\log\left(\dfrac{\theta_1}{1-\theta_1} \cdot \dfrac{1-\theta_0}{\theta_0}\right)} \right) \geq -\frac{\log \alpha}{\log\left(\dfrac{\theta_1}{1-\theta_1} \cdot \dfrac{1-\theta_0}{\theta_0}\right)}$$

Accordingly, in an upper Page's control scheme, with binomial data, we use

$$K^+ = \frac{n \log\left(\dfrac{1-\theta_0}{1-\theta_1}\right)}{\log\left(\dfrac{\theta_1}{1-\theta_1} \cdot \dfrac{1-\theta_0}{\theta_0}\right)} \tag{11.5.7}$$

and

$$h^+ = -\frac{\log \alpha}{\log\left(\dfrac{\theta_1}{1-\theta_1} \cdot \dfrac{1-\theta_0}{\theta_0}\right)} \tag{11.5.8}$$

Poisson Distributions   When the statistics X_t have Poisson distribution with mean λ, then for specified levels λ_0 and λ_1, $0 < \lambda_0 < \lambda_1 < \infty$,

$$\sum_{i=1}^{t} \log \frac{f(X_i; \lambda_1)}{f(X_i; \lambda_0)} = \log\left(\frac{\lambda_1}{\lambda_0}\right) \sum_{i=1}^{t} X_i - t(\lambda_1 - \lambda_0) \tag{11.5.9}$$

It follows that the control parameters are

$$K^+ = \frac{\lambda_1 - \lambda_0}{\log(\lambda_1/\lambda_0)} \tag{11.5.10}$$

and

$$h^+ = -\frac{\log \alpha}{\log(\lambda_1/\lambda_0)} \tag{11.5.11}$$

11.5.3   Lower and Two-Sided Page's Scheme

To test whether a significant drop has occurred in the process level (mean), we can use a lower Page scheme. According to this scheme, we set $S_0^- \equiv 0$ and

$$S_t^- = \min\{S_{t-1}^- + (X_t - K^-), 0\} \qquad t = 1, 2, \ldots \tag{11.5.12}$$

Here the CUSUM values S_t^- are either zero or negative. We decide that a shift down in the process level from θ_0 to $\theta_1, \theta_1 < \theta_0$ occurred as soon as $S_t^- < h^-$. The control parameters K^- and h^- are determined by the formulas given in Section 11.5.2 by setting $\theta_1 < \theta_0$, and $\alpha = 0$. In this case only the lower boundary is considered.

EXAMPLE **11.5**   In file COAL.DAT are data on the number of coal mine disasters (explosions) in England per year for the period 1850 to 1961. These data are plotted in Figure 11.7. It seems that the average number of disasters per year dropped after 40 years from 3 to 2 and later settled around an average of 1 per year. We apply the lower Page's scheme to see when we detect this change for the first time. It is plausible to assume that the number of disasters per year, X_t, is a random variable having a Poisson distribution. We therefore set $\lambda_0 = 3$ and $\lambda_1 = 1$. The formulas of the previous section, with K^+ and h^+ replaced by K^- and h^-, yield, for $\beta = .01$, $K^- = (\lambda_1 - \lambda_0)/\log(\lambda_1/\lambda_0) = 1.82$ and $h^- = -[\log(0.01)/\log(1/3)] = -4.19$.

F I G U R E   **11.7**

Number of yearly coal mine disasters in England

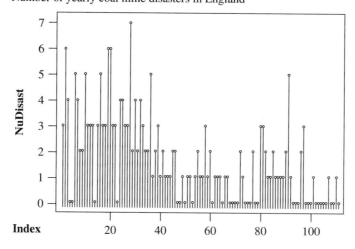

In Table 11.9, we find the values of X_t, $X_t - K^-$, and S_t^- for $t = 1, \ldots, 50$. We see that $S_t^- < h^-$ for the first time at $t = 47$. The graph of S_t^- versus t is plotted in Figure 11.8.

If we wish to control simultaneously against changes in the process level in either the upward or downward direction, we use upper and lower Page's schemes together, which triggers an alarm as soon as either $S_t^+ > h^+$ or $S_t^- < h^-$. Such a two-sided scheme is denoted by the four control parameters (K^+, h^+, K^-, h^-).   ■

EXAMPLE **11.6**   Yashchin (1991a) illustrates the use of a two-sided Page's control scheme on data, that represent the difference between the thickness of a grown silicon layer and its

TABLE   11.9
Lower Page's control scheme for mine disaster data

t	X_t	$X_t - K^-$	S_t^-	t	X_t	$X_t - K^-$	S_t^-
1	3	1.179	0	26	3	1.179	0
2	6	4.179	0	27	3	1.179	0
3	4	2.179	0	28	7	5.179	0
4	0	−1.820	−1.820	29	2	0.179	0
5	0	−1.820	−3.640	30	4	2.179	0
6	5	3.179	−0.461	31	2	0.179	0
7	4	2.179	0	32	4	2.179	0
8	2	0.179	0	33	3	1.179	0
9	2	0.179	0	34	2	0.179	0
10	5	3.179	0	35	2	0.179	0
11	3	1.179	0	36	5	3.179	0
12	3	1.179	0	37	1	−0.820	−0.820
13	3	1.179	0	38	2	0.179	−0.640
14	0	−1.820	−1.820	39	3	1.179	0
15	3	1.179	−0.640	40	1	−0.820	−0.820
16	5	3.179	0	41	2	0.179	−0.640
17	3	1.179	0	42	1	−0.820	−1.461
18	3	1.179	0	43	1	−0.820	−2.281
19	6	4.179	0	44	1	−0.820	−3.102
20	6	4.179	0	45	2	0.179	−2.922
21	3	1.179	0	46	2	0.179	−2.743
22	3	1.179	0	47	0	−1.820	−4.563
23	0	−1.820	−1.820	48	0	−1.820	−6.384
24	4	2.179	0	49	1	−0.820	−7.204
25	4	2.179	0	50	0	−1.820	−9.025

FIGURE   11.8
Lower Page's CUSUM control chart

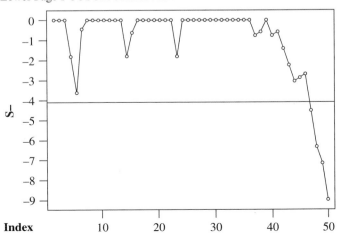

target value. The units of the measurements are in $\overset{\circ}{A}$. He applies the control scheme $(K^+ = 3, h^+ = 9, K^- = -2, h^- = -5)$. The values of X_t, S_t^+, and S_t^- are given in Table 11.10. Note that $S_t^+ > h^+$ for the first time at $t = 40$. This is an indication that a significant drift upward occurred in the level of thickness.

TABLE 11.10

Computation of (S_t^+, S_t^-) in a two-sided control scheme

t	X_t	$X_t - K^+$	$X_t - K^-$	S_t^+	S_t^-
1	-4	-7	-2	0	-2
2	-1	-4	1	0	-1
3	3	0	5	0	0
4	-2	-5	0	0	0
5	-2.5	-5.5	-0.5	0	-0.5
6	-0.5	-3.5	1.5	0	0
7	1.5	-1.5	3.5	0	0
8	-3	-6	-1	0	-1
9	4	1	6	1	0
10	3.5	0.5	5.5	1.5	0
11	-2.5	-5.5	-0.5	0	-0.5
12	-3	-6	-1	0	-1.5
13	-3	-6	-1	0	-2.5
14	-0.5	-3.5	1.5	0	-1
15	-2.5	-5.5	-0.5	0	-1.5
16	1	-2	3	0	0
17	-1	-4	1	0	0
18	-3	-6	-1	0	-1
19	1	-2	3	0	0
20	4.5	-2	6.5	1.5	0
21	-3.5	-6.5	-1.5	0	-1.5
22	-3	-6	-1	0	-2.5
23	-1	-4	1	0	-1.5
24	4	1	6	1	0
25	-0.5	-3.5	1.5	0	0
26	-2.5	-5.5	-0.5	0	-0.5
27	4	1	6	1	0
28	-2	-5	0	0	0
29	-3	-6	-1	0	-1
30	-1.5	-4.5	0.5	0	-0.5
31	4	1	6	1	0
32	2.5	-0.5	4.5	0.5	0
33	-0.5	-3.5	1.5	0	0
34	7	4	9	4	0
35	5	2	7	6	0
36	4	1	6	7	0
37	4.5	1.5	6.5	8.5	0
38	2.5	-0.5	4.5	8	0
39	2.5	-0.5	4.5	7.5	0
40	5	2	3	9.5	0

Figure 11.9 shows the two-sided control chart for the data of Table 11.10, as obtained from MINITAB.   ■

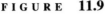

FIGURE 11.9
CUSUM two-sided control chart for thickness difference, control parameters ($K^+ = 3, h^+ = 9, K^- = -3, h^- = -9$)

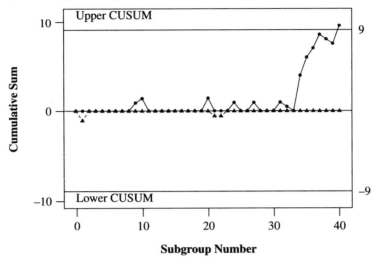

The two-sided Page's control scheme can be boosted by changing the values of S_0^+ and S_0^- to nonzero. These are called *headstart values*. The introduction of nonzero headstarts was suggested by Lucas and Crosier (1982) in order to bring the history of the process into consideration and accelerate the initial response of the scheme. Lucas (1982) also suggested combining the CUSUM scheme with the Shewhart control chart. If any X_t value exceeds an upper limit UCL, or falls below a lower limit LCL, an alarm should be triggered. A more advanced theory for the evaluation of CUSUM with headstarts is given in Yaschin (1985b).

11.5.4 Average Run Length, Probability of False Alarm, and Conditional Expected Delay

The run length (RL) is defined as the number of time units until either $S_t^+ > h_t^+$ or $S_t^- < h^-$ for the first time. We have seen already that the average run length (ARL) is an important characteristic of a control procedure when there is either no change in the mean level (ARL(0)) or the mean level shifted to $\mu_1 = \mu_0 + \delta\sigma$ before the control procedure started (ARL(δ)). When the shift from μ_0 to μ_1 occurs at some change point τ, $\tau > 0$, then we would like to know the probability of false alarm—that is, the probability that the run length is smaller than τ—and the conditional expected run length given that RL > τ. If RL < τ we say that a false alarm occurred. The probability of this is denoted by PFA. The conditional expected delay is $E\{RL - \tau \mid RL > \tau\}$ This is denoted by CED. It is difficult to compute these characteristics of the Page control scheme analytically. The theory required for such an analysis is quite complicated (see Yashchin, 1985b). We provide computer programs that approximate these characteristics numerically by simulation (see subdirectory \CHA11\).

Programs CUSARLN.EXE, CUSARLB.EXE, and CUSARLP.EXE compute the ARL function for normal, binomial, and Poisson distributions, respectively. Programs CUSPERN.EXE, CUSPERB.EXE, and CUSPERP.EXE compute the **probability of false alarm (PFA)** and the **conditional expected delay (CED)** for the normal, binomial, and Poisson distributions.

Table 11.11 lists estimates of the $ARL(\delta)$ for the normal distribution, as obtained from program CUSARLN.EXE, with $nr = 100$ runs (SE = standard deviation(RL)/$\sqrt{nr}$). Notice that these simulation programs yield the estimate of $ARL(0) \pm 2^*SE\{RL\}$.

TABLE 11.11

$ARL(\delta)$ estimates for the normal distribution, $\mu = \delta$, $\sigma = 1$, NR $= 100$ ($K^+ = 1$, $h^+ = 3$, $K^- = -1$, $h^- = -3$)

δ	ARL	2*SE
0	1225.0	230.875
0.5	108.0	22.460
1.0	18.7	3.393
1.5	7.1	0.748

FIGURE 11.10

Histogram of RL for $\mu = 10$, $\sigma = 5$, $K^+ = 12$, $h^+ = 29$, $K^- = 8$, $h^- = 29$

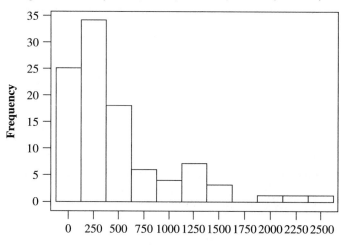

ARL (0)

Figure 11.10 is a histogram of the run lengths corresponding to the two-sided control scheme ($K^+ = 1$, $h^+ = 3$, $K^- = -1$, $h^- = -3$) in the normal case with $\mu = 0$ and $\sigma = 1$. The distribution of RL is very skewed.

Program CUSARLN.EXE can also be used to determine the values of the control parameters h^+ and h^- so that a certain ARL(0) is attained. For example, if we use the Shewhart 3-σ control charts for the sample means in the normal case, the probability that, under no shift in the process level, a point will fall outside the control limits is .0026 and ARL(0) = 385. Suppose we wish to devise a two-sided CUSUM control scheme when $\mu_0 = 10$, $\sigma = 5$, $\mu_1^+ = 14$, $\mu_1^- = 6$. Here μ_1^+ and μ_1^- designate the upper or lower alternatives to μ_0. We obtain $K^+ = 12$ and $K^- = 8$. If we take $\alpha = .01$, we obtain $h^+ = [-25 \times \log(0.01)]/4 = 28.78$. For the parameters $\mu = 10$, $\sigma = 5$, $K^+ = 12$, $h^+ = 29$, $K^- = 8$, and $h^- = -29$, program CUSARLN.EXE yields the estimate ARL(0) $= 464 \pm 99.3$. If we use $\alpha = .05$, we obtain $h^+ = 18.72$. Under the control parameters (12, 18.7, 8, −18.7), we obtain ARL(0) $= 67.86 \pm 13.373$. We can now run the program for several $h^+ = -h^-$ values to obtain an ARL(0) estimate close to 385 (see Table 11.12). Thus, $h^+ = 29$ would yield a control scheme having an ARL(0) close to that of a Shewhart 3-σ scheme.

TABLE 11.12

ARL(0) estimates for $\mu = 10$, $\sigma = 5$, $K^+ = 12$, $K^- = 8$, $h^+ = -h^-$

h^+	18.72	25	27.5	30
ARL(0)	67.86 ± 13.37	186 ± 35.22	319 ± 65.16	412.96 ± 74.65

Program CUSARLB.EXE computes the estimates of the ARL(δ) for the binomial distribution. To illustrate, consider the case of the binomial distribution $B(n, \theta)$ with $n = 100$ and $\theta = .05$. A two-sided Page's control scheme, protecting against a shift above, $\theta_1^+ = .07$, or below, $\theta_1^- = .03$, can use the control parameters $K^+ = 5.95$, $h^+ = 12.87$, $K^- = 3.92$, $h^- = -8.66$. This program yields, for nr $= 100$ runs, the estimate ARL(0) $= 371.2 \pm 63.884$. Furthermore, for $\delta = \theta_1/\theta_0$, we obtain for the same control scheme:

$$\text{ARL}\left(\frac{.06}{.05}\right) = 40.5 \pm 7.676 \qquad \text{ARL}\left(\frac{.07}{.05}\right) = 11.4 \pm 1.303$$

Similarly, program CUSARLP.EXE can be used to estimate the ARL(δ) in the Poisson case. For example, suppose that X_t has a Poisson distribution with mean $\lambda_0 = 10$. We wish to control the process against shifts in λ greater than $\lambda_1^+ = 15$ or smaller than $\lambda_1^- = 7$. Using the control parameters $K^+ = 12.33$, $h^+ = 11.36$, $K^- = 8.41$, and $h^- = -12.91$, we obtain the estimate ARL(0) $= 284.2 \pm 54.648$.

We can use program CUSPERN.EXE to estimate the probability of false alarm (PFA) and the conditional expected delay (CED) if a change in the mean of magnitude $\delta\sigma$ occurs at time τ. Table 11.13 gives some estimates obtained from this program. Programs CUSPERB.EXE and CUSPERP.EXE perform similar simulations for binomial and Poisson variables, respectively.

TABLE 11.13

Estimates of PFA and CED, normal distribution $\mu_0 = 0$, $\sigma = 1$, control parameters ($K^+ = 1$, $h^+ = 3$, $K^- = -1$, $h^- = -3$), $\tau = 100$, nr $= 500$

δ	PFA	CED
0.5	0.07	15.5 ± 19.82
1	0.07	16.23 ± 10.87
1.5	0.06	6.57 ± 9.86

11.6
Bayesian Detection

The Bayesian approach to the problem of detecting changes in distributions can be described in the following terms. Suppose we decide to monitor the stability of a process with a statistic T having a distribution with p.d.f. $f_T(t; \theta)$, where θ designates the parameters on which the distribution depends (process mean, variance, and so on). The statistic T could be the mean, $\overline{X}$, of a random sample of size n; the sample standard deviation, S; or the proportion of defectives in the sample. A sample of size n is drawn from the process at predetermined epochs. Let T_i ($i = 1, 2, \ldots$) denote the monitoring statistic at the ith time points. Suppose that m such samples are drawn and the statistics $T_1, T_2, \ldots, T_m$ are independent. Let $\tau = 0, 1, 2, \ldots$ denote the location of the point of change in the process parameter θ_0 to $\theta_1 = \theta_0 + \Delta$. τ is called the *change-point of* θ_0. The event $\{\tau = 0\}$ signifies that all the m samples have been drawn after the change point. The event $\{\tau = i\}$, for $i = 1, \ldots, m - 1$, signifies that the change-point occurred between the ith and $(i + 1)$st sampling times. Finally, the event $\{\tau = m^+\}$ signifies that the change-point has not occurred before the first m sampling times.

Given $T_1, \ldots, T_m$, the likelihood function of τ for specified values of θ_0 and θ_1 is defined as

$$
L_m(\tau; T_1, \ldots, T_m) = \begin{cases} \displaystyle\prod_{i=1}^{m} f(T_i; \theta_1) & \tau = 0 \\[2ex] \displaystyle\prod_{i=1}^{\tau} f(T_i; \theta_0) \prod_{j=\tau+1}^{m} f(T_j; \theta_1) & 1 \le \tau \le m - 1 \\[2ex] \displaystyle\prod_{i=1}^{m} f(T_i; \theta_0) & \tau = m^+ \end{cases} \quad \textbf{(11.6.1)}
$$

A maximum likelihood estimator of τ, given $T_1, \ldots, T_m$, is the argument maximizing $L_m(\tau; T_1, \ldots, T_m)$.

In the Bayesian framework, the statistician gives various possible values of τ nonnegative weights that reflect his belief as to where the change-point could occur. High weight expresses higher confidence. To standardize the approach, we will assume that the sum of all weights is 1, and we call these weights the *prior prob-abilities* of τ. Let $\pi(\tau)$ $(\tau = 0, 1, 2, \ldots)$ denote the prior probabilities of τ. If the occurrence of the change-point is a realization of some random process, the following modified-geometric prior distribution can be used:

$$
\pi_m(\tau) = \begin{cases} \pi & \text{if } \tau = 0 \\ (1 - \pi)p(1 - p)^{i-1} & \text{if } \tau = i \ (i = 1, \ldots, m - 1) \\ (1 - \pi)(1 - p)^{m-1} & \text{if } \tau = m^+ \end{cases} \tag{11.6.2}
$$

where $0 < \pi < 1$, $0 < p < 1$ are prior parameters. Applying Bayes' formula, we convert the prior probabilities after observing $T_1, \ldots, T_m$, to *posterior probabilities*. Let π_m denote the posterior probability of the event $\{\tau \leq m\}$ given $T_1, \ldots, T_m$. Using the above modified-geometric prior distribution, and employing Bayes' theorem, we obtain the formula

$$
\pi_m = \tag{11.6.3}
$$

$$
\frac{\dfrac{\pi}{(1 - \pi)(1 - p)^{m-1}} \prod_{j=1}^{m} R_j + \dfrac{p}{(1 - p)^{m-1}} \sum_{i=1}^{m-1} (1 - p)^{i-1} \prod_{j=i+1}^{m} R_j}{\dfrac{\pi}{(1 - \pi)(1 - p)^{m-1}} \prod_{j=1}^{m} R_j + \dfrac{p}{(1 - p)^{m-1}} \sum_{i=1}^{m-1} (1 - p)^{i-1} \prod_{j=i+1}^{m} R_j + 1}
$$

where

$$
R_j = \frac{f(T_j; \boldsymbol{\theta}_1)}{f(T_j; \boldsymbol{\theta}_0)} \qquad j = 1, 2, \ldots \tag{11.6.4}
$$

The **Bayesian detection** of a change-point is a procedure that detects a change as soon as $\pi_m \geq \pi^*$, where π^* is a value in $(0, 1)$ close to 1.

The preceding procedure can be simplified. If we believe that the monitoring starts when $\boldsymbol{\theta} = \boldsymbol{\theta}_0$ (that is, $\pi = 0$) and p is very small, we can then represent π_m, approximately, by

$$
\tilde{\pi}_m = \frac{\displaystyle\sum_{i=1}^{m-1} \prod_{j=i+1}^{m} R_j}{\displaystyle\sum_{i=1}^{m-1} \prod_{j=i+1}^{m} R_j + 1} \tag{11.6.5}
$$

The statistic

$$
W_m = \sum_{i=1}^{m-1} \prod_{j=i+1}^{m} R_j \tag{11.6.6}
$$

is called the **Shiryayev-Roberts (SR) statistic**. Notice that $\tilde{\pi}_m \geq \pi^*$ if $W_m \geq \pi^*/(1 - \pi^*)$. The term $\pi^*/(1 - \pi^*)$ is called the *stopping threshold*. If the Bayes' procedure is to flag a change as soon as $\tilde{\pi}_m \geq 0.95$, for example, the procedure that flags as soon as $W_m \geq 19$ is equivalent.

We now illustrate the use of the SR statistic in the special case of monitoring the mean, θ_0, of a process. The statistic T is the sample mean, $\overline{X}_n$, based on a sample of

n observations. We will assume that $\overline{X}_n$ has a normal distribution $N(\theta_0, \sigma/\sqrt{n})$, and at the change-point, θ_0 shifts to $\theta_1 = \theta_0 + \delta\sigma$. It is straightforward to verify that the likelihood ratio is

$$R_j = \exp\left\{-\frac{n\delta^2}{2\sigma^2} + \frac{n\delta}{\sigma^2}(\overline{X}_j - \theta_0)\right\} \qquad j = 1, 2, \ldots \qquad (11.6.7)$$

Accordingly, the SR statistic is

$$W_m = \sum_{i=1}^{m-1} \exp\left\{\frac{n\delta}{\sigma^2}\sum_{j=i+1}^{m}(\overline{X}_j - \theta_0) - \frac{n\delta^2(m-i)}{2\sigma^2}\right\} \qquad (11.6.8)$$

EXAMPLE 11.7 We illustrate the procedure numerically. Suppose that $\theta_0 = 10$, $n = 5$, $\delta = 2$, $\pi^* = .95$, and $\sigma = 3$. The stopping threshold is 19. Suppose that $\tau = 10$. The values of $\overline{X}_j$ have the normal distribution $N\left(10, 3/\sqrt{5}\right)$ for $j = 1, \ldots, 10$ and $N\left(10 + \delta\sigma, 3/\sqrt{5}\right)$ for $j = 11, 12, \ldots$. In Table 11.14, we present the values of W_m.

T A B L E   11.14
Values of W_m for $\delta = 0.5(0.5)2.0$, $n = 5$, $\tau = 10$, $\sigma = 3$, $\pi^* = .95$

m	$\delta = 0.5$	$\delta = 1.0$	$\delta = 1.5$	$\delta = 2.0$
2	0.3649	0.0773	0.0361	0.0112
3	3.1106	0.1311	0.0006	0.0002
4	3.2748	0.0144	0.0562	0.0000
5	1.1788	0.0069	0.0020	0.0000
6	10.1346	0.2046	0.0000	0.0291
7	14.4176	0.0021	0.0527	0.0000
8	2.5980	0.0021	0.0015	0.0000
9	0.5953	0.6909	0.0167	0.0000
10	0.4752	0.0616	0.0007	0.0001
11	1.7219	5.6838	848.6259	1538.0943
12	2.2177	73.8345		
13	16.3432			
14	74.9618			

We see that the SR statistic detects the change point quickly if δ is large. ∎

The larger the critical level $w^* = \pi^*/(1 - \pi^*)$, the smaller the frequency of detecting the change-point before it happens (false alarm). Two characteristics of the procedure are of interest:

1   Probability of false alarm (PFA)

2 Conditional expected delay (CED), given that the alarm is given after the change-point.

Programs SHROARLN.EXE and SHROARLP.EXE estimate the ARL(0) of the procedure for the normal and Poisson cases, respectively. Programs SHROPERN.EXE and SHROPERP.EXE estimate the PFA and CED of these procedures. Table 11.15 gives simulation estimates of the PFA and CED for several values of δ. The estimates are based on 100 simulation runs.

T A B L E   11.15
Estimates of PFA and CED for $\mu_0 = 10$, $\sigma = 3$, $n = 5$, $\tau = 10$, stopping threshold = 99

	$\delta = 0.5$	$\delta = 1.0$	$\delta = 1.5$	$\delta = 2.0$
PFA	0.00	0.04	0.05	0.04
CED	21.02	7.19	4.47	3.41

If the amount of shift δ is large ($\delta > 1$), then the conditional expected delay (CED) is small. The estimates of PFA are small due to the large threshold value. Another question of interest is: What is the average run length (ARL) when there is no change in the mean? We estimated the ARL(0) for the same example of normally distributed sample means using program SHROARLN.EXE and performing 100 independent simulation runs. Table 11.16 gives the estimated values of ARL(0) as a function of the stopping threshold.

T A B L E   11.16
Average run length of Shiryayev-Roberts procedure, $\mu_0 = 10$, $\delta = 2$, $\sigma = 3$, $n = 5$

Stopping threshold	ARL(0)
19	49.37 ± 10.60
50	100.81 ± 19.06
99	224.92 ± 41.11

Thus, although the procedure based on the Shiryayev-Roberts detection is sensitive to changes, in a stable situation (no changes), it is expected to run till an alarm is given. Figure 11.11 shows a box-whiskers plot of the run length with stopping threshold of 99 when there is no change. For more details on analytic data aspects of the Shiryayev-Roberts procedure, see Kenett and Pollak (1996). See also Zacks (1991).

FIGURE **11.11**

Box and whiskers plot of 100 run lengths of the Shiryayev-Roberts procedure normal distribution, $\mu_0 = 10, \delta = 2, \sigma = 3, n = 5$, stopping threshold 99

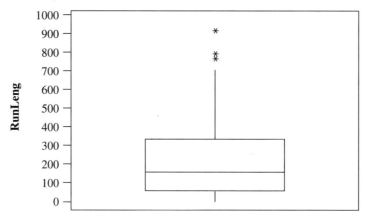

11.7

Process Tracking

Process tracking is a procedure that repeatedly estimates certain characteristics of the process being monitored. The CUSUM detection procedure, as well as that of Shiryayev-Roberts, are designed to provide a warning immediately after changes occur. However, whenever the process is stopped, these procedures do not provide direct information on the current location of the process mean (or the process variance). In the Shewhart $\bar{X}$ control chart, each point provides an estimate of the process mean at that specific time. However, the precision of these estimates is generally low, because the estimates are based on small samples. As long as there is no evidence that a change in the process mean has occurred, we may want to use an average of all previous sample means as an estimator of the current value of the process mean. Indeed, after observing m samples, each of size n, the grand average $\bar{\bar{X}}_m = 1/m(\bar{X}_1 + \cdots + \bar{X}_m)$ has the standard error $\sigma/\sqrt{nm}$, whereas the standard error of the last mean, $\bar{X}_m$, is only $\sigma/\sqrt{n}$. It is well established by statistical estimation theory that as long as the process mean μ_0 does not change, $\bar{\bar{X}}_m$ is the best (minimum variance) unbiased estimator of $\mu_m = \mu_0$. However, if μ_0 has changed to $\mu_1 = \mu_0 + \delta\sigma$, between the τ-th and the $(\tau + 1)$st sample, where $\tau < m$, then the grand mean $\bar{\bar{X}}_m$ is a biased estimator of μ_1 (the current mean). The expected value of $\bar{\bar{X}}_m$ is $1/m(\tau\mu_0 + (m - \tau)\mu_1) = \mu_1 - (\tau/m)\delta\sigma$. Thus, if the change-point, τ, is close to m, the bias of $\bar{\bar{X}}$ can be considerable. The bias of the estimator of the current mean, when $1 < \tau < m$, can be reduced by considering different types of estimators. In this chapter, we examine three procedures for tracking and monitoring the process mean: the exponentially weighted moving average (EWMA) procedure, the Bayes' estimation of the current mean (BECM), and the Kalman filter.

11.7.1 The EWMA Procedure

The **exponentially weighted moving averages (EWMAB)** chart is a control chart for the process mean that, at time t ($t = 1, 2, \ldots$) plots the statistic

$$\hat{\mu}_t = (1 - \lambda)\hat{\mu}_{t-1} + \lambda \overline{X}_t \tag{11.7.1}$$

where $0 < \lambda < 1$, and $\hat{\mu}_0 = \mu_0$ is the initial process mean. The Shewhart $\overline{X}$ control chart is the limiting case of $\lambda = 1$. Small values of λ give high weight to the past data. It is customary to use the values of $\lambda = 0.2$ or $\lambda = 0.3$.

By repeated application of the recursive formula, we obtain

$$\begin{aligned}
\hat{\mu}_t &= (1 - \lambda)^2 \hat{\mu}_{t-2} + \lambda(1 - \lambda)\overline{X}_{t-1} + \lambda \overline{X}_t \\
&= \cdots \\
&= (1 - \lambda)^t \mu_0 + \lambda \sum_{i=1}^{t} (1 - \lambda)^{t-i} \overline{X}_i
\end{aligned} \tag{11.7.2}$$

In this formula, $\hat{\mu}_t$ is a weighted average of the first t means $\overline{X}_1, \ldots, \overline{X}_t$ and μ_0, with weights that decrease geometrically as $t - i$ grows.

Let τ denote the epoch of change from μ_0 to $\mu_1 = \mu_0 + \delta\sigma$. As in the previous section, $\{\tau = i\}$ implies that

$$E\{\overline{X}_j\} = \begin{cases} \mu_0 & \text{for } j = 1, \ldots, i \\ \mu_1 & \text{for } j = i+1, i+2, \ldots \end{cases} \tag{11.7.3}$$

Accordingly, the expected value of the statistic $\hat{\mu}_t$ (an estimator of the current mean μ_t) is

$$E\{\hat{\mu}_t\} = \begin{cases} \mu_0 & \text{if } t \leq \tau \\ \mu_1 - \delta\sigma(1 - \lambda)^{t-\tau} & \text{if } t > \tau \end{cases} \tag{11.7.4}$$

The bias of $\hat{\mu}_t$, $-\delta\sigma(1 - \lambda)^{t-\tau}$ decreases to zero geometrically fast as t grows above τ. This is a faster decrease in bias than that of the grand mean, $\overline{\overline{X}}_t$, which was discussed earlier.

The variance of $\hat{\mu}_t$ can be easily determined because $\overline{X}_1, \overline{X}_2, \ldots, \overline{X}_t$ are independent and $\text{var}\{\overline{X}_j\} = \sigma^2/n, j = 1, 2, \ldots$. Hence,

$$\begin{aligned}
\text{var}\{\hat{\mu}_t\} &= \frac{\sigma^2}{n}\lambda^2 \sum_{i=1}^{t}(1 - \lambda)^{2(t-i)} \\
&= \frac{\sigma^2}{n}\lambda^2 \frac{1 - (1 - \lambda)^{2t}}{1 - (1 - \lambda)^2}
\end{aligned} \tag{11.7.5}$$

This variance converges to

$$\text{avar}\{\hat{\mu}_t\} = \frac{\sigma^2}{n}\frac{\lambda}{2 - \lambda} \tag{11.7.6}$$

as $t \to \infty$.

An EWMA control chart for monitoring shifts in the mean is constructed as follows. Starting at $\hat{\mu}_0 = \mu_0$, the points $(t, \hat{\mu}_t)$, $t = 1, 2, \ldots$ are plotted. As soon as these points cross either one of the control limits,

$$CL = \mu_0 \pm L\frac{\sigma}{\sqrt{n}}\sqrt{\frac{\lambda}{2 - \lambda}} \qquad\qquad \textbf{(11.7.7)}$$

an alarm is given that the process mean has shifted.

Figure 11.12 shows an EWMA chart with $\mu_0 = 10$, $\sigma = 3$, $n = 5$, $\lambda = 0.2$, and $L = 2$. The values of $\hat{\mu}_t$ indicate that a shift in the mean took place after the eleventh sampling epoch. An alarm for change is given after the fourteenth sample.

As in the previous sections, we characterize the efficacy of the EWMA chart in terms of PFA and CED when a shift occurs, and in terms of the ARL when there is no shift. Table 11.17 gives estimates of PFA and CED based on 1000 simulation runs. The simulations were from normal distributions, with $\mu_0 = 10$, $\sigma = 3$, $n = 5$. The change-point was at $\tau = 10$. The shift was from μ_0 to $\mu_1 = \mu_0 + \delta\sigma$. The estimates $\hat{\mu}_t$ were determined with $\lambda = 0.2$. We see in this table that if we construct control limits with the value of $L = 3$, then the PFA is very small and the CED is not large.

The estimated ARL values for this example are

L	2	2.5	3.0
ARL	48.7	151.36	660.9

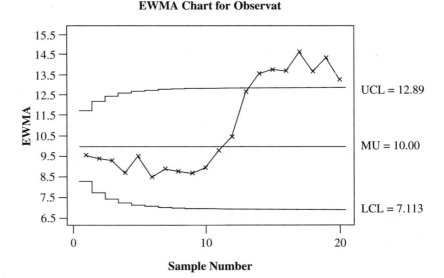

FIGURE 11.12
EWMA chart, $\mu_0 = 10$, $\sigma = 3$, $\delta = 0$, $n = 5$, $\lambda = 0.2$, and $L = 2$

TABLE **11.17**
Simulation estimates of PFA and CED of an EWMA chart

		CED			
L	PFA	$\delta = 0.5$	$\delta = 1.0$	$\delta = 1.5$	$\delta = 2.0$
2	0.168	3.93	2.21	1.20	1.00
2.5	0.043	4.35	2.67	1.41	1.03
3	0.002	4.13	3.36	1.63	1.06

11.7.2 The BECM Procedure

In this section, we examine a Bayesian procedure for estimating the current mean μ_t $(t = 1, 2, \ldots)$ called *Bayes' estimation of the current mean (BECM)*. Let $\overline{X}_1, \overline{X}_2, \ldots, \overline{X}_t, t = 1, 2, \ldots$ be means of samples of size n. The distribution of $\overline{X}_i$ is $N(\mu_i, \sigma/\sqrt{n})$, where σ is the process standard deviation. To simplify the exposition, we will assume that σ is known and fixed throughout all sampling epochs. In actual cases, we have to monitor also whether σ changes with time.

If the process mean stays stable throughout the sampling periods, then

$$\mu_1 = \mu_2 = \cdots = \mu_t = \mu_0$$

Let us consider this case first and present the Bayes' estimator of μ_0. In the Bayesian approach, the model assumes that μ_0 itself is random with some prior distribution. If we assume that the prior distribution of μ_0 is normal—say, $N(\mu^*, \tau)$—then using Bayes' theorem we can show that the posterior distribution of μ_0, given the t sample means, is normal with mean

$$\hat{\mu}_{B,t} = \left(1 - \frac{nt\tau^2}{\sigma^2 + nt\tau^2}\right)\mu^* + \frac{nt\tau^2}{\sigma^2 + nt\tau^2}\overline{\overline{X}}_t \tag{11.7.8}$$

and variance

$$w_t^2 = \tau^2\left(1 - \frac{nt\tau^2}{\sigma^2 + nt\tau^2}\right) \tag{11.7.9}$$

where $\overline{\overline{X}}_t = 1/t\sum_{i=1}^{t}\overline{X}_i$. The mean $\hat{\mu}_{B,t}$ of the posterior distribution is commonly taken as the Bayes' estimator of μ_0 (see Chapter 6).

It is interesting to notice that $\hat{\mu}_{B,t}, t = 1, 2, \ldots$ can be determined recursively by the formula

$$\hat{\mu}_{B,t} = \left(1 - \frac{nw_{t-1}^2}{\sigma^2 + nw_{t-1}^2}\right)\hat{\mu}_{B,t-1} + \frac{nw_{t-1}^2}{\sigma^2 + nw_{t-1}^2}\overline{X}_t \tag{11.7.10}$$

where $\hat{\mu}_{B,0} = \mu^*$, $w_0^2 = \tau^2$, and

$$w_t^2 = \frac{\sigma^2 w_{t-1}^2}{\sigma^2 + n w_{t-1}^2} \qquad\qquad (11.7.11)$$

This recursive formula resembles that of the EWMA estimator. The difference here is that the weight λ is a function of time—that is,

$$\lambda_t = \frac{n w_{t-1}^2}{\sigma^2 + n w_{t-1}^2} \qquad\qquad (11.7.12)$$

From this recursive formula for w_t^2, we find that $w_t^2 = \lambda_t(\sigma^2/n)$, or $\lambda_t = \lambda_{t-1}/(1 + \lambda_{t-1})$, $t = 2, 3, \ldots$, where $\lambda_1 = n\tau^2/(\sigma^2 + n\tau^2)$. The procedures become more complicated if change-points are introduced.

11.7.3　The Kalman Filter

In this section, we discuss a model of dynamic changes in the observed sequence of random variables and a Bayesian estimator of the current mean called the **Kalman filter**.

At time t, let Y_t denote an observable random variable with mean μ_t. We assume that μ_t may change at random from one time epoch to another, according to the model

$$\mu_t = \mu_{t-1} + \Delta_t \qquad t = 1, 2, \ldots$$

where Δ_t, $t = 1, 2, \ldots$ is a sequence of i.i.d. random variables having a normal distribution $N(\delta, \sigma_2)$. Furthermore, we assume that μ_0 is distributed like $N(\mu_0^*, w_0)$, and the observation equation is

$$Y_t = \mu_t + \epsilon_t \qquad t = 1, 2, \ldots$$

where ϵ_t are i.i.d., $N(0, \sigma_\epsilon)$. According to this dynamic model, the mean at time t (the current mean) is normally distributed with mean

$$\hat{\mu}_t = B_t(\hat{\mu}_{t-1} + \delta) + (1 - B_t)Y_t \qquad\qquad (11.7.13)$$

where

$$B_t = \frac{\sigma_e^2}{\sigma_\epsilon^2 + \sigma_2^2 + w_{t-1}^2} \qquad\qquad (11.7.14)$$

$$w_t^2 = B_t(\sigma_2^2 + w_{t-1}^2) \qquad\qquad (11.7.15)$$

The posterior variance of μ_t is w_t^2. The term $\hat{\mu}_t$ is the Kalman filter.

If the prior parameters σ_ϵ^2, σ_2^2, and δ are unknown, we could use a small number of data to estimate these parameters.

According to the dynamic model, we can write

$$y_t = \mu_0 + \delta t + \epsilon_t^* \qquad t = 1, 2, \ldots$$

where $\epsilon_t^* = \sum_{i=1}^{t}[(\Delta_i - \delta) + \epsilon_i]$. Notice that $E\{\epsilon_t^*\} = 0$ for all t and $V\{\epsilon_t^*\} = t(\sigma_2^2 + \sigma_\epsilon^2)$. Let $U_t = y_t/\sqrt{t}, t = 1, 2, \ldots$; then we can write the regression model

$$U_t = \mu_0 x_{1t} + \delta x_{2t} + \eta_t \qquad t = 1, 2, \ldots$$

where $x_{1t} = 1/\sqrt{t}$, and $x_{2t} = \sqrt{t}$ and $\eta_t, t = 1, 2\ldots$ are independent random variables, with $E\{\eta_t\} = 0$ and $V\{\eta_t\} = (\sigma_2^2 + \sigma_\epsilon^2)$.

Using the first m points of (t, y_t) and fitting, by the method of least squares (see Chapter 3), the regression equation of U_t against (x_{1t}, x_{2t}), we obtain estimates of μ_0, δ, and $(\sigma_2^2 + \sigma_\epsilon^2)$. Estimates of σ_e^2 can be obtained, if y_t are group means, by estimating within-group variance; otherwise, we assume a value for σ_ϵ^2 smaller than the least-squares estimate of $\sigma_2^2 + \sigma_\epsilon^2$. We illustrate this now by example.

EXAMPLE 11.8   Figure 11.13 shows the Dow-Jones financial index for the 300 business days of 1935 (file DOJO1935.DAT). The Kalman filter estimates of the current means are plotted in this figure as well. These estimates were determined by the formula

$$\hat{\mu}_t = B_t(\hat{\mu}_{t-1} + \delta) + (1 - B_t)y_t \tag{11.7.16}$$

where the prior parameters were computed as just suggested, on the basis of the first $m = 20$ data points. The least-squares estimates of μ_0, δ, and $\sigma_2^2 + \sigma_\epsilon^2$ are, respectively, $\hat{\mu}_0 = 127.484$, $\hat{\delta} = 0.656$, and $\hat{\sigma}_2^2 + \hat{\sigma}_\epsilon^2 = 0.0731$. For $\hat{\sigma}_\epsilon^2$, we have chosen the value 0.0597 and for w_0^2, the value 0.0015. The first 50 values of the data, y_t, and the estimate $\hat{\mu}_t$ are given in Table 11.18.   ■

FIGURE 11.13

The daily Dow-Jones financial index for 1935

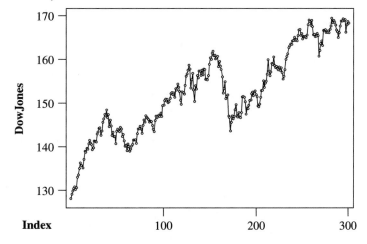

TABLE 11.18

The Dow-Jones index for the first 50 days of 1935, and the Kalman filter estimates

t	y_t	$\hat{\mu}_t$	t	y_t	$\hat{\mu}_t$
1	128.06	128.0875	26	141.31	141.5413
2	129.05	128.8869	27	141.2	141.8236
3	129.76	129.6317	28	141.07	141.9515
4	130.35	130.3119	29	142.9	142.7171
5	130.77	130.8927	30	143.4	143.3831
6	130.06	130.9880	31	144.25	144.1181
7	130.59	131.2483	32	144.36	144.6189
8	132.99	132.3113	33	142.56	144.2578
9	133.56	133.1894	34	143.59	144.4178
10	135.03	134.2893	35	145.59	145.2672
11	136.26	135.4378	36	146.32	146.0718
12	135.68	135.9388	37	147.31	146.9459
13	135.57	136.2108	38	147.06	147.3989
14	135.13	136.2161	39	148.44	148.1991
15	137.09	136.9537	40	146.65	148.0290
16	138.96	138.1156	41	147.37	148.1923
17	138.77	138.7710	42	144.61	147.2604
18	139.58	139.4843	43	146.12	147.2434
19	139.42	139.8704	44	144.72	146.7082
20	140.68	140.5839	45	142.59	145.5755
21	141.47	141.3261	46	143.38	145.1632
22	140.78	141.5317	47	142.34	144.5157
23	140.49	141.5516	48	142.35	144.1145
24	139.35	141.1370	49	140.72	143.2530
25	139.74	141.0238	50	143.58	143.7857

11.7.4 Hoadley's QMP

Hoadley (1981, 1986, 1988) introduced at Bell Laboratories a **quality measurement plan (QMP)** that uses Bayesian methods of estimating the current mean of a process. QMPs provide reporting capabilities for large data sets and, in a certain sense, are an improvement over the Shewhart 3-sigma control chart. These plans were implemented throughout Western Electric Co. in the late 1980s. The main idea is that the process mean does not remain at a constant level but changes at random every time period according to some distribution.

This framework is similar to that of the Kalman filter, but it was developed for observations X_t having Poisson distributions with means λ_t ($t = 1, 2, \ldots$), and where $\lambda_1, \lambda_2, \ldots$ are independent random variables having a common gamma distribution $G(\nu, \Lambda)$. The parameters ν and Λ are unknown and are estimated from the data. At the end of each period, a box-whiskers plot is put on a chart. The center line of the box-whiskers plot represents the posterior mean of λ_t, given past observations. The lower and upper sides of the box represent the .05th and .95th quantiles of the posterior distribution of λ_t. The lower whisker starts at the .01th quantile of the posterior distribution and the upper whisker ends at the .99th quantile of that distribution. These box-whiskers plots are compared to a desired quality level.

We saw in Section 6.8.3 that if X_t has a Poisson distribution $P(\lambda_t)$, and λ_t has a gamma distribution $G(\nu, \Lambda)$, then the posterior distribution of λ_t, given X_t, is the gamma distribution $G(\nu + X_t, \Lambda/(1 + \Lambda))$. Thus, the Bayes' estimate of λ_t, for a squared error loss, is the posterior expectation

$$\hat{\lambda}_t = (\nu + X_t)\frac{\Lambda}{1 + \Lambda} \tag{11.7.17}$$

Similarly, the pth quantile of the posterior distribution is

$$\lambda_{t,p} = \frac{\Lambda}{1 + \Lambda}G_p(\nu + X_t, 1) \tag{11.7.18}$$

where $G_p(\nu + X_t, 1)$ is the pth quantile of the standard gamma distribution $G(\nu + X_t, 1)$. If ν is an integer, then

$$G_p(\nu + X_t, 1) = \frac{1}{2}\chi_p^2[2(\nu + X_t)] \tag{11.7.19}$$

We assumed that $\lambda_1, \lambda_2, \ldots$ are independent and identically distributed. This implies that $X_1, X_2, \ldots$ are independent and have the same negative-binomial (NB) predictive distribution, with predictive expectation

$$E\{X_t\} = \nu\Lambda \tag{11.7.20}$$

and predictive variance

$$V\{X_t\} = \nu\Lambda(1 + \Lambda) \tag{11.7.21}$$

We therefore can estimate the prior parameters ν and Λ by the consistent estimators

$$\hat{\Lambda}_T = \left(\frac{S_T^2}{\overline{X}_T} - 1\right)^+ = \max\left(0, \frac{S_T^2}{\overline{X}_T} - 1\right) \tag{11.7.22}$$

and

$$\hat{\nu}_T = \frac{\overline{X}_T}{\hat{\Lambda}_T} \tag{11.7.23}$$

where $\overline{X}_T$ and S_T^2 are the sample mean and sample variance, respectively, of $X_1, X_2, \ldots, X_T$. For determining $\hat{\lambda}_t$ and $\lambda_{t,p}$, we can substitute $\hat{\Lambda}_T$ and $\hat{\nu}_T$ in these equations, with $T = t - 1$. We illustrate this estimation method, called *parametric empirical Bayes' method*, in the following example.

EXAMPLE **11.9** File SOLDEF.DAT gives the results of testing batches of circuit boards for defects in solder points after wave solderings. The batches include boards of similar design. There are close to 1000 solder points on each board. The results X_t are number of defects in parts per million (ppm) points. The quality standard is $\lambda^0 = 100$ (ppm). The λ_t values below λ^0 represent high-quality soldering. In this data file, there are $N = 380$ test results. Only 78 batches had an X_t value greater than $\lambda^0 = 100$.

If we take UCL $= \lambda^0 + 3\sqrt{\lambda^0} = 130$, we see that only 56 batches had X_t values greater than the UCL. All runs of consecutive X_t values greater than 130 are of length not greater than 3. We conclude, therefore, that the occurrence of low-quality batches is sporadic, caused by common causes. These batches are excluded from the analysis. Table 11.19 gives the X_t values and the associated values of $\overline{X}_{t-1}$, S^2_{t-1}, $\hat{\Lambda}_{t-1}$ and $\hat{v}_{t-1}$, with $t = 10, \ldots, 20$. The statistics $\overline{X}_{t-1}$ and so on are functions of $X_1, \ldots, X_{t-1}$.

Table 11.20 lists the values of $\hat{\lambda}_t$ and the quantiles $\lambda_{t,.01}$ for $p = .01, .05, .95,$ and .99. ■

TABLE 11.19

Number of defects (ppm) and associated statistics for the SOLDEF data

t	X_t	$\overline{X}_{t-1}$	S^2_{t-1}	$\hat{\Lambda}_{t-1}$	$\hat{v}_{t-1}$
10	29	23.66666	75.55555	2.192488	10.79443
11	16	24.20000	70.56000	1.915702	12.63244
12	31	23.45454	69.70247	1.971811	11.89492
13	19	24.08333	68.24305	1.833621	13.13429
14	18	23.69230	64.82840	1.736263	13.64556
15	20	23.28571	62.34693	1.677475	13.88140
16	103	23.06666	58.86222	1.551830	14.86416
17	31	28.06250	429.5585	14.30721	1.961423
18	33	28.23529	404.7681	13.33553	2.117296
19	12	28.50000	383.4722	12.45516	2.288207
20	46	27.63157	376.8642	12.63889	2.186233

TABLE 11.20

Empirical Bayes' estimates of λ_t and $\lambda_{t,p}$, $p = .01, .05, .95, .99$

t	$\hat{\lambda}_t$	Quantiles of posterior distributions			
		.01	.05	.95	.99
10	27.32941	18.40781	21.01985	34.45605	37.40635
11	18.81235	12.34374	14.23760	24.59571	26.98991
12	28.46099	19.59912	22.19367	35.60945	38.56879
13	20.79393	13.92388	15.93527	26.82811	29.32615
14	20.08032	13.52340	15.44311	25.95224	28.38310
15	21.22716	14.44224	16.42871	27.22614	29.70961
16	71.67607	57.21276	61.44729	82.53656	87.03260
17	30.80809	16.17445	20.45885	39.63539	43.28973
18	32.66762	17.93896	22.25118	41.73586	45.48995
19	13.22629	2.578602	5.696005	18.98222	21.36506
20	44.65323	28.14758	32.98006	55.23497	59.61562

11.8
Automatic Process Control

Certain production lines—such as those in the chemical industry, paper industry, and automobile industry—are fully automated. In such production lines, it is often possible to build in feedback and control mechanisms—**automatic process control**—so that if the process mean or standard deviation changes significantly, a correction will be made automatically via the control mechanism. If μ_t denotes the level of the process mean at time t, and u_t denotes the control level at time t, the **dynamic linear model (DLM)** of the process mean is

$$\mu_t = \mu_{t-1} + \Delta_t + bu_{t-1} \qquad t = 1, 2, \ldots \tag{11.8.1}$$

and the observations equation is, as before:

$$Y_t = \mu_t + \epsilon_t \qquad t = 1, 2, \ldots \tag{11.8.2}$$

Δ_t is a random disturbance in the process evolution. The recursive equation of the DLM is linear in the sense that the effect on μ_t of u_{t-1} is proportional to u_{t-1}. The control could be on a vector of several variables whose level at time t is given by a vector $\mathbf{u}_t$. The question is: How do we determine the levels of the control variables? This question of optimal control of systems when the true level μ_t of the process mean is not known exactly but is only estimated from the observed values of Y_t, is a subject of studies in the field of *stochastic control*. We refer the reader to the book by Aoki (1989) and to the paper by Box and Kramer (1992).

It is common practice in many industries to use the *proportional rule* for control. That is, if the process level (mean) is targeted at μ_0 and the estimated level at time t is $\hat{\mu}_t$, then

$$u_t = -p(\hat{\mu}_t - \mu_0) \tag{11.8.3}$$

where p is some factor determined by the DLM, by cost factors, and so on. This rule is not necessarily optimal. It depends on the objectives of the optimization. For example, suppose that the DLM with control is

$$\mu_t = \mu_{t-1} + bu_{t-1} + \Delta_t \qquad t = 1, 2, \ldots \tag{11.8.4}$$

where the process mean is set at μ_0 at time $t = 0$. Δ_T is a random disturbance having a normal distribution $N(\delta, \sigma)$. The process level μ_t is estimated by the Kalman filter, which was described in a previous section. We have the option adjusting the mean at each time period at a cost of $c_A u^2$ (\$). However, at the end of T periods, we pay a penalty of \$ $c_d(\mu_T - \mu_0)^2$ for the deviation of μ_T from the target level. In this example, the optimal levels of u_t, for $t = 0, \ldots, T-1$, are given by

$$u_t^0 = -\frac{bq_{t+1}}{c_A + q_{t+1}b^2}(\hat{\mu}_t - \mu_0) \tag{11.8.5}$$

where

$$q_T = c_d$$

and, for $t = 0, \ldots, T - 1$

$$q_t = \frac{c_A q_{t+1}}{c_A + q_{t+1} b^2} \qquad (11.8.6)$$

These formulas are obtained as special cases from the general result given in Aoki (1989, p. 128). Thus, we see that the values that u_t obtains under the optimal scheme are proportional to $-(\hat{\mu}_t - \mu_0)$ but with a varying factor of proportionality,

$$p_t = b q_{t+1}/(c_A + q_{t+1} b^2) \qquad (11.8.7)$$

Table 11.21 gives the optimal values of p_t for the case of $c_A = 100$, $c_d = 1000$, $b = 1$, and $T = 15$.

T A B L E   11.21

Factors of proportionality in optimal control

t	q_t	p_t
15	——	——
14	90.909	.909
13	47.619	.476
12	32.258	.323
11	24.390	.244
10	19.608	.196
9	16.393	.164
8	14.085	.141
7	12.346	.124
6	10.989	.110
5	9.901	.099
4	9.009	.090
3	8.264	.083
2	7.634	.076
1	7.092	.071

If the penalty for deviation from the target is cumulative, we want to minimize the total expected penalty function—namely,

$$J_t = c_d \sum_{t=1}^{T} E\{(\mu_t - \mu_0)^2\} + c_A \sum_{t=0}^{T-1} u_t^2 \qquad (11.8.8)$$

The optimal solution in this case is somewhat more complicated than the preceding rule, and it is also not one with a fixed factor of proportionality p. To obtain this solution, we use *dynamic programming*. This optimization procedure is not discussed here, but the interested reader is referred to Aoki (1989). We note only that the optimal solution using this method yields, for example, that the last control is at the level (when $b = 1$) of

$$u_{T-1}^0 = -\frac{c_d}{c_A + c_d}(\hat{\mu}_{T-1} - \mu_0) \qquad (11.8.9)$$

FIGURE   **11.14**

EWMA chart for average film speed in subgroups of $n = 5$ film rolls, $\mu_0 = 105$, $\sigma = 6.53$, $\lambda = 0.2$

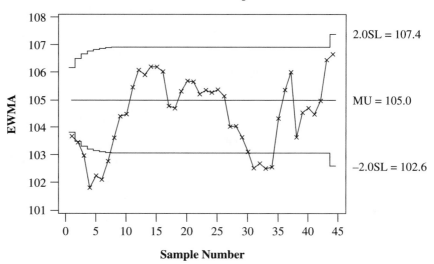

The optimal control at $t = T - 2$ is

$$u_{T-2}^0 = -\frac{cd}{c_A + 2c_d}(\hat{\mu}_{T-2} - \mu_0)$$ **(11.8.10)**

and so on.

Finally, a simple but reasonable method of automatic process control is to use the EWMA chart, and whenever the trend estimates, $\hat{\mu}_t$, are above or below the upper or lower control limits, then a control of size

$$u_t = -(\hat{\mu}_t - \mu_0)$$ **(11.8.11)**

is applied. Figure 11.14 illustrates the results of such a control procedure on film speed (file FILMSP.DAT) in a production process of coating film rolls. This EWMA chart was constructed with $\mu_0 = 105$, $\sigma = 6.53$, $\lambda = 0.2$, $L = 2$, and $n = 5$. Notice that at the beginning the process was out of control. After a remedial action, the process returned to a state of control. At time 30 it drifted downward but was corrected again.

11.9
Chapter Highlights

Tests of randomness, called run tests, are discussed. These tests are required as a first step in checking the statistical stability of a process.

Modifications of the 3-sigma control limits that include warning limits are presented as are control limits based on economic considerations.

Particular attention is given to the determination of the frequency and size of subgroups used in the process control system. The theory of cumulative sum control charts, CUSUM, is introduced and the main results are given. Special computer programs are given for estimating the probability of false alarm, conditional expected delay, and expected run length. Special sections on modern topics including Bayesian detection, process tracking, multivariate control charts, and automatic process control introduce readers to nonstandard techniques and applications.

The main concepts and definitions introduced in this chapter include:

- Run tests
- Average run length
- Operating characteristic function
- Multivariate control chart
- Cumulative sum control chart
- Probability of false alarm
- Conditional expected delay
- Bayesian detection
- Shiryayev-Roberts statistic
- Process tracking
- Exponentially weighted moving average
- Kalman filter
- Quality measurement plan
- Automatic process control
- Dynamic linear model

Exercises

11.1.1   Use program DISTRUNS.EXE to tabulate the distribution of the number of runs in a sample of size $n = 25$ if the number of elements above the sample mean is $m_2 = 10$.

a   What are Q_1, M_e, and Q_3 of this distribution?

b   Compute the expected value, μ_R, and the standard deviation, σ_R.

c   What is $\Pr\{10 \leq R \leq 16\}$?

d   Determine the normal approximation to $\Pr\{10 \leq R \leq 16\}$.

11.1.2   Use MINITAB to perform a run test on the simulated cycle times from the pistons, which are in data file CYCLT.DAT. Is the number of runs above the mean cycle time significantly different from its expected value?

11.1.3

a   What is the expected number of runs up or down in a sample of size 50?

b   Compute the number of runs up or down in the cycle time data (CYCLT.DAT).

c   Is this number significantly different than expected?

d   What is the probability that a random sample of size 50 will have at least one run of size greater or equal to 5?

11.1.4   Analyze the observations in YARNSTRG.-DATA for runs.

11.1.5   Run TURB2.EXE at the upper level of the seven control parameters and generate 50 samples of size 5. Analyze the output file OTURB2.DAT for runs (both $\overline{X}$- and S-charts).

11.1.6

a   Run TURB21.EXE at the upper level of the seven control parameters and generate 50 samples of size 5 (both $\overline{X}$ and S charts).

b Repeat the exercise with TURB22.EXE.

c Repeat the exercise with TURB23.EXE.

d Compare the results in parts a, b, and c with those of Exercise 11.1.5.

11.1.7 Construct a *p*-chart for the fraction of defective substrates received at a particular point in the production line. One thousand ($n = 1000$) substrates are sampled each week. Remove data for any week for which the process is not in control. Be sure to check for runs as well as points outside the control limits. Construct a revised *p*-chart and check for runs again.

Week	Number of defectives	Week	Number of defectives
1	18	16	38
2	14	17	29
3	9	18	35
4	25	19	24
5	27	20	20
6	18	21	23
7	21	22	17
8	16	23	20
9	18	24	19
10	24	25	17
11	20	26	16
12	19	27	10
13	22	28	8
14	22	29	10
15	20	30	9

11.1.8 Substrates were inspected weekly for defects on two different production lines. The weekly sample sizes and the number of defectives are indicated in the data set that follows. Plot these data and indicate which of the lines is not in a state of statistical control. On what basis do you make your decision? Use MINITAB to construct control charts for the two production lines. *Note*: When the sample size is not the same for each sampling period, use variable control limits. If $X(i)$ and $n(i)$ represent the number of defects and sample size, respectively, for sampling period i, then the upper and lower control limits for the *i*th period are

$$\text{UCL}_i = \bar{p} + 3(\bar{p}(1 - \bar{p})/n_i)^{1/2}$$

and

$$\text{LCL}_i = \bar{p} - 3(\bar{p}(1 - \bar{p})/n_i)^{1/2}$$

where

$$\bar{p} = \sum X(i) / \sum n(i)$$

is the center line for the control chart.

Week	Line 1 X_i	Line 1 n_i	Line 2 X_i	Line 2 n_i
1	45	7920	135	2640
2	72	6660	142	2160
3	25	6480	16	240
4	25	4500	5	120
5	33	5840	150	2760
6	35	7020	156	2640
7	42	6840	140	2760
8	35	8460	160	2980
9	50	7020	195	2880
10	55	9900	132	2160
11	26	9180	76	1560
12	22	7200	85	1680

11.1.9 In designing a control chart for the fraction of defectives *p*, a random sample of size *n* is drawn from the daily production output (very large lot). How large should *n* be so that the probability of detecting a shift from $p_0 = .01$ to $p_t = .05$ within a 5-day period will not be smaller than .8?

11.1.10 The following data represent dock-to-stock cycle times for a certain type of shipment (class *D*). Incoming shipments are classified according to their type, which is determined by the size of the item and the shipment, the type of handling required, and the destination of the shipment. Samples of five shipments per day are tracked from their initial arrival to their final destination, and the time it takes for completion of this cycle is noted. The samples are selected as follows: at five preselected times during the day, the next class *D* shipment to arrive is tagged and the arrival time and identity of the shipment are recorded. When the shipment reaches its final destination, the time is again recorded. The difference between these times is the cycle time in hours. The cycle time is always recorded for the day of arrival.

Day	\multicolumn{5}{c}{Dock-to-stock cycle times Times}

Dock-to-stock cycle times

Day			Times		
1	27	43	49	32	36
2	34	29	34	31	41
3	36	32	48	35	33
4	31	41	51	51	34
5	43	35	30	32	31
6	28	42	35	40	37
7	38	37	41	34	44
8	28	44	44	34	50
9	44	36	38	44	35
10	30	43	37	29	32
11	36	40	50	37	43
12	35	36	44	34	32
13	48	49	44	27	32
14	45	46	40	35	33
15	38	36	43	38	34
16	42	37	40	42	42
17	44	31	36	42	39
18	32	28	42	39	27
19	41	41	35	41	44
20	44	34	39	30	37
21	51	43	36	50	54
22	52	50	50	44	49
23	52	34	38	41	37
24	40	41	40	23	30
25	34	38	39	35	33

a Construct $\overline{X}$- and S-charts from the data. Are any points out of control? Are there any trends in the data? If any points are beyond the control limits, assume that we can determine special causes for them and recalculate the control limits excluding those points.

b Use a t-test to decide whether the mean cycle time for days 21 and 22 was significantly greater than 45.

c Make some conjectures about possible causes of unusually long cycle times. Might other appropriate data have been collected, such as the times at which the shipments reached intermediate points in the cycle? Why would such data be useful?

11.2.1 Consider the modified Shewhart control chart for sample means with $a = 3$, $w = 2$, and $r = 4$. What is the ARL of this procedure when $\delta = 0, 1, 2$ and the sample size is $n = 10$?

11.2.2 Repeat Exercise 11.2.1 for $a = 3$, $w = 1$, and $r = 15$ when $n = 5$ and $\delta = 0.5$.

11.3.1 Suppose that a shift in the mean is occurring at random according to an exponential distribution with mean of 1 hour. The hourly cost is $100 per shift of size $\delta = (\mu_1 - \mu_0)/\sigma$. The cost of sampling and testing is $d = \$10$ per item. How often should samples of size $n = 5$ be taken when lifts of size $\delta \geq 1.5$ should be detected?

11.3.2 Compute the OC(p) function for a Shewhart 3-sigma control chart for p based on samples of size $n = 20$ when $p_0 = 0.10$. [Use the formula for exact computations.]

11.3.3 How large should the sample size n be for a 3-sigma control chart for p if we want the probability of detecting a shift from $p_0 = .01$ to $p_t = .05$ to be $1 - \beta = .90$?

11.3.4 Suppose that a measurement X of hardness of brackets after heat treatment has a normal distribution. Every hour a sample of n units is drawn and an $\overline{X}$-chart with control limits $\mu_0 \pm 3\sigma/\sqrt{n}$ is used. Here μ_0 and σ are the assumed process mean and standard deviation. The OC function is

$$OC(\delta) = \Phi(3 - \delta\sqrt{n}) + \Phi(3 + \delta\sqrt{n}) - 1$$

where $\delta = (\mu - \mu_0)/\sigma$ is the standardized deviation of the true process mean from the assumed one.

a How many hours on average would it take to detect a shift in the process mean of size $\delta = 1$ when $n = 5$?

b What should be the smallest sample size n so that a shift in the mean of size $\delta = 1$ would be on average detected in less than 3 hours?

c Assume you have two options: to sample $n_1 = 5$ elements every hour or to sample $n_2 = 10$ elements every two hours. Which one would you choose? State your criterion for choosing between the two options and make the necessary computations.

11.4.1 In data file TSQ.DAT we find 368 T^2 values corresponding to the vectors (x, y, θ) in the PLACE.DAT file. The first $n = 48$ vectors in the PLACE.DAT file were used as a base sample to compute the vector of means **m** and the covariance matrix S. The T^2 values are for the other individual vectors ($m = 1$). Plot the T^2 values in the file TSQ.DAT. Compute the UCL and describe from the plot what might have happened in the placement process generating the (x, y, θ) values.

11.5.1 Electric circuits are designed to have an output of 220 (volts DC). If the mean output is above 222 (volts DC), you want to detect the shift as soon as possible. Examine the sample of data file OELECT.DAT and construct a CUSUM chart using MINITAB under the menu Stat\Control Charts\CUSUM. For this purpose, compute K^+ and h^+ (with $\alpha = .001$). In MINITAB, substitute K^+ for the "Target," h^+/S for "h," where S is the Std of OELECT. For "Sigma," substitute S under :Historical." For "Subgroup size," put 1. Is there an indication of a shift in the mean?

11.5.2 File OELECT1.DAT gives the output voltages of 25 electric circuits. The nominal value for these circuits is 220 volts DC. Corrective actions are required when the average output voltage increases above 222 volts or drops below 218 volts.

a Construct a CUSUM chart for these data with $\alpha = .001$, and $\sigma^2 = 25.534$.

b Use CUSARLN.EXE to estimate the average run length when the average output voltage takes the values 220, 222, 225, 235 volts, $\sigma = 9.5$ and the control parameters as in a.

c Use CUSPERN.EXE to determine the probability of false alarm (PFA) and conditional expected delay (CED) when the average output voltage takes the values 222, 225, 235 volts past the first 30 observations.

11.5.3 Estimate the average run length ARL(0) for the CUSUM scheme designed in Exercise 11.5.1.

11.5.4 Estimate the probability of false alarm and the conditional expected delay in the binomial case when a two-sided CUSUM scheme is used to detect a change in θ from $\theta_0 = .05$ to $\theta_1 = .10$ or from θ_0 to $\theta_1 = .03$, with $\alpha = .001$. The sample size is $n = 100$. [First determine K^+, h^+, K^-, and h^-, and then use program CUSPERB.EXE; with $\tau = 30$.]

11.5.5 Estimate the probability of false alarm and the conditional expected delay in the Poisson case with a two-sided CUSUM scheme. The parameters are $\lambda_0 = 15$ and $\lambda_1 = 25$ or $\lambda_1 = 7$. Use $\alpha = .001$, $\tau = 30$.

11.5.6 Construct a modified Shewhart (a, w, r)-control chart that matches as closely as possible the CUSUM of Exercise 11.5.2a.

11.5.7 A two-sided CUSUM control scheme is based on sample means.

a Determine the control parameters K^+, h^+, K^-, and h^- when $\mu_0 = 100$, $\mu_1 = 110$ or $\mu_1 = 90$, $\sigma = 20$, $n = 5$, and $\alpha = .001$.

b Estimate the PFA and CED when the change-point is at $\tau = 10, 20, 30$.

c How would the properties of the CUSUM change if each sample size were increased from 5 to 20.

11.6.1 Show that the Shiryayev-Roberts statistic W_n for detecting a shift in a Poisson distribution from a mean λ_0 to a mean $\lambda_1 = \lambda_0 + \delta$ is

$$W_m = (1 + W_{m-1})R_m$$

where $W_0 \equiv 0$, $R_m = \exp\{-\delta + X_m \log(\rho)\}$, and $\rho = \lambda_1/\lambda_0$.

11.6.2 Macro SHYROB.MTB computes the Shiryayev-Roberts statistic W_m for Poisson data. Apply this program to the data file SOLDEF.DAT, which contains $n = 380$ values. The data should have a mean value of $\lambda_0 = 6.5$ (defects/10^5 points). A change to $\lambda_1 = 10$ (defects/10^5 points) should be detected. Declare that a shift in λ occurred as soon as $W_m > 99$. (This corresponds to $\pi^* = 0.99$). When should you give an alarm signal? To run the macro (path: C:\ISTAT\CHA11), put the data in column $C1$. Execute the following MINITAB commands

```
MTB> LET C1 = C1/10
MTB> LET C2 = EXPO(-3.5) *
EXPO(LOGE(10/6.5)*C1).
MTB> LET C3(1) = C2(1)
MTB> LET k1 = 1
MTB> EXEC
'C:\ISTAT\CHA11\SHYROB.MTB' 99
```

Column $C3$ will have the values of W_m after you execute the macro 99 times.

11.6.3 Use program SHROARLP.EXE to estimate ARL(0) of the Shiryayev-Roberts procedure when the observations come from a Poisson distribution with $\lambda_0 = 6.5$. For a positive shift of $\delta = 3.5$ and stopping threshold of $W_m > 99$, draw the box and whisker plot of the run length (see file SHROARLP.DAT).

11.7.1 Analyze the data in file OELECT1.DAT with an EWMA control chart with $\lambda = 0.2$.

11.7.2 Analyze the variable diameter1 in the data file ALMPIN.DAT with an EWMA control chart with $\lambda = 0.2$. Explain how you would apply the automatic process control technique described in Section 11.8.

11.7.3 Construct the Kalman filter for the Dow-Jones daily index, which is given in the data file DOW1941.DAT.

12

Classical Design and Analysis of Experiments

Industry uses experiments to improve productivity, reduce variability, and obtain robust products and manufacturing processes. In this chapter, we study how to design and analyze experiments aimed at testing scientific or technological hypotheses. These hypotheses are concerned with the effects of procedures or treatments on the yield, the relationship among variables, the conditions under which a production process yields maximum output or other optimum result, and so on. We also look at the classical methods of experimental design.

Guiding Principles and Basic Steps

The following are guiding principles used in designing experiments.

1 The objectives of the study are clearly stated, and criteria have been established to test whether they have been met.

2 **Response variable(s)** are clearly defined.

3 All factors that might affect the response variable(s) are listed and specified. We call these **controllable factors**.

4 The type of measurements or observations on all variables is specified.

5 Levels of the controllable factors are determined.

6 A **statistical model** of the relationship between pertinent variables and their error distributions is formulated.

7 An **experimental layout** is designed so that inferences made from the gathered data are:

a	Valid	b	Precise
c	Generalizable	d	Easy to obtain

8 Trials are performed in random order, if possible, to avoid bias by outside factors.

9 A protocol of execution is prepared, as is the method of analysis. The method of analysis depends on the design.

10 The execution of the experiment carefully follows the protocol.

11 The results of the experiments are carefully analyzed and reported.

12 Confirmatory experiments are conducted to validate the inferences (conclusions) of the main experiments.

We illustrate these principles with two examples.

EXAMPLE **12.1** The first example deals with a problem of measuring the weights of objects. It illustrates an experimental layout (design) and explains why choosing an optimal design is important.

Step 1. **Formulation of objectives**
The objective is to devise a measurement plan that will yield weight estimates with maximal precision under a fixed number of four weighing operations.

Step 2. **Description of response**
The weight measurement device is a chemical balance with right and left pans. One or more objects can be put on either pan. The response variable Y is the measurement read on the scale of the chemical balance. This is equal to the total weight of objects on the right pan ($+$) minus the total weight of objects on the left pan ($-$), plus a measurement error.

Step 3. **Controllable variables**
Suppose we have four objects O_1, O_2, O_3, O_4 with unknown weights w_1, w_2, w_3, w_4. The controllable (influencing) variables are

$$
X_{ij} = \begin{cases} 1 & \text{if } j\text{th object is put on } + \text{ pan} \\ & \text{in the } i\text{th measurement} \\ \\ -1 & \text{if } j\text{th object is put on } - \text{ pan} \\ & \text{in the } i\text{th measurement} \end{cases} \qquad i, j = 1, 2, 3, 4
$$

Step 4. **Type of measurements**
The response Y is measured on a continuous scale in an interval (y^*, y^{**}). The observations are a realization of continuous random variables.

Step 5. **Levels of controllable variables**
$X_{ij} = \pm 1$, as in Step 3.

Step 6. **A Statistical Model**
The measurement model is linear—that is,

$$
Y_i = w_1 X_{i1} + w_2 X_{i2} + w_3 X_{i3} + w_4 X_{i4} + e_i \qquad i = 1, \ldots, 4
$$

where e_1, e_2, e_3, e_4 are independent random variables with $E\{e_i\} = 0$ and $V\{e_i\} = \sigma^2, i = 1, 2, \ldots, 4$.

Step 7.   Experimental layout

An experimental layout is represented by a 4-×-4 matrix

$$(X) = (X_{ij}; i, j = 1, \ldots, 4)$$

Such a matrix is called a *design matrix*.

Given a design matrix (X) and a vector of measurements $\mathbf{Y} = (Y_1, \ldots, Y_4)'$, we estimate $\mathbf{w} = (w_1, \ldots, w_4)'$ by

$$\hat{\mathbf{W}} = (L)\mathbf{Y}$$

where (L) is a 4-×-4 matrix. We say that the design is valid if there exists a matrix L such that $E\{\hat{\mathbf{W}}\} = \mathbf{w}$. Any nonsingular design matrix (X) represents a valid design with $(L) = (X)^{-1}$.

Indeed, $E\{\mathbf{Y}\} = (X)\mathbf{w}$. Hence

$$E\{\hat{\mathbf{W}}\} = (X)^{-1}E\{\mathbf{Y}\} = \mathbf{w}$$

The precision of the design matrix (X) is measured by $\left(\sum_{i=1}^{4} V\{\hat{W}_i\}\right)^{-1}$.

The problem is to find a design matrix $(X)^0$ that maximizes the precision.

It can be shown that an optimal design is given by the orthogonal array

$$(X)^0 = \begin{bmatrix} 1 & -1 & -1 & 1 \\ 1 & 1 & -1 & -1 \\ 1 & -1 & 1 & -1 \\ 1 & 1 & 1 & 1 \end{bmatrix}$$

or any row (or column) permutation of this matrix. Notice that in this design, in each one of the first three weighing operations (rows), two objects are put on the left pan $(-)$ and two on the right. Also, each object except the first is put twice on $(-)$ and twice on $(+)$. The weight estimates under this design are

$$\hat{\mathbf{W}} = \frac{1}{4} \begin{bmatrix} 1 & 1 & 1 & 1 \\ -1 & 1 & -1 & 1 \\ -1 & -1 & 1 & 1 \\ 1 & -1 & -1 & 1 \end{bmatrix} \begin{bmatrix} Y_1 \\ Y_2 \\ Y_3 \\ Y_4 \end{bmatrix}$$

Moreover,

$$\sum_{i=1}^{4} V\{\hat{W}_i\} = \sigma^2$$

The order of measurements is random.   ∎

EXAMPLE **12.2**   The second example illustrates a complex process with a large number of factors that may affect the yield variables.

Wave soldering of circuit pack assemblies (CPAs) is an automated process of soldering that, if done in an optimal fashion, can raise quality and productivity. The process, however, involves three phases and many variables. Here we analyze the various steps required for designing an experiment to learn the effects of the various factors on the process. We follow the process description of Lin and Kacker (1989).

If the soldering process yields good results, the CPAs can proceed directly to automatic testing. This is a big savings in direct labor cost and an increase in productivity. The wave-soldering process (WSP) occurs in three phases. In phase I, called *fluxing*, the solder joint surfaces are cleaned by the soldering flux, which also protects them against reoxidation. Fluxing lowers the surface tension for better solder wetting and solder joint formation.

Phase II of the WSP is the *soldering assembly*. This is performed in a wave soldering machine. After preheating the solution, the noncomponent side of the assembly is immersed in a solder wave for 1 to 2 seconds. All solder points are completed when the CPA exits the wave. Preheating must be gradual. Correct heating is essential to effective soldering. Also important is the conveyor speed and the conveyor's angle.

Phase III, the last phase of the process, is *detergent cleaning*. The assembly is first washed in detergent solution, then rinsed in water, and finally dried with hot air. The temperature of the detergent solution is raised to achieve effective cleaning and prevent excessive foaming. The rinse water is heated to obtain effective rinsing.

The design steps are as follows:

1 **Objectives**: To find the effects of the various factors on the quality of wave soldering, and to optimize the process.

2 **Response variables**: There are four yield variables:
   a Insulation resistance
   b Cleaning characterization
   c Soldering efficiency
   d Solder mask cracking

3 **Controllable variables**: There are 17 variables (factors) associated with the three phases of the process:

I. Flux formulation	II. Wave soldering	III. Detergent cleaning
A. Type of activator	H. Amount of flux	N. Detergent concentration
B. Amount of activator	I. Preheat time	O. Detergent temperature
C. Type of surfactant	J. Solder temperature	P. Cleaning conveyor speed
D. Amount of surfactant	K. Conveyor speed	Q. Rinse water temperature
E. Amount of antioxidant	L. Conveyor angle	
F. Type of solvent	M. Wave height setting	
G. Amount of solvent		

4 **Measurements**
   a Insulation resistance: Test at 30 minutes, 1 and 4 days after soldering at $-35°C$, 90% relative humidity (RH), no bias voltage, and $-65°C$, 90% RH, no bias voltage (continuous variable)

    b Cleaning characterization: Amounts of residues on the board (continuous variable)

    c Soldering efficiency: Visual inspection of no solder, insufficient solder, good solder, excess solder, and other defects (discrete variables)

    d Solder mask cracking: Visual inspection of cracked spots on the solder mask (discrete variables)

5 **Levels of controllable factors**

Factor	Number of Levels	Factor	Number of Levels
A	2	J	3
B	3	K	3
C	2	L	3
D	3	M	2
E	3	N	2
F	2	O	2
G	3	P	3
H	3	Q	2
I	3		

6 **Statistical model**: Response variables are related to controllable variables by linear models having "main effects" and "interaction" parameters, as will be explained in Section 13.3.

7 **Experiment layout**: A fractional factorial experiment is designed, as explained in Section 13.9. Such a design is needed because a full factorial design contains $3^{10}2^7 = 7,558,272$ possible combinations of factor levels. A fractional replication design chooses a manageable fraction of the full factorial in a manner that allows valid inference and precise estimates of the parameters of interest.

8 **Protocol of execution**: Suppose it is decided to perform a fraction of $3^3 2^2 = 108$ trials at certain levels of the 17 factors. However, the setup of the factors takes time, and one cannot perform more than 4 trials a day. The experiment will last 27 days. It is important to construct the design so the important effects to be estimated will not be confounded with possible differences between days (blocks). The order of the trials within each day is randomized as are the trials assigned to different days. Randomization is an important component of the design: it enhances its validity. The execution protocol should clearly specify the order of execution of the trials. ∎

12.2
Blocking and Randomization

Blocking and randomization are devices in the planning of experiments that are aimed at increasing the precision of the outcome and ensuring the validity of the inference. **Blocking** is used to reduce errors. A block is a portion of the experimental material that is expected to be more homogeneous than the whole. For example, if the experiment is designed to test the effect of polyester coating on the current output of electronic circuits, the variability between circuits could be considerably larger than the effect of the coating on the current output. To reduce this component of variance, we can block by circuit. Each circuit will be tested under two treatments: no-coating and coating. We first test the current output of a circuit without coating. Then we coat the circuit and test again. Such a comparison of before and after treatments of the same unit is called *paired comparison.*

Another example of blocking is the famous boys' shoes example of Box, Hunter, and Hunter (1978, p. 97). Two kinds of materials for shoe soles are to be tested by fixing the soles on *n* pairs of boys' shoes and measuring the amount of wear on the soles after a period of active wear. Because there is high variability in the activities of boys, it will not be clear whether any difference in degree of wear is due to differences between the sole materials or to differences among the boys. However, by blocking, we can reduce much of the variability. Each pair of shoes is assigned both types of soles. The comparison within each block (pair of shoes) is free of the variability among boys. Furthermore, because boys use their right and left feet differently, each type of sole material is assigned randomly to the left or right shoe. Thus, the treatments (two types of soles) are assigned within each block at random.

Other examples of blocks are machines, shifts of production, days of the week, operators, and so on.

Generally, if there are *t* treatments to compare, and *b* blocks, and if all *t* treatments can be performed within a single block, we assign all the *t* treatments to each block. The order of applying the treatments within each block should be **randomized**. Such a **block design** is called a **randomized complete block design (RCBD)**. We will see later how a proper analysis of the yield can validly test for the effects of the treatments.

If not all treatments can be applied within each block, it is desirable to assign treatments to blocks in some balanced fashion. Such designs, to be discussed later, are called *balanced incomplete block designs (BIBDs).*

Randomization within each block is important also to validate the assumption that the error components in the statistical model are independent. This assumption may not be valid if treatments are not assigned at random to the experimental units within each block.

12.3
Additive and Nonadditive Linear Models: Main Effects and Interactions

Seventeen factors that might influence the outcome in WSP are listed in Example 12.2. Some of these factors, such as type of activator (*A*) or type of surfactant (*C*) are categorical variables. The number of levels listed for these factors was 2. That is, the study compares the effects of two types of activators and two types of surfactants.

If the variables are continuous, such as amount of activator (B), we can use a regression linear model to represent the effects of the factors on the yield variables. Such models will be discussed later (Section 12.7.3). In this section, we examine linear models that are valid for both categorical and continuous variables.

We start with a simple case in which the response depends on one factor only. Thus, A designates some factor that is applied at different levels, $A_1, \ldots, A_a$. These could be a categories. The levels of A are also called *treatments*.

Suppose that at each level of A we make n independent repetitions (replicas) of the experiment. Let Y_{ij}, $i = 1, \ldots, a$ and $j = 1, \ldots, n$ denote the observed yield at the jth replication of level A_i. We model the random variables Y_{ij} as

$$Y_{ij} = \mu + \tau_i^A + e_{ij} \qquad i = 1, \ldots, a, \ j = 1, \ldots, n \tag{12.3.1}$$

where μ and $\tau_1^A, \ldots, \tau_a^A$ are unknown parameters satisfying

$$\sum_{i=1}^{a} \tau_i^A = 0 \tag{12.3.2}$$

and e_{ij}, $i = 1, \ldots, a, j = 1, \ldots, n$, are independent random variables such that

$$E\{e_{ij}\} = 0 \qquad \text{and} \qquad V\{e_{ij}\} = \sigma^2 \tag{12.3.3}$$

for all $i = 1, \ldots, a; j = 1, \ldots, n$.

Let

$$\overline{Y}_i = \frac{1}{n} \sum_{j=1}^{n} Y_{ij} \qquad i = 1, \ldots, a$$

The expected values of these means are

$$E\{\overline{Y}_i\} = \mu + \tau_i^A \qquad i = 1, \ldots, a \tag{12.3.4}$$

Let

$$\overline{\overline{Y}} = \frac{1}{a} \sum_{i=1}^{a} \overline{Y}_i \tag{12.3.5}$$

This is the mean of all $N = a \times n$ observations (the grand mean). Because $\sum_{i=1}^{a} \tau_i^A = 0$, we find that

$$E\{\overline{\overline{Y}}\} = \mu \tag{12.3.6}$$

The parameter τ_i^A is called the **main effect** of A at level i.

If there are two factors, A and B, at a and b levels, respectively, there are $a \times b$ *treatment combinations* (A_i, B_j), $i = 1, \ldots, a, j = 1, \ldots, b$. Suppose also that n independent replicas are made at each one of the treatment combinations. The yield at the kth replication of treatment combination (A_i, B_j) is given by

$$Y_{ijk} = \mu + \tau_i^A + \tau_j^B + \tau_{ij}^{AB} + e_{ijk} \tag{12.3.7}$$

The error terms e_{ijk} are independent random variables satisfying

$$E\{e_{ijk}\} = 0 \qquad \text{and} \qquad V\{e_{ijk}\} = \sigma^2 \tag{12.3.8}$$

for all $i = 1, \ldots, a, j = 1, \ldots, b, k = 1, \ldots, n$.

We further assume that

$$\sum_{j=1}^{b} \tau_{ij}^{AB} = 0 \qquad i = 1, \ldots, a$$

$$\sum_{i=1}^{a} \tau_{ij}^{AB} = 0 \qquad j = 1, \ldots, b$$

$$\sum_{i=1}^{a} \tau_i^{A} = 0 \tag{12.3.9}$$

$$\sum_{j=1}^{b} \tau_j^{B} = 0$$

τ_i^A is the main effect of A at level i, τ_j^B is the main effect of B at level j, and τ_{ij}^{AB} is the **interaction effect** at (A_i, B_j).

If all the interaction effects are zero, then the model reduces to

$$Y_{ijk} = \mu + \tau_i^A + \tau_j^B + e_{ijk} \tag{12.3.10}$$

Such a model is called *additive*. If not all the interaction components are zero, then the model is called *nonadditive*.

This model is generalized in a straightforward manner to include a larger number of factors. Thus, for three factors, there are three types of main effect terms, τ_i^A, τ_j^B, and τ_k^C; three types of interaction terms, τ_{ij}^{AB}, τ_{ik}^{AC}, and τ_{jk}^{BC}; and one type of interaction, τ_{ijk}^{ABC}.

Generally, if there are p factors, there are 2^p types of parameters,

$$\mu, \tau_i^A, \tau_j^B, \ldots, \tau_{ij}^{AB}, \tau_{ik}^{AC}, \ldots, \tau_{ijk}^{ABC}, \ldots$$

and so on. Interaction parameters between two factors are called first-order interactions. Interaction parameters among three factors are called second-order interactions, and so on. In design modeling for experiments, it is often assumed that all interaction parameters of higher than first-order are zero.

12.4
The Analysis of Randomized Complete Block Designs
12.4.1 Several Blocks, Two Treatments per Block: Paired Comparison

We now examine paired comparisons more closely. As in the shoe soles example or the example of the effect of polyester coating on circuits output, two treatments are applied to each one of n blocks. The linear model can be written as

$$Y_{ij} = \mu + \tau_i + \beta_j + e_{ij} \qquad i = 1, 2; \ j = 1, \ldots, n \tag{12.4.1}$$

where τ_i is the effect of the ith treatment and β_j is the effect of the jth block. Variable e_{ij} is an independent random variable representing the experimental random error or deviation. It is assumed that $E\{e_{ij}\} = 0$ and $V\{e_{ij}\} = \sigma_e^2$. Because we are interested in testing whether the two treatments have different effects, the analysis is based on within-block differences

$$D_j = Y_{2j} - Y_{1j} = \tau_2 - \tau_1 + e_j^* \qquad j = 1, \ldots, n \tag{12.4.2}$$

The error terms e_j^* are independent random variables with $E\{e_j^*\} = 0$ and $V\{e_j^*\} = \sigma_d^2, j = 1, \ldots, n$, where $\sigma_d^2 = 2\sigma_e^2$.

An unbiased estimator of σ_d^2 is

$$S_d^2 = \frac{1}{n-1} \sum_{j=1}^{n} (D_j - \overline{D}_n)^2 \tag{12.4.3}$$

where $\overline{D}_n = 1/n \sum_{j=1}^{n} D_j$. The hypotheses to be tested are:

$$H_0 : \delta = \tau_2 - \tau_1 = 0$$

against

$$H_1 : \delta \neq 0$$

12.4.1.1   The t-Test

Most commonly used is the t-test, in which H_0 is tested by computing the test statistic

$$t = \frac{\sqrt{n}\,\overline{D}_n}{S_d} \tag{12.4.4}$$

If $e_1^*, \ldots, e_n^*$ are i.i.d., normally distributed, then, under the null hypothesis, t has a t distribution with $n - 1$ degrees of freedom. In this case, H_0 is rejected if

$$|t| > t_{1-\alpha/2}[n-1]$$

where α is the selected level of significance.

12.4.1.2   Randomization Tests

A randomization test for paired comparison constructs a *reference distribution* of all possible averages of the differences that can be obtained by randomly assigning the sign $+$ or $-$ to the value of D_j. It then computes an average difference $\overline{D}$ for each one of the 2^n sign assignments.

The *P*-value of the test, for the two-sided alternative, is determined according to this reference distribution by

$$P = \Pr\{|\overline{D}| \geq |\text{Observed } \overline{D}|\}$$

For example, suppose we have four differences with values 1.1, 0.3, −0.7, −0.1. The mean is $\overline{D}_4 = 0.15$. There are $2^4 = 16$ possible ways of assigning a sign to $|D_i|$. These signs and the associated average differences are shown in Table 12.1.

T A B L E 12.1

Sign Assignments and Values of *D*

Signs				*D*
−1	−1	−1	−1	−0.55
1	−1	−1	−1	0
−1	1	−1	−1	−0.4
1	1	−1	−1	0.15
−1	−1	1	−1	−0.20
1	−1	1	−1	0.35
−1	1	1	−1	−0.05
1	1	1	−1	0.50
−1	−1	−1	1	−0.50
1	−1	−1	1	0.05
−1	1	−1	1	−0.35
1	1	−1	1	0.2
−1	−1	1	1	−0.15
1	−1	1	1	0.40
−1	1	1	1	0
1	1	1	1	0.55

Under the reference distribution, all these possible means are equally probable. The *P*-value associated with the observed $\overline{D} = 0.15$ is $P = \frac{12}{16} = 0.75$. If the number of pairs (blocks) *n* is large, the procedure becomes cumbersome because we have to determine all the 2^n sign assignments. If $n = 20$, there are $2^{20} = 1,048,576$ such assignments. We can, however, estimate the *P*-value by taking an RSWR from this reference distribution. This can be easily done by using MINITAB with the macro RPCOMP.MTB. This macro is

```
Random k1 C2;
Integer 1 2.
Let C3 = 2 * (C2 − 1.5)
Let k2 = mean(C1 * C3)
stack C4 k2 C4
end
```

In order to execute it, we first set $k1 = n$ by the command

```
MTB> Let k1 = n
```

where n is the sample size. In column $C1$, we set the n values of the observed differences, $D_1, \ldots, D_n$. Initiate column $C4$ by

```
MTB> Let C4(1) = mean(C1).
```

After executing this macro M times, we estimate the P-value by the proportion of cases in $C4$ whose absolute value is greater or equal to that of the mean ($C1$).

EXAMPLE **12.3**   In this example, we analyze the results of the shoe soles experiment, as reported in Box, Hunter, and Hunter (1978, p. 100). The observed differences in the wear of the soles between type B and type A for $n = 10$ children are:

$$0.8, \quad 0.6, \quad 0.3, \quad -0.1, \quad 1.1 \quad -0.2, \quad 0.3, \quad 0.5, \quad 0.5, \quad 0.3$$

The average difference is $\overline{D}_{10} = 0.41$.

A t-test of H_0, after setting the differences in column $C3$, is obtained by the MINITAB command,

```
MTB> T Test 0.0 C3;
SUBC> Alternative 0.
```

The result of this t-test is

TEST OF MU = 0.000 VS MU N.E. 0.000

	N	MEAN	STDEV	SE MEAN	T	P VALUE
C3	10	0.410	0.387	0.122	3.35	0.0086

Executing macro RPCOMP.MTB $M = 200$ times on the data in column $C1$ gave 200 values of $\overline{D}$, whose stem-and-leaf plot is shown in Figure 12.1.

According to the figure, the P-value is estimated by

$$\hat{P} = \frac{2}{200} = .01$$

This estimate is almost the same as the P-value of the t-test.   ∎

F I G U R E   12.1
Stem-and-leaf plot of 200 random difference averages

Character Stem-and-Leaf Display
Stem-and-leaf of $C4$ $N = 200$
Leaf Unit $= 0.010$

```
    3  -3  955
    6  -3  111
   17  -2  99755555555
   22  -2  33311
   44  -1  999999999777775555555
   63  -1  3333333333111111111
   88  -0  99999999977777777755555555
  (17)  -0  33333333333311110
   95   0  11111111111333333333
   75   0  555555555577777777799999999
   50   1  111133333
   41   1  555555557777999999999
   19   2  113
   16   2  555777
   10   3  11133
    5   3  779
    2   4  13
```

12.4.2   Several Blocks, t Treatments per Block

As noted earlier, randomized complete block designs (RCBDs) are those in which each block contains all the t treatments. The treatments are assigned to the experimental units in each block at random. Let b denote the number of blocks. The linear model for these designs is

$$Y_{ij} = \mu + \tau_i + \beta_j + e_{ij} \qquad i = 1, \ldots, t \; j = 1, \ldots, b \qquad \text{(12.4.5)}$$

where Y_{ij} is the yield of the ith treatment in the jth block. The main effect of the ith treatment is τ_i, and the main effect of the jth block is β_j. It is assumed that the effects are additive (no interaction). Under this assumption, each treatment is tried only once in each block. The different blocks serve the role of replicas. However, because the blocks may have additive effects, β_j, we have to adjust for the effects of blocks in estimating σ^2. This is done as shown in the ANOVA table (Table 12.2). Further assume that e_{ij} are the error random variables with $E\{e_{ij}\} = 0$ and $V\{e_{ij}\} = \sigma^2$ for all (i, j).

Here,

$$\text{SST} = \sum_{i=1}^{t} \sum_{j=1}^{b} (Y_{ij} - \overline{\overline{Y}})^2, \qquad \text{(12.4.6)}$$

TABLE   12.2

ANOVA table for RCBD

Source of variation	DF	SS	MS	E{MS}
Treatments	$t-1$	SSTR	MSTR	$\sigma^2 + \dfrac{b}{t-1}\displaystyle\sum_{i=1}^{t}\tau_i^2$
Blocks	$b-1$	SSBL	MSBL	$\sigma^2 + \dfrac{t}{b-1}\displaystyle\sum_{j=1}^{b}\beta_j^2$
Error	$(t-1)(b-1)$	SSE	MSE	σ^2
Total	$tb-1$	SST	—	

$$\text{SSTR} = b\sum_{i=1}^{t}(\overline{Y}_{i.} - \overline{\overline{Y}})^2, \tag{12.4.7}$$

$$\text{SSBL} = t\sum_{j=1}^{b}(\overline{Y}_{.j} - \overline{\overline{Y}})^2, \tag{12.4.8}$$

and

$$\text{SSE} = \text{SST} - \text{SSTR} - \text{SSBL}$$

$$\overline{Y}_{i.} = \frac{1}{b}\sum_{j=1}^{b} Y_{ij} \qquad \overline{Y}_{.j} = \frac{1}{t}\sum_{i=1}^{t} Y_{ij} \tag{12.4.9}$$

and $\overline{\overline{Y}}$ is the grand mean.

The significance of the treatment effects is tested by the F statistic

$$F_t = \frac{\text{MSTR}}{\text{MSE}} \tag{12.4.10}$$

The significance of the block effects is tested by

$$F_b = \frac{\text{MSBL}}{\text{MSE}} \tag{12.4.11}$$

These statistics are compared with the corresponding $(1 - \alpha)$th fractiles of the F distribution. Under the assumption that $\sum_{i=1}^{t}\tau_i = 0$, the main effects of the treatments are estimated by

$$\hat{\tau}_i = \overline{Y}_{i.} - \overline{\overline{Y}} \qquad i = 1, \ldots, t \tag{12.4.12}$$

These are least-squares estimates. Each such estimate is a linear contrast

$$\hat{\tau}_i = \sum_{i'=1}^{t} c_{ii'}\overline{Y}_{.i'} \tag{12.4.13}$$

where

$$c_{ii'} = \begin{cases} 1 - \dfrac{1}{t} & \text{if } i = i' \\ -\dfrac{1}{t} & \text{if } i \neq i' \end{cases} \qquad \textbf{(12.4.14)}$$

Hence,

$$V\{\hat{\tau}_i\} = \frac{\sigma^2}{b} \sum_{i'=1}^{t} c_{ii'}^2$$

$$= \frac{\sigma^2}{b} \left(1 - \frac{1}{t}\right) \qquad i = 1, \dots, t \qquad \textbf{(12.4.15)}$$

An unbiased estimator of σ^2 is given by the MSE. Thus, simultaneous confidence intervals for τ_i $(i = 1, \dots, t)$, according to the Scheffé method (see Section 8.9), are

$$\hat{\tau}_i \pm S_\alpha \left(\frac{\text{MSE}}{b} \left(1 - \frac{1}{t}\right)\right)^{1/2} \qquad i = 1, \dots, t \qquad \textbf{(12.4.16)}$$

where

$$S_\alpha = ((t-1)F_{1-\alpha}[t-1, (t-1)(b-1)])^{1/2}$$

EXAMPLE 12.4 In Example 8.10, we estimated the effects of hybrids on the resistance in cards. We have $t = 6$ hybrids (treatments) on a card, and 32 cards. We can now test whether there are significant differences between the cards by considering them as blocks and using the ANOVA for RCBD. In this case, $b = 32$. Using MINITAB and data file HADPAS.DAT, we obtain Table 12.3. Because $F_{.99}[5, 155] = 3.1375$ and $F_{.99}[31, 155] = 1.8105$, both the treatment effects and the card effects are significant.

TABLE 12.3
ANOVA for hybrid data

Source	DF	SS	MS	F
Hybrids	5	1,780,741	356148	105.7
Cards	31	2,804,823	90478	26.9
Error	155	522,055	3368	—
Total	191	5,107,619	—	—

The estimator of σ, $\hat{\sigma} = (\text{MSW})^{1/2}$, according to this ANOVA, is $\hat{\sigma} = 58.03$. Notice that this estimator is considerably smaller than the one of Example 8.9. This is due to the variance reduction effect of the blocking.

The simultaneous confidence intervals, at level of confidence .95, for the treatment effects are:

Hybrid 1: 177.67 ± 31.57

Hybrid 6: 51.17 ± 31.57

Hybrid 5: 14.82 ± 31.57

Hybrid 2: -62.93 ± 31.57

Hybrid 4: -65.33 ± 31.57

Hybrid 3: -115.40 ± 31.57

Accordingly, the effects of hybrid 2 and hybrid 4 are not significantly different, and that of hybrid 5 is not significantly different from zero.   ■

12.5
Balanced Incomplete Block Designs

As mentioned before, it is often the case that blocks are not sufficiently large to accommodate all the t treatments. For example, in testing the durability of fabric, a special machine (a Martindale wear tester) is used that can accommodate only four pieces of cloth simultaneously. Here the block size is fixed at $k = 4$, whereas the number of treatments t is the number of types of cloths to be compared. **Balanced incomplete block designs (BIBD)** are designs that assign t treatments to b blocks of size k $(k < t)$ as follows:

1   Each treatment is assigned only once to any one block.

2   Each treatment appears in r blocks; r is the number of replicas.

3   Every pair of two different treatments appears in λ blocks.

4   The order of treatments within each block is randomized.

5   The order of blocks is randomized.

According to these requirements, there are altogether $N = tr = bk$ trials. Moreover, the following equality should hold

$$\lambda(t - 1) = r(k - 1) \tag{12.5.1}$$

The question is how to design a BIBD for a given t and k. We can obtain a BIBD by the complete combinatorial listing of the $\binom{t}{k}$ selections without replacement of k out of t letters. In this case, the number of blocks is

$$b = \binom{t}{k} \tag{12.5.2}$$

The number of replicas is $r = \binom{t - 1}{k - 1}$, and $\lambda = \binom{t - 2}{k - 2}$. The total number of trials is

$$N = tr = t\binom{t-1}{k-1} = \frac{t!}{(k-1)!(t-k)!} = k\binom{t}{k} \qquad (12.5.3)$$
$$= kb$$

Such designs of BIBD are called **combinatoric designs**. However, they may be too big. For example, if $t = 8$ and $k = 4$, we are required to have $\binom{8}{4} = 70$ blocks. Thus, the total number of trials is $N = 70 \times 4 = 280$ and $r = \binom{7}{3} = 35$. Here $\lambda = \binom{6}{2} = 15$.

Advanced algebraic methods can yield smaller designs for $t = 8$ and $k = 4$. Box, Hunter, and Hunter (1978, p. 272), for example, lists a BIBD of $t = 8$, $k = 4$ in $b = 14$ blocks. Here $N = 14 \times 4 = 56$, $r = 7$, and $\lambda = 3$.

It is not always possible to have a BIBD smaller than a complete combinatoric design. This is the case with $t = 8$ and $k = 5$. Here the smallest number of blocks possible is $\binom{8}{5} = 56$, and $N = 56 \times 5 = 280$.

Box, Hunter, and Hunter (1978, pp. 270–274) lists some useful BIBDs for $k = 2, \ldots, 6$, $t = k, \ldots, 10$. Let B_i denote the set of treatments in the ith block. For example, if block 1 contains the treatments 1, 2, 3, 4, then $B_1 = \{1, 2, 3, 4\}$. Let Y_{ij} be the yield of treatment $j \in B_i$. The effects model is

$$Y_{ij} = \mu + \beta_i + \tau_j + e_{ij} \qquad i = 1, \ldots, b; \; j \in B_i \qquad (12.5.4)$$

$\{e_{ij}\}$ are random experimental errors with $E\{e_{ij}\} = 0$ and $V\{e_{ij}\} = \sigma^2$ for all (i, j). The block and treatment effects, $\beta_1, \ldots, \beta_b$ and $\tau_1, \ldots, \tau_t$, satisfy the constraints

$$\sum_{j=1}^{t} \tau_j = 0 \text{ and } \sum_{i=1}^{b} \beta_i = 0.$$

Let T_j be the set of all indices of blocks containing the jth treatment. The least-squares estimates of the treatment effects are obtained in the following manner.

Let $W_j = \sum_{i \in T_j} Y_{ij}$ be the sum of all Y values under the jth treatment. Let W_j^* be the sum of the values in all the r blocks that contain the jth treatment—that is, $W_j^* = \sum_{i \in T_j} \sum_{l \in B_i} Y_{il}$. Compute

$$Q_j = kW_j - W_j^* \qquad j = 1, \ldots, t \qquad (12.5.5)$$

The LSE of τ_j is

$$\hat{\tau}_j = \frac{Q_j}{t\lambda} \qquad j = 1, \ldots, t \qquad (12.5.6)$$

Notice that $\sum_{j=1}^{t} Q_j = 0$. Thus, $\sum_{j=1}^{t} \hat{\tau}_j = 0$. Let $\overline{\overline{Y}} = 1/N \sum_{i=1}^{b} \sum_{l \in B_i} Y_{il}$. The adjusted treatment average is defined as $\overline{Y}_j^* = \overline{\overline{Y}} + \hat{\tau}_j, j = 1, \ldots, t$. The ANOVA for a BIBD is given in the Table 12.4.

TABLE   12.4
ANOVA for a BIBD

Source of variation	DF	SS	MS	E{MS}
Blocks	$b-1$	SSBL	MSBL	$\sigma^2 + \dfrac{t}{b-1}\displaystyle\sum_{i=1}^{b}\beta_i^2$
Treatments (adjusted)	$t-1$	SSTR	MSTR	$\sigma^2 + \dfrac{b}{t-1}\displaystyle\sum_{j=1}^{t}\tau_j^2$
Error	$N-t-b+1$	SSE	MSE	σ^2
Total	$N-1$	SST	—	—

Here,

$$\text{SST} = \sum_{i=1}^{b}\sum_{l\in B_i} Y_{il}^2 - \left(\sum_{i=1}^{b}\sum_{l\in B_i} Y_{il}\right)^2 / N \tag{12.5.7}$$

$$\text{SSBL} = \frac{1}{k}\sum_{i=1}^{b}\left(\sum_{l\in B_i} Y_{il}\right)^2 - N\overline{\overline{Y}}^2 \tag{12.5.8}$$

$$\text{SSTR} = \frac{1}{\lambda kt}\sum_{j=1}^{t} Q_j^2 \tag{12.5.9}$$

and

$$\text{SSE} = \text{SST} - \text{SSBL} - \text{SSTR} \tag{12.5.10}$$

The significance of the treatment effects is tested by the statistic

$$F = \frac{\text{MSTR}}{\text{MSE}} \tag{12.5.11}$$

EXAMPLE 12.5   Six different adhesives ($t = 6$) are tested for their bonding strength in a lamination process under curing pressure of 200 (psi). Lamination can be done in blocks of size $k = 4$.

A combinatoric design will have $\binom{6}{4} = 15$ blocks, with $r = \binom{5}{3} = 10$, $\lambda = \binom{4}{2} = 6$, and $N = 60$. The treatment indices of the 15 blocks are given in Table 12.5. The observed bond strength in these trials is given in Table 12.6.

The grand mean of the bond strength is $\overline{\overline{Y}} = 27.389$. The sets T_j and the sums W_j, W_j^* are listed in Table 12.7. The ANOVA is given in Table 12.8, and the adjusted mean effects of the adhesives are listed in Table 12.9.

The variance of each adjusted mean effect is

$$V\{\overline{Y}_j^*\} = \frac{k\sigma^2}{t\lambda} \qquad j = 1, \ldots, t \tag{12.5.12}$$

TABLE 12.5
Block sets

i	B_i	i	B_i
1	1, 2, 3, 4	9	1, 3, 5, 6
2	1, 2, 3, 5	10	1, 4, 5, 6
3	1, 2, 3, 6	11	2, 3, 4, 5
4	1, 2, 4, 5	12	2, 3, 4, 6
5	1, 2, 4, 6	13	2, 3, 5, 6
6	1, 2, 5, 6	14	2, 4, 5, 6
7	1, 3, 4, 5	15	3, 4, 5, 6
8	1, 3, 4, 6		

TABLE 12.6
Values of Y_{il}, $l \in B_i$

i	Y_{il}	i	Y_i
1	24.7, 20.8, 29.4, 24.9	8	23.1, 29.3, 27.1, 34.4
2	24.1, 20.4, 29.8, 30.3	9	22.0, 29.8, 31.9, 36.1
3	23.4, 20.6, 29.2, 34.4	10	22.8, 22.6, 33.2, 34.8
4	23.2, 20.7, 26.0, 30.8	11	21.4, 29.6, 24.8, 31.2
5	21.5, 22.1, 25.3, 35.4	12	21.3, 28.9, 25.3, 35.1
6	21.4, 20.1, 30.1, 34.1	13	21.6, 29.5, 30.4, 33.6
7	23.2, 28.7, 24.9, 31.0	14	20.1, 25.1, 32.9, 33.9
		15	30.1, 24.0, 30.8, 36.5

TABLE 12.7
The set T_j and the statistics W_j, W_j^*, Q_j

j	T_j	W_j	W_j^*	Q_j
1	1, 2, 3, 4, 5, 6, 7, 8, 9, 10	229.54	1077.7	−159.56
2	1, 2, 3, 4, 5, 6, 11, 12, 13, 14	209.02	1067.4	−231.31
3	1, 2, 3, 7, 8, 9, 11, 12, 13, 15	294.12	1107.6	68.90
4	1, 4, 5, 7, 8, 10, 11, 12, 14, 15	250.00	1090.9	−90.90
5	2, 4, 6, 7, 9, 10, 11, 13, 14, 15	312.49	1107.5	142.47
6	3, 5, 6, 8, 9, 10, 12, 13, 14, 15	348.18	1123.8	268.90

TABLE 12.8
ANOVA for BIBD

Source	DF	SS	MS	F
Blocks	14	161.78	11.556	23.99
Treatment (adjusted)	5	1282.76	256.552	532.54
Error	40	19.27	0.48175	—
Total	59	1463.81		

TABLE 12.9

Mean effects and their SE

Treatment	$\overline{Y}_j^*$	SE$\{\overline{Y}_j^*\}$
1	22.96	1.7445
2	20.96	1.7445
3	29.33	1.7445
4	24.86	1.7445
5	31.35	1.7445
6	34.86	1.7445

Thus, the SE of $\overline{Y}_i^*$ is

$$\text{SE}\{\overline{Y}_j^*\} = \left(\frac{k\,\text{MSE}}{t\lambda}\right)^{1/2} \qquad j = 1, \ldots, t \tag{12.5.13}$$

It seems that there are two homogeneous groups of treatments: $\{1, 2, 4\}$ and $\{3, 5, 6\}$. ∎

12.6
Latin Square Designs

Latin square designs are designs that we can block for two error-inducing factors in a balanced fashion, and yet save ourselves a considerable number of trials.

Suppose we have t treatments to test, and we wish to block for two factors. We assign the blocking factors t levels (the number of treatments) in order to obtain squared designs. For example, suppose we wish to study the effects of four new designs of keyboards for desktop computers. The design of the keyboard might have an effect on the speed of typing or on the number of typing errors. The factors Typist or Type of Job may confound the results. Thus, we can block by Typist and by Job. We should pick at random 4 typists and 4 different jobs. We construct a square with 4 rows and 4 columns for the blocking factors. (See Table 12.10.)

Let A, B, C, D denote the 4 keyboard designs. We assign the letters to the cells of this square so that

TABLE 12.10

A 4-×-4 Latin square

	Job 1	Job 2	Job 3	Job 4
Typist 1	*A*	*B*	*C*	*D*
Typist 2	*B*	*A*	*D*	*C*
Typist 3	*C*	*D*	*A*	*B*
Typist 4	*D*	*C*	*B*	*A*

1 Each letter appears exactly once in a row

2 Each letter appears exactly once in a column

Finally, the order of performing these trials is random. Notice that a design which contains all the combinations of typist, job, and keyboard spans over $4 \times 4 \times 4 = 64$ combinations. Thus, the Latin square design saves many trials. However, it is based on the assumption of no interactions between the treatments and the blocking factors. That is, in order to obtain a valid analysis, the model relating the response to the factor effects should be additive:

$$Y_{ijk} = \mu + \beta_i + \gamma_j + \tau_k + e_{ijk} \qquad i, j, k = 1, \ldots, t \tag{12.6.1}$$

where μ is the grand mean, β_i are the row effects, γ_j are the column effects, and τ_k are the treatment effects. The experimental error variables are $\{e_{ijk}\}$, with $E\{e_{ijk}\} = 0$ and $V\{e_{ijk}\} = \sigma^2$ for all (i, j). Furthermore,

$$\sum_{i=1}^{t} \beta_i = \sum_{j=1}^{t} \gamma_j = \sum_{k=1}^{t} \tau_k = 0 \tag{12.6.2}$$

The Latin square presented in Table 12.10 is not unique. There are other 4-×-4 Latin squares. For example,

$$
\begin{array}{cccc}
D & C & B & A \\
B & A & D & C \\
C & D & A & B \\
\end{array}
\qquad
\begin{array}{cccc}
C & D & A & B \\
D & C & B & A \\
B & A & D & C \\
\end{array}
$$

A few Latin square designs, for $t = 3, \ldots, 9$ are given in Box, Hunter, and Hunter (1978, pp. 261–262).

If we perform only one replication of the Latin square, the ANOVA for testing the main effects is as shown in Table 12.11.

T A B L E 12.11

ANOVA for a Latin square, one replication

Source	DF	SS	MS	F
Treatments	$t-1$	SSTR	MSTR	MSTR/MSE
Rows	$t-1$	SSR	MSR	MSR/MSE
Columns	$t-1$	SSC	MSC	MSC/MSE
Error	$(t-1)(t-2)$	SSE	MSE	—
Total	t^2-1	SST	—	—

Formulas for the various SS terms are given below. At this time, we wish to emphasize that if t is small—say, $t = 3$—then the number of DF for SSE is only 2. This is too small. The number of DFs for the error SS can be increased by performing replicas. One way to do this is to perform the same Latin square r times independently, making all as similar as possible. However, significant differences between replicas may emerge nevertheless. The ANOVA for r identical replicas is as given in Table 12.12. Notice that now we have rt^2 observations. Let $T_{....}$ and $Q_{....}$ be the sum and sum of squares of all observations. Then,

$$\text{SST} = Q_{....} - T_{....}^2/rt^2 \tag{12.6.3}$$

Let $T_{i...}$ denote the sum of rt observations in the ith row of all r replications. Then

$$\text{SSR} = \frac{1}{tr}\sum_{i=1}^{t} T_{i...}^2 - \frac{T_{....}^2}{rt^2} \tag{12.6.4}$$

Similarly, let $T_{.j..}$ and $T_{..k.}$ be the sums of all rt observations in column j of all replicas and treatment k of all r replicas; then

$$\text{SSC} = \frac{1}{rt}\sum_{j=1}^{t} T_{.j..}^2 - \frac{T_{....}^2}{rt^2} \tag{12.6.5}$$

and

$$\text{SSTR} = \frac{1}{rt}\sum_{k=1}^{t} T_{..k.}^2 - \frac{T_{....}^2}{rt^2} \tag{12.6.6}$$

Finally, let $T_{...l}$ ($l = 1, \ldots, r$) denote the sum of all t^2 observations in the lth replication. Then,

$$\text{SSREP} = \frac{1}{t^2}\sum_{l=1}^{r} T_{...l}^2 - \frac{T_{....}^2}{rt^2} \tag{12.6.7}$$

The pooled sum of squares for error is obtained by

$$\text{SSE} = \text{SST} - \text{SSR} - \text{SSC} - \text{SSTR} - \text{SSREP} \tag{12.6.8}$$

Notice that if $t = 3$ and $r = 3$, the number of DF for SSE increases from 2 (when $r = 1$) to 18 (when $r = 3$).

The most important hypothesis is that connected with the main effects of the treatments. This we test with the statistic

$$F = \text{MSTR}/\text{MSE} \tag{12.6.9}$$

TABLE 12.12

ANOVA for Replicated Latin Square

Source	DF	SS	MS	F
Treatments	$t - 1$	SSTR	MSTR	MSTR/MSE
Rows	$t - 1$	SSR	MSR	—
Columns	$t - 1$	SSC	MSC	—
Replicas	$r - 1$	SSREP	MSREP	—
Error	$(t - 1)[r(t + 1) - 3])$	SSE	MSE	—
Total	$rt^2 - 1$	SST	—	—

EXAMPLE **12.6**   Five models of keyboards (treatments) were tested in a Latin square design in which the blocking factors are typist and job. Five typists were randomly selected from a pool of typists of similar capabilities. Five typing jobs were selected. Each typing job had 4000 characters. The yield, Y_{ijk}, is the number of typing errors found at the ith typist, jth job under the kth keyboard. The Latin square design used is presented in Table 12.13.

TABLE **12.13**
Latin square design, $t = 5$

	\multicolumn{5}{c}{Job}
Typist	**1**	**2**	**3**	**4**	**5**
1	A	B	C	D	E
2	B	C	D	E	A
3	C	D	E	A	B
4	D	E	A	B	C
5	E	A	B	C	D

The five keyboards are denoted by the letters A, B, C, D, E. The experiment spanned 5 days. Each day, a typist was assigned a job at random (from those not yet tried). The keyboard used is the one associated with the job. Only one job was tried in a given day. The observed number of typing errors (per 4000 characters) is given in Table 12.14.

Figures 12.2–12.4 present boxplots of the error rates for the three different factors. The only influencing factor is the typist effect, as shown in Figure 12.4.

The total sum of squares is

$$Q = 30636$$

Thus,

$$SST = 30636 - \frac{798^2}{25} = 5163.84$$

Similarly,

$$SSR = \frac{1}{5}(100^2 + 277^2 + \cdots + 187^2) - \frac{798^2}{25}$$
$$= 4554.64$$
$$SSC = \frac{1}{5}(178^2 + 138^2 + \cdots + 156^2) - \frac{798^2}{25}$$
$$= 270.641$$

and

$$SSTR = \frac{1}{5}(162^2 + 166^2 + \cdots + 170^2) - \frac{798^2}{25}$$
$$= 69.4395$$

The analysis of variance, following that in Table 12.12, is given in Table 12.15.

T A B L E **12.14**
Number of typing errors

Typist	Job 1	Job 2	Job 3	Job 4	Job 5	Row sum
1	A 20	B 18	C 25	D 17	E 20	100
2	B 65	C 40	D 55	E 58	A 59	277
3	C 30	D 27	E 35	A 21	B 27	140
4	D 21	E 15	A 24	B 16	C 18	94
5	E 42	A 38	B 40	C 35	D 32	187
Column sum	178	138	179	147	156	798

Sums Keyboard	A 162	B 166	C 148	D 152	E 170

F I G U R E **12.2**
Effect of keyboard on error rate

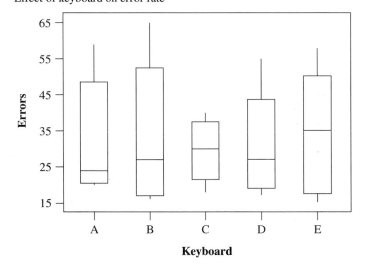

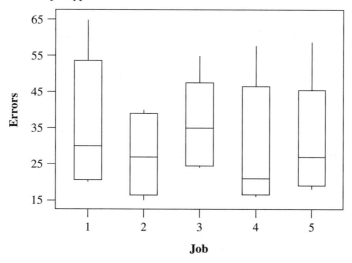

F I G U R E **12.3**

Effect of job type on error rate

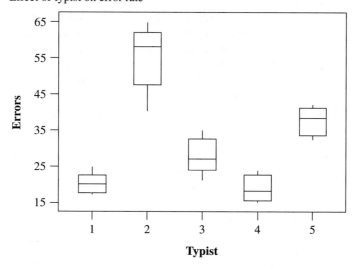

F I G U R E **12.4**

Effect of typist on error rate

The null hypothesis that the main effects of the keyboards are zero cannot be rejected. The largest source of variability in this experiment was the typists. The different jobs contributed also to the variability. The *P*-value for the *F*-test of Jobs is .062. ∎

TABLE   12.15
ANOVA for keyboard Latin square experiment

Source	DF	SS	MS	F
Typist	4	4554.640	1138.66	50.772
Job	4	270.641	67.66	3.017
Keyboard	4	69.439	17.3598	0.774
Error	12	269.120	22.4267	—
Total	24	5163.840	—	—

12.7
Full Factorial Experiments

12.7.1   The Structure of Factorial Experiments

Full factorial experiments are those in which complete trials are performed of all combinations of the various factors at all levels. For example, if there are five factors, each one tested at three levels, there are altogether $3^5 = 243$ treatment combinations. All 243 treatment combinations are tested. The full factorial experiment may also be replicated several times. The order in which the trials are performed is random.

In full factorial experiments, the number of levels of different factors do not have to be the same. Some factors might be tested at two levels and others at three or four levels. Full factorial, or certain fractional factorials (which will be discussed later), are necessary if the statistical model is not additive. We need to perform factorial experiments, full or fractional, in order to estimate or test the effects of interactions. In a full factorial experiment, all the main effects and interactions can be tested or estimated. Recall that if there are p factors $A, B, C, \ldots$, there are p types of main effects, $\binom{p}{2}$ types of pairwise interactions $AB, AC, BC, \ldots$, $\binom{p}{3}$ interactions between three factors, $ABC, ABD, \ldots$, and so on. On the whole, together with the grand mean μ, there are 2^p types of parameters.

In the following section we discuss the structure of the ANOVA for testing the significance of main effects and interactions. This is followed by a section on estimation. In Sections 12.7.4 and 12.7.5 we discuss the structure of full factorial experiments with two and three levels per factor, respectively.

12.7.2   The ANOVA for Full Factorial Designs

The analysis of variance for **full factorial designs** is done to test the hypothesis that main effects or interaction parameters are equal to zero. We present the ANOVA for a two-factor situation: factor A at a levels and factor B at b levels. The method can be generalized to any number of factors.

The structure of the experiment is such that all $a \times b$ treatment combinations are tested. Each treatment combination is repeated n times. The model is

$$Y_{ijk} = \mu + \tau_i^A + \tau_j^B + \tau_{ij}^{AB} + e_{ijk} \tag{12.7.1}$$

$$i = 1, \ldots, a; \ j = 1, \ldots, b; \ k = 1, \ldots, n$$

e_{ijk} are independent random variables with $E\{e_{ijk}\} = 0$ and $V\{e_{ijk}\} = \sigma^2$ for all i, j, k.
Let

$$\overline{Y}_{ij} = \frac{1}{n} \sum_{k=1}^{n} Y_{ijk} \tag{12.7.2}$$

$$\overline{Y}_{i.} = \frac{1}{b} \sum_{j=1}^{b} \overline{Y}_{ij} \qquad i = 1, \ldots, a \tag{12.7.3}$$

$$\overline{Y}_{.j} = \frac{1}{a} \sum_{i=1}^{a} \overline{Y}_{ij} \qquad j = 1, \ldots, b \tag{12.7.4}$$

and

$$\overline{\overline{Y}} = \frac{1}{ab} \sum_{i=1}^{a} \sum_{j=1}^{b} \overline{Y}_{ij} \tag{12.7.5}$$

The ANOVA first partitions the total sum of squares of deviations from $\overline{\overline{Y}}$:

$$\text{SST} = \sum_{i=1}^{a} \sum_{j=1}^{b} \sum_{k=1}^{n} (Y_{ijk} - \overline{\overline{Y}})^2 \tag{12.7.6}$$

into two components

$$\text{SSW} = \sum_{i=1}^{a} \sum_{j=1}^{b} \sum_{k=1}^{n} (Y_{ijk} - \overline{Y}_{ij})^2 \tag{12.7.7}$$

and

$$\text{SSB} = n \sum_{i=1}^{a} \sum_{j=1}^{b} (\overline{Y}_{ij} - \overline{\overline{Y}})^2 \tag{12.7.8}$$

It is straightforward to show that

$$\text{SST} = \text{SSW} + \text{SSB} \tag{12.7.9}$$

In the second stage, the sum of squares of deviations, SSB, is partitioned into three components—SSI, SSMA, SSMB—where

$$\text{SSI} = n \sum_{i=1}^{a} \sum_{j=1}^{b} (\overline{Y}_{ij} - \overline{Y}_{i.} - \overline{Y}_{.j} + \overline{\overline{Y}})^2 \tag{12.7.10}$$

$$\text{SSMA} = nb \sum_{i=1}^{a} (\overline{Y}_{i.} - \overline{\overline{Y}})^2 \tag{12.7.11}$$

and

$$\text{SSMB} = na \sum_{j=1}^{b} (\overline{Y}_{.j} - \overline{\overline{Y}})^2 \tag{12.7.12}$$

That is,

$$\text{SSB} = \text{SSI} + \text{SSMA} + \text{SSMB} \tag{12.7.13}$$

All these terms are collected in a table of ANOVA, as shown in Table 12.16.

TABLE   12.16
Table of ANOVA for a 2-factor factorial experiment

Source of variation	DF	SS	MS	F
A	$a-1$	SSMA	MSA	F_A
B	$b-1$	SSMB	MSB	F_B
AB	$(a-1)(b-1)$	SSI	MSAB	F_{AB}
Between	$ab-1$	SSB	—	—
Within	$ab(n-1)$	SSW	MSW	—
Total	$N-1$	SST	—	—

Thus,

$$\text{MSA} = \frac{\text{SSMA}}{a-1} \tag{12.7.14}$$

$$\text{MSB} = \frac{\text{SSMB}}{b-1} \tag{12.7.15}$$

and

$$\text{MSAB} = \frac{\text{SSI}}{(a-1)(b-1)} \tag{12.7.16}$$

$$\text{MSW} = \frac{\text{SSW}}{ab(n-1)} \tag{12.7.17}$$

Finally, we compute the F-statistics

$$F_A = \frac{\text{MSA}}{\text{MSW}} \tag{12.7.18}$$

$$F_B = \frac{\text{MSB}}{\text{MSW}} \tag{12.7.19}$$

and

$$F_{AB} = \frac{\text{MSAB}}{\text{MSW}} \tag{12.7.20}$$

F_A, F_B and F_{AB} are statistics that test, respectively, the significance of the main effects of A, the main effects of B, and the interactions AB.
   If $F_A < F_{1-\alpha}[a-1, ab(n-1)]$, the null hypothesis

$$H_0^A : \tau_1^A = \cdots = \tau_k^A = 0$$

cannot be rejected.

If $F_B < F_{1-\alpha}[b-1, ab(n-1)]$, the null hypothesis

$$H_0^B : \tau_1^B = \cdots = \tau_b^B = 0$$

cannot be rejected.

Also, if

$$F_{AB} < F_{1-\alpha}[(a-1)(b-1), ab(n-1)]$$

we cannot reject the null hypothesis

$$H_0^{AB} : \tau_{11}^{AB} = \cdots = \tau_{ab}^{AB} = 0$$

The ANOVA for two factors can be performed by MINITAB. We illustrate this estimation and test in the following example.

EXAMPLE **12.7**   In Chapter 10, we introduced the piston example. Seven prediction factors for the piston cycle time were listed. These are

  A. Piston weight, 30–60 (kg)
  B. Piston surface area, 0.005–0.020 (m^2)
  C. Initial gas volume, 0.002–0.010 (m^3)
  D. Spring coefficient, 1000–5000 (N/m)
  E. Atmospheric pressure, 90,000–100,000 (N/m^2)
  F. Ambient temperature, 290–296 (K)
  G. Filling gas temperature, 340–360 (K)

We are interested in testing the effects of the piston weight (*A*) and the spring coefficient (*D*) on the cycle times (in sec). For this purpose, we designed a factorial experiment at three levels of *A* and three levels of *D*. The levels are $A_1 = 30$ (kg), $A_2 = 45$ (kg) and $A_3 = 60$ (kg). The levels of factor *D* (spring coefficient) are $D_1 = 1500$ (N/m), $D_2 = 3000$ (N/m), and $D_3 = 4500$ (N/m). Five replicas were performed at each treatment combination ($n = 5$). The data were obtained by using the computer simulator TURBEXP.EXE. The five factors not under study were kept at the levels $B = 0.01$ (m^2), $C = 0.005$ (m^3), $E = 95,000$ (N/m^2), $F = 293$ (K) and $G = 350$ (K). The data, which include 45 observations, are given in file TURBEXP.DAT. (*Note*: Running TURBEXP.EXE produces an output file labeled OTURBEXP.DAT.)

We start with the two-way ANOVA. Table 12.17 is output from MINITAB, following the command:

```
MTB> ANOVA C3 = C1 C2 C1 * C2.
```

TABLE   12.17

Two-way ANOVA for cycle time

Source	DF	SS	MS	F	p
Spr_Cof	2	0.9926	0.496	10.77	0.000
Pist_Wg	2	0.1931	0.097	2.10	0.138
Spr_Cof*Pist_Wg	4	0.1514	0.038	0.82	0.520
Error	36	1.6587	0.046		
Total	44	2.9958			

Figure 12.5 shows the effect of the factor spring coefficient on cycle time. Spring coefficient at 4500 (N/m) reduces the mean cycle time and its variability. The boxplot at the right margin is for the combined samples.

Figure 12.6 is an interaction plot showing the effect of piston weight on the mean cycle time at each level of the spring coefficient. Generally, if the lines in the interaction plot are not parallel, the model is not additive. However, according to the ANOVA table, the indicated interaction is not significant.

The *P*-values are computed with the appropriate *F* distributions. We see in the ANOVA table that only the main effects of the spring coefficient (*D*) are significant. Because the effects of the piston weight (*A*) and that of the interaction are not significant, we can estimate σ^2 by a pooled estimator, which is

$$\hat{\sigma}^2 = \frac{SSW + SSI + SSMA}{36 + 4 + 2} = \frac{2.2711}{42}$$

$$= 0.0541$$

FIGURE   12.5

Effect of spring coefficient on cycle time

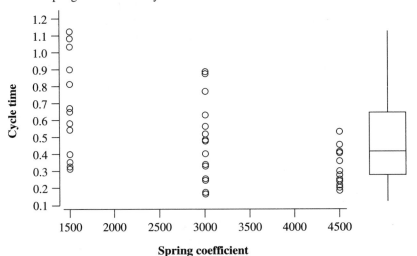

FIGURE   **12.6**

Interaction plot of piston weight by spring coefficient

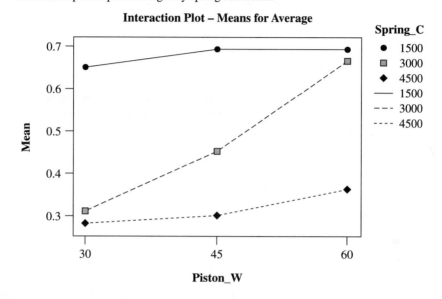

To estimate the main effects of D, we pool all data from samples having the same level of D. We obtain pooled samples of size $n_p = 15$. The means of the cycle time for these samples are

	D_1	D_2	D_3	Grand
$\overline{Y}$	0.679	0.477	0.316	0.491
Main effects	0.188	−0.014	−0.175	—

The standard error of these main effects is $\text{SE}\{\hat{\tau}_j^D\} = 0.21448/\sqrt{45}\sqrt{\tfrac{1}{2}} = 0.0425$.

Because we estimate on the basis of the pooled samples, and the main effects $\hat{\tau}_j^D$ ($j = 1, 2, 3$) are contrasts of three means, the coefficient S_α for the simultaneous confidence intervals has the formula

$$S_\alpha = (2F_{.95}[2, 42])^{1/2}$$

$$= \sqrt{2 \times 3.22} = 2.538.$$

The simultaneous confidence intervals for τ_j^D, at $\alpha = .05$, are

	Lower limit	Upper limit
τ_1^D:	0.0801	0.2959
τ_2^D:	−0.1219	0.0939
τ_3^D:	−0.2829	−0.0671

We see that the confidence interval for τ_2^D covers zero. Thus, $\hat{\tau}_2^D$ is not significant. The significant main effects are $\hat{\tau}_1^D$ and $\hat{\tau}_3^D$. ∎

12.7.3 Estimating Main Effects and Interactions

In discussing the estimation of the main effects and interaction parameters, we confine our presentation to the case of two factors A and B, which are at a and b levels, respectively. The number of replicas of each treatment combination is n. We further assume that the errors $\{e_{ijk}\}$ are i.i.d., having a normal distribution $N(0, \sigma^2)$.

Let

$$\overline{Y}_{ij} = \frac{1}{n} \sum_{l=1}^{n} Y_{ijl} \tag{12.7.21}$$

and

$$Q_{ij} = \sum_{l=1}^{n} (Y_{ijl} - \overline{Y}_{ij})^2 \tag{12.7.22}$$

for $i = 1, \ldots, a; j = 1, \ldots, b$. It can be shown that the least-squares estimators of τ_i^A, τ_j^B, and τ_{ij}^{AB} are, respectively,

$$\hat{\tau}_{i.}^A = \overline{Y}_{i.} - \overline{\overline{Y}} \qquad i = 1, \ldots, a$$
$$\hat{\tau}_j^B = \overline{Y}_{.j} - \overline{\overline{Y}} \qquad j = 1, \ldots, b \tag{12.7.23}$$

and

$$\hat{\tau}_{ij}^{AB} = \overline{Y}_{ij} - \overline{Y}_{i.} - \overline{Y}_{.j} + \overline{\overline{Y}} \tag{12.7.24}$$

where

$$\overline{Y}_{i.} = \frac{1}{b} \sum_{j=1}^{b} \overline{Y}_{ij} \tag{12.7.25}$$

and

$$\overline{Y}_{.j} = \frac{1}{a} \sum_{i=1}^{a} \overline{Y}_{ij} \tag{12.7.26}$$

Furthermore, an unbiased estimator of σ^2 is

$$\hat{\sigma}^2 = \frac{\sum_{i=1}^{a} \sum_{j=1}^{b} Q_{ij}}{ab(n-1)} \tag{12.7.27}$$

The standard errors of the estimators of the interactions are

$$\text{SE}\{\hat{\tau}_{ij}^{AB}\} = \frac{\hat{\sigma}}{\sqrt{n}} \left(\left(1 - \frac{1}{a}\right)\left(1 - \frac{1}{b}\right) \right)^{1/2} \tag{12.7.28}$$

for $i = 1, \ldots, a; j = 1, \ldots, b$. The standard errors of the estimators of the main effects are

$$\text{SE}\{\hat{\tau}_i^A\} = \frac{\hat{\sigma}}{\sqrt{nb}}\left(1 - \frac{1}{a}\right)^{1/2} \qquad i = 1, \ldots, a \qquad \text{(12.7.29)}$$

and

$$\text{SE}\{\hat{\tau}_j^B\} = \frac{\hat{\sigma}}{\sqrt{na}}\left(1 - \frac{1}{b}\right)^{1/2} \qquad j = 1, \ldots, b \qquad \text{(12.7.30)}$$

Simultaneous confidence limits at level $1 - \alpha$ for such parameters are obtained by

$$\hat{\tau}_i^A \pm S_\alpha \cdot \text{SE}\{\hat{\tau}_i^A\}$$
$$\hat{\tau}_j^B \pm S_\alpha \cdot \text{SE}\{\hat{\tau}_j^B\} \qquad \text{(12.7.31)}$$

and

$$\hat{\tau}_{ij}^{AB} \pm S_\alpha \text{SE}\{\hat{\tau}_{ij}^{AB}\} \qquad \text{(12.7.32)}$$

where

$$S_\alpha = ((ab - 1)F_{1-\alpha}[ab - 1, ab(n - 1)])^{1/2}$$

Any confidence interval that covers the value zero implies that the corresponding parameter is not significantly different from zero.

12.7.4 2^m Factorial Designs

2^m factorial designs are full factorials of m factors, each at two levels. The levels are labeled Low and High or 1 and 2. If the factors are categorical, then the labeling of the levels is arbitrary and the values of the main effects and interaction parameters depend on this arbitrary labeling. Here we discuss experiments in which the factor levels are measured on a continuous scale, as in the case of the factors affecting piston cycle time. The levels of the ith factor ($i = 1, \ldots, m$) are fixed at x_{i1} and x_{i2}, where $x_{i1} < x_{i2}$.

By simple transformation, all factor levels can be reduced to

$$c_i = \begin{cases} +1 & \text{if } x = x_{i2} \\ \\ -1 & \text{if } x = x_{i1} \end{cases} \qquad i = 1, \ldots, m$$

In such a factorial experiment, there are 2^m possible treatment combinations. Let $(i_1, \ldots, i_m)$ denote a treatment combination, where $i_1, \ldots, i_m$ are indices, such that

$$i_j = \begin{cases} 0 & \text{if } c_i = -1 \\ 1 & \text{if } c_i = 1 \end{cases}$$

Thus, if there are $m = 3$ factors, the number of possible treatment combinations is $2^3 = 8$. These are given in Table 12.18.

TABLE 12.18

Treatment combinations of a 2^3 experiment

ν	i_1	i_2	i_3
0	0	0	0
1	1	0	0
2	0	1	0
3	1	1	0
4	0	0	1
5	1	0	1
6	0	1	1
7	1	1	1

The index ν of the standard order is given by the formula

$$\nu = \sum_{j=1}^{m} i_j 2^{j-1} \tag{12.7.33}$$

Notice that ν ranges from 0 to $2^m - 1$. Tables of the treatment combinations for a 2^m factorial design, arranged in standard order, can be obtained by using program GENEDSN.EXE. This program is designed to yield fractional replications of 2^m designs, which will be explained in Section 12.8. For now, to obtain a 2^m design, insert the values $k = 0$ for "Degree of Fractionation," and the value $kb = 0$ for the "Fractional Block to Print." In Table 12.19, we present the output of this program for the case of a 2^5 factorial experiment. The program prints the labels, $l_j = i_j + 1$ $(j = 1, \ldots, m)$. In performing such an experiment, *the order in which the treatment combinations are applied should be randomized.*

A random order of the treatments can be obtained by MINITAB using the commands

```
MTB> Set C1
DATA> 1(0 : 31/1)1
DATA> End.
MTB> Sample 32 C1 C2.
```

The first command puts the integer 0 to 31 in column $C1$. The next command draws an RSWOR of size 32 from $C1$ and puts it in $C2$. The following random order was obtained:

9, 19, 6, 30, 11, 10, 17, 15, 18, 21, 1, 12, 14, 8, 13, 28

5, 16, 29, 24, 25, 27, 23, 30, 4, 20, 7, 22, 26, 2, 3, 31

TABLE 12.19

Labels in standard order for a 2^5 factorial design

v	l_1	l_2	l_3	l_4	l_5	v	l_1	l_2	l_3	l_4	l_5
0	1	1	1	1	1	16	1	1	1	1	2
1	2	1	1	1	1	17	2	1	1	1	2
2	1	2	1	1	1	18	1	2	1	1	2
3	2	2	1	1	1	19	2	2	1	1	2
4	1	1	2	1	1	20	1	1	2	1	2
5	2	1	2	1	1	21	2	1	2	1	2
6	1	2	2	1	1	22	1	2	2	1	2
7	2	2	2	1	1	23	2	2	2	1	2
8	1	1	1	2	1	24	1	1	1	2	2
9	2	1	1	2	1	25	2	1	1	2	2
10	1	2	1	2	1	26	1	2	1	2	2
11	2	2	1	2	1	27	2	2	1	2	2
12	1	1	2	2	1	28	1	1	2	2	2
13	2	1	2	2	1	29	2	1	2	2	2
14	1	2	2	2	1	30	1	2	2	2	2
15	2	2	2	2	1	31	2	2	2	2	2

Thus, in the first trial ($v = 9$), factors 2, 3, and 5 are at level 1 and factors 1 and 4 are at level 2. In the second trial ($v = 19$), factors 3 and 4 are at level 1 and factors 1, 2, and 5 are at level 2.

Let Y_v, $v = 0, 1, \ldots, 2^m - 1$, denote the yield of the vth treatment combination. We now discuss the estimation of main effects and interaction parameters. Starting with the simple case of two factors, we present the variables schematically in Table 12.20.

TABLE 12.20

Treatment means in a 2^2 design

Factor B	Factor A		Row
	1	**2**	**means**
1	$\overline{Y}_0$	$\overline{Y}_1$	$\overline{Y}_{1.}$
2	$\overline{Y}_2$	$\overline{Y}_3$	$\overline{Y}_{2.}$
Column means	$\overline{Y}_{.1}$	$\overline{Y}_{.2}$	$\overline{\overline{Y}}$

According to our previous definition, there are four main effects, τ_1^A, τ_2^A, τ_1^B, τ_2^B, and four interaction effects, τ_{11}^{AB}, τ_{12}^{AB}, τ_{21}^{AB}, τ_{22}^{AB}. But because $\tau_1^A + \tau_2^A = \tau_1^B + \tau_2^B = 0$, we can represent the main effects of A and B by τ_2^A and τ_2^B. Similarly, because

$\tau_{11}^{AB} + \tau_{12}^{AB} = 0 = \tau_{11}^{AB} + \tau_{21}^{AB}$ and $\tau_{12}^{AB} + \tau_{22}^{AB} = 0 = \tau_{21}^{AB} + \tau_{22}^{AB}$, we can represent the interaction effects by τ_{22}^{AB}.

The main effect τ_2^A is estimated by

$$\hat{\tau}_2^A = \overline{Y}_{.2} - \overline{\overline{Y}}$$
$$= \frac{1}{2}(\overline{Y}_1 + \overline{Y}_3) - \frac{1}{4}(\overline{Y}_0 + \overline{Y}_1 + \overline{Y}_2 + \overline{Y}_3)$$
$$= \frac{1}{4}(-\overline{Y}_0 + \overline{Y}_1 - \overline{Y}_2 + \overline{Y}_3)$$

The estimator of τ_2^B is

$$\hat{\tau}_2^B = \overline{Y}_{2.} - \overline{\overline{Y}}$$
$$= \frac{1}{2}(\overline{Y}_2 + \overline{Y}_3) - \frac{1}{4}(\overline{Y}_0 + \overline{Y}_1 + \overline{Y}_2 + \overline{Y}_3)$$
$$= \frac{1}{4}(-\overline{Y}_0 - \overline{Y}_1 + \overline{Y}_2 + \overline{Y}_3)$$

Finally, the estimator of τ_{22}^{AB} is

$$\hat{\tau}_{22}^{AB} = \overline{Y}_3 - \overline{Y}_{2.} - \overline{Y}_{.2} + \overline{\overline{Y}}$$
$$= \overline{Y}_3 - \frac{1}{2}(\overline{Y}_2 + \overline{Y}_3) - \frac{1}{2}(\overline{Y}_1 + \overline{Y}_3) + \frac{1}{4}(\overline{Y}_0 + \overline{Y}_1 + \overline{Y}_2 + \overline{Y}_3)$$
$$= \frac{1}{4}(\overline{Y}_0 - \overline{Y}_1 - \overline{Y}_2 + \overline{Y}_3)$$

The parameter μ is estimated by the grand mean $\overline{\overline{Y}} = \frac{1}{4}(\overline{Y}_0 + \overline{Y}_1 + \overline{Y}_2 + \overline{Y}_3)$. All these estimators can be presented in a matrix form as follows:

$$
\begin{bmatrix} \hat{\mu} \\ \hat{\tau}_2^A \\ \hat{\tau}_2^B \\ \hat{\tau}_{22}^{AB} \end{bmatrix} = \frac{1}{4} \begin{bmatrix} 1 & 1 & 1 & 1 \\ -1 & 1 & -1 & 1 \\ -1 & -1 & 1 & 1 \\ 1 & -1 & -1 & 1 \end{bmatrix} \cdot \begin{bmatrix} \overline{Y}_0 \\ \overline{Y}_1 \\ \overline{Y}_2 \\ \overline{Y}_3 \end{bmatrix}
$$

The indices in a 2^2 design are given in the following 4-×-2 matrix:

$$
D_{2^2} = \begin{bmatrix} 1 & 1 \\ 2 & 1 \\ 1 & 2 \\ 2 & 2 \end{bmatrix}
$$

The corresponding C coefficients are the second and third columns in the matrix:

$$
C_{2^2} = \begin{bmatrix} 1 & -1 & -1 & 1 \\ 1 & 1 & -1 & -1 \\ 1 & -1 & 1 & -1 \\ 1 & 1 & 1 & 1 \end{bmatrix}
$$

The fourth column of this matrix is the product of the elements in the second and third columns. Notice also that the linear model for the yield vector is

$$
\begin{bmatrix} \bar{Y}_0 \\ \bar{Y}_1 \\ \bar{Y}_2 \\ \bar{Y}_3 \end{bmatrix} = \begin{bmatrix} 1 & -1 & -1 & 1 \\ 1 & 1 & -1 & -1 \\ 1 & -1 & 1 & -1 \\ 1 & 1 & 1 & 1 \end{bmatrix} \begin{bmatrix} \mu \\ \tau_2^A \\ \tau_2^B \\ \tau_{22}^{AB} \end{bmatrix} + \begin{bmatrix} e_1 \\ e_2 \\ e_2 \\ e_4 \end{bmatrix}
$$

where e_1, e_2, e_3, and e_4 are independent random variables, with $E\{e_i\} = 0$ and $V\{e_i\} = \sigma^2, i = 1, 2, \ldots, 4$.

Let $\mathbf{Y}^{(4)} = (Y_0, Y_1, Y_2, Y_3)'$, $\boldsymbol{\theta}^{(4)} = (\mu, \tau_2^A, \tau_2^B, \tau_{22}^{AB})'$, and $\mathbf{e}^{(4)} = (e_1, e_2, e_3, e_4)'$. Then the model is

$$
\mathbf{Y}^{(4)} = C_{2^2}\boldsymbol{\theta}^{(4)} + \mathbf{e}^{(4)}
$$

This is the usual linear model for multiple regression. The least-squares estimator of $\boldsymbol{\theta}^{(4)}$ is

$$
\hat{\boldsymbol{\theta}}^{(4)} = [C'_{2^2}C_{2^2}]^{-1}C'_{2^2}\mathbf{Y}^{(4)}
$$

The matrix C_{2^2} has orthogonal column (row) vectors and

$$
C'_{2^2}C_{2^2} = 4I_4
$$

where I_4 is the identity matrix of rank 4. Therefore,

$$
\hat{\boldsymbol{\theta}}^{(4)} = \frac{1}{4}C'_{2^2}\mathbf{Y}^{(4)}
$$

$$
= \frac{1}{4}\begin{bmatrix} 1 & 1 & 1 & 1 \\ -1 & 1 & -1 & 1 \\ -1 & -1 & 1 & 1 \\ 1 & -1 & -1 & 1 \end{bmatrix}\begin{bmatrix} \bar{Y}_0 \\ \bar{Y}_1 \\ \bar{Y}_2 \\ \bar{Y}_3 \end{bmatrix}
$$

This is identical to the solution obtained earlier.

The estimators of the main effects and interactions are the least-squares estimators, as was mentioned before.

This can now be generalized to the case of m factors. For a model with m factors, there are 2^m parameters. The mean μ, m main effects $\tau^1, \ldots, \tau^m$, $\binom{m}{2}$ first-order interactions $\tau^{ij}, i \neq j = 1, \ldots, m$, $\binom{m}{3}$ second-order interactions $\tau^{ijk}, i \neq j \neq k$, and so on. We can now order the parameters in a standard manner as follows. Each of the 2^m parameters can be represented by a binary vector $(j_1, \ldots, j_m)$, where $j_i = 0, 1$ $(i = 1, \ldots, m)$. The vector $(0, 0, \ldots, 0)$ represents the grand mean μ. A vector $(0, 0, \ldots, 1, 0, \ldots, 0)$ where the 1 is the ith component represents the main effect of the ith factor $(i = 1, \ldots, m)$. A vector with two 1s, at the ith and jth components $(i = 1, \ldots, m - 1; j = i + 1, \ldots, m)$, represents the first-order interaction between factor i and factor j. A vector with three 1s, at the i, j, and k components, represents the second-order interaction between factors i, j, and k, and so on.

Let $\omega = \sum_{i=1}^{m} j_i 2^{i-1}$ and β_ω be the parameter represented by the vector with index ω. For example, β_3 corresponds to $(1, 1, 0, \ldots, 0)$, which represents the first-order interaction between factors 1 and 2.

Let $\mathbf{Y}^{(2^m)}$ be the yield vector, whose components are arranged in the standard order, with index $v = 0, 1, 2, \ldots, 2^m - 1$. Let C_{2^m} be the matrix of coefficients obtained recursively by the equations

$$C_2 = \begin{bmatrix} 1 & -1 \\ 1 & 1 \end{bmatrix} \tag{12.7.34}$$

and

$$C_{2^l} = \begin{bmatrix} C_{2^{l-1}} & -C_{2^{l-1}} \\ C_{2^{l-1}} & C_{2^{l-1}} \end{bmatrix} \tag{12.7.35}$$

$l = 2, 3, \ldots, m$. Then the linear model relating $\mathbf{Y}^{(2^m)}$ to $\boldsymbol{\beta}^{(2^m)}$ is

$$\mathbf{Y}^{(2^m)} = C_{2^m} \cdot \boldsymbol{\beta}^{(2^m)} + \mathbf{e}^{(2^m)} \tag{12.7.36}$$

where

$$\boldsymbol{\beta}^{(2^m)} = (\beta_0, \beta_1, \ldots, \beta_{2^m-1})'$$

Because the column vectors of C_{2^m} are orthogonal, $(C_{2^m})'C_{2^m} = 2^m I_{2^m}$, the least-squares estimator (LSE) of $\boldsymbol{\beta}^{(2^m)}$ is

$$\hat{\boldsymbol{\beta}}^{(2^m)} = \frac{1}{2^m}(C_{2^m})'\mathbf{Y}^{(2^m)} \tag{12.7.37}$$

Accordingly, the LSE of β_ω is

$$\hat{\beta}_\omega = \frac{1}{2^m} \sum_{v=0}^{2^m-1} c^{(2^m)}_{(v+1),(\omega+1)} Y_v \tag{12.7.38}$$

where $c^{(2^m)}_{ij}$ is the ith row and jth column element of C_{2^m}; that is, we multiply the components of $\mathbf{Y}^{(2^m)}$ by those of the column of C_{2^m}, corresponding to the parameter β_ω, and divide the sum of products by 2^m.

We do not have to estimate all 2^m parameters; we can limit our attention to only those parameters that are of interest, as will be shown in the following example.

Because $c^{(2^m)}_{ij} = \pm 1$, the variance of $\hat{\beta}_\omega$ is

$$V\{\hat{\beta}_\omega\} = \frac{\sigma^2}{2^m} \quad \text{for all } \omega = 0, \ldots, 2^m - 1 \tag{12.7.39}$$

Finally, if every treatment combination is repeated n times, estimation of the parameters is based on the means $\overline{Y}_v$ of n replications. The variance of $\hat{\beta}_\omega$ becomes

$$V\{\hat{\beta}_\omega\} = \frac{\sigma^2}{n2^m} \tag{12.7.40}$$

The variance σ^2 can be estimated by the pooled variance estimator obtained from the between-replication variance within each treatment combination. That is, if Y_{vj}, $j = 1, \ldots, n$, are the observed values at the vth treatment combination, then

$$\hat{\sigma}^2 = \frac{1}{(n-1)2^m} \sum_{v=0}^{2^m-1} \sum_{j=1}^{n} (Y_{vj} - \overline{Y}_v)^2 \tag{12.7.41}$$

EXAMPLE **12.8**   In Example 12.7, we studied the effects of two factors on the cycle time of a piston in a gas turbine, keeping the other five factors fixed. In this example, we perform a 2^5 experiment with the piston, varying factors A, B, C, D, and F at two levels, and keeping the atmospheric pressure (factor E) fixed at 90,000 (N/m^2) and the filling gas temperature (factor G) at 340 (K). The two levels of each factor are those specified in Example 12.7 as the limits of the experimental range. Thus, for example, the low level of piston weight (factor A) is 30 (kg) and its high level is 60 (kg). The treatment combinations are listed in Table 12.21.

The simulation program according to which the cycle times are generated is TURB5.EXE. The number of replications is $n = 5$. In file OTURB5.DAT, we find

TABLE   **12.21**
Labels of treatment combinations and average response

A	B	C	D	F	$\overline{Y}$
1	1	1	1	1	0.929
2	1	1	1	1	1.111
1	2	1	1	1	0.191
2	2	1	1	1	0.305
1	1	2	1	1	1.072
2	1	2	1	1	1.466
1	2	2	1	1	0.862
2	2	2	1	1	1.318
1	1	1	2	1	0.209
2	1	1	2	1	0.340
1	2	1	2	1	0.123
2	2	1	2	1	0.167
1	1	2	2	1	0.484
2	1	2	2	1	0.690
1	2	2	2	1	0.464
2	2	2	2	1	0.667
1	1	1	1	2	0.446
2	1	1	1	2	0.324
1	2	1	1	2	0.224
2	2	1	1	2	0.294
1	1	2	1	2	1.067
2	1	2	1	2	1.390
1	2	2	1	2	0.917
2	2	2	1	2	1.341
1	1	1	2	2	0.426
2	1	1	2	2	0.494
1	2	1	2	2	0.271
2	2	1	2	2	0.202
1	1	2	2	2	0.482
2	1	2	2	2	0.681
1	2	2	2	2	0.462
2	2	2	2	2	0.649

the means $\overline{Y}_\nu$ and standard deviations S_ν of the five observations in each treatment combination. We now use MINITAB to analyze the results, and estimate the parameters. We can retrieve the file OTURB5.MTW to follow the steps of the analysis. There are 22 columns of 32 rows in the file. Columns $C1$–$C5$ contain the treatment combination labels, as in Table 12.21. Columns $C6$–$C10$ contain the coefficients $c_{\nu\omega}^{(32)}$, corresponding to the main effects of the five factors. These were obtained from columns $C1$–$C5$ by the transformation $C = 2\ (i - 1.5)$. The following MINITAB command shows this transformation for column $C6$:

```
MTB> let C6 = 2 * (C1 − 1.5).
```

Columns $C11$ and $C12$ contain $\overline{Y}_\nu$ and S_ν, respectively. An estimate of σ^2 can be obtained from column $C12$ by the MINITAB commands

```
MTB> let k1 = mean(C12 **2)
MTB> print k1.
```

The constant $k1$ represents $\hat{\sigma}^2$. We obtain the value $\hat{\sigma}^2 = 0.02898$. Columns $C13$–$C22$ contain the $\binom{5}{2} = 10$ products of pairs of columns from $C6$ to $C10$— for example, $C13 = C6 * C7$, $C14 = C6 * C8$, and so on. We need these columns to estimate the first-order interactions. The estimated variance of all LSE of the parameters is

$$\hat{V}\{\hat{\beta}_\omega\} = \frac{\hat{\sigma}^2}{5 \times 32} = 0.0001811$$

Correspondingly, the square root of the variance is the standard error of the estimates—that is, $\mathrm{SE}\{\hat{\beta}_\omega\} = 0.01346$. To estimate the main effect of A, we use the commands:

```
MTB> let k2 = mean(C6 * C11)
MTB> print k2
```

We obtain the value $\hat{\beta}_1 = 0.0871$. Table 12.22 gives the LSEs of all 5 main effects and 10 first-order interactions. The SE values in the table are the standard errors of the estimates and the t values are $t = \mathrm{LSE}/\mathrm{SE}$.

Values of t that are greater in magnitude than 2.6 are significant at $\alpha = .02$. However, if we want all 15 tests to have simultaneously a level of significance of $\alpha = .05$, we should use as a critical value the Scheffé coefficient $\sqrt{31 \times F_{.95}[31, 128]} = 6.91$ because all LSEs are contrasts of 32 means. In Table 12.22, we marked with one $*$ the t values greater in magnitude than 2.6, and with $**$, those greater than 7. These

TABLE 12.22

LSE of main effects and interactions

Effect	LSE	SE	t
A	0.08781	0.01346	6.52*
B	−0.09856	0.01346	−7.32**
C	0.24862	0.01346	18.47**
D	−0.20144	0.01346	14.97**
F	−0.02275	0.01346	−1.69
AB	0.00150	0.01346	0.11
AC	0.06169	0.01346	4.58*
AD	−0.02725	0.01346	−2.02
AF	−0.02031	0.01346	−1.51
BC	0.05781	0.01346	4.29*
BD	0.04850	0.01346	3.60*
BF	0.03919	0.01346	2.91*
CD	−0.10194	0.01346	−7.57**
CF	0.02063	0.01346	1.53
DF	0.05544	0.01346	4.12*

estimates of the main effects and first-order interactions can also be obtained by running a multiple regression of C11 on 15 predictors in C6–C10 and C13–C22.

When we execute this regression, we obtain $R^2 = .934$. The variance around the regression surface is $s_{y|x}^2 = 0.02053$. This is significantly greater than $\hat{\sigma}^2/5 = 0.005795$. Thus, there might be significant high-order interactions that have not been estimated.

Figure 12.7 is a graphical display of the main effects of factors $A, B, C, D,$ and F. The left limit of a line shows the average response at a low level and the right limit

FIGURE 12.7

Main effects plot

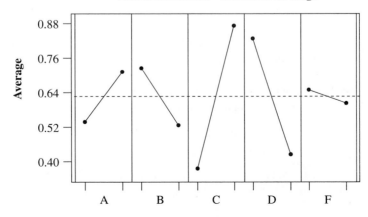

that at a high level. Factors C and D seem to have the highest effect, as is shown by the t values in Table 12.22. Figure 12.8 shows the two-way interactions of the various factors. Interaction $C * D$ is the most pronounced.   ∎

FIGURE   12.8
Two-way interaction plots

Interaction Plot – Means for Average

A				
	B			
		C		
			D	
				F

12.7.5 3^m Factorial Designs

In this section, we discuss estimation and testing of model parameters when the design is full factorial, of m factors, each at $p = 3$ levels. We assume that the levels are measured on a continuous scale and are labeled Low, Medium, and High. We introduce the indices i_j $(j = 1, \ldots, m)$, which assume the values 0, 1, 2 for the Low, Medium, and High levels, correspondingly, of each factor. Thus, we have 3^m treatment combinations, represented by vectors of indices $(i_1, i_2, \ldots, i_m)$. The index v of the standard order of treatment combinations is

$$v = \sum_{j=1}^{m} i_j 3^{j-1} \tag{12.7.42}$$

This index ranges from 0 to $3^m - 1$. Let $\overline{Y}_v$ denote the yield of n replicas of the vth treatment combination, $n \geq 1$.

Because we obtain the yield at three levels of each factor, we can estimate not only the linear effects but also the quadratic effects of each factor. For example, if we have $m = 2$ factors, we can use a multiple regression method to fit the model

$$Y = \beta_0 + \beta_1 x_1 + \beta_2 x_1^2 + \beta_3 x_2 + \beta_4 x_1 x_2 + \\ \beta_5 x_1^2 x_2 + \beta_6 x_2^2 + \beta_7 x_1 x_2^2 + \beta_8 x_1^2 x_2^2 + e \tag{12.7.43}$$

This is a quadratic model in two variables. β_1 and β_3 represent the linear effects of x_1 and x_2. β_2 and β_6 represent the quadratic effects of x_1 and x_2. The other coefficients represent interaction effects. β_4 represents the linear × linear interaction, β_5 represents the quadratic × linear interaction, and so on. We have two main effects for each factor (linear and quadratic) and four interaction effects.

Generally, if there are m factors, we have, in addition to β_0, $2m$ parameters for main effects (linear and quadratic), $2^2 \binom{m}{2}$ parameters for interactions between 2 factors, $2^3 \binom{m}{3}$ interactions between 3 factors, and so on. Generally, we have 3^m parameters, where

$$3^m = \sum_{j=0}^{m} 2^j \binom{m}{j}$$

As in the case of 2^m models, each parameter in a 3^m model is represented by a vector of m indices $(\lambda_1, \lambda_2, \ldots, \lambda_m)$, where $\lambda_j = 0, 1, 2$. Thus, for example, the vector $(0, 0, \ldots, 0)$ represents the grand mean $\mu = \beta_0$. A vector $(0, \ldots, 0, 1, 0, \ldots, 0)$ with 1 at the ith component represents the linear effect of the ith factor. Similarly, $(0, 0, \ldots, 0, 2, 0, \ldots, 0)$ represents the quadratic effect of the ith factor. Two indices equal to 1 and all the rest zero represent the linear × linear interaction of the ith and jth factors, and so on. The standard order of the parameters is

$$\omega = \sum_{j=1}^{m} \lambda_j 3^{j-1} \qquad \omega = 0, \ldots, 3^m - 1$$

If m is not too large, it is also customary to label the factors by the letters $A, B, C, \ldots$ and the parameters by $A^{\lambda_1} B^{\lambda_2} C^{\lambda_3} \ldots$. In this notation, a letter to the zero power is omitted. Table 12.23 lists the parameters of a 3^3 system.

T A B L E   12.23
Main effects and interactions of a 3^3 factorial

ω	Parameter	Indices	ω	Parameter	Indices
0	Mean	(0,0,0)	15	B^2C	(0,2,1)
1	A	(1,0,0)	16	AB^2C	(1,2,1)
2	A^2	(2,0,0)	17	A^2B^2C	(2,2,1)
3	B	(0,1,0)	18	C^2	(0,0,2)
4	AB	(1,1,0)	19	AC^2	(1,0,2)
5	A^2B	(2,1,0)	20	A^2C^2	(2,0,2)
6	B^2	(0,2,0)	21	BC^2	(0,1,2)
7	AB^2	(1,2,0)	22	ABC^2	(1,1,2)
8	A^2B^2	(2,2,0)	23	A^2BC^2	(2,1,2)
9	C	(0,0,1)	24	B^2C^2	(0,2,2)
10	AC	(1,0,1)	25	AB^2C^2	(1,2,2)
11	A^2C	(2,0,1)	26	$A^2B^2C^2$	(2,2,2)
12	BC	(0,1,1)			
13	ABC	(1,1,1)			
14	A^2BC	(2,1,1)			

It is easy to transform the x values of each factor to

$$X_j = \begin{cases} -1 & \text{if } i_j = 0 \\ 0 & \text{if } i_j = 1 \\ 1 & \text{if } i_j = 2 \end{cases}$$

However, the matrix of coefficients X that is obtained when we have quadratic and interaction parameters is not orthogonal. Thus, we need to use the computer, with the usual multiple regression program, to obtain the least-squares estimators. Another approach is to redefine the effects so that the statistical model will be linear, with a matrix having coefficients obtained by the method of orthogonal polynomials (see Draper and Smith, 1981, p. 166). Thus, consider the model

$$\mathbf{Y}^{(3^m)} = \mathbf{\Psi}_{(3^m)} \boldsymbol{\gamma}^{(3^m)} + \mathbf{e}^{(3^m)} \tag{12.7.44}$$

where

$$\mathbf{Y}^{(3^m)} = (\overline{Y}_0, \ldots, \overline{Y}_{3^m-1})'$$

and $\mathbf{e}^{(3^m)} = (e_0, \ldots, e_{3^m-1})'$ is a vector of random variables with $E\{e_\nu\} = 0$ and $V\{e_\nu\} = \sigma^2$ for all $\nu = 0, \ldots, 3^m - 1$. Moreover,

$$\mathbf{\Psi}_{(3)} = \begin{bmatrix} 1 & -1 & 1 \\ 1 & 0 & -2 \\ 1 & 1 & 1 \end{bmatrix} \tag{12.7.45}$$

and for $m \geq 2$,

$$\mathbf{\Psi}_{(3^m)} = \begin{bmatrix} \mathbf{\Psi}_{(3^{m-1})} & -\mathbf{\Psi}_{(3^{m-1})} & \mathbf{\Psi}_{(3^{m-1})} \\ \mathbf{\Psi}_{(3^{m-1})} & 0 & -2\mathbf{\Psi}_{(3^{m-1})} \\ \mathbf{\Psi}_{(3^{m-1})} & \mathbf{\Psi}_{(3^{m-1})} & \mathbf{\Psi}_{(3^{m-1})} \end{bmatrix} \tag{12.7.46}$$

The matrices $\mathbf{\Psi}_{(3^m)}$ have orthogonal column vectors and

$$(\mathbf{\Psi}_{(3^m)})'(\mathbf{\Psi}_{(3^m)}) = \mathbf{\Delta}_{(3^m)} \tag{12.7.47}$$

where $\mathbf{\Delta}_{(3^m)}$ is a diagonal matrix whose diagonal elements are equal to the sum of squares of the elements in the corresponding column of $\mathbf{\Psi}_{(3^m)}$. For example, for $m = 1$,

$$\mathbf{\Delta}_{(3)} = \begin{pmatrix} 3 & 0 & 0 \\ 0 & 2 & 0 \\ 0 & 0 & 6 \end{pmatrix}$$

For $m = 2$, we obtain

$$\mathbf{\Psi}_{(9)} = \begin{bmatrix} 1 & -1 & 1 & -1 & 1 & -1 & 1 & -1 & 1 \\ 1 & 0 & -2 & -1 & 0 & 2 & 1 & 0 & -2 \\ 1 & 1 & 1 & -1 & -1 & -1 & 1 & 1 & 1 \\ 1 & -1 & 1 & 0 & 0 & 0 & -2 & 2 & -2 \\ 1 & 0 & -2 & 0 & 0 & 0 & -2 & 0 & 4 \\ 1 & 1 & 1 & 0 & 0 & 0 & -2 & -2 & -2 \\ 1 & -1 & 1 & 1 & -1 & 1 & 1 & -1 & 1 \\ 1 & 0 & -2 & 1 & 0 & -2 & 1 & 0 & -2 \\ 1 & 1 & 1 & 1 & 1 & 1 & 1 & 1 & 1 \end{bmatrix}$$

and

$$\Delta_{(9)} = \begin{bmatrix} 9 & & & & & & & & \\ & 6 & & & & & & & \\ & & 18 & & & & 0 & & \\ & & & 6 & & & & & \\ & & & & 4 & & & & \\ & & & & & 12 & & & \\ & & 0 & & & & 18 & & \\ & & & & & & & 12 & \\ & & & & & & & & 36 \end{bmatrix}$$

Thus, the LSE of $\boldsymbol{\gamma}^{(3^m)}$ is

$$\hat{\boldsymbol{\gamma}}^{(3^m)} = \Delta_{(3^m)}^{-1}(\Psi_{(3^m)})'\mathbf{Y}^{(3^m)} \tag{12.7.48}$$

These LSEs are the best linear unbiased estimators and

$$V\{\hat{\gamma}_\omega\} = \frac{\sigma^2}{n\sum_{i=1}^{3^m}(\Psi_{i,\omega+1}^{(3^m)})^2}, \qquad \omega = 0, 1, \ldots, 3^m - 1. \tag{12.7.49}$$

If the number of replicas, n, is greater than 1, then σ^2 can be estimated by

$$\hat{\sigma}^2 = \frac{1}{3^m(n-1)} \sum_{v=0}^{3^m-1} \sum_{l=1}^{n}(Y_{vl} - \bar{Y}_v)^2 \tag{12.7.50}$$

If $n = 1$, we can estimate σ^2 if it is known a priori that some parameters γ_ω are zero. Let Λ_0 be the set of all parameters that can be assumed to be negligible. Let K_0 be the number of elements of Λ_0. If $\omega \in \Lambda_0$, then $\hat{\gamma}_\omega^2 \left(\sum_{j=1}^{3^m}(\Psi_{i,\omega+1}^{(3^m)})^2\right)$ is distributed like $\sigma^2\chi^2[1]$. Therefore, an unbiased estimator of σ^2 is

$$\hat{\sigma}^2 = \frac{1}{K_0} \sum_{\omega \in \Lambda_0} \hat{\gamma}_\omega^2 \left(\sum_{j=1}^{3^m}(\Psi_{j,\omega+1}^{(3^m)})^2\right) \tag{12.7.51}$$

EXAMPLE 12.9 Oikawa and Oka (1987) reported the results of a 3^3 experiment to investigate the effects of three factors, A, B, C, on the stress levels of a membrane Y. The data are given in file STRESS.DAT. The first three columns of the data file provide the levels of the three factors, and column 4 presents the stress values. The matrix $\Psi_{(3^3)}$ and the stress data are given in the MINITAB file STRESS.MTW. The first 27 columns are those of the matrix of coefficients $\Psi_{(3^3)}$. Column $C28$ contains the stress values of the corresponding treatment combinations.

The LSE of the parameters can be easily obtained by multiplying column $C28$ by the column of $\Psi_{(3^3)}$ representing the parameter, and dividing by the corresponding

$\Delta_{(3^3)}$ value. For example, to estimate the linear effect of A, we use the MINITAB command

MTB> let k1 = sum(C2 * C28)/sum(C2 ** 2).

The values of the LSE and their standard errors are given in Table 12.24. The formula for the variance of an LSE is

$$V\{\hat{\gamma}_\omega\} = \sigma^2/n \left(\sum_{i=1}^{3^3} (\Psi_{i,\omega+1}^{(3^3)})^2 \right) \qquad (12.7.52)$$

where $\psi_{i,j}^{(3^3)}$ is the coefficient of $\Psi_{(3^3)}$ in the ith row and jth column.

TABLE 12.24
The LSE of the parameters of the 3^3 system

Parameter	LSE	SE	Significance
A	44.917	2.309	
A^2	−1.843	1.333	n.s.
B	−42.494	2.309	
AB	−16.558	2.828	
A^2B	−1.897	1.633	n.s.
B^2	6.557	1.333	
AB^2	1.942	1.633	n.s.
A^2B^2	−0.171	0.943	n.s.
C	26.817	2.309	
AC	22.617	2.828	
A^2C	0.067	1.633	n.s.
BC	−3.908	2.828	n.s.
ABC	2.013	3.463	n.s.
A^2BC	1.121	1.999	n.s.
B^2C	−0.708	1.633	n.s.
AB^2C	0.246	1.099	n.s.
A^2B^2C	0.287	1.154	n.s.
C^2	−9.165	1.333	
AC^2	−4.833	1.633	
A^2C^2	0.209	0.943	n.s.
BC^2	2.803	1.633	n.s.
ABC^2	−0.879	1.999	n.s.
A^2BC^2	0.851	1.154	n.s.
B^2C^2	−0.216	0.943	n.s.
AB^2C^2	0.287	1.154	n.s.
$A^2B^2C^2$	0.059	0.666	n.s.

Suppose that for technological considerations we decide that all interaction parameters involving quadratic components are negligible (zero). In this case, we can estimate σ^2 by $\hat{\hat{\sigma}}^2$. In this example, the set Λ_0 contains 16 parameters:

$$\Lambda_0 = \{A^2B, AB^2, A^2B^2, A^2C, A^2BC, B^2C, AB^2C, A^2B^2C, AC^2,$$
$$A^2C^2, BC^2, ABC^2, A^2BC^2, B^2C^2, AB^2C^2, A^2B^2C^2\}$$

Thus, $K_0 = 16$ and the estimator $\hat{\hat{\sigma}}^2$ has 16 degrees of freedom. The estimate of σ^2 is $\hat{\hat{\sigma}}^2 = 95.95$. The estimates of the standard errors (SE) in Table 12.24 use this estimate. All nonsignificant parameters are denoted by n.s. Figures 12.9 and 12.10 show the main effects and interaction plots. ∎

FIGURE **12.9**
Main effects plot for a 3^3 design

Main Effects Plot – Means for STRESS

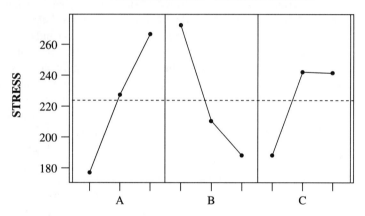

FIGURE **12.10**
Interaction plots for a 3^3 design

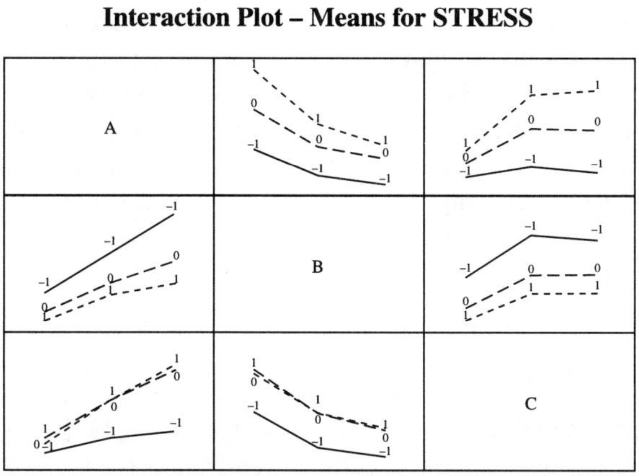

Interaction Plot – Means for STRESS

12.8

Blocking and Fractional Replications of 2^m Factorial Designs

Full factorial experiments with a large number of factors might be impractical. For example, if there are $m = 12$ factors, even at $p = 2$ levels, the total number of treatment combinations is $2^{12} = 4096$. This size for an experiment is generally not necessary because most high-order interactions may be negligible, and there is no need to estimate 4096 parameters. If only main effects and first-order interactions are considered important, and all the rest are believed to be negligible, we have to estimate and test only $1 + 12 + \binom{12}{2} = 79$ parameters. A fraction of the experiment, of size $2^7 = 128$, would be sufficient. Such a fraction can even be replicated several times. The question is: How do we choose a fraction of the full factorial in such a way that desirable properties of orthogonality, equal variances of estimators, and so on will be kept, and the parameters of interest will be estimable on an unbiased basis?

The problem of fractioning the full factorial experiment arises also when the full factorial cannot be performed in one block—that is, when several blocks are required to accommodate all the treatment conditions. For example, a 2^5 experiment is designed, but only $8 = 2^3$ treatment combinations can be performed in any given block (day, machine, and so on). We have to design the fractions that will be assigned to each block in such a way that if there are significant differences between the blocks, the block effects will not confound or obscure parameters of interest. Such fractions are called **fractional factorial designs**. We start with a simple illustration of the fractionization procedure and the properties of the ensuing estimators.

Consider 3 factors—*A*, *B*, and *C*—at 2 levels. We wish to partition the $2^3 = 8$ treatment combinations to two fractions of size $2^2 = 4$. Let $\lambda_i = 0, 1$ ($i = 1, 2, 3$) and let $A^{\lambda_1} B^{\lambda_2} C^{\lambda_3}$ represent the 8 parameters. One way of representing the treatment combinations, when the number of factors is not large is by using lowercase letters $a, b, c, \ldots$. The letter a indicates that factor A is at the High level ($i_1 = 1$) and so on with the other factors. The absence of a letter indicates that the corresponding factor is at the Low level. The symbol (1) indicates that all levels are Low. Treatment combinations and associated coefficients $c_{ij}^{(2^3)}$ are shown in Table 12.25.

Suppose that the treatment combinations should be partitioned into two fractional replications (blocks) of size 4. We have to choose a parameter, called a *defining parameter*, according to which the partition will be done. This defining parameter is, in a sense, sacrificed because its effects will be either confounded with the block effects or inestimable if only one block of trials is performed. Let us choose the parameter ABC as a defining parameter. We partition the treatment combinations into two blocks according to the signs of the coefficients corresponding to ABC. These are the products of the coefficients in the A, B, and C columns. Thus, two blocks are obtained:

$$B_- = \{(1), ab, ac, bc\}$$
$$B_+ = \{a, b, c, abc\}$$

If 2^m treatment combinations are partitioned to $2^k = 2$ blocks, we say that the *degree of fractionation* is $k = 1$, the fractional replication is of size 2^{m-k}, and the design

TABLE   12.25

A 2^3 factorial

Treatments	Main effects			Defining parameter
	A	*B*	*C*	*ABC*
(1)	−1	−1	−1	−1
a	1	−1	−1	1
b	−1	1	−1	1
ab	1	1	−1	−1
c	−1	−1	1	1
ac	1	−1	1	−1
bc	−1	1	1	−1
abc	1	1	1	1

is $1/2^k$ fraction of a full factorial. For example, if $m = 5$ factors and we want to partition into 4 blocks of 8, the degree of fractionization is $k = 2$. We select $k = 2$ parameters to serve as defining parameters—say, *ACE* and *BDE*—and we partition the treatment combinations according to the signs ± 1 of the coefficients in the *ACE* and *BDE* columns. This becomes very cumbersome if m and k are large. Program GENEDSN.EXE performs this partitioning and prints into a file the block requested. We will return to this later.

Let's now check the properties of estimators in the 2^{3-1} fractional replication if only the block B_- is performed. The defining parameter is *ABC*. Let $Y(1)$ be the response of treatment combination (1); this is Y_0 in the standard order notation. Let $Y(a)$ be the response of a and so on. The results of performing B_-, with the associated coefficients of parameters of interest, can be presented as in Table 12.26.

We see that the six columns of coefficients are orthogonal to each other, and each column has two −1s and two +1s. The LSEs of the parameters are orthogonal contrasts given by

$$\hat{A} = \frac{1}{4}(-Y(1) + Y(ab) + Y(ac) - Y(bc))$$

$$\hat{B} = \frac{1}{4}(-Y(1) + Y(ab) - Y(ac) + Y(bc))$$

TABLE   12.26

Coefficients and response for several treatment combinations (t.c.)

t.c.	*A*	*B*	*C*	*AB*	*AC*	*BC*	*Y*
(1)	−1	−1	−1	1	1	1	$Y(1)$
ab	1	1	−1	1	−1	−1	$Y(ab)$
ac	1	−1	1	−1	1	−1	$Y(ac)$
bc	−1	1	1	−1	−1	1	$Y(bc)$

and so on. The variances of all these estimators, when $n = 1$, are equal to $\sigma^2/4$. However, the estimators might be biased. The expected value of the first estimator is

$$E\{\hat{A}\} = \frac{1}{4}(-E\{Y(1)\} + E\{Y(ab)\} + E\{Y(ac)\} - E\{Y(bc)\})$$

Now,

$$E\{Y(1)\} = \mu - A - B - C + AB + AC + BC - ABC$$
$$E\{Y(ab)\} = \mu + A + B - C + AB - AC - BC - ABC$$
$$E\{Y(ac)\} = \mu + A - B + C - AB + AC - BC - ABC$$

and

$$E\{Y(bc)\} = \mu - A + B + C - AB - AC + BC - ABC$$

Collecting all these terms, the result is

$$E\{\hat{A}\} = A - BC$$

Similarly, one can show that

$$E\{\hat{B}\} = B - AC$$
$$E\{\hat{C}\} = C - AB$$
$$E\{\hat{AB}\} = AB - C$$

and so on. The LSEs of all the parameters are biased unless $AB = AC = BC = 0$. The bias terms are called *aliases*. Aliases are obtained by multiplying the parameter of interest by the defining parameter when any letter raised to the power 2 is eliminated—for example,

$$A \otimes ABC = A^2 BC = BC$$

The sign of the alias is the sign of the block. Because we used the block B_-, all the aliases appear with a negative sign.

The general rules for finding the aliases in 2^{m-k} designs is as follows. To obtain a 2^{m-k} fractional replication, one needs k defining parameters. The multiplication operation of parameters was just illustrated. The k defining parameters should be independent in the sense that none can be obtained as a product of the others. Such independent defining parameters are called *generators*. For example, to choose four defining parameters when the factors are A, B, C, D, E, F, G, and H, first choose two parameters, such as $ABCH$ and $ABEFG$. The product of these two is $CEFGH$. Next, for the third defining parameter, choose any one that is different from $\{ABCH, ABEFG, CEFGH\}$. Suppose one chooses $BDEFH$. The three independent parameters, $ABCH$, $ABEFG$, and $BDEFH$, generate a subgroup of eight parameters, including the mean μ. These are:

μ	$BDEFH$
$ABCH$	$ACDEF$
$ABEFG$	$ADGH$
$CEFGH$	$BCDG$

Finally, to choose a fourth independent defining parameter, we can choose any parameter not listed among these eight.

Suppose that the parameter *BCEFH* is chosen. Now we obtain a subgroup of $2^4 = 16$ defining parameters by adding to the 8 just listed their products with *BCEFH*. Thus, this subgroup is

μ	*BCEFH*
ABCH	*AEF*
ABEFG	*ACGH*
CEFGH	*BG*
BDEFH	*CD*
ACDEF	*ABDH*
ADGH	*ABCDEFG*
BCDG	*DEFGH*

Notice that the subgroup, excluding the mean, includes two first-order interactions: *CD* and *BG*. This shows that the choice of defining parameters was not a good one because the aliases that will be created by these defining parameters will include main effects and other low-order interactions.

Given a subgroup of defining parameters, the aliases of a given parameter are obtained by multiplying the parameter by the defining parameters. Table 12.27 lists the aliases of the eight main effects with respect to the subgroup of 2^4 defining parameters.

In this table, most of the aliases to the main effects are high-order interactions, which are generally negligible. However, among the aliases to *A* there is *EF*. Among the aliases to *B*, there is the main effect *G*. Among the aliases to *C* there is *D*, and

TABLE 12.27

Aliases to the main effects in a 2^{8-4} design, with generators *ABCH*, *ABEFG*, *BDEFH*, and *BCEFH*

Main effects	Aliases
A	*BCH, BEFG, ACEFGH, ABDEFH, CDEF, DGH, ABCDG, ABCEFH,* **EF***, CGH, ABG, ACD, BDH, BCDEFG, ADEFGH*
B	*ACH, AEFG, BCEFGH, DEFH, ABCDEF, ABDGH, CDG, CEFH, ABEF, ABCGH,* **G***, BCD, ADH, ACDEFG, BDEFGH*
C	*ABH, ABCEFG, EFGH, BCDEFH, ADEF, ACDGH, BDG, BEFH, ACEF, AGH, BCG,* **D***, ABCDH, ABDEFG, CDEFGH*
D	*ABCDH, ABDEFG, CDEFGH, BEFH, ACEF, AGH, BCG, BCDEFH, ADEF, ACDGH, BDG,* **C***, ABH, ABCEFG, EFGH*
E	*ABCEH, ABFG, CFGH, BDFH, ACDF, ADEGH, BCDEG, BCFH,* **AF***, ACEGH, BEG, CDE, ABDEH, ABCDFG, DFGH*
F	*ABCFH, ABEG, CEGH, BDEH, ACDE, ADFGH, BCDFG, BCEH,* **AE***, ACFGH, BFG, CDF, ABDFH, ABCDEG, DEGH*
G	*ABCGH, ABEF, CEFH, BDEFGH, ACDEFG, ADH, BCD, BCEFGH, AEFG, ACH,* **B***, CDG, ABDGH, ABCDEF, DEFH*
H	*ABC, ABEFGH, CEFG, BDEF, ACDEFH, ADG, BCDGH, BCEF, AEFH, ACG, BGH, CDH, ABD, ABCDEFGH, DEFG*

so on. This design is not good because it may yield strongly biased estimators. The resolution of a 2^{m-k} design is the length of the smallest word (excluding μ) in the subgroup of defining parameters. For example, if in a 2^{8-4} design we use the 4 generators $BCDE$, $ACDF$, $ABCG$ and $ABDH$, we obtain the 16 defining parameters $\{\mu,$ $BCDE, ACDF, ABEF, ABCG, ADEG, BDFG, CEFG, ABDH, ACEH, BDFH, DEFH,$ $CDGH, BEGH, AFGH, ABCDEFGH\}$. The length of the smallest word, excluding μ, among these defining parameters is four. Thus, the present 2^{8-4} design is a *resolution IV* design. In this design, all aliases of main effects are second-order interactions or higher (words of length greater or equal to 3). Aliases to first-order interactions are interactions of first order or higher. The present design is obviously better in terms of resolution than the previous one (which is of resolution II). We should always try to get resolution IV or higher. If the degree of fractionation is too high, resolution IV designs may not exist. For example, in $2^{6-3}, 2^{7-4}, 2^{9-5}, 2^{10-6}$, and 2^{11-7}, we have only resolution III designs. One way to reduce the bias is to choose several fractions at random. For example, in a 2^{11-7}, we have $2^7 = 128$ blocks of size $2^4 = 16$. If we execute only one block, the best we can have is a resolution III. In this case, some main effects are biased (confounded) with some first-order interactions. If we choose n blocks at random (RSWOR) out of the 128 possible ones, and compute the average estimate of the effects, the bias is reduced to zero but the variance of the estimators is increased.

To illustrate this, suppose we have a 2^{6-2} design with generators $ABCE$ and $BCDF$. This will yield a resolution IV design. There are four blocks, and the corresponding bias terms of the LSE of A are

	Block
0	$-BCE - ABCDF + DEF$
1	$BCE - ABCDF - DEF$
2	$-BCE + ABCDF - DEF$
3	$BCE + ABCDF + DEF$

If we choose one block at random, the expected bias is the average of these four terms, which is zero. The total variance of $\hat{A}$ is $\sigma^2/16 +$ Variance of conditional bias $= \sigma^2/16 + [(BCE)^2 + (ABCDF)^2 + (DEF)^2]/4$.

As mentioned earlier, one can construct blocks of 2^{m-k} using program GENEDSN.EXE.

EXAMPLE **12.10**   In this example, we illustrate the construction of fractional replications using program GENEDSN.EXE. The case that is illustrated is a 2^{8-4} design. Here we can construct 16 fractions, each of size 16. As discussed before, four generating parameters should be specified. Let these be $BCDE$, $ACDF$, $ABCG$, and $ABDH$. These parameters generate resolution IV designs. The first question that appears on the screen is "How many factors?" Type in the answer 8. To answer the question "What is the degree of fractionation, k?" type in 4. There are $2^4 = 16$ blocks, indexed $0, 1, \ldots, 15$. Each index is determined by the signs of the four generators that determine the block. Thus, the signs $(-1, -1, 1, 1)$ correspond to $(0, 0, 1, 1)$, which yield the index $\sum_{j=1}^{4} i_j 2^{j-1} = 12$. The next question is: "Which fractional block, kb should be printed?" The answer is an integer $0, 1, \ldots, 15$.

After this, we have to insert the indices of the generators. Generator 1 is $BCDE = A^0 B^1 C^1 D^1 E^1 F^0 G^0 H^0$. We therefore type successively on new lines 0, 1, 1, 1, 1, 0, 0, 0. For generator 2, we type 1, 0, 1, 1, 0, 1, 0, 0; for generator 3, we type 1, 1, 1, 0, 0, 0, 1, 0; and for generator 4, we type 1, 1, 0, 1, 0, 0, 0, 1. The design is printed in the output file FRACFAC.DAT. In Table 12.28, two blocks, for $kb = 0$ and $kb = 1$, are printed. ■

TABLE 12.28
Blocks of 2^{8-4} designs

Block 0								Block 1							
1	1	1	1	1	1	1	1	1	2	2	2	1	1	1	1
1	2	2	2	2	1	1	1	1	1	1	1	2	1	1	1
2	1	2	2	1	2	1	1	2	2	1	1	1	2	1	1
2	2	1	1	2	2	1	1	2	1	2	2	2	2	1	1
2	2	2	1	1	1	2	1	2	1	1	2	1	1	2	1
2	1	1	2	2	1	2	1	2	2	2	1	2	1	2	1
1	2	1	2	1	2	2	1	1	1	2	1	1	2	2	1
1	1	2	1	2	2	2	1	1	2	1	2	2	2	2	1
2	2	1	2	1	1	1	2	2	1	2	1	1	1	1	2
2	1	2	1	2	1	1	2	2	2	1	2	2	1	1	2
1	2	2	1	1	2	1	2	1	1	1	2	1	2	1	2
1	1	1	2	2	2	1	2	1	2	2	1	2	2	1	2
1	1	2	2	1	1	2	2	1	2	1	1	1	1	2	2
1	2	1	1	2	1	2	2	1	1	2	2	2	1	2	2
2	1	1	1	1	2	2	2	2	2	2	2	1	2	2	2
2	2	2	2	2	2	2	2	2	1	1	1	2	2	2	2

Box, Hunter, and Hunter (1978, p. 410) recommends generators for 2^{m-k} designs. Some of these generators are given in Table 12.29.

The LSE of the parameters is performed by first writing the columns of coefficients $c_{i,j} = \pm 1$ corresponding to the design, multiplying the coefficients by the Y values, and dividing by 2^{m-k}.

12.9
Exploration of Response Surfaces

The functional relationship between the yield variable Y and the experimental variables $(x_1, \ldots, x_k)$ is modeled as

$$Y = f(x_1, \ldots, x_k) + e$$

where e is a random variable with zero mean and a finite variance σ^2. The set of points $\{f(x_1, \ldots, x_k), x_i \in D_i, i = 1, \ldots, k\}$, where $(D_1, \ldots, D_k)$ is the experimental domain of the x variables, is called a *response surface*. Two types of response

TABLE 12.29

Some generators for 2^{m-k} designs

			m	
k	5	6	7	8
1	ABCDE	ABCDEF	ABCDEFGH	ABCDEFGH
2	ABD	ABCE	ABCDF	ABCDG
	ACE	BCDF	ABDEG	ABEFH
3		ABD	ABCE	ABCF
		ACD	BCDF	ABDG
		BCF	ACDG	BCDEH
4			ABD	BCDE
			ACE	ACDF
			BCF	ABCG
			ABCG	ABDH

surfaces were discussed before: the linear

$$f(x_1, \ldots, x_k) = \beta_0 + \sum_{i=1}^{k} \beta_i x_i \tag{12.9.1}$$

and the quadratic

$$f(x_1, \ldots, x_k) = \beta_0 + \sum_{i=1}^{k} \beta_i x_i + \sum_{i=1}^{k} \beta_{ii} x_i^2 + \sum \sum_{i \neq j} \beta_{ij} x_i x_j \tag{12.9.2}$$

Response surfaces may be of complicated functional form. We assume here that in local domains of interest, they can be approximated by linear or quadratic models.

Researchers are interested in studying or exploring the nature of response surfaces in certain domains of interest for the purpose of predicting future yield, and, in particular, for optimizing a process, by choosing the x values to maximize (or minimize) the expected yield (or the expected loss). In this section, we present special designs for the exploration of quadratic surfaces and for the determination of optimal domains (conditions). Designs for quadratic models are called *second-order designs*. We start with the theory of second-order designs and conclude with the optimization process. For additional reading on exploration of response surfaces and their relation to quality improvement, see Myer and Montgomery (1995).

12.9.1 Second-Order Designs

Second-order designs are constructed to estimate the parameters of the quadratic response function:

$$E\{Y\} = \beta_0 + \sum_{i=1}^{k} \beta_i x_i + \sum_{i=1}^{k} \beta_{ii} x_i^2 + \sum_{i=1}^{k-1} \sum_{j=i+1}^{k} \beta_{ij} x_i x_j \tag{12.9.3}$$

In this case, the number of regression coefficients is $p = 1 + 2k + \binom{k}{2}$. We will arrange the vector β in the form

$$\beta' = (\beta_0, \beta_{11}, \ldots, \beta_{kk}, \beta_1, \ldots, \beta_k, \beta_{12}, \ldots, \beta_{1k}, \beta_{23}, \ldots, \beta_{2k}, \ldots, \beta_{k-1,k})$$

Let N be the number of x points. The design matrix takes the form

$$(X) = \begin{bmatrix} 1 & x_{11}^2 & \cdots & x_{1k}^2 & x_{11} & \cdots & x_{1k} & x_{11}x_{12} & \cdots & x_{1,k-1}x_{1,k} \\ 1 & x_{21}^2 & & x_{2k}^2 & x_{21} & & x_{2k} & x_{21}x_{22} & & \\ 1 & x_{31}^2 & & x_{3k}^2 & x_{31} & & x_{3k} & & & \\ \vdots & \vdots & & \vdots & \vdots & & \vdots & \vdots & & \vdots \\ 1 & x_{N1}^2 & & x_{Nk}^2 & x_{N1} & \cdots & x_{Nk} & x_{N1}x_{N2} & \cdots & x_{N,k-1}x_{N,k} \end{bmatrix}$$

Impose on the x values the conditions:

i $\displaystyle\sum_{j=1}^{N} x_{ji} = 0, i = 1, \ldots, k$

ii $\displaystyle\sum_{j=1}^{N} x_{ji}^3 = 0, i = 1, \ldots, k$

iii $\displaystyle\sum_{j=1}^{N} x_{ji}^2 x_{jl} = 0, i \neq l$

iv $\displaystyle\sum_{j=1}^{N} x_{ji}^2 = b, i = 1, \ldots, k$ (12.9.4)

v $\displaystyle\sum_{j=1}^{N} x_{ji}^2 x_{jl}^2 = c, i \neq l$

vi $\displaystyle\sum_{j=1}^{N} x_{ji}^4 = c + d$

The matrix $(S) = (X)'(X)$ can be written in the form

$$(S) = \begin{bmatrix} (U) & 0 \\ 0 & (B) \end{bmatrix} \tag{12.9.5}$$

where (U) is the $(k + 1) \times (k + 1)$ matrix

$$(U) = \begin{bmatrix} N & b\mathbf{1}_k' \\ b\mathbf{1}_k & d\mathbf{I}_k + c\mathbf{J}_k \end{bmatrix} \tag{12.9.6}$$

and (B) is a diagonal matrix of order $\dfrac{k(k+1)}{2}$

$$
(B) = \begin{bmatrix} b & & & & & & \\ & \ddots & & & & & \\ & & b & & & 0 & \\ & & & c & & & \\ & & & & c & & \\ & 0 & & & & \ddots & \\ & & & & & & c \end{bmatrix}.
$$

(12.9.7)

Moreover, $\mathbf{1}_k$ is a vector of k 1's and J_k is a $k \times k$ matrix of 1's. One can verify that

$$
(U)^{-1} = \begin{bmatrix} p & q\mathbf{1}'_k \\ q\mathbf{1}_k & tI_k + sJ_k \end{bmatrix}
$$

(12.9.8)

where

$$
p = \frac{d + kc}{N(d + kc) - b^2 k}
$$

$$
q = -\frac{b}{N(d + kc) - b^2 k}
$$

$$
t = \frac{1}{d}
$$

(12.9.9)

$$
s = \frac{b^2 - Nc}{d[N(d + kc) - b^2 k]}
$$

Notice that U is singular if $N(d + kc) = b^2 k$. Therefore, we say that the design is nonsingular if

$$
N \neq \frac{b^2 k}{d + kc}
$$

Furthermore, if $N = b^2/c$, then $s = 0$. In this case, the design is called *orthogonal*.

Let $\mathbf{x}^{0'} = (x_1^0, \ldots, x_k^0)$ be a point in the experimental domain, and

$$
\boldsymbol{\xi}^{0'} = (1, (x_1^0)^2, \ldots, (x_k^0)^2, x_1^0, \ldots, x_k^0, x_1^0 x_2^0, x_1^0 x_3^0, \ldots, x_{k-1}^0 x_k^0)
$$

The variance of the predicted response at $\mathbf{x}^0$ is

$$
V\{\hat{Y}(\mathbf{x}^0)\} = \sigma^2 \boldsymbol{\xi}^{0'} \begin{bmatrix} (U)^{-1} & & & 0 & & \\ & b^{-1} & & & & 0 \\ & & \ddots & & & \\ 0 & & & b^{-1} & & \\ & & & & c^{-1} & \\ & 0 & & & & \ddots \\ & & & & & c^{-1} \end{bmatrix} \boldsymbol{\xi}^0
$$

$$
= \sigma^2 \left[p + \frac{1}{b} \sum_{i=1}^{k} (x_i^0)^2 + (t + s) \sum_{i=1}^{k} (x_i^0)^4 \right]
$$

(12.9.10)

$$+ \frac{1}{c} \sum \sum_{h<j} (x_h^0)^2 (x_j^0)^2 + 2b \sum_{i=1}^{k} (x_i^0)^2$$

$$+ 2s \sum \sum_{h<j} (x_h^0)^2 (x_j^0)^2 \Bigg]$$

$$= \sigma^2 \Bigg[p + \rho^2 \left(2b + \frac{1}{b} \right) + (t+s) \sum_{i=1}^{k} (x_i^0)^4$$

$$+ 2 \left(s + \frac{1}{2c} \right) \sum \sum_{h<j} (x_k^0)^2 (x_j^0)^2 \Bigg]$$

where $\rho^2 = \sum_{i=1}^{k} (x_i^0)^2$. Notice that

$$\rho^4 = \left(\sum_{i=1}^{k} (x_i^0)^2 \right)^2 = \sum_{i=1}^{k} (x_i^0)^4 + 2 \sum \sum_{h<j} (x_h^0)^2 (x_j^0)^2$$

Thus, if $d = 2c$, then $t + s = s + 1/2c$ and

$$V\{\hat{Y}(\mathbf{x}^0)\} = \sigma^2 \left[p + \rho^2 \left(2b + \frac{1}{b} \right) + (t+s)\rho^4 \right] \qquad \text{(12.9.11)}$$

Such a design ($d = 2c$) is called *rotatable*, because $V\{\hat{Y}(\mathbf{x}^0)\}$ is constant for all points $\mathbf{x}^0$ on the circumference of a circle of radius ρ, centered at the origin.

12.9.2 Some Specific Second-Order Designs

12.9.2.1 3^k Designs

Consider a factorial design of k factors, each one at three levels: $-1, 0, 1$. In this case, the number of points is $N = 3^k$. Obviously $\sum_{j=1}^{3^k} x_{ji} = 0$ for all $i = 1, \ldots, k$. Also, $\sum_{j=1}^{3^k} kx_{ji}^3 = 0$ and $\sum_{j=1}^{3^k} x_{ji}^2 x_{jl} = 0$, $i \neq l$. In terms of the parameters defined in equations 12.9.4, we get here

$$b = \sum_{j=1}^{3^k} x_{ji}^2 = \frac{2}{3} 3^k = 2 \cdot 3^{k-1}$$

$$\qquad \text{(12.9.12)}$$

$$c = \sum_{j=1}^{3^k} x_{ji}^2 x_{jl}^2 = \frac{2}{3} b = 4 \cdot 3^{k-2}$$

Hence

$$d = b - c = 2 \cdot 3^{k-1} - 4 \cdot 3^{k-2} = 2 \cdot 3^{k-2} \qquad \text{(12.9.13)}$$

Moreover, $b^2 = 4 \cdot 3^{2k-2}$ and $N \cdot c = 3^k \cdot 4 \cdot 3^{k-2} = 4 \cdot 3^{2k-2}$. Thus, $Nc = b^2$. The design is therefore orthogonal. However, $d \neq 2 \cdot c$. Thus, the design is not rotatable.

12.9.2.2   Central Composite Designs

A *central composite design* is one in which we start with $n_c = 2^k$ points of a factorial design in which each factor is at levels -1 and $+1$. To these points we add $n_a = 2k$ axial points, which are at a fixed distance α from the origin. These are the points

$$(\pm\alpha, 0, \ldots, 0), (0, \pm\alpha, 0, \ldots, 0), \ldots, (0, 0, \ldots, 0, \pm\alpha)$$

Finally, put n_0 points at the origin. These n_0 observations yield an estimate of the variance σ^2. Thus, the total number of points is $N = 2^k + 2k + n_0$. In such a design, the parameters of equations 12.9.4 are

$$b = 2^k + 2\alpha^2$$
$$c = 2^k$$
$$c + d = 2^k + 2\alpha^4 \tag{12.9.14}$$

or

$$d = 2\alpha^4$$

The rotatability condition is $d = 2c$. Thus, the design is rotatable if

$$\alpha^4 = 2^k \qquad \text{or} \qquad \alpha = 2^{k/4} \tag{12.9.15}$$

For this reason, in central composite designs with $k = 2$ factors, we use $\alpha = \sqrt{2} = 1.414$. For $k = 3$ factors, we use $\alpha = 2^{3/4} = 1.6818$. For rotatability and orthogonality, the following should be satisfied:

$$n_0 + 2^k = \frac{4\alpha^2(2^k + \alpha^2)}{2^k} \tag{12.9.16}$$

and because $\alpha^2 = 2^{k/2}$ (for rotatability)

$$n_0 = 4(2^{k/2} + 1) - 2^k \tag{12.9.17}$$

Thus, if $k = 2$, the number of points at the origin is $n_0 = 8$. For $k = 4$, we need $n_0 = 4$ points at the origin. For $k = 3$, there is no rotatability because $4(2^{3/2} + 1) - 8 = 7.313$.

EXAMPLE **12.11**   In Example 12.7, the piston simulation experiment was considered. We tested there, via a 3^2 experiment, the effects of piston weight (factor A) and the spring coefficient (factor D). It was found that the effects of the spring coefficient were significant, whereas the piston weight had no significant effect on cycle time. Now we will conduct a central composite design of four factors in order to explore the response surface. The factors chosen are: piston surface area (factor B), initial gas volume (factor C), spring coefficient (factor D), and filling gas temperature (factor G). The experiment is performed with the simulator program \CHA12\TURB4.EXE. The factors and associated levels in this experiment are given in Table 12.30.

TABLE **12.30**

Factors and level in Turb4 experiment

Factor	Levels				
Piston surface area	.0075	.01	.0125	.015	.0175
Initial gas volume	.0050	.00625	.0075	.00875	.0100
Spring coefficient	1000	2000	3000	4000	5000
Filling gas temperature	340	345	350	355	360
Code	-2	-1	0	1	2

It is desirable to have both orthogonality and rotatability. Because $k = 4$, we have $\alpha = 2$ and $n_0 = 4$. The number of replications is $n_r = 30$. The central composite design is read into the program from the input file \DATA\DESMATQ.DAT. This program is given on the left-hand side of Table 12.31. The results are given in the output file \DATA\OTURB4.DAT. Figure 12.11 shows the main effects plot for the four factors. The spring coefficient and the filling gas temperature have similar main effects. With increasing levels of these two factors, the cycle time average is monotonically decreasing.

FIGURE **12.11**

Main effects plot

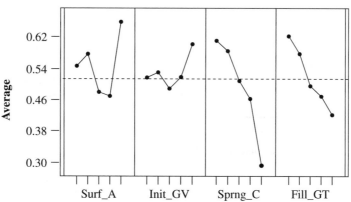

Main Effects Plot – Means for Average

TABLE **12.31**

Central composite design and the mean and standard deviations of cycle time

	Code	Levels			
B	C	D	G	$\overline{Y}$	STD
−1.00	−1.00	−1.00	−1.00	0.671	0.2328
1.00	−1.00	−1.00	−1.00	0.445	0.1771
−1.00	1.00	−1.00	−1.00	0.650	0.2298
1.00	1.00	−1.00	−1.00	0.546	0.2228
−1.00	−1.00	1.00	−1.00	0.534	0.1650
1.00	−1.00	1.00	−1.00	0.410	0.1688
−1.00	1.00	1.00	−1.00	0.534	0.1257
1.00	1.00	1.00	−1.00	0.495	0.1388
−1.00	−1.00	−1.00	1.00	0.593	0.2453
1.00	−1.00	−1.00	1.00	0.542	0.2266
−1.00	1.00	−1.00	1.00	0.602	0.2185
1.00	1.00	−1.00	1.00	0.509	0.1977
−1.00	−1.00	1.00	1.00	0.480	0.1713
1.00	−1.00	1.00	1.00	0.411	0.1658
−1.00	1.00	1.00	1.00	0.435	0.1389
1.00	1.00	1.00	1.00	0.438	0.1482
2.00	0.00	0.00	0.00	0.458	0.1732
−2.00	0.00	0.00	0.00	0.635	0.1677
0.00	2.00	0.00	0.00	0.570	0.1569
0.00	−2.00	0.00	0.00	0.481	0.1757
0.00	0.00	2.00	0.00	0.428	0.1064
0.00	0.00	−2.00	0.00	0.742	0.3270
0.00	0.00	0.00	2.00	0.496	0.2029
0.00	0.00	0.00	−2.00	0.549	0.1765
0.00	0.00	0.00	0.00	0.490	0.1802
0.00	0.00	0.00	0.00	0.468	0.1480
0.00	0.00	0.00	0.00	0.481	0.1636
0.00	0.00	0.00	0.00	0.557	0.1869

In Table 12.32, we present the results of regression analysis of the mean cycle time $\overline{Y}$ on 14 predictors $x_1^2, x_2^2, x_3^2, x_4^2, x_1, x_2, x_3, x_4, x_1x_2, x_1x_3, x_1x_4, x_2x_3, x_2x_4, x_3x_4$, where x_1 corresponds to factor B, x_2 to factor C, x_3 to factor D, and x_4 to factor G.

We see from the table that only factor D (spring coefficient) has a significant quadratic effect. Factors B, D, and G have significant linear effects. The interaction effects of B with C, D, and G will also be added. Thus, the response surface can be approximated by the equation,

$$Y = .499 - 0.0440x_1 - 0.0604x_3 - 0.0159x_4 + 0.0171x_3^2$$
$$+ 0.0148x_1x_2 + 0.0153x_1x_3 + 0.0177x_1x_4$$

The contour lines for the mean cycle time corresponding to $x_2 = x_4 = 0$ are shown in Figure 12.12. ∎

TABLE   12.32

MINITAB regression analysis

	Predictor	Coef	Stdev	*t*-ratio	*p*
μ	constant	0.499000	0.018530	26.92	0.000
β_{11}	x_1^2	0.007469	0.007567	0.99	0.342
β_{22}	x_2^2	0.002219	0.007567	0.29	0.774
β_{33}	x_3^2	0.017094	0.007567	2.26	0.042
β_{44}	x_4^2	0.001469	0.007567	0.19	0.849
β_1	x_1	−0.044042	0.007567	−5.82	0.000
β_2	x_2	0.012542	0.007567	1.66	0.121
β_3	x_3	−0.060375	0.007567	−7.98	0.000
β_4	x_4	−0.015875	0.007567	−2.10	0.056
β_{12}	x_1x_2	0.014813	0.009267	1.60	0.134
β_{13}	x_1x_3	0.015313	0.009267	1.65	0.122
β_{14}	x_1x_4	0.017687	0.009267	1.91	0.079
β_{23}	x_2x_3	0.000688	0.009267	0.07	0.942
β_{24}	x_2x_4	−0.012937	0.009267	−1.40	0.186
β_{34}	x_3x_4	−0.008937	0.009267	−0.96	0.352

$$s = 0.03707 \quad R\text{-sq} = 90.4\% \quad R\text{-sq(adj)} = 80.0\%$$

FIGURE   12.12

Contour lines of the response surface ($Y = .499 - 0.0440x_1 - 0.0604x_3 + 0.0171x_3^2 + 0.0153x_1x_3$)

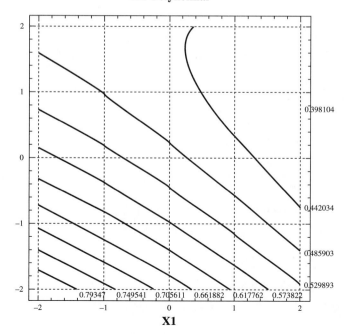

XY Polynomial

12.9.3 Approaching the Region of Optimal Yield

Often, the purpose for fitting a response surface is to locate the levels of the factors that yield optimal results.

Initially, one might be far from the optimal regions, and a series of small experiments may be performed to move toward the region. Thus, we start with simple first-order experiments, like 2^k factorial, and fit to the results a linear model of the form

$$\hat{Y} = b_0 + b_1 x_1 + \cdots + b_k x_k$$

We now wish to determine a new point—$\boldsymbol{\xi}^*$, say—whose distance from the center of the first-stage experiment (say 0) is R, and with maximal (or minimal) predicted yield. The predicted yield at $\boldsymbol{\xi}^*$ is

$$\hat{y}^* = b_0 + \sum_{i=1}^{k} b_i \xi_i^*$$

To find $\boldsymbol{\xi}^*$, we differentiate the Lagrangian

$$L = b_0 + \sum_{i=1}^{k} b_i \xi_i^* + \lambda \left(R^2 - \sum_{i=1}^{k} (\xi_i^*)^2 \right)$$

with respect to ξ_i^* ($i = 1, \ldots, k$) and λ. The solution is

$$\xi_i^* = R \frac{b_i}{\sqrt{\sum_{i=1}^{k} b_i^2}} \qquad i = 1, \ldots, k$$

The direction of the steepest ascent (descent) is in the direction of the normal (perpendicular) to the contours of equal response.

At the second stage, we perform experiments at a few points along the direction of the steepest ascent (at $R_1, R_2, \ldots$) until there is no further increase in the mean yield. We then enter the third stage, at which we perform a second-order design centered at a new region, where the optimal conditions seem to prevail.

12.9.4 Canonical Representation

The quadratic response function

$$\hat{Y} = b_0 + \sum_{i=1}^{k} b_i x_i + \sum_{i=1}^{k} b_{ii} x_i^2 + 2 \sum \sum_{i<j} b_{ij} x_i x_j \tag{12.9.18}$$

can be written in the matrix form

$$\hat{Y} = b_0 + \mathbf{b}'\mathbf{x} + \mathbf{x}'\mathbf{B}\mathbf{x} \tag{12.9.19}$$

where $\mathbf{x}' = (x_1, \ldots, x_k)$, $\mathbf{b}' = (b_1, \ldots, b_k)$, and

$$\mathbf{B} = \begin{bmatrix} b_{11} & b_{12} & \cdots & b_{1k} \\ b_{12} & \ddots & & \vdots \\ \vdots & & \ddots & \vdots \\ b_{1k} & \cdots & \cdots & b_{kk} \end{bmatrix}$$

Let $\nabla \hat{Y}$ be the gradient of $\hat{Y}$—that is,

$$\nabla \hat{Y} = \frac{\partial}{\partial \mathbf{x}} \hat{Y} = \mathbf{b} + 2\mathbf{Bx}$$

Let $\mathbf{x}^0$ be a point at which the gradient is zero—namely,

$$\mathbf{x}^0 = -\frac{1}{2}\mathbf{B}^{-1}\mathbf{b} \tag{12.9.20}$$

assuming that the matrix $\mathbf{B}$ is nonsingular. Making the transformation (change of origin to $\mathbf{x}^0$) $\mathbf{z} = \mathbf{x} - \mathbf{x}^0$, we obtain

$$\begin{aligned} \hat{Y} &= b_0 + (\mathbf{x}^0 + \mathbf{z})'\mathbf{b} + (\mathbf{x}^0 + \mathbf{z})'\mathbf{B}(\mathbf{x}^0 + \mathbf{z}) \\ &= \hat{Y}_0 + \mathbf{z}'\mathbf{Bz} \end{aligned} \tag{12.9.21}$$

where $\hat{Y}_0 = b_0 + \mathbf{b}'\mathbf{x}^0$.

The matrix B is real symmetric. Thus, there exists an orthogonal matrix $\mathbf{H}$ (see Appendix I) that consists of the normalized eigenvectors of B such that

$$\mathbf{HBH}' = \begin{pmatrix} \lambda_1 & & \\ 0 & \ddots & 0 \\ & & \lambda_k \end{pmatrix}$$

where λ_i $(i = 1, \ldots, k)$ are the eigenvalues of B. We make now a new transformation (rotation)—namely,

$$\mathbf{w} = \mathbf{HZ}$$

Because H is orthogonal, $\mathbf{z} = \mathbf{H}'\mathbf{w}$ and

$$\begin{aligned} \mathbf{z}'\mathbf{Bz} &= \mathbf{w}'\mathbf{HBH}'\mathbf{w} \\ &= \sum_{i=1}^{k} \lambda_i w_i^2 \end{aligned}$$

In these new coordinates,

$$\hat{Y} = \hat{Y}_0 + \sum_{i=1}^{k} \lambda_i w_i^2 \tag{12.9.22}$$

This representation of the quadratic surface is called the *canonical form*. We see immediately that if $\lambda_i > 0$ for all $i = 1, \ldots, k$, then $\hat{Y}_0$ is a point of minimum. If $\lambda_i < 0$ for all $i = 1, \ldots, k$, then $\hat{Y}_0$ is a maximum. If some eigenvalues are positive and some are negative, then $\hat{Y}_0$ is a saddle point.

The following examples of second-order equations are taken from Box, Hunter, and Hunter (1978, pp. 527–530).

1 Simple maximum:

$$\hat{Y} = 83.57 + 9.39x_1 + 7.12x_2 - 7.44x_1^2 - 3.71x_2^2 - 5.80x_1x_2$$
$$= 87.69 - 9.02w_1^2 - 2.13w_2^2$$

2 Minimax:

$$\hat{Y} = 84.29 + 11.06x_1 + 4.05x_2 - 6.46x_1^2 - 0.43x_2^2 - 9.38x_1x_2$$
$$= 87.69 - 9.02w_1^2 + 2.13w_2^2$$

3 Stationary ridge:

$$\hat{Y} = 83.93 + 10.23x_1 + 5.59x_2 - 6.95x_1^2 - 2.07x_2^2 - 7.59x_1x_2$$
$$= 87.69 - 9.02w_1^2 + 0.00w_2^2$$

4 Rising ridge:

$$\hat{Y} = 82.71 + 8.80x_1 + 8.19x_2 - 6.95x_1^2 - 2.07x_2^2 - 7.59x_1x_2$$
$$= 87.69 - 9.02w_1^2 + 2.97w_2^2$$

12.10
Chapter Highlights

This chapter covers the range of classic experimental designs, including complete block designs, Latin squares, and full and fractional factorial designs with factors at two and three levels. The basic approach to the analysis is through modeling the response variable and computing ANOVA tables. Particular attention is also given to the generation of designs, and a special software module GENEDSN.EXE is used to generate designs.

The main concepts and definitions introduced in this chapter include:

- Response variable
- Controllable factor
- Statistical model

- Experimental layout
- Blocking
- Randomization
- Block designs
- Randomized complete block design
- Main effects
- Interaction effect
- Balanced incomplete block design
- Combinatoric designs
- Latin square design
- Full factorial design
- Fractional factorial design

12.11
Exercises

12.1.1 Describe a production process familiar to you, such as baking cakes or manufacturing concrete. List the pertinent variables. What is (are) the response variable(s)? Classify the variables that affect the response to noise variables and control variables. How many levels would you consider for each variable?

12.1.2 Different types of adhesive are used in a lamination process in manufacturing a computer card. The card is tested for bond strength. In addition to the type of adhesive, another factor that might influence bond strength is curing pressure (currently at 200 psi). Follow the basic steps of experimental design to set a possible experiment for testing the effects of adhesives and curing pressure on bond strength.

12.2.1 Provide an example where blocking can reduce the variability of a product.

12.3.1 Three factors—A, B, C—are tested in a given experiment that is designed to assess their effects on the response variable. Each factor is tested at three levels. List all the main effects and interactions.

12.3.2 Let x_1, x_2 be two quantitative factors and Y a response variable. A regression model $Y = \beta_0 + \beta_1 x_1 + \beta_2 x_2 + \beta_{12} x_1 x_2 + e$ is fitted to the data. Explain why β_{12} can be used as an interaction parameter.

12.4.1 Consider the ISC values for times t_1, t_2, and t_3 in data file SOCELL.DAT (see Table 3.8). Use a t-test to make a paired comparison to see whether the mean ISC in time t_2 is different from that in time t_1.

12.4.2 Use macro RPCOMP.MTB to perform a randomization test for the differences in the ISC values of the solar cells at times t_2 and t_3 (data file SOCELL.DAT).

12.4.3 Box, Hunter, and Hunter (1978, p. 209) gives the following results of four treatments A, B, C, D in penicillin manufacturing in five different blends (blocks).

	Treatments			
Blends	*A*	*B*	*C*	*D*
1	89	88	97	94
2	84	77	92	79
3	81	87	87	85
4	87	92	89	84
5	79	81	80	88

Perform an ANOVA to test whether there are significant differences among the treatments or among the blends.

12.5.1 Eight treatments A, B, C, ..., H were tested in a BIBD of 28 blocks, $k = 2$ treatments per block, $r = 7$, and $\lambda = 1$. The results of the experiments are given in the table. Make an ANOVA to test the significance of the block effects and treatment effects. If the treatment effects are significant, make multiple comparisons of the treatments.

Block	Treatments				Block	Treatments			
1	A	38	B	30	15	D	11	G	24
2	C	50	D	27	16	F	37	H	39
3	E	33	F	28	17	A	23	F	40
4	G	62	H	30	18	B	20	D	14
5	A	37	C	25	19	C	18	H	10
6	B	38	H	52	20	E	22	G	52
7	D	89	E	89	21	A	66	G	67
8	F	27	G	75	22	B	23	F	46
9	A	17	D	25	23	C	22	E	28
10	B	47	G	63	24	D	20	H	40
11	C	32	F	39	25	A	27	H	32
12	E	20	H	18	26	B	10	E	40
13	A	5	E	15	27	C	32	G	33
14	B	45	C	38	28	D	18	F	23

12.6.1 Four different methods of preparing concrete mixtures A, B, C, D were tested. These methods consisted of two different mixture ratios of cement to water and two blending durations. The four methods (treatments) were blocks in four batches and four days, according to a Latin square design. The concrete was poured into cubes and tested for compressive strength (kg/cm^2) after 7 days of storage in special rooms with 20°C temperature and 50% relative humidity. The results were as follows:

	Batches			
Days	**1**	**2**	**3**	**4**
1	A	B	C	D
	312	299	315	290
2	C	A	D	B
	295	317	313	300
3	B	D	A	C
	295	298	312	315
4	D	C	B	A
	313	314	299	300

Are the differences between the strength values of different treatments significant? [Perform the ANOVA.]

12.7.1 Repeat the experiments described in Example 12.7 at the low levels of factors B, C, E, F, and G.

Perform the ANOVA for the main effects and interaction of spring coefficient and piston weight on the cycle time. Are your results different from those obtained in the example?

12.7.2   File OTURB3.DAT contains the coefficient matrix of a 2^7 factorial experiment in which the seven factors of the piston cycle time experiment were tested (see Example 12.7). The two response columns are the means and standard deviations of samples of size $n = 5$. Compute the least-squares estimates of the main effects on the means and on the standard deviations.

12.7.3   A 2^4 factorial experiment gave the following response values, arranged in standard order: 72, 60, 90, 80, 65, 60, 85, 80, 60, 50, 88, 82, 58, 50, 84, 75.

a   Estimate all possible main effects.

b   Estimate σ^2 under the assumption that all interaction parameters are zero.

c   Determine a confidence interval for σ^2 at level of confidence .99.

12.7.4   A 3^2 factorial experiment with $n = 3$ replications gave the following observations:

	A_1	A_2	A_3
	18.3	17.9	19.1
B_1	17.9	17.6	19.0
	18.5	16.2	18.9
	20.5	18.2	22.1
B_2	21.1	19.5	23.5
	20.7	18.9	22.9
	21.5	20.1	22.3
B_3	21.7	19.5	23.5
	21.9	18.9	23.3

Perform an ANOVA to test the main effects and interactions. Break the between-treatments sum of squares to 1-DF components. Use the Scheffé S_α coefficient to determine which effects are significant.

12.8.1   Use program GENEDSN.EXE to construct a 2^{8-2} fractional replication using generators *ABCDG* and *ABEFH*. What is the resolution of this design? Write the aliases to the main effects and to the first-order interactions with factor *A*.

12.8.2   Consider a full factorial experiment of $2^6 = 64$ runs. We need to partition the runs into 8 blocks of 8. We do this by selecting three generators and using

program GENEDSN.EXE. The parameters in the group of defining parameters are confounded with the effects of blocks and are not estimable. Show which parameters are not estimable if the blocks are generated by *ACE*, *ABEF*, and *ABCD*.

12.9.1   A 2^2 factorial design is expanded by using four observations at 0. The design matrix and the response are:

X_1	X_2	Y
-1	-1	55.8
-1	-1	54.4
1	-1	60.3
1	-1	60.9
-1	1	63.9
-1	1	64.4
1	1	67.9
1	1	68.5
0	0	61.5
0	0	62.0
0	0	61.9
0	0	62.4

a   Fit a response function of the form: $Y = \beta_0 + \beta_1 X_1 + \beta_2 X_2 + \beta_{12} X_1 X_2 + e$ and plot its contour lines.

b   Estimate the variance σ^2 and test the goodness of fit of this model.

12.9.2   The following from Myers (1976, p. 175) represents a design matrix and the response for a control composite design

X_1	X_2	Y
1.0	0.000	95.6
0.5	0.866	77.9
-0.5	0.866	76.2
-1.0	0	54.5
-0.5	-0.866	63.9
0.5	-0.866	79.1
0	0	96.8
0	0	94.8
0	0	94.4

a   Estimate the response function and its stationary point.

b   Plot contours of equal response in two dimensions.

c   Conduct an ANOVA.

13

Quality by Design

The factorial designs discussed in Chapter 12 were developed in the 1930s by R. A. Fisher and F. Yates at the Rothemstad agricultural station in Britain. Fractional replications were developed in the 1940s by D. Finney, also at Rothemstad. After World War II, researchers at the Imperial Chemical Laboratories (ICL) in Britain applied these experimental design methods to industrial problems. Fractional replication and orthogonal array designs were reintroduced in the 1980s to improve the quality of manufacturing processes. The objectives of the classic designs and those of contemporary quality improvement designs are, however, different. For agronomists, the objective is to find treatment combinations that will lead to maximal yield in the growth of an agricultural product. Chemical engineers want to find the right combination of pressure, temperature, and other factors that will lead to the maximal amount of product coming from a reactor. For manufacturing engineers, the objective is to design a process so that products will be as close as possible to some specified target without much fluctuation over time.

The following is an example of a flaw in an engineering design, that caused severe problems over time. A small British electronics company called Encrypta designed an ingenious electronic seal for trucks and storerooms. Industrial versions of the D-size batteries were used to drive the circuit and numeric display. When Encrypta started to receive defective seals back from customers, they conducted a failure mode analysis that revealed that, when dropped on a hard surface, the batteries would heat up and cause a short circuit. Encrypta won £30,000 in compensation from the battery manufacturer and switched to Vidor batteries made by Fuji, which passed the test. Encrypta found that the D batteries had failed when dropped because a clothlike separation inside ruptured and allowed active chemicals to mix and discharge the battery. Fuji used a tough, rolled separator that eliminated the problem. [From *New Scientist* (15 September 1988), 119, 39.)

In this chapter, we discuss a comprehensive **quality engineering** approach developed in the 1950s by Japanese engineer Genichi Taguchi. Taguchi labeled his methodology **off-line quality control**. The basic ideas of off-line quality control originated while Taguchi was working at the Electrical Communications Laboratory (ECL) of the Nippon Telephone and Telegraph Company (NTT). Taguchi's task was to help Japanese engineers develop high-quality products with raw materials of poor quality, outdated manufacturing equipment, and an acute shortage of skilled engineers. Central to his approach was the application of statistically designed experiments. Over time, Taguchi's impact on Japan expanded to a wide range of industries. He won the Deming Prize for application of quality in 1960 and three Deming Prizes for literature on quality in 1951, 1953, and 1984. In 1959, the Japanese company NEC followed Taguchi's methods and ran 402 such experiments. In 1976, Nippon Denso, which is a 20,000 employee company producing electrical parts for automobiles, reported that it had run 2700 experiments using the Taguchi off-line quality-control method. Off-line quality control was first applied in the West to integrated circuit manufacturing (see Phadke et al., 1983). Applications of off-line quality control now range from the design of automobiles, copiers, and electronic systems to cash flow optimization in banking, improvements in computer response times, and runway utilization in an airport.

13.1
Off-Line Quality Control, Parameter Design, and the Taguchi Method

Kackar (1985, 1986), Dehand (1989), Phadke (1989), John (1990, ch. 19), and others all provide explanations of the Taguchi methodology for off-line experimentation. We provide here a concise summary of this approach.

The performance of products or processes is typically quantified by performance measures. Examples include measures such as piston cycle time, yield of a production process, output voltage of an electronic circuit, noise level of a compressor, and response times of a computer system. These performance measures might be affected by several factors that have to be set at specific levels to get desired results. For example, the piston simulator introduced in previous chapters has seven factors that can be used to control piston cycle times. The aim of off-line quality control is to determine the factor–level combination that gives the least variability to the appropriate performance measure, while keeping the mean value of the measure on target. The goal is to control both accuracy and variability. In the next section, we discuss an optimization strategy that solves this problem by minimizing various loss functions.

13.1.1 Product and Process Optimization Using Loss Functions

Optimization problems of products or processes can take many forms, depending on the objectives to be reached. These objectives are typically derived from customer requirements. Performance parameters such as dimensions, pressure, and velocity usually have a target or nominal value. The objective is to reach the target within a range bounded by upper and lower specification limits. We call such cases "*nominal is best.*" Noise levels, shrinkage factors, amount of wear, and deterioration usually need to be as low as possible. We call such cases "*the smaller the better.*" When we measure strength, efficiency, yields, or time to failure, our goal is, in most cases, to reach the maximum possible levels. These cases are called "*the larger the better.*" These three types of cases require different objective (target) functions to optimize. Taguchi introduced the concept of **loss function** to help determine the appropriate optimization procedure.

When "nominal is best," specification limits are typically two-sided, with an upper specification limit (USL) and a lower specification limit (LSL). These limits are used to differentiate between conforming and nonconforming products. Nonconforming products are usually fixed, retested, and sometimes downgraded or simply scrapped. In all cases, defective products carry a loss to the manufacturer. Taguchi argues that only products on target should carry no loss. Any deviation carries a loss that is not always immediately perceived by the customer or production personnel. Taguchi proposes a quadratic function as a simple approximation to a graduated loss function that measures loss on a continuous scale. A quadratic loss function has the form

$$L(y, M) = K(y - M)^2 \tag{13.1.1}$$

where y is the value of the performance characteristic of a product, M is the target value of this characteristic, and K is a positive constant that yields monetary or other utility value to the loss. For example, suppose that $(M - \Delta, M + \Delta)$ is the **customer's tolerance interval** around the target. When y falls out of this interval, the product has to be repaired or replaced at a cost of A. Then, for this product,

$$A = K\Delta^2 \tag{13.1.2}$$

or

$$K = A/\Delta^2 \tag{13.1.3}$$

The **manufacturer's tolerance interval** is generally tighter than that of the customer—namely, $(M - \delta, M + \delta)$, where $\delta < \Delta$. We can obtain the value of δ in the following manner. Suppose the cost to the manufacturer to repair a product that exceeds the customer's tolerance, before shipping the product, is B, $B < A$. Then

$$B = \left(\frac{A}{\Delta^2} \right)(Y - M)^2$$

or

$$Y = M \pm \Delta \left(\frac{B}{A} \right)^{1/2} \tag{13.1.4}$$

Thus,

$$\delta = \Delta \left(\frac{B}{A} \right)^{1/2} \tag{13.1.5}$$

The manufacturer should reduce the variability in the product performance characteristic so that process capability C_{pk} for the tolerance interval $(M - \delta, M + \delta)$ should be high. Figure 13.1 presents these relationships schematically.

F I G U R E   13.1

Quadratic loss and tolerance intervals

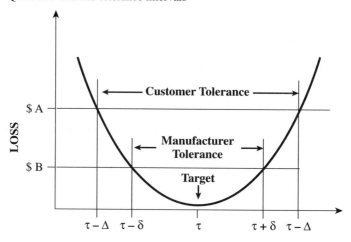

Notice that the expected loss is

$$E\{L(Y, M)\} = K(\text{bias}^2 + \text{variance}) \tag{13.1.6}$$

where bias $= \mu - M$, $\mu = E\{Y\}$, and variance $= E\{(Y - \mu)^2\}$. Thus, the objective is to have a manufacturing process with μ as close as possible to the target M, and variance, σ^2, as small as possible ($\sigma < \frac{\delta}{3}$ so that $C_{pk} > 1$). Recall that variance $+$ bias^2 is the mean-squared error, MSE. Thus, when "normal is best," the objective should be to minimize the MSE.

Objective functions for cases of "the bigger the better" or "the smaller the better" depend on the case under consideration. In cases where the performance measure is the lifetime of a product, the objective might be to design the product to maximize its expected lifetime. In the literature we may find the objective of minimizing $1/n \sum \frac{1}{y_j}$, which is an estimator of $E\{1/Y\}$. This parameter, however, may not always exist (for example, when Y has an exponential distribution), and this objective function might be senseless.

13.1.2   Major Stages in Product and Process Design

A major challenge to industry is to reduce variability in products and processes. The previous section dealt with measuring the impact of such variability. In this section, we discuss methods for actually reducing variability. The design of products or processes involves two main steps: designing the system and setting tolerances. System design is the stage where engineering skills, innovation, and technology are pooled together to create a basic design. Once the design is ready to go into production, one has to specify tolerances of parts and subassemblies so that the product or process meets its requirements. Loose tolerances are typically less expensive than tight tolerances. Taguchi proposed to change the classical approach to the design of products and processes and add an intermediate stage of **parameter design**. Thus, the three major stages in designing a product or a process are:

I   **System design**—This is when the product architecture and technology are determined.

II   **Parameter design**—At this stage, a planned optimization program is carried out to minimize variability and costs.

III   **Tolerance design**—Once the optimum performance is determined, tolerances should be specified so the product or process stays within specifications. The setting of optimum values of tolerance factors is called **tolerance design**.

Table 13.1 (adapted from Phadke, 1989) shows the relationships between the type of problems experienced in industrial products and processes and the various design phases. For definition of noise factors see next paragraph.

T A B L E   13.1

Noise factors and design phases

Activity	Design phase	Leverage on noise factors			Comments
		External	Manufacturing imperfections	Natural deterioration	
Product design	a. System design	High	High	High	Involves innovation to reduce sensitivity to all noise factors.
	b. Parameter design	High	High	High	Most important step for reducing sensitivity to all noise factors.
	c. Tolerance design	High	High	High	Method for selecting most economical grades of materials, components, and manufacturing equipment, and best operating environment for the product.
Manufacturing process design	a. Concept design	Low	High	Low	Involves innovation to reduce the effect of manufacturing imperfections.
	b. Parameter design	Low	High	Low	Important for reducing sensitivity of unit-to-unit variation to manufacturing variations.
	c. Tolerance design	Low	High	Low	Method for determining tolerance on manufacturing process parameters.
Manufacturing	a. Concept design	Low	High	Low	Method of detecting problems when they occur and correcting them.
	b. Parameter design	Low	High	Low	Method of compensating for known problems.
	c. Tolerance design	Low	High	Low	Last alternative—useful when process capability is poor.
Customer usage	warranty and repair	Low	Low	Low	

13.1.3   Design Parameters and Noise Factors

Taguchi puts the variables that affect performance characteristics into two categories: *design parameters* and *source of noise*. All factors that cause variability are included in the source of noise. Sources of noise are classified as external and internal. External sources are those external to the product, such as environmental conditions (temperature, humidity, dust, and so on); human variations in operating the product; and other similar factors. Internal sources of variability are those connected with manufacturing imperfections and product degradation or natural deterioration.

Design parameters, are controllable factors that can be set at predetermined values (level). The product designer has to specify the values of the design parameters to achieve the objectives. This is done by running an experiment called *parameter design*. Under manufacturing conditions, the values of these parameters may vary slightly from values determined in the parameter design stage (the nominal ones). In *tolerance designs*, we test the effects of such variability and determine tolerances that yield the desired results at lower cost.

EXAMPLE **13.1**   An *RL* circuit is an electrical circuit of alternating current that obtains an input of voltage 100 V AC and frequency 55 Hz. The output current of the circuit is aimed at 10 A, with tolerances of $\Delta = \pm 4$ A. Four factors influence the output y:

   1   V: Input voltage (V)
   2   f: Input frequency (Hz)
   3   R: Resistance (ohm)
   4   L: Self-inductance (H)

R and *L* are controllable factors; *V* and *f* are noise factors. Assume that *V* has a distribution between 90 and 110 V and *f* has a distribution between 55 and 65 Hz. *R* and *L* are design parameters. What should the values of R (Ω) and L (H) be to obtain an output y distributed around the target of $M = 10$ (A), with minimal mean-squared error and lowest cost? In Section 13.1.4, we study how we can take advantage of the nonlinear relationship between the above factors to attain lower variability and high accuracy.   ■

13.1.4   Parameter Design Experiments

In a parameter design experiment, we test the effects of controllable factors and noise factors on the performance characteristics of the product in order to:

   1   Make the product robust (insensitive) to environmental conditions
   2   Make the product insensitive to component variation
   3   Minimize the mean-squared error about a target value

We distinguish between two types of experiments: physical experiments and computer-based simulation experiments. In Section 13.4, we discuss computer-based simulation experiments. Let $\boldsymbol{\theta} = (\theta_1, \ldots, \theta_k)$ be the vector of design parameters. The vector of noise variables is denoted by $\mathbf{x} = (x_1, \ldots, x_m)$.

In many situations, the response function $y \in f(\boldsymbol{\theta}, \mathbf{x})$ involves the factors $\boldsymbol{\theta}$ and $\mathbf{x}$ in a nonlinear fashion. The RL circuit described in Example 13.1 involves the four factors $V, f, R,$ and L and the output y according to the nonlinear response function

$$y = \frac{V}{(R^2 + (2\pi fL)^2)^{1/2}}$$

If the noise factors V and f had no variability, we could determine the values of R and L to always obtain a target value $y_0 = M$. However, the variability of V and f around their nominal values turns y into a random variable, Y, with expected value μ and variance σ^2, which depend on the settings of the design parameters R and L and on the variances of V and f. The effects of nonlinearity on the distribution of Y will be studied in Section 13.2. The objective of parameter design experiments is to take advantage of the effects of the nonlinear relationship. The strategy is to perform a factorial experiment to investigate the effects of the design parameters (controllable factors). If we learn from the experiments that certain design parameters affect the mean of Y but not its variance and that other design factors affect the variance but not the mean, we can use the latter group to reduce the variance of Y as much as possible, and then adjust the levels of the parameters in the first group to set μ close to the target M. We illustrate this approach in the following example.

EXAMPLE **13.2**   The data for this example are taken from John (1990, p. 335). Three factors A, B, C (controllable) affect the output, Y, of a system. To estimate the main effects of A, B, C, a 2^3 factorial experiment was conducted. Each treatment combination was repeated four times, at the low and high levels of two noise factors. The results are given in Table 13.2.

TABLE   **13.2**
Response at a 2^3 factorial experiment

Factor levels			Response			
A	B	C	y_1	y_2	y_3	y_4
−1	−1	−1	60.5	61.7	60.5	60.8
1	−1	−1	47.0	46.3	46.7	47.2
−1	1	−1	92.1	91.0	92.0	91.6
1	1	−1	71.0	71.7	71.1	70.0
−1	−1	1	65.2	66.8	64.3	65.2
1	−1	1	49.5	50.6	49.5	50.5
−1	1	1	91.2	90.5	91.5	88.7
1	1	1	76.0	76.0	78.3	76.4

The mean, $\overline{Y}$, and standard deviation, S, of Y at the 8 treatment combinations are as follows:

v	$\overline{Y}$	S
0	60.875	0.4918
1	46.800	0.3391
2	91.675	0.4323
3	70.950	0.6103
4	65.375	0.9010
5	50.025	0.5262
6	90.475	1.0871
7	76.675	0.9523

Regressing the column $\overline{Y}$ on the three orthogonal columns under A, B, C in Table 13.2, we obtain

$$\text{Mean} = 69.1 - 7.99A + 13.3B + 1.53C$$

with $R^2 = .991$. Moreover, the coefficient 1.53 of C is not significant (P-value of .103). Thus, the significant main effects on the mean yield are of factors A and B only. Regressing the column of S on A, B, C, we obtain the equation

$$\text{STD} = 0.655 - 0.073A + 0.095B + 0.187C$$

with $R^2 = .805$. Only the main effect of C is significant. Factors A and B have no effect on the standard deviation. The strategy is therefore to set the value of C at -1 (as small as possible) and the values of A and B to adjust the mean response to be equal to the target value M. If $M = 85$, we find A and B to solve the equation

$$69.1 - 7.99A + 13.3B = 85$$

Letting $B = 0.75$, then $A = -0.742$. The optimal setting of the design parameters A, B, C is at $A = -0.742$, $B = 0.75$, and $C = -1$. ∎

13.1.5 Performance Statistics

As we just saw, the performance characteristic y at various combinations of the design parameters is represented by the mean, $\overline{Y}$, and standard deviation, S, of the y values observed at various combinations of noise factors. We performed the analysis first on $\overline{Y}$ and then on S to detect which design parameters influence $\overline{Y}$ but not S, and which influence S but not $\overline{Y}$. Let $\eta(\boldsymbol{\theta})$ denote the expected value of Y as a function of the design parameters $\theta_1, \ldots, \theta_k$. Let $\sigma^2(\boldsymbol{\theta})$ denote the variance of Y as a function of $\boldsymbol{\theta}$. This situation corresponds to the case where $\eta(\boldsymbol{\theta})$ and $\sigma^2(\boldsymbol{\theta})$ are independent. The objective in setting the values of $\theta_1, \ldots, \theta_k$ is to minimize the mean-squared error

$$\text{MSE}(\boldsymbol{\theta}) = B^2(\boldsymbol{\theta}) + \sigma^2(\boldsymbol{\theta}) \tag{13.1.7}$$

where $B(\boldsymbol{\theta}) = \eta(\boldsymbol{\theta}) - M$. The *performance statistic* is an estimator of $\mathrm{MSE}(\boldsymbol{\theta})$—namely,

$$\widehat{\mathrm{MSE}}(\boldsymbol{\theta}) = (\overline{Y}(\boldsymbol{\theta}) - M)^2 + S^2(\boldsymbol{\theta}) \qquad \text{(13.1.8)}$$

If $\overline{Y}(\boldsymbol{\theta})$ and $S^2(\boldsymbol{\theta})$ depend on different design parameters, we perform the minimization in two steps. First we minimize $S^2(\boldsymbol{\theta})$ and then $\hat{B}^2(\boldsymbol{\theta}) = (\overline{Y}(\boldsymbol{\theta}) - M)^2$. If $\eta(\boldsymbol{\theta})$ and $\sigma^2(\boldsymbol{\theta})$ are not independent, the problem is more complicated.

Taguchi recommends devising a function of $\eta(\boldsymbol{\theta})$ and $\sigma(\boldsymbol{\theta})$, called a *signal-to-noise ratio (SN)*, and maximizing an estimator of this SN function. Taguchi devised a large number of such performance statistics. In particular, he recommended maximizing the performance statistic

$$\eta = 10 \log \left(\frac{\overline{Y}^2}{S^2} - \frac{1}{n} \right) \qquad \text{(13.1.9)}$$

which is used in many studies. It is difficult to justify the performance statistic $\eta(\mathrm{SN}$ ratio). We have to be careful here because maximizing this SN might achieve bad results if $\eta(\boldsymbol{\theta})$ is far from the target M. Thus, if the objective is to set the design parameters to obtain means close to M and small standard deviations, we should minimize the mean-squared error and not necessarily maximize the SN ratio.

13.2
The Effects of Nonlinearity

As mentioned in Section 13.1, the response function $f(\boldsymbol{\theta}, \mathbf{x})$ might be nonlinear in $\boldsymbol{\theta}$ and $\mathbf{x}$. An example was given for the case of the output current of an *RL* circuit—namely,

$$Y = \frac{V}{(R^2 + (2\pi fL)^2)^{1/2}}$$

This is a nonlinear function of the design parameters R and L and the noise factor f. We have assumed that V and f are random variables and that R and L are constant parameters. The output current is the random variable Y. What is the expected value and variance of Y? Generally, we can estimate the expected value of Y and its variance by simulation, using the function $f(\boldsymbol{\theta}, \mathbf{X})$ and the assumed joint distribution of $\mathbf{X}$. An approximation to the expected value and the variance of Y can be obtained by the following method.

Let the random variables $X_1, \ldots, X_k$ have expected values $\xi_1, \ldots, \xi_k$ and a variance–covariance matrix

$$V = \begin{bmatrix} \sigma_1^2 & \sigma_{12} & \cdots & \sigma_{1k} \\ \sigma_{21} & & & \\ \vdots & & \ddots & \vdots \\ \sigma_{k1} & \sigma_{k2} & \cdots & \sigma_k^2 \end{bmatrix}$$

Assuming that $f(\boldsymbol{\theta}, \mathbf{X})$ can be expanded into a Taylor series around the means $\boldsymbol{\xi}_1 = (\xi_1, \ldots, \xi_k)$, we obtain the approximation

$$f(\boldsymbol{\theta}, \mathbf{X}) \cong f(\boldsymbol{\theta}, \boldsymbol{\xi}) + \sum_{i=1}^{k} (x_i - \xi_i) \frac{\partial}{\partial x_i} f(\boldsymbol{\theta}, \boldsymbol{\xi}) \tag{13.2.1}$$

$$+ \frac{1}{2} (\mathbf{X} - \boldsymbol{\xi})' H(\boldsymbol{\theta}, \boldsymbol{\xi})(\mathbf{X} - \boldsymbol{\xi})$$

where $H(\boldsymbol{\theta}, \boldsymbol{\xi})$ is a $k \times k$ matrix of second-order partial derivatives evaluated at ξ_i with (i, j)th element equal to

$$H_{ij}(\boldsymbol{\theta}, \boldsymbol{\xi}) = \frac{\partial^2}{\partial x_i \partial x_j} f(\boldsymbol{\theta}, \boldsymbol{\xi}) \qquad i, j = 1, \ldots, k \tag{13.2.2}$$

Thus, the expected value of $f(\boldsymbol{\theta}, \mathbf{X})$ is approximated by

$$E\{f(\boldsymbol{\theta}, \mathbf{X})\} \cong f(\boldsymbol{\theta}, \boldsymbol{\xi}) + \frac{1}{2} \sum_{i=1}^{k} \sum_{j=1}^{k} \sigma_{ij} H_{ij}(\boldsymbol{\theta}, \boldsymbol{\xi}) \tag{13.2.3}$$

and the variance of $f(\boldsymbol{\theta}, \mathbf{X})$ is approximated by

$$V\{f(\boldsymbol{\theta}, \mathbf{X})\} \cong \sum_{i=1}^{k} \sum_{j=1}^{k} \sigma_{ij} \frac{\partial}{\partial x_i} f(\boldsymbol{\theta}, \boldsymbol{\xi}) \frac{\partial}{\partial x_j} f(\boldsymbol{\theta}, \boldsymbol{\xi}) \tag{13.2.4}$$

As seen in these approximations, if the response variable Y is a nonlinear function of the random variables $X_1, \ldots, X_m$, its expected value depends also on the variances and covariances of the Xs. This is not the case if Y is a linear function of the x's. Moreover, in the linear case, the formula for $V\{Y\}$ is exact.

EXAMPLE 13.3   Consider the function

$$Y = \frac{V}{(R^2 + (2\pi f L)^2)^{1/2}}$$

where $R = 5.0$ (Ω) and $L = 0.02$ (H). V and f are independent random variables having normal distributions:

V is distributed like $N(100, 3)$

f is distributed like $N(55, 5/3)$

Notice that

$$\frac{\partial y}{\partial v} = \frac{1}{(R^2 + (2\pi f L)^2)^{1/2}}$$

$$\frac{\partial y}{\partial f} = -4V(R^2 + (2\pi f L)^2)^{-3/2} \pi^2 L^2 f$$

$$\frac{\partial^2 y}{\partial v^2} = 0$$

$$\frac{\partial^2 y}{\partial v \partial f} = -4(R^2 + (2\pi f L)^2)^{-3/2}\pi^2 L^2 f$$

Also,

$$\frac{\partial^2 y}{\partial f^2} = 4V(R^2 + (2\pi f L)^2)^{-5/2}\pi^2 L^2(8\pi^2 L^2 f^2 - R^2)$$

Substituting in these derivatives the values of R and L and the expected values of V and f, we obtain

$$\frac{\partial y}{\partial v} = \frac{1}{8.5304681} = 0.11723$$

$$\frac{\partial y}{\partial f} = -0.13991$$

$$\frac{\partial^2 y}{\partial v^2} = 0$$

$$\frac{\partial^2 y}{\partial v \partial f} = -0.0013991$$

$$\frac{\partial^2 y}{\partial f^2} = 0.002466$$

Accordingly, an approximation for $E\{Y\}$ is given by

$$E\{Y\} \cong 11.722686 + \frac{1}{2}(9 \times 0 + 2.7778 \times 0.002466) = 11.7261$$

The variance of Y is approximated by

$$V\{Y\} \cong 9 \times (0.11723)^2 + \frac{25}{9} \times (-0.13991)^2 = 0.17806$$

To check the goodness of these approximations, we do the following simulation using MINITAB. We simulate $N = 500$ normal random variables having mean 100 and standard deviation 3 into $C1$. Similarly, 500 normal random variables having mean 55 and standard deviation 1.67 are simulated into $C2$. In $C3$, we put the values of Y. This is done by the following program:

```
MTB> Random 500 C1;
SUBC> Normal 100 3.
MTB> Random 500 C2;
SUBC> Normal 55 1.67.
MTB> let k1 = 8 * ATAN(1)
MTB> let C3 = C1/sqrt(25 + (k1 * 0.02 * C2) **2)
MTB> mean C3
MTB> stan C3
```

The results obtained are $\overline{Y}_{500} = 11.687$ and $S^2_{500} = 0.17123$. The analytical approximations are very close to the simulation estimates. Actually, a .95 confidence interval for $E\{Y\}$ is given by $\overline{Y}_{500} \pm 2(S_{500}/\sqrt{500})$, which is (11.650, 11.724). The result of the analytical approximation, 11.7261, is only slightly above the upper confidence limit. The approximation is quite good.

It is interesting to estimate the effects of the design parameters R and L on $E\{Y\}$ and $V\{Y\}$. We conduct a small experiment on the computer for estimating $E\{Y\}$ and $V\{Y\}$ by a 3^2 factorial experiment. The levels of R and L as recommended by Taguchi in his review paper (see Ghosh, 1990, pp. 1–34) are

	0	1	2
R	0.05	5.00	9.50
L	0.01	0.02	0.03

In each treatment combination, we simulate 500 y values. The results are given in Table 13.3.

The objective is to find the combinations of R and L yield the minimum MSE $= (E\{Y\} - M)^2 + V\{Y\}$, where M is the target of 10 (Ω). It seems that the best setting of the design parameters is $R = 9.5$ (Ω) and $L = 0.01$ (H). The combination of $R = 0.05$ (Ω) and $L = 0.03$ (H) also yields a very small MSE. One should choose the least expensive setting. ∎

T A B L E 13.3
Means, variances, and MSE of Y in a 3^2 experiment

R	L	$\overline{Y}_{500}$	S^2_{500}	MSE
0	0	28.943	1.436	360.27
0	1	14.556	0.387	21.14
0	2	9.628	0.166	0.30
1	0	16.441	0.286	41.77
1	1	11.744	0.171	3.21
1	2	8.607	0.109	1.88
2	0	9.891	0.087	0.10
2	1	8.529	0.078	2.24
2	2	7.119	0.064	8.36

13.3
Taguchi's Designs

To simulate the effect of noise factors, Taguchi advocates the combination of two experimental arrays, an inner array and an outer array. The inner array is used to determine factor–level combinations of factors that can be controlled by the designer of the product or process. The outer array is used to generate the variability due to noise factors that is experienced by the product or process under optimization in its day to day operation.

The experimental arrays used by Taguchi are *orthogonal array* designs. An example of a design with 15 factors at 2 levels each, using 16 experiments is given in Table 13.4. The levels are indicated by 1 and 2, and the first row consists of all factors at level 1. This experimental array was introduced in Chapter 12 as a 2^{15-11} fractional factorial design, which is a design with 15 factors and 16 experimental runs. The corresponding full factorial design consists of $2^{15} = 32,768$ experiments. Taguchi labeled several experimental arrays using a convenient notation and reproduced them in tables that were widely distributed among engineers (see Taguchi and Taguchi 1987). The availability of these tables made it convenient for practitioners to design and run such experiments.

Note from Table 13.4 that if we run the experiment using the order of the experiment array, we will find it convenient to assign to column 1 a factor that is difficult to change from level 1 to level 2. For example, if changing the temperature of a solder bath requires 5 hours, the assignment of temperature to column 1 would require only one change of temperature. Column 15, however, has 10 changes between levels. Taguchi recommends that in some cases randomization be abandoned for the benefit of simplicity and cost. If we choose to run the experiment in the order of the experimental array, we can reduce the practical difficulties of running the experiment by the proper assignment of factors to columns. A factor with easily changed levels can

TABLE 13.4

Factor–level combinations of 15 factors at 2 levels, each in an $L_{16}(2^{15})$ orthogonal array

	Columns														
Trial	**1**	**2**	**3**	**4**	**5**	**6**	**7**	**8**	**9**	**10**	**11**	**12**	**13**	**14**	**15**
1	1	1	1	1	1	1	1	1	1	1	1	1	1	1	1
2	1	1	1	1	1	1	1	2	2	2	2	2	2	2	2
3	1	1	1	2	2	2	2	1	1	1	1	2	2	2	2
4	1	1	1	2	2	2	2	2	2	2	2	1	1	1	1
5	1	2	2	1	1	2	2	1	1	2	2	1	1	2	2
6	1	2	2	1	1	2	2	2	2	1	1	2	2	1	1
7	1	2	2	2	2	1	1	1	1	2	2	2	2	1	1
8	1	2	2	2	2	1	1	2	2	1	1	1	1	2	2
9	2	1	2	1	2	1	2	1	2	1	2	1	2	1	2
10	2	1	2	1	2	1	2	2	1	2	1	2	1	2	1
11	2	1	2	2	1	2	1	1	2	1	2	2	1	2	1
12	2	1	2	2	1	2	1	2	1	2	1	1	2	1	2
13	2	2	1	1	2	2	1	1	2	2	1	1	2	2	1
14	2	2	1	1	2	2	1	2	1	1	2	2	1	1	2
15	2	2	1	2	1	1	2	1	2	2	1	2	1	1	2
16	2	2	1	2	1	1	2	2	1	1	2	1	2	2	1

be assigned to column 15 with low penalty. However, assigning a factor to column 15 whose levels are difficult to change might make the whole experiment impractical.

To simulate the noise factors, we can design a second experiment using an external array. In some cases, noise cannot be directly simulated, and the external array consists of replicating the internal array experiments over a specified length of time or amount of material. In Section 13.6 we describe two such experiments. The first experiment deals with a speedometer cable, where the percentage of shrinkage is measured on several pieces of cable taken from various parts of a spool by running a heat test. The external array simply consists of the sampled parts of the spool. The second experiment deals with optimizing the response time of a computer system. Here the inner array experiment was carried out on an operational system so that the variability induced by various users was not specifically simulated. A retrospective study verified that there was no bias in user methods.

The design given in Table 13.4 can be used for up to 15 factors. It allows us to compute estimates of main effects provided there are no interactions of any order between them. However, if all first-order interactions are potentially significant, this design cannot be used with more than 5 factors. (The resolution of this design is III.) To assist the engineer with the correct choice of columns from the table of orthogonal arrays, Taguchi devised a graphical method of presenting the columns of an orthogonal array table that are confounded with first-order interactions of some factors. These graphs are called *linear graphs*.

Figure 13.2 shows two linear graphs associated with Table 13.4. Linear graph LG_1 corresponds to the case where all interactions might be significant. The graph has 5 vertices and 10 lines connecting the vertices. The factors A, B, C, D, E are assigned to the columns with numbers at the vertices. Thus, according to this linear graph, the assignment of factors to columns is as follows:

Factor	A	B	C	D	E
Column	1	2	4	8	15

We see that column 3 can be used to estimate the interaction AB; column 6 the interaction BC; column 5, the interaction AC; and so on.

F I G U R E   13.2
Two linear graphs for $L_{16}(2^{15})$

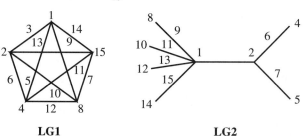

LG1   LG2

The second linear graph, LG_2, represents the case where only some interactions are significant. These are *AB, AC, AD, AE, FG,* and *FH*. In this case, we can perform the 16-trial experiment with 8 factors and assign them to columns as follows:

Factor	A	B	C	D	E	F	G	H
Column	1	8	10	12	14	2	4	5

The columns that can be used to estimate the interactions are 6, 7, 9, 11, 13, 15.

Although the emphasis in Taguchi's methodology is on estimating main effects, we should not forget that interactions might exist. It is better to be cautious and not to oversaturate a small design with too many factors. Recall that when fractional replications are used, the estimates of main effects might be confounded. We wish to choose a design with sufficient resolution (see Chapter 12), and this may require sufficiently large fractions. The table that we presented (Table 13.4) is a fraction of a 2^{15} factorial experiment. Taguchi also prepared tables of orthogonal arrays for 3^k factorial experiments and for mixtures of factors with 2 and 3 levels ($2^m \times 3^k$ factorials). The reader is referred to the tables of Taguchi and Taguchi (1987) and also to Appendix C of Phadke (1989).

13.4
Computer-Aided Designs

We mentioned earlier that when the response function $y = f(\theta, \mathbf{X})$ is known, as is the case in circuit designs, we can test the effects of the design parameters, θ, and of the noise variables, $\mathbf{X}$, by performing experiments on the computer.

The simulators TURB1–TURB5, TURBEXP, and ELECT1–ELECT3 (see Appendix II) are examples of such computer experiments. These experiments are designed to replace the difficult methods of sensitivity analysis that engineers have been using. Sensitivity analysis by methods of applied mathematics requires differentiation of complicated functions of many variables, followed by long and tedious analysis to solve large systems of differential equations. However, if we apply fractional replication designs, response surface designs, Latin squares, or other methods that were discussed in the previous chapter, we can obtain very good results relatively cheaply and fast. A complicated response function $f(\theta, \mathbf{X})$ is replaced by a relatively simple linear or quadratic regression function that is fitted to the values obtained. If the R^2 of the regression is close to 1, we have a satisfactory, simple approximation to a complex situation. We illustrate this with an example.

EXAMPLE **13.4**   Earlier we presented the piston turbine simulator, which computes the cycle speed as a function of seven factors. The functional relationship between the response Y and the seven factors is complicated but known (see Appendix II). Thus, in the parameter design stage, before a model or a prototype of the piston is built, we can run experiments on the computer to test the effects of the various parameters (controllable factors) on the outcome. Such a factorial experiment was described in Example 12.8. The goodness of fit in that example is $R^2 = .934$.

To test how well a linear regression function of main effects and first-order interaction terms can approximate the exact nonlinear response function, we rerun program TURB5.EXE here. For our current purpose, we set all standard deviations of the input factors to be equal to zero. The response now is completely deterministic. Table 13.5 is the matrix of coefficients of the 2^5 factorial experiment and the deterministic cycle times associated with parameter settings. The factor levels corresponding to these coefficients are given in Example 12.7.

TABLE **13.5**

Coefficients' matrix and exact cycle times in a 2^5 factorial experiment

1	−1	−1	−1	−1	−1	0.531
2	1	−1	−1	−1	−1	0.751
3	−1	1	−1	−1	−1	0.184
4	1	1	−1	−1	−1	0.261
5	−1	−1	1	−1	−1	1.077
6	1	−1	1	−1	−1	1.523
7	−1	1	1	−1	−1	0.914
8	1	1	1	−1	−1	1.293
9	−1	−1	−1	1	−1	0.295
10	1	−1	−1	1	−1	0.417
11	−1	1	−1	1	−1	0.116
12	1	1	−1	1	−1	0.163
13	−1	−1	1	1	−1	0.486
14	1	−1	1	1	−1	0.687
15	−1	1	1	1	−1	0.468
16	1	1	1	1	−1	0.662
17	−1	−1	−1	−1	1	0.533
18	1	−1	−1	−1	1	0.754
19	−1	1	−1	−1	1	0.186
20	1	1	−1	−1	1	0.263
21	−1	−1	1	−1	1	1.076
22	1	−1	1	−1	1	1.522
23	−1	1	1	−1	1	0.913
24	1	1	1	−1	1	1.291
25	−1	−1	−1	1	1	0.296
26	1	−1	−1	1	1	0.418
27	−1	1	−1	1	1	0.117
28	1	1	−1	1	1	0.165
29	−1	−1	1	1	1	0.486
30	1	−1	1	1	1	0.687
31	−1	1	1	1	1	0.468
32	1	1	1	1	1	0.662

The multiple regression of the cycle time Y on the coefficients of factors A, B, C, D, E is

$$Y = 0.615 + 0.105X_1 - 0.107X_2 + 0.274X_3 - 0.202X_4 + 0.00028X_5$$
$$- 0.0183X_1X_2 + 0.0470X_1X_3 - 0.0348X_1X_4 + 0.0521X_2X_3$$
$$+ 0.0472X_2X_4 - 0.110X_3X_4$$

The variables $X_1, \ldots, X_4$ correspond to factors $A, \ldots, D$, and X_5 corresponds to factor E. In this case, the goodness of fit is given by $R^2 = .997$, which is almost perfect. The standard errors, the t values, and the P-values of the regression coefficients are given in Table 13.6.

The example shows that we can approximate complicated response functions very well, within the experimental domain, by linear regression. ∎

TABLE 13.6

Standard errors and significance of regression coefficients

	Coef	Stdev	t-ratio	p
Intercept	0.614531	0.005066	121.29	.000
X_1	0.105406	0.005066	20.80	.000
X_2	−0.106656	0.005066	−21.05	.000
X_3	0.273906	0.005066	54.06	.000
X_4	−0.202469	0.005066	−39.96	.000
X_5	0.000281	0.005066	0.06	.956
X_1X_2	−0.018281	0.005066	−3.61	.002
X_1X_3	0.047031	0.005066	9.28	.000
X_1X_4	−0.034844	0.005066	−6.88	.000
X_2X_3	0.052094	0.005066	10.28	.000
X_2X_4	0.047219	0.005066	9.32	.000
X_3X_4	−0.110219	0.005066	−21.75	.000

Computer-simulated experiments can be run in the design stage of a new system whenever the functional relationship between the design parameters and the response is known. For the noise factors, one can simulate values according to some reasonable distribution or use an outer array according to the methodology of Taguchi. For additional reading on computer experiments for parameter design, see Welch, Yu, Kang, and Sacks (1990).

13.5

Tolerance Designs

Usually parts that are installed in systems—resistors, capacitors, transistors, and other parts of a mechanical nature—have some deviations in their characteristics from the nominal ones. For example, a resistor with a nominal resistance of 8200 ohms will have an actual resistance value that is a random deviate around that nominal value. Parts are classified according to their tolerances. Grade A could have a tolerance interval $\pm 1\%$ of the nominal value, grade B $\pm 5\%$, grade C $\pm 10\%$, and so on. Parts with high-grade tolerances are more expensive than those with low-grade ones. Due to the nonlinear dependence of the system output (performance characteristic) on the input values of its components, not all component variances contribute equally to the variance of the output. We also saw that the variances of the components affect the means of the output characteristics. It is therefore important to perform experiments to determine which tolerance grade should be assigned to each component. We illustrate such a problem in the following example.

EXAMPLE **13.5**   Taguchi (1987, Vol. 1, p. 379) describes a tolerance design for a circuit that converts alternating current of 100 V AC to a direct current of 220 V DC. This example is based on an experiment performed in 1974 at the Shin Nippon Denki Company.

The output of the system, Y, depends, in a complicated manner, on 17 factors. We present this function in Appendix II. Simulators ELECT1–ELECT3 were constructed to experiment with the system. In this example, we use program ELECT3.EXE to execute a fractional replication of 2^{13-8} to investigate the effects of two tolerance grades of 13 components: 10 resistors, and 3 transistors, on the output of the system. The two design levels for each factor are the two tolerance grades. For example, if we specify for a given factor a tolerance of 10%, then the experiment at level 1 will use a tolerance of 5% and at level 2 a tolerance of 10%. The value of a given factor is simulated according to a normal distribution with mean at the nominal value of that factor. The standard deviation is one-sixth the length of the tolerance interval. For example, if the nominal value for factor A is 8200 (ohm) and the tolerance level is 10%, the standard deviation for level 1 is 136.67 (ohm) and for level 2 is 273.33 (ohm).

As mentioned earlier, the control factors are 10 resistors, labeled A–J, and 3 transistors, labeled K–M. The nominal levels of these factors are:

$$A = 8200, \ B = 220000, \ C = 1000, \ D = 33000, \ E = 56000, \ F = 5600,$$
$$G = 3300, \ H = 58.5, \ I = 1000, \ J = 120, \ K = 130, \ L = 100, \ M = 130$$

The nominal levels and the tolerance levels can be modified by editing the file PARELCT.DAT. The levels of the 13 factors in the 2^{13-8} fractional replication are given in Table 13.7.

We perform this experiment on the computer using program ELECT3.EXE. We want to find a treatment combination (run) that yields a small MSE at a low cost per circuit. We assume that grade B parts (5% tolerance) cost \$1 and grade C parts (10% tolerance) cost \$0.50. To obtain sufficiently precise estimates of the MSE, we perform at each run a simulated sample of size $n = 100$. The results of this experiment are given in Table 13.8.

We see that the runs with small mean-squared errors (MSE) are 1, 11, 19, and 25. Among these, the runs with the smallest total cost (TC) are 11 and 19. The MSE of run 11 is somewhat smaller than that of run 19. The difference, however, is not significant. We can choose either combination of tolerance levels for the manufacture of the circuits.   ∎

TABLE 13.7
Factor levels for the 2^{13-8} design

Run	A	B	C	D	E	F	G	H	I	J	K	L	M
1	1	1	1	1	1	1	1	1	1	1	1	1	1
2	2	2	2	2	2	2	2	2	1	1	1	1	1
3	2	1	2	2	1	1	2	1	2	2	1	1	1
4	1	2	1	1	2	2	1	2	2	2	1	1	1
5	1	1	2	1	2	2	2	1	2	1	2	1	1
6	2	2	1	2	1	1	1	2	2	1	2	1	1
7	2	1	1	2	2	2	1	1	1	2	2	1	1
8	1	2	2	1	1	1	2	2	1	2	2	1	1
9	2	2	2	1	2	1	1	1	2	1	1	2	1
10	1	1	1	2	1	2	2	2	2	1	1	2	1
11	1	2	1	2	2	1	2	1	1	2	1	2	1
12	2	1	2	1	1	2	1	2	1	2	1	2	1
13	2	2	1	1	1	2	2	1	1	1	2	2	1
14	1	1	2	2	2	1	1	2	1	1	2	2	1
15	1	2	2	2	1	2	1	1	2	2	2	2	1
16	2	1	1	1	2	1	2	2	2	2	2	2	1
17	2	2	2	1	2	1	1	1	2	1	1	1	2
18	1	1	1	2	1	2	2	2	2	1	1	1	2
19	1	2	1	2	2	1	2	1	1	2	1	1	2
20	2	1	2	1	1	2	1	2	1	2	1	1	2
21	2	2	1	1	1	2	2	1	1	1	2	1	2
22	1	1	2	2	2	1	1	2	1	1	2	1	2
23	1	2	2	2	1	2	1	1	2	2	2	1	2
24	2	1	1	1	2	1	2	2	2	2	2	1	2
25	1	1	1	1	1	1	1	1	1	1	1	2	2
26	2	2	2	2	2	2	2	2	1	1	1	2	2
27	2	1	2	2	1	1	2	1	2	2	1	2	2
28	1	2	1	1	2	2	1	2	2	2	1	2	2
29	1	1	2	1	2	2	2	1	2	1	2	2	2
30	2	2	1	2	1	1	1	2	2	1	2	2	2
31	2	1	1	2	2	2	1	1	1	2	2	2	2
32	1	2	2	1	1	1	2	2	1	2	2	2	2

13.6
Case Studies

13.6.1 The Quinlan Experiment at Flex Products

Our first case study is an experiment carried out at Flex Products in Midvale, Ohio (Quinlan, 1985). Flex Products, a subcontractor of General Motors, manufactures mechanical speedometer cables. The basic cable design had not changed for 15 years and General Motors had made many unsuccessful attempts to reduce the

TABLE **13.8**

Performance characteristics of tolerance design experiment

Run	$\overline{Y}$	STD	MSE	TC
1	219.91	3.6420	13.2723	13
2	219.60	7.5026	56.4490	9
3	220.21	5.9314	35.2256	10
4	220.48	7.3349	54.0312	10
5	219.48	4.8595	23.8851	10
6	219.82	6.3183	39.9533	10
7	219.61	6.0647	36.9327	10
8	219.40	5.2205	27.6136	10
9	220.29	5.6093	31.5483	10
10	218.52	6.5635	45.2699	10
11	219.71	4.0752	16.6914	10
12	220.27	5.6723	32.2479	10
13	220.74	5.8068	34.2665	10
14	219.93	5.4065	29.2351	10
15	219.92	5.6605	32.0477	9
16	219.71	6.9693	48.6552	9
17	219.93	5.1390	26.4142	10
18	221.49	6.6135	45.9585	10
19	219.98	4.1369	17.1143	10
20	220.10	6.5837	43.3551	10
21	220.65	6.0391	36.8932	10
22	219.38	5.7089	32.9759	10
23	220.26	6.2068	38.5920	9
24	219.97	6.3469	40.2840	9
25	220.53	4.0378	16.5847	12
26	220.20	6.6526	44.2971	8
27	220.22	5.4881	30.1676	9
28	219.48	6.1564	38.1717	9
29	219.60	5.1583	26.7681	9
30	220.50	6.3103	40.0699	9
31	221.43	5.8592	36.3751	9
32	220.22	5.2319	27.4212	9

speedometer errors. To correct the problem, Flex products decided to apply off-line quality control and involve in the project customers, production personnel, and engineers with experience in the product and manufacturing process. A large experiment involving 15 factors was designed and completed. The data showed that much improvement could be gained by a few simple changes. The results were dramatic, and the loss per unit was reduced from \$2.12 to \$0.13 by changing the braid type, the liner material, and the braiding tension.

The experiment was as follows:

1 **Problem definition**: The product under investigation is an extruded thermo-plastic speedometer casing used to cover the mechanical speedometer cable on automobiles. Excessive shrinkage of the casing is causing noise in the mechanical speedometer cable assembly.

2 **Response variable**: The performance characteristic in this problem is the post-extrusion shrinkage of the casing. The percent shrinkage is obtained by measuring approximately 600 mm of casing that has been properly conditioned (*A*), placing that casing in a 2-hour heat soak in an air-circulating oven, reconditioning the sample, and measuring the length (*B*). Shrinkage is computed as: Shrinkage $= 100 \times (A - B)/A$.

3 **Control factors**:

Liner process:

   A : Liner outer diameter

   B : Liner die

   C : Liner material

   D : Liner line speed

Wire braiding:

   E : Wire braid type

   F : Braiding tension

   G : Wire diameter

   H : Liner tension

   I : Liner temperature

Coating process:

   J : Coating material

   K : Coating dye type

   L : Melt temperature

   M : Screen pack

   N : Cooling method

   O : Line speed

4 **Factor levels**: Existing (1)–Changed (2)

5 **Experimental array**: $L_{16}(2^{15})$ orthogonal array (Table 13.4)

6 **Number of replications**: Four random samples of 600 mm from the 3000 feet manufactured at each experimental run

7 **Data analysis**: Signal-to-noise ratios (SN) are computed for each experimental run and analyzed using main effect plots and an ANOVA. Savings are derived from loss function computations. The signal-to-noise formula used by Quinlan is

$$\eta = -10 \log_{10} \left(\frac{1}{n} \sum_{i=1}^{n} y_i^2 \right)$$

For example, experimental run number 1 produced shrinkage factors of: 0.49, 0.54, 0.46, and 0.45. The SN is 6.26. The objective is to maximize the SN by properly setting up the 15 controllable factors. Table 13.9 shows the factor levels and the SN values for all 16 experimental runs.

TABLE 13.9

Factor levels and SN values

							Factor									
Run	*A*	*B*	*C*	*D*	*E*	*F*	*G*	*H*	*I*	*J*	*K*	*L*	*M*	*N*	*O*	*SN*
1	1	1	1	1	1	1	1	1	1	1	1	1	1	1	1	6.26
2	1	1	1	1	1	1	1	2	2	2	2	2	2	2	2	4.80
3	1	1	1	2	2	2	2	1	1	1	1	2	2	2	2	21.04
4	1	1	1	2	2	2	2	2	2	2	2	1	1	1	1	15.11
5	1	2	2	1	1	2	2	1	1	2	2	1	1	2	2	14.03
6	1	2	2	1	1	2	2	2	2	1	1	2	2	1	1	16.69
7	1	2	2	2	2	1	1	1	1	2	2	2	2	1	1	12.91
8	1	2	2	2	2	1	1	2	2	1	1	1	1	2	2	15.05
9	2	1	2	1	2	1	2	1	2	1	2	1	2	1	2	17.67
10	2	1	2	1	2	1	2	2	1	2	1	2	1	2	1	17.27
11	2	1	2	2	1	2	1	1	2	1	2	2	1	2	1	6.82
12	2	1	2	2	1	2	1	2	1	2	1	1	2	1	2	5.43
13	2	2	1	1	2	2	1	1	2	2	1	1	2	2	1	15.27
14	2	2	1	1	2	2	1	2	1	1	2	2	1	1	2	11.2
15	2	2	1	2	1	1	2	1	2	2	1	2	1	1	2	9.24
16	2	2	1	2	1	1	2	2	1	1	2	1	2	2	1	4.68

Notice that Quinlan, by using the orthogonal array $L_{16}(2^{15})$ for all 15 factors, assumes that there are no significant interactions. If this assumption is correct, then the main effects of the 15 factors are as follows:

Factor	*A*	*B*	*C*	*D*	*E*	*F*	*G*	*H*
Main effect	−1.145	0.29	1.14	−0.86	3.60	1.11	2.37	−0.82

Factor	*I*	*J*	*K*	*L*	*M*	*N*	*O*
Main effect	0.49	−0.34	−1.19	0.41	0.22	0.28	0.22

Figure 13.3 shows the main effects plot for this experiment. Factors *E* and *G* seem to be most influential. These main effects, as defined in Chapter 12, are the regression coefficients of SN on the design coefficients ±1. Only the effects of factors *E* and *G* are significant. If the assumption of no-interaction

FIGURE **13.3**

Main effects plot for Quinlan experiment

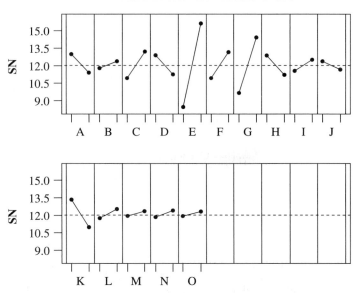

is wrong and all first-order interactions are significant, then, as shown in the linear graph LG_1 in Figure 13.2, only the effects of factors *A*, *B*, *D*, *H*, and *O* are not confounded. The effects of the other factors are confounded with first-order interactions. The main effect of factor *E* is confounded with interaction *AD*, that of *G* is confounded with *HO*. In order to confirm the first hypothesis—that all interactions are negligible—an additional experiment should be performed in which factors *E* and *G* are assigned to columns that do not represent possible interactions (columns 1 and 2 of Table 13.4, for example). The results of the additional experiment should confirm the conclusions of the original experiment.

8 **Results**: As a result of Quinlan's analysis, factors *E* and *G* were properly changed. This reduced the average shrinkage index from 26% to 5%. The shrinkage standard deviation was also reduced—from 0.05 to 0.025. This was considered a substantial success in quality improvement.

13.6.2 Computer Response Time Optimization

The next case study is an experiment that was part of an extensive effort to optimize a UNIX operating system running on a VAX 11-780 machine (Pao, Phadke, and Sherrerd, 1985). The machine had 48 user terminal ports, 2 remote job entry links,

4 megabytes of memory, and 5 disk drives. The typical number of users logged on at a given time was between 20 and 30.

1 **Problem definition**: Users complained that system performance was very poor, especially in the afternoon. The objective of the improvement effort was to both minimize response time and reduce variability in response.

2 **Response variable**: To get an objective measurement of response time, two specific representative commands called **STANDARD** and **TRIVIAL** were used. The **STANDARD** command consisted of creating, editing, and removing a file. The **TRIVIAL** command was the UNIX system "date" command. Response times were measured by submitting these commands every 10 minutes and clocking the time the system took to complete their execution.

3 **Control factors**:

  A : Disk drives

  B : File distribution

  C : Memory size

  D : System buffers

  E : Sticky bits

  F : KMCs used

  G : INODE table entries

  H : Other system tables

4 **Factor levels**:

Factor	Levels		
A: RM05 & RP06	4& 1		4 & 2
B: File distribution	a	b	c
C: Memory size (MB)	4	3	3.5
D: System buffers	1/5	1/4	1/3
E: Sticky bits	0	3	8
F: KMCs used	2		0
G: INODE table entries	400	500	600
H: Other system tables	a	b	c

5 **Experimental array**: The design was an orthogonal array $L_{18}(3^8)$. This and the mean response are given in Table 13.10. Each mean response in the table is over $n = 96$ measurements.

TABLE 13.10

Factor levels and mean response time[a]

	F	B	C	D	E	A	G	H	Mean (sec)	SN
1	1	1	1	1	1	1	1	1	4.65	−14.66
2	1	1	2	2	2	2	2	2	5.28	−16.37
3	1	1	3	3	3	3	3	3	3.06	−10.49
4	1	2	1	1	2	2	3	3	4.53	−14.85
5	1	2	2	2	3	3	1	1	3.26	−10.94
6	1	2	3	3	1	1	2	2	4.55	−14.96
7	1	3	1	2	1	3	2	3	3.37	−11.77
8	1	3	2	3	2	1	3	1	5.62	−16.72
9	1	3	3	1	3	2	1	2	4.87	−14.67
10	2	1	1	3	3	2	2	1	4.13	−13.52
11	2	1	2	1	1	3	3	2	4.08	−13.79
12	2	1	3	2	2	1	1	3	4.45	−14.19
13	2	2	1	2	3	1	3	2	3.81	−12.89
14	2	2	2	3	1	2	1	3	5.87	−16.75
15	2	2	3	1	2	3	2	1	3.42	−11.65
16	2	3	1	3	2	3	1	2	3.66	−12.23
17	2	3	2	1	3	1	2	3	3.92	−12.81
18	2	3	3	2	1	2	3	1	4.42	−13.71

[a] Factor *A* had only 2 levels. All level 3s in the table were changed to level 2.

6 **Data analysis**: The measure of performance characteristic used was the SN ratio:

$$\eta = -10 \log_{10} \left(\frac{1}{n} \sum_{i=1}^{n} y_i^2 \right)$$

where y_i is the *i*th response time.

Figure 13.4 is the main effects plot of the eight factors. The linear and quadratic effects of these factors were found to be as follows:

Factor	Linear	Quadratic
A	0.97	—
B	0.19	−0.15
C	−1.24	−1.32
D	−0.37	−1.23
E	1.72	1.86
F	0.44	—
G	0.17	−0.63
H	0.05	1.29

Factors having substantial effects were *A*, *C*, *D*, *E*, and *H*. As a result of this experiment, the number of disk drives was changed to 4 and 2. The system buffers were

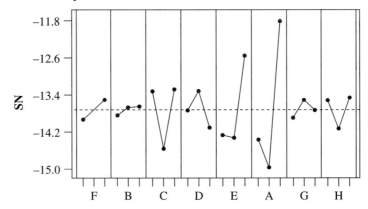

FIGURE   13.4
Main effects plot

changed from 1/3 to 1/4, and the number of sticky bits was changed from 0 to 8. After introducing these changes, the average response time dropped from 6.15 sec to 2.37 sec, with a substantial reduction in response time variability.

Chapter Highlights

Quality is largely determined by decisions made in the early planning phases of products and processes. A particularly powerful technique for making optimal design decisions is the statistically designed experiment. This chapter covers the basics of experimental designs in the context of engineering and economic optimization problems. Taguchi's loss function, signal-to-noise ratios, factorial models, and orthogonal arrays are discussed using case studies and simple examples.

The main concepts and definitions introduced in this chapter include:

- Quality engineering
- Off-line quality control
- Loss function
- Customer's tolerance interval
- Manufacturer's tolerance interval
- Parameter design
- Tolerance design

Exercises

13.1.1   The objective is to find the levels of the factors of the turbo piston that yield an average cycle time of

0.45 sec. Execute program TURB5.EXE, with sample size $n = 100$.

a Use the output file OTURB5.DAT to determine which treatment combination yields the smallest MSE $= (\overline{Y} - 0.45)^2 + S^2$.

b Which treatment combination yields the largest SN ratio, $\eta = 10 \; \log_{10}[(Y^2/S^2) - (1/100)]$? What is the MSE at this treatment combination? The five factors that are varied are: piston weight, piston surface area, initial gas volume, spring coefficient, and ambient temperature. The factors atmospheric pressure and filling gas temperature are kept constant at the midrange level.

13.1.2 Run program TURB3.EXE with sample size of $n = 100$. The output file OTURB3.DAT contains the sample means and standard deviation of the $2^7 = 128$ treatment combinations of a full factorial experiment for the effects on the piston cycle time. Perform regression analysis to find which factors have significant effects on the signal-to-noise ratio SN $= \log((\overline{X}/S)^2)$.

13.2.1 Let (X_1, X_2) have joint distribution with means (ξ_1, ξ_2) and covariance matrix

$$V = \begin{pmatrix} \sigma_1^2 & \sigma_{12} \\ \sigma_{12} & \sigma_2^2 \end{pmatrix}$$

Find approximations to the expected values and variances of:

i $Y = X_1/X_2$

ii $Y = \log(X_1^2/X_2^2)$

iii $Y = (X_1^2 + X_2^2)^{1/2}$

13.2.2 The relationship between the absorption ratio Y of a solid image in a copied paper and the light intensity X is given by the function

$$Y = 0.0782 + \frac{0.90258}{1 + 0.6969X^{-1.4258}}$$

Assuming that X has the gamma distribution $G(1, 1.5)$, approximate the expected value and variance of Y.

13.2.3 Let $\overline{X}_n$ and S_n^2 be the mean and variance of a random sample of size n from a normal distribution $N(\mu, \sigma)$. We know that $\overline{X}_n$ and S_n^2 are independent, $\overline{X}_n$ is distributed like $N\left(\mu, \sigma/\sqrt{n}\right)$ and S_n^2 is distributed like $[\sigma^2/(n-1)]\chi^2[n-1]$. Find an approximation to the expected value and variance of $Y = \log\left(\overline{X}_n/S_n^2\right)$.

13.3.1 An experiment based on an L_{18} orthogonal array involving eight factors gave the following results

(see Phadke, et al. 1983). Each run had $n = 5$ replications.

	Factors									
Run	**1**	**2**	**3**	**4**	**5**	**6**	**7**	**8**	$\overline{X}$	S
1	1	1	1	1	1	1	1	1	2.500	0.0827
2	1	1	2	2	2	2	2	2	2.684	0.1196
3	1	1	3	3	3	3	3	3	2.660	0.1722
4	1	2	1	1	2	2	3	3	1.962	0.1696
5	1	2	2	2	3	3	1	1	1.870	0.1168
6	1	2	3	3	1	1	2	2	2.584	0.1106
7	1	3	1	2	1	3	2	3	2.032	0.0718
8	1	3	2	3	2	1	3	1	3.267	0.2101
9	1	3	3	1	3	2	1	2	2.829	0.1516
10	2	1	1	3	3	2	2	1	2.660	0.1912
11	2	1	2	1	1	3	3	2	3.166	0.0674
12	2	1	3	2	2	1	1	3	3.323	0.1274
13	2	2	1	2	3	1	3	2	2.576	0.0850
14	2	2	2	3	1	2	1	3	2.308	0.0964
15	2	2	3	1	2	3	2	1	2.464	0.0385
16	2	3	1	3	2	3	1	2	2.667	0.0706
17	2	3	2	1	3	1	2	3	3.156	0.1569
18	2	3	3	2	1	2	3	1	3.494	0.0473

Analyze the effects of the factors on the SN ratio $\eta = \log(\overline{X}/S)$.

13.4.1 Computer programs TURB3.EXE, TURB31.EXE, TURB32.EXE, and TURB33.EXE perform a full factorial (2^7), a $1/8$ (2^{7-3}), $1/4$ (2^{7-2}), and $1/2$ (2^{7-1}) fractional replications of the cycle time experiment. Execute these programs, estimate the main effects of the seven factors with respect to SN $= \log(\overline{X}/S)$, and compare the results obtained from these experiments.

13.4.2 To see the effect of the variances of random variables on the expected response in nonlinear cases, execute program TURB5.EXE, with $n = 20$, and compare the output means (file OTURB5.EXE) to the values in Table 13.5.

13.5.1 Run program ELECT3.EXE with 1% and 2% tolerances, and compare the results to those of Table 13.8. [First edit file C:\ISTAT\DATA\PARELCT.DAT by changing 5 to 1 and 10 to 2.]

14

Reliability Analysis

Industrial products are considered to be of high quality if they conform to their design specifications and appeal to the customer. However, products can fail after a while due to degradation over time or to some instantaneous shock. A system or a component of a system is said to be *reliable* if it continues to function according to specifications for a long time. Reliability of a product is a dynamic notion, over time. We say that a product is highly reliable if the probability that it will function properly for a specified long period is close to 1. As will be defined later, the reliability function, $R(t)$, is the probability that a product will continue functioning at least t units of time.

We distinguish between the reliability of systems that are unrepairable and the reliability of those that are repairable. A repairable system, after failure, goes through a period of repair and then returns to function normally. Highly reliable systems need less repair. Repairable systems that need less repair are more available to operate, and are therefore more desirable. **Availability** of a system at time t is the probability that the system will be up and running at time t. To increase the availability of repairable systems, maintenance procedures are devised. Maintenance is designed to prevent failures of a system by periodic replacement of parts, tuning, cleaning, and so on. It is important to develop maintenance procedures based on the reliability properties of system components that are cost effective and helpful to the availability of systems.

One of the intriguing features of component and system failure is its random nature. Therefore, we consider the length of time that a part functions till failure as a random variable, called the *life length* of the component or system. The distribution functions of life length variables are called **life distributions**. The role of statistical reliability theory is to develop methods of estimating the characteristics of life distributions from failure data, and to design experiments called *life tests*. Associated with life testing is **accelerated life testing**. Highly reliable systems may take a long time till failure. In accelerated life tests, early failures are induced by subjecting the systems to higher than normal stress. When analyzing the results of such experiments, we need to know how to relate failure distributions under stressful conditions to those under normal operating conditions. This chapter provides the foundation for the theoretical and practical treatment of these subjects. For additional readings, see Zacks (1992).

The following examples illustrate the importance of reliability analysis and modifications (improvements) for industry.

1 **Florida Power and Light** A reduction in the power plant outage rate from 14% in 1986 to less than 4% in 1989 has generated $300 million in savings to customers on an investment of $5 million for training and consulting. Customer service interruptions dropped from 100 minutes per year in 1982 to 42 minutes per year in 1989. [*From: Forbes* (March 1991), 18, 112–114.]

2 **Tennessee Valley Authority (TVA)** The Athens Utilities Board is one of 160 power distributors supplied by TVA, with a service region of 100 square miles, 10,000 customers, and a peak load of 80 MW. One year's worth of trouble service data was examined in three South Athens feeders. The primary circuit failure rate was 15.3 failures/year/mile; restoring service using automatic equipment took, on average, 3 minutes per switch, whereas manual switching required approximately 20 minutes. Line repair generally takes 45 minutes. The average outage cost for an industrial customer in the United States is $11.87/kWh (in 1986 dollars). Without automation, the yearly outage cost for a 6000-kW load per year would be, on average, $540,000. The automation required to restore service in 3 minutes in South Athens costs about $35,000. Automation has reduced outage costs to $340,000. These improvements in the reliability of the power supply have therefore produced an average return on investment of $9.7 for every dollar invested in automation. [*From: IEEE Transactions on Power Delivery* (January 1989), 770–777.]

3 **AT&T** An original plan for a transatlantic telephone cable called for three spares to back up each transmitter in the 200 repeaters that would relay calls across the seabed. A detailed reliability analysis with SUPER (System Used for Prediction and Evaluation of Reliability) indicated that one spare would be enough. This reduced the cost of the project by 10%—and AT&T won the job with a bid just 5% less than that of its nearest competitor. [*From: Business Week* (8 June 1987), 140–141.]

4 **AVX** The level of reliability now achieved by tantalum capacitors, along with their small size and high stability, is promoting their use in many applications that are electrically and environmentally more aggressive than in the past. The 1990 failure rates were 0.67 FIT (failures in 10^9 component hours), several orders of magnitude lower than that in 1980. Moreover, shorts cause approximately 67% of the total failures. [*From: Electronic Engineering* (March 1991), 43–51.]

5 **Siemens** Broadband transmission systems use a significant number of microwave components that are expected to work without failure right from the first switch-on. The 565-Mbit coaxial repeater uses 30 diodes and transistor functions in each repeater, which adds up to 7000 SP87-11 transistors along the 250-km link. The link must not fail within 15 years. Redundant circuits are not possible because of the complex circuitry. Accelerated life testing has

demonstrated that the expected failure rate of the SP87-11 transistor is less than 1 FIT, thus meeting the 15-year requirement. [*From: Microwave and RF Engineering* (April 1989), S29–S34.]

6 National Semiconductor   A single-bit error in a microelectronic device can cause an entire system to crash. In developing the BiCmos III component, one-third of the design team was assigned the job of improving the component's reliability. Accelerated life tests under high temperature and high humidity (145°C, 85% relative humidity and under bias) showed the improved device to have a failure rate below 100 FIT. In a system using 256-kbit BiCmos III static random-access memories, this translates into less than one failure in 18 years. [*From: Electronics* (4 February 1988), 61–62.]

7 Lockheed   Some 60% of the cost of a military aircraft now goes for its electronic systems, and many military contracts require the manufacturer to provide service at a fixed price for product defects that occur during the warranty period. Lockheed Corporation produces switching logic units used in the U.S. Navy S-3A antisubmarine aircraft to distribute communications within and outside the aircraft. These units were high on the Pareto of component failures. They were therefore often removed for maintenance, which damaged the chassis. The mean time between failures (MTBF) for the switching logic units was approximately 100 hours. Changes in the design and improved screening procedures increased the MTBF to 500 hours. The average number of units removed each week from nine aircraft dropped from 1.8 to 0.14. [*From: IEEE Spectrum* (December 1986), 37–42.]

Basic Notions

14.1.1 Time Categories

The following **time categories** play an important role in the theory of reliability, availability, and maintainability of systems.

I Usage-Related Time Categories
  1 *Operating time* is the time interval during which the system is in actual operation.
  2 *Scheduled operating time* is the time interval during which the system is required to operate properly.
  3 *Free time* is the time interval during which the system is scheduled to be off duty.
  4 *Storage time* is the time interval during which the system is stored as a spare part.

II Equipment Condition Time Categories
  1 **Up time** is the time interval during which the system is operating or ready for operation.

2 **Down time** is the time interval out of the scheduled operating time during which the system is in a state of failure (inoperable). Down time is the sum of (a) administrative time, (b) active repair time and (c) logistic time (repair suspension due to lack of parts).

III Indices

$$\text{Scheduled operating time} = \text{Operating time} + \text{down time}$$

$$\textbf{Intrinsic availability} = \frac{\text{Operating time}}{\text{Operating time} + \text{active repair time}}$$

$$\text{Availability} = \frac{\text{Operating time}}{\text{Operating time} + \text{down time}}$$

$$\textbf{Operational readiness} = \frac{\text{Up time}}{\text{Total calendar time}}$$

EXAMPLE **14.1**   A machine is scheduled to operate for two shifts a day (8 hours each shift), five days a week. During the last 48 weeks, the machine was down five times. The average down time is partitioned as follows:

1   Average administrative time = 9 hr
2   Average repair time = 30 hr
3   Average logistic time = 7.6 hr

Thus, the total down time in the 48 weeks is

$$\text{Down time} = 5 \times (9 + 30 + 7.6) = 233 \text{ hr}$$

The total scheduled operating time is $48 \times 16 \times 5 = 3840$ hr. Thus, total operating time = 3607 hr. The indices of availability and intrinsic availability are

$$\text{Availability} = \frac{3607}{3840} = 0.9393$$

$$\text{Intrinsic availability} = \frac{3607}{3607 + 150} = 0.9601$$

Finally, the operational readiness of the machine is

$$\text{Operational readiness} = \frac{8064 - 233}{8064} = 0.9711 \quad \blacksquare$$

14.1.2   Reliability and Related Functions

The length of life (lifetime) of a (product) system is the length of the time interval T from its initial activation till its failure. If a system is switched on and off, we consider the total active time of the system till its failure. T is a nonnegative random variable. The distribution of T is a life distribution. We generally assume that T is a continuous random variable, having a p.d.f. $f_T(t)$ and c.d.f. $F_T(t)$. The **reliability function** of a (product) system is defined as

$$
\begin{aligned}
R(t) &= \Pr\{T \geq t\} \\
&= 1 - F_T(t), \quad t \geq 0.
\end{aligned}
$$

(14.1.1)

The expected lifetime of a product is called its **mean time to failure (MTTF)**. This quantity is given by

$$
\begin{aligned}
\mu &= \int_0^\infty t f_T(t)\,dt \\
&= \int_0^\infty R(t)\,dt
\end{aligned}
$$

(14.1.2)

The instantaneous **hazard function** of a product, also called the **failure rate function** is defined as

$$
h(t) = \frac{f(t)}{R(t)} \qquad t \geq 0
$$

(14.1.3)

Notice that $h(t)$ and $f(t)$ have the dimension of $1/T$. That is, if T is measured in hours, the dimension of $h(t)$ is [1/hr].

Notice that $h(t) = -d/dt \log(R(t))$. Accordingly,

$$
R(t) = \exp\left\{-\int_0^t h(u)\,du\right\}
$$

(14.1.4)

The function

$$
H(t) = \int_0^t h(u)\,du
$$

(14.1.5)

is called the *cumulative hazard rate function*.

EXAMPLE 14.2   In many applications of reliability theory, the exponential distribution with mean μ is used for T. In this case

$$
f_T(t) = \frac{1}{\mu}\exp\{-t/\mu\} \qquad t \geq 0
$$

and

$$
R(t) = \exp\{-t/\mu\} \qquad t \geq 0
$$

In this model, the reliability function diminishes from 1 to 0 exponentially fast, relative to μ.

The hazard rate function is

$$
h(t) = \frac{\frac{1}{\mu}\cdot\exp\{-t/\mu\}}{\exp\{-t/\mu\}} = \frac{1}{\mu} \qquad t \geq 0
$$

That is, the exponential model is valid for cases where the hazard rate function is a constant, independent of time. If the MTTF is $\mu = 100$ (hr), we expect 1 failure per 100 (hr)—that is,

$$
h(t) = \frac{1}{100}\left[\frac{1}{hr}\right] \qquad \blacksquare
$$

14.2
System Reliability

In this section, we discuss how to compute the reliability function of a system as a function of the reliability of its components (modules). Thus, if we have a system composed of k subsystems (components or modules) with reliability functions $R_1(t), \ldots, R_k(t)$, the reliability of the system is given by

$$R_{\text{sys}}(t) = \psi(R_1(t), \ldots, R_k(t)) \qquad t \geq 0 \qquad \text{(14.2.1)}$$

The function $\psi(\cdot)$ is called a **structure function**. It reflects the functional relationship between the subsystems and the system. Here we discuss some structure functions of simple systems. We assume that the random variables $T_1, \ldots, T_k$ representing the life length of the subsystems are independent.

Consider a system having two subsystems (modules) C_1 and C_2. We say that the subsystems are *connected in series* if a failure of either one of the subsystems causes immediate failure of the system. We represent this series connection by a *block diagram*, as in Figure 14.1. Let I_i ($i = 1, \ldots, k$) be indicator variables that assume the value 1 if the component C_i does not fail during a specified time interval $(0, t_0)$. If C_i fails during $(0, t_0)$, then $I_i = 0$. A *series structure function* of k components is

$$\psi_s(I_1, \ldots, I_k) = \prod_{i=1}^{k} I_i \qquad \text{(14.2.2)}$$

The expected value of I_i is

$$E\{I_i\} = \Pr\{I_i = 1\} = R_i(t_0) \qquad \text{(14.2.3)}$$

Thus, if the system is connected in series, then, because $T_1, \ldots, T_k$ are independent,

$$R_{\text{sys}}^{(s)}(t_0) = E\{\psi_s(I_1, \ldots, I_k)\} = \prod_{i=1}^{k} R_i(t_0)$$

$$\qquad \qquad \qquad \qquad \text{(14.2.4)}$$

$$= \psi_s(R_1(t_0), \ldots, R_k(t_0))$$

Thus, the system reliability function for subsystems connected in series is given by $\psi_s(R_1, \ldots, R_k)$, where $R_1, \ldots, R_k$ are the reliability values of the components.

A system composed of k subsystems is said to be *connected in parallel* if the system fails the instant all subsystems fail. In a parallel connection, it is sufficient that one of the subsystems function for the whole system to function.

The structure function for parallel connection is (see Figure 14.1)

$$\psi_p(I_1, \ldots, I_k) = 1 - \prod_{i=1}^{k} (1 - I_i) \qquad \text{(14.2.5)}$$

The reliability function for a system in parallel is, in the case of independence,

$$R_{\text{sys}}^{(p)}(t_0) = E\{\psi_p(I_1, \ldots, I_k)\}$$

$$\qquad \qquad \qquad \text{(14.2.6)}$$

$$= 1 - \prod_{i=1}^{k} (1 - R_i(t_0))$$

F I G U R E   14.1

Block diagrams for systems in series and in parallel

Components in Series

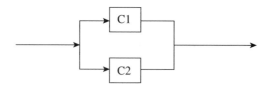

Components in Parallel

EXAMPLE 14.3   A computer card has 200 components that should function correctly. The reliability of each component for a period of 200 hours of operation is $R = .9999$. The components are independent of each other. What is the reliability of the card for this time period? Because all the components should function, we consider a series structure function. Thus, the system reliability for $t_0 = 200$ (hr) is

$$R_{sys}^{(s)}(t_0) = (.9999)^{200} = .9802$$

Despite the fact that each component is unlikely to fail, there is nevertheless a probability of .02 that the card will fail within 200 hours. If each of the components has a reliability of only .99, the card reliability is

$$R_{sys}^{(s)}(t_0) = (.99)^{200} = .134$$

This shows why it is so essential in the electronics industry to demand the most reliable products from the vendors of components.

Suppose there is room for some redundancy on the card. In this case, it is decided to use parts with reliability $R = .99$ and duplicate each component in a parallel structure. The parallel structure of duplicated components is considered a module. The reliability of each module is $R_M = 1 - (1 - .99)^2 = .9999$. The reliability of the whole system is again

$$R_{sys}^{(s)} = (R_M)^{200} = .9802$$

Thus, by changing the structure of the card, we can achieve the .98 reliability with 200 pairs of components, each with reliability value of .99.   ∎

Systems may have more complicated structures. In Figure 14.2, we see the block diagram of a system consisting of five components. Let $R_1, R_2, \ldots, R_5$ denote the reliability values of the five components $C_1, \ldots, C_5$, respectively. Let M_1 be the

FIGURE   14.2

A parallel series structure

A Parallel Series System

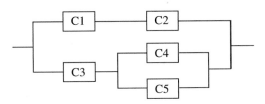

module consisting of components C_1 and C_2, and let M_2 be the module consisting of the other components. The reliability of M_1 for some specified time interval is

$$R_{M_1} = R_1 R_2$$

The reliability of M_2 is

$$
\begin{aligned}
R_{M_2} &= R_3(1 - (1 - R_4)(1 - R_5)) \\
&= R_3(R_4 + R_5 - R_4 R_5) \\
&= R_3 R_4 + R_3 R_5 - R_3 R_4 R_5
\end{aligned}
$$

Finally, the system reliability for that block diagram is

$$
\begin{aligned}
R_{sys} &= 1 - (1 - R_{M_1})(1 - R_{M_2}) \\
&= R_{M_1} + R_{M_2} - R_{M_1} R_{M_2} \\
&= R_1 R_2 + R_3 R_4 + R_3 R_5 - R_3 R_4 R_5 \\
&\quad - R_1 R_2 R_3 R_4 - R_1 R_2 R_3 R_5 + R_1 R_2 R_3 R_4 R_5
\end{aligned}
$$

Another important structure function is the k out of n subsystems function. In other words, if a system consists of n subsystems, it is required that at least k, $1 \le k \le n$ subsystems function throughout the specified time period in order for the system to function. Assuming independence of the lifetimes of the subsystems, we can construct the reliability function of the system by simple probabilistic considerations. For example, if we have 3 subsystems with reliability values for the given time period of R_1, R_2, R_3 and at least 2 of the 3 should function, then the system reliability is

$$
\begin{aligned}
R_{sys}^{2(3)} &= 1 - (1 - R_1)(1 - R_2)(1 - R_3) - R_1(1 - R_2)(1 - R_3) \\
&\quad - R_2(1 - R_1)(1 - R_3) - R_3(1 - R_1)(1 - R_2) \\
&= R_1 R_2 + R_1 R_3 + R_2 R_3 - 2 R_1 R_2 R_3
\end{aligned}
$$

If all the subsystems have the same reliability value R for a specified time period, then the reliability function of the system, in a k out of n structure, can be computed by using the binomial c.d.f. $B(j; n, R)$—that is,

$$R_{sys}^{k(n)} = 1 - B(k - 1; n, R) \qquad \textbf{(14.2.7)}$$

EXAMPLE **14.4**   A cooling system for a reactor has 3 identical cooling loops. Each cooling loop has 2 identical pumps connected in parallel. The cooling system requires that 2 of the 3 cooling loops operate successfully. The reliability of a pump over the life span of the plant is $R = .6$. Let's now compute the reliability of the cooling system.

First, the reliability of a cooling loop is

$$R_{cl} = 1 - (1 - R)^2 = 2R - R^2$$
$$= 1.2 - .36 = .84$$

Finally, the system reliability is

$$R_{sys}^{2(3)} = 1 - B(1; 3, .84) = .9314$$

This reliability can be increased by choosing pumps with higher reliability. If pump reliability is .9, the loop's reliability is .99 and the system's reliability is .9997. ∎

The reader is referred to Zacks (1992, ch. 3) for additional methods of computing system reliability.

14.3
Availability of Repairable Systems

Repairable systems alternate during their functional lives through cycles of up phase and down phase. During the up phase, the system functions as required till it fails. At the moment of failure, the system enters the down phase. The system remains in this down phase until it is repaired and activated again. The length of time the system is in the up phase is called the **time till failure (TTF)**. The length of time the system is in the down phase is called the **time till repair (TTR)**. Both TTF and TTR are modeled as random variables T and S, respectively. We assume here that T and S are independent. The **cycle time** is the random variable $C = T + S$.

The process in which the system goes through these cycles is called a **renewal process**. Let $C_1, C_2, C_3, \ldots$ be a sequence of cycles of a repairable system. We assume that $C_1, C_2, \ldots$ are i.i.d. random variables.

Let $F(t)$ be the c.d.f. of the TTF and $G(t)$, the c.d.f. of the TTR. Let $f(t)$ and $g(t)$ be the corresponding p.d.f.'s. Let $K(t)$ denote the c.d.f. of C. Because T and S are independent random variables,

$$K(t) = \Pr\{C \le t\}$$

$$= \int_0^t f(x) P\{S \le t - x\} dx \qquad \text{(14.3.1)}$$

$$= \int_0^t f(x) G(t - x) dx$$

Assuming that $G(0) = 0$, differentiation of $K(t)$ yields the p.d.f. of the cycle time, $k(t)$—namely,

$$k(t) = \int_0^t f(x) g(t - x) dx \qquad \text{(14.3.2)}$$

The operation of getting $k(t)$ from $f(t)$ and $g(t)$ is called a *convolution*.

The *Laplace transform* of an integrable function $f(t)$, on $0 < t < \infty$, is defined as

$$f^*(s) = \int_0^\infty e^{-ts} f(t) dt \qquad s \geq 0 \tag{14.3.3}$$

Notice that if $f(t)$ is a p.d.f. of a nonnegative continuous random variable, then $f^*(s)$ is its moment-generating function (m.g.f.) at $-s$. Because $C = T + S$, and T, S are independent, the m.g.f. of C is $M_C(u) = M_T(u)M_S(u)$ for all $u \leq u^*$ at which these m.g.f. exist. In particular, if $k^*(s)$ is the Laplace transform of $k(t)$,

$$k^*(s) = f^*(s)g^*(s) \qquad s \geq 0 \tag{14.3.4}$$

EXAMPLE **14.5**   Suppose that T is exponentially distributed like $E(\beta)$, and S is exponentially distributed like $E(\gamma)$ for $0 < \beta, \gamma < \infty$—that is,

$$f(t) = \frac{1}{\beta} \exp\{-t/\beta\}$$

$$g(t) = \frac{1}{\gamma} \exp\{-t/\gamma\}$$

The p.d.f. of C is

$$k(t) = \int_0^t f(x)g(t-x)dx$$

$$= \begin{cases} \dfrac{1}{\beta - \gamma}(e^{-t/\beta} - e^{-t/\gamma}) & \text{if } \beta \neq \gamma \\ \dfrac{t}{\beta^2}e^{-t/\beta} & \text{if } \beta = \gamma \end{cases}$$

The corresponding Laplace transforms are

$$f^*(s) = (1 + s\beta)^{-1}$$
$$g^*(s) = (1 + s\gamma)^{-1}$$
$$k^*(s) = (1 + s\beta)^{-1}(1 + s\gamma)^{-1} \quad \blacksquare$$

Let $N_F(t)$ denote the number of failures of a system during the time interval $(0, t]$. Let $W(t) = E\{N_F(t)\}$. Similarly, let $N_R(t)$ be the number of repairs during $(0, t]$ and $V(t) = E\{N_R(t)\}$. Obviously, $N_R(t) \leq N_F(t)$ for all $0 < t < \infty$.

Let $A(t)$ denote the probability that the system is up at time t. $A(t)$ is the **availability function** of the system. In unrepairable systems, $A(t) = R(t)$.

Let us assume that $W(t)$ and $V(t)$ are differentiable, and let $w(t) = W'(t)$, $v(t) = V'(t)$.

The **failure intensity function** of repairable systems is defined as

$$\lambda(t) = \frac{w(t)}{A(t)} \qquad t \geq 0 \tag{14.3.5}$$

Notice that if the system is unrepairable, then $W(t) = F(t)$, $w(t) = f(t)$, $A(t) = R(t)$, and $\lambda(t)$ is the hazard function $h(t)$. Let $Q(t) = 1 - A(t)$. The **repair intensity function** is

$$\mu(t) = \frac{v(t)}{Q(t)} \qquad t \geq 0 \tag{14.3.6}$$

The function $V(t) = E\{N_R(t)\}$ is called the **renewal function**. Notice that

$$\Pr\{N_R(t) \geq n\} = \Pr\{C_1 + \cdots + C_n \leq t\}$$
$$= K_n(t) \qquad t \geq 0 \tag{14.3.7}$$

where $K_n(t)$ is the c.d.f. of $C_1 + \cdots + C_n$.

Because $N_R(t)$ is a nonnegative random variable, the renewal function is

$$V(t) = \sum_{n=1}^{\infty} \Pr\{N_R(t) \geq n\}$$
$$= \sum_{n=1}^{\infty} K_n(t) \tag{14.3.8}$$

EXAMPLE 14.6   Suppose that TTF is distributed like $E(\beta)$ and that the repair is instantaneous. Then C is distributed like $E(\beta)$ and $K_n(t)$ is the c.d.f. of $G(n, \beta)$—that is,

$$K_n(t) = 1 - P\left(n - 1; \frac{t}{\beta}\right) \qquad n = 1, 2, \ldots$$

where $P(j; \lambda)$ is the c.d.f. of a Poisson random variable with mean λ. Thus, in the present case,

$$V(t) = \sum_{n=1}^{\infty} \left(1 - P\left(n - 1; \frac{t}{\beta}\right)\right)$$
$$= E\left\{\text{Pois}\left(\frac{t}{\beta}\right)\right\} = \frac{t}{\beta}, \qquad t \geq 0$$

Here $\text{Pois}\left(\frac{t}{\beta}\right)$ designates a random variable having a Poisson distribution with mean t/β. ∎

At time t, $0 < t < \infty$, there are two possible events:

E_1: The first cycle is not yet terminated.

E_2: The first cycle terminated at some time before t.

Accordingly, $V(t)$ can be written as

$$V(t) = K(t) + \int_0^t k(x)V(t - x)dx \tag{14.3.9}$$

The derivative of $V(t)$ is called the **renewal density**. Because $V(0) = 0$, we find, by differentiating this equation, that

$$v(t) = k(t) + \int_0^t k(x)v(t-x)dx \qquad \text{(14.3.10)}$$

Let $v^*(s)$ and $k^*(s)$ denote the Laplace transforms of $v(t)$ and $k(t)$, respectively. Then, from equation 14.3.10,

$$v^*(s) = k^*(s) + k^*(s)v^*(s) \qquad \text{(14.3.11)}$$

or, because $k^*(s) = f^*(s)g^*(s)$,

$$v^*(s) = \frac{f^*(s)g^*(s)}{1 - f^*(s)g^*(s)} \qquad \text{(14.3.12)}$$

The renewal density $v(t)$ can be obtained by inverting $v^*(s)$, i.e., finding a function $v(t)$ whose Laplace transform is $v^*(s)$.

EXAMPLE 14.7   As before, suppose that the TTF is $E(\beta)$ and that the TTR is $E(\gamma)$. Let $\lambda = 1/\beta$ and $\mu = 1/\gamma$ with

$$f^*(s) = \frac{\lambda}{\lambda + s}$$

and

$$g^*(s) = \frac{\mu}{\mu + s}$$

Then

$$v^*(s) = \frac{\lambda\mu}{s^2 + (\lambda + \mu)s}$$

$$= \frac{\lambda\mu}{\lambda + \mu}\left(\frac{1}{s} - \frac{1}{s + \lambda + \mu}\right)$$

Now $1/s$ is the Laplace transform of 1, and $(\lambda + \mu)/(s + \lambda + \mu)$ is the Laplace transform of $E(1/(\lambda + \mu))$. Hence

$$v(t) = \frac{\lambda\mu}{\lambda + \mu} - \frac{\lambda\mu}{\lambda + \mu}e^{-t(\lambda + \mu)} \qquad t \geq 0$$

Integrating $v(t)$, we obtain the renewal function

$$V(t) = \frac{\lambda\mu}{\lambda + \mu}t - \frac{\lambda\mu}{(\lambda + \mu)^2}(1 - e^{-t(\lambda + \mu)}) \qquad 0 \leq t < \infty$$

In a similar fashion we can show that

$$W(t) = \frac{\lambda\mu}{\lambda + \mu}t + \frac{\lambda^2}{(\lambda + \mu)^2}(1 - e^{-t(\lambda + \mu)}) \qquad 0 \leq t < \infty$$

Because $W(t) > V(t)$ if and only if the last cycle is still incomplete and the system is down, the probability, $Q(t)$, that the system is down at time t is

$$Q(t) = W(t) - V(t)$$

$$= \frac{\lambda}{\lambda + \mu} - \frac{\lambda}{\lambda + \mu}e^{-t(\lambda + \mu)} \qquad t \geq 0$$

Thus, the availability function is

$$A(t) = 1 - Q(t)$$

$$= \frac{\mu}{\lambda + \mu} + \frac{\lambda}{\lambda + \mu} e^{-t(\lambda + \mu)} \qquad t \geq 0$$

Notice that the availability at large values of t is approximately

$$\lim_{t \to \infty} A(t) = \frac{\mu}{\lambda + \mu} = \frac{\beta}{\beta + \gamma} \qquad \blacksquare$$

The availability function $A(t)$ can be determined from $R(t)$ and $v(t)$ by solving the equation

$$A(t) = R(t) + \int_0^t v(x) R(t - x) dx \tag{14.3.13}$$

The Laplace transform of this equation is

$$A^*(s) = \frac{R^*(s)}{1 - f^*(s) g^*(s)} \qquad 0 < s < \infty \tag{14.3.14}$$

This theory can be useful in assessing different system structures with respect to their availability. The following asymptotic (large t approximations) results are very useful. Let μ and σ^2 be the mean and variance of the cycle time.

1 $\displaystyle \lim_{t \to \infty} \frac{V(t)}{t} = \frac{1}{\mu}$ $\tag{14.3.15}$

2 $\displaystyle \lim_{t \to \infty} (V(t + a) - V(t)) = \frac{a}{\mu} \qquad a > 0$ $\tag{14.3.16}$

3 $\displaystyle \lim_{t \to \infty} \left(V(t) - \frac{t}{\mu} \right) = \frac{\sigma^2}{2\mu^2} - \frac{1}{2}$ $\tag{14.3.17}$

If the p.d.f of C, $k(t)$, is continuous, then

4 $\displaystyle \lim_{t \to \infty} v(t) = \frac{1}{\mu}$ $\tag{14.3.18}$

5 $\displaystyle \lim_{t \to \infty} \Pr \left\{ \frac{N_R(t) - t/\mu}{(\sigma^2 t/\mu^3)^{1/2}} \leq z \right\} = \Phi(z)$ $\tag{14.3.19}$

6 $\displaystyle A_\infty = \lim_{T \to \infty} \frac{1}{T} \int_0^T A(t) dt = \frac{E\{TTF\}}{E\{TTF\} + E\{TTR\}}$ $\tag{14.3.20}$

According to equation 14.3.15, the expected number of renewals, $V(t)$, is approximately t/μ for large t. According to equation 14.3.16, we expect approximately a/μ renewals in a time interval $(t, t + a)$ when t is large. The third result (equation 14.3.17) says that t/μ is an underestimate (overestimate) for large t if the squared coefficient of variation, σ^2/μ^2, of the cycle time is greater (smaller) than 1. The last three properties can be interpreted in a similar fashion. We illustrate these asymptotic properties with examples.

EXAMPLE **14.8**   Consider a repairable system. The TTF (hr) has a gamma distribution like $G(2, 100)$. The TTR (hr) has a Weibull distribution $W(2, 2.5)$. Thus, the expected TTF is $\mu_T = 200$ (hr), and the expected TTR is $\mu_S = 2.5 \times \Gamma \left(\frac{3}{2} \right) = 1.25 \sqrt{\pi} = 2.2$ (hr). The asymptotic availability is

$$A_\infty = \frac{200}{202.2} = .989$$

That is, in the long run, the proportion of total availability time is 98.9%.

The expected cycle time is $\mu_C = 202.2$, and the variance of the cycle time is

$$\sigma_C^2 = 2 \times 100^2 + 6.25 \left[\Gamma(2) - \Gamma^2 \left(\frac{3}{2} \right) \right]$$

$$= 20,000 + 1.34126 = 20,001.34126$$

Thus, during 2000 hr of scheduled operation, we expect close to $2000/202.2 \cong 10$ renewal cycles. The probability that $N_R(2000)$ will be less than 11 is

$$\Pr\{N_R(2000) \le 11\} \cong \Phi \left(\frac{1.11}{2.20} \right) = \Phi(0.505) = .6932 \quad \blacksquare$$

An important question is: What is the probability, for large values of t, that we will find the system operating and that it will continue to operate without a failure for at least u additional time units. This function is called the **asymptotic operational reliability**, and is given by

$$R_\infty(u) = A_\infty \cdot \frac{\displaystyle\int_u^\infty R(u)du}{\mu_T} \qquad 0 \le u \tag{14.3.21}$$

where $R(u) = 1 - F_T(u)$.

EXAMPLE **14.9**   We continue discussing Example 14.8. In this case,

$$F_T(u) = \Pr\{G(2, 100) \le u\}$$

$$= \Pr \left\{ G(2, 1) \le \frac{u}{100} \right\}$$

$$= 1 - P \left(1; \frac{u}{100} \right) = 1 - e^{-u/100} - \frac{u}{100} e^{-u/100}$$

and

$$R(u) = e^{-u/100} + \frac{u}{100} e^{-u/100}$$

Furthermore, $\mu_T = 200$ and $A_\infty = .989$. Hence,

$$R_\infty(u) = .989 \cdot \frac{\displaystyle\int_u^\infty \left(1 + \frac{x}{100} \right) e^{-x/100} dx}{200}$$

$$= \frac{98.9}{200} \left(2 + \frac{u}{100} \right) e^{-u/100}$$

Thus, $R_\infty(0) = .989$, $R_\infty(100) = .546$, and $R_\infty(200) = .268$ ∎

Before concluding this section, we introduce two computer programs, AVAILDIS.EXE and RENEWDIS.EXE, which provide the EBD of the number of renewals in a specified time interval and the EBD of the asymptotic availability index A_∞ based on observed samples of failure times and repair times. These programs provide computer-aided estimates of the renewal distribution and of the precision of A_∞. We illustrate these programs in the following example.

EXAMPLE **14.10**   Consider again the renewal process described in Example 14.8. Suppose that file SAMPLE1.DAT contains $n = 50$ observed values of i.i.d. TTF from $G(2, 100)$, and SAMPLE2.DAT contains $n = 50$ observed repair times. We run program RENEWDIS.EXE 1000 times to obtain an EBD of the number of renewals in 1000 hr. The output of the program is in file C:/ISTAT/DATA/RENEWDIS.DAT. The program yields that the mean number of renewals for 1000 hours of operation is 5.261. This is the bootstrap estimate of $V(1000)$. The asymptotic approximation is $1000/202.2 = 4.946$. The bootstrap confidence interval for $V(1000)$ at .95 level of confidence is $(3,8)$. This confidence interval covers the asymptotic approximation. Accordingly, the bootstrap estimate of 5.26 is not significantly different from the asymptotic approximation.   ∎

Other topics of interest are maintenance, repairability, and availability. The objective is to increase the availability by instituting maintenance procedures and by adding standby systems and repair personnel. The question is what the optimal maintenance period is and how many standby systems and repair personnel to add. The interested reader is referred to Zacks (1992, ch. 4) and Gertsbakh (1989).

In the following sections, we discuss statistical problems associated with reliability assessment when one does not know definitely the model and the values of its parameters.

Types of Observations on TTF

Proper analysis of data depends on the type of observations available. In dealing with TTF and TTR random variables, we want to have observations that give us the exact length of time interval from activation (failure) of a system (component) till its failure (repair). However, we may find that proper records have not been kept, and we may find only the number of failures (repairs) in a given period of time. These are discrete random variables and not the continuous ones under investigation. Another type of problem typical of reliability studies is that some observations are **censored**. For example, if n identical systems are tested for a specified length of time t^*, we may observe only a random number, K_n, of failures in the time interval $(0, t^*]$. On the other $n - K_n$ systems, which did not fail, we have only partial information—that is, their TTF is greater than t^*. The observations on these systems are called *right censored*. In this example, n units are being tested at the same time. The censoring time t^* is a fixed time. Sometimes we have observations with **random censoring**. This is the case when we carry a study for a fixed length of time t^* (years) but the units (systems) enter the study at random times between 0 and t^*, according to some distribution.

Suppose a unit enters the study at the random time τ, $0 < \tau < t^*$, and its TTF is T. We can observe only $W = \min(T, t^* - \tau)$. Here the censoring time is the random variable $t^* - \tau$. An example of such a situation is when we sell a product under warranty. The units of this product are sold to different customers at random times during the study period $(0, t^*)$. Products that fail are brought back for repair. If this happens during the study period, we have an uncensored observation on the TTF of that unit; otherwise, the observation is censored: $W = t^* - \tau$.

The censored observations just described are **time censored**. Another type of censoring is **frequency censoring**. This occurs when n units are put on test at the same time, but the test is terminated the instant the rth failure occurs. In this case, the length of the test is the rth-order statistic of failure times $T_{n,r}$ $(r = 1, \ldots, n)$. Notice that $T_{n,r} = T_{(r)}$, where $T_{(1)} < T_{(2)} < \cdots < T_{(n)}$ are the order statistics of n i.i.d. TTFs. If T is distributed exponentially—$E(\beta)$, for example—the expected length of the experiment is

$$E\{T_{n,r}\} = \beta \left(\frac{1}{n} + \frac{1}{n-1} + \cdots + \frac{1}{n-r+1} \right)$$

There may be a substantial time saving if we terminate the study at the rth failure, when $r < n$. For example, in the exponential case, with $E\{T\} = \beta = 1000$ (hr) and $n = 20$,

$$E\{T_{20,20}\} = 1,000 \times \left(1 + \frac{1}{2} + \frac{1}{3} + \cdots + \frac{1}{20} \right) = 3597.7 \text{ (hr)}$$

However, for $r = 10$, we have $E\{T_{20,10}\} = 668.8$ (hr). Thus, a frequency censored experiment, with $r = 10$ and $n = 20$, $\beta = 1000$, lasts on average only 19% of the length of time of an uncensored experiment. We will see later how one can determine the optimal n and r for estimating the mean TTF (MTTF) in the exponential case.

14.5
Graphical Analysis of Life Data

In this section, we discuss some graphical procedures to fit a life distribution to failure data, and obtain estimates of the parameters from the graphs.

Let $t_1, t_2, \ldots, t_n$ be n uncensored observation on i.i.d. random variables, $T_1, \ldots, T_n$, having some life distribution $F(t)$.

The empirical c.d.f., given $t_1, \ldots, t_n$, is defined as

$$F_n(t) = \frac{1}{n} \sum_{i=1}^{n} I\{t_i \leq t\} \tag{14.5.1}$$

where $I\{t_i \le t\}$ is the indicator variable assuming the value 1 if $t_i \le t$ and the value 0 otherwise. A theorem in probability theory states that the empirical c.d.f. $F_n(t)$ converges to $F(t)$ as $n \to \infty$.

Figure 14.3 shows the empirical c.d.f. of a random sample of 100 variables having the Weibull distribution $W(1.5, 100)$. Because $F_n(t_{(i)}) = i/n$ for $i = 1, 2, \ldots, n$, the (i/n)th quantile of $F_n(t)$ is the ordered statistic $t_{(i)}$. Accordingly, if $F(t)$ has some specific distribution, the scattergram of $(F^{-1}(i/n), t_{(i)})$ $(i = 1, \ldots, n)$ should be around a straight line with slope 1. The plot of $t_{(i)}$ versus $F^{-1}(i/n)$ is called a *QQ-plot* (quantile versus quantile). The QQ plot is the basic graphical procedure to test whether a given sample of failure times is generated by a specific life distribution. Because $F^{-1}(1) = \infty$ for the interesting life distribution, the quantile of F is taken at $i/(n+1)$ or at some other $(i+\alpha)/(n+\beta)$, which gives better plotting position for a specific distribution. For the normal distribution, $(i - 3/8)/(n + 1/4)$ is used.

FIGURE   14.3

The empirical c.d.f. of a random sample of 100 variables from $W(1.5, 100)$

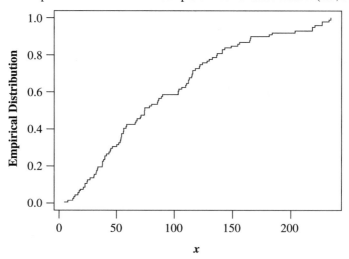

If the distribution depends on location and scale parameters, we plot $t_{(i)}$ against the quantiles of the standard distribution. The intercept and the slope of the line fitted through the points yield estimates of these location and scale parameters. For example, suppose that $t_1, \ldots, t_n$ are values of a sample from a $N(\mu, \sigma^2)$ distribution. Thus, $t_{(i)} \cong \mu + \sigma \Phi^{-1}(i/n)$. Thus, if we plot $t_{(i)}$ against $\Phi^{-1}(i/n)$, we should have points around a straight line whose slope is an estimate of σ and intercept an estimate of μ.

We focus attention here on three families of life distributions:

1   The exponential or shifted exponential

2   The Weibull

3   The log-normal

The *shifted-exponential* c.d.f. has the form

$$F(t; \mu, \beta) = \begin{cases} 1 - \exp\left\{-\dfrac{t-\mu}{\beta}\right\} & t \geq \mu \\ 0 & t < \mu \end{cases} \tag{14.5.2}$$

The starting point of the exponential distribution $E(\beta)$ is shifted to a point μ. Location parameters of interest in reliability studies are $\mu \geq 0$. Notice that the pth quantile, $0 < p < \infty$, of the shifted exponential is

$$t_p = \mu + \beta(-\log(1-p)) \tag{14.5.3}$$

Accordingly, for exponential QQ plots, we plot $t_{(i)}$ versus $E_{i,n} = -\log[1 - [i/(n+1)]]$. Notice that in this plot, the intercept estimates the location parameter μ, and the slope estimates β. In the Weibull case, $W(\nu, \beta)$, the c.d.f. is

$$F(t; \nu, \beta) = 1 - \exp\left\{-\left(\frac{t}{\beta}\right)^{\nu}\right\} \qquad t \geq 0 \tag{14.5.4}$$

Thus, if t_p is the pth quantile,

$$\log t_p = \log \beta + \frac{1}{\nu}\log(-\log(1-p)) \tag{14.5.5}$$

For this reason, we plot $\log t_{(i)}$ versus

$$W_{i,n} = \log\left(-\log\left(1 - \frac{i}{n+1}\right)\right) \qquad i = 1, \ldots, n \tag{14.5.6}$$

The slope of the straight line estimates $1/\nu$ and the intercept estimates $\log \beta$. In the log-normal case, we plot $\log t_{(i)}$ against $\Phi^{-1}[(i - 3/8)/(n + 1/4)]$.

EXAMPLE 14.11   Figure 14.4 shows the QQ plot of 100 values generated at random from an exponential distribution $E(5)$. We fit a straight line through the origin to the points by the method of least squares. A linear regression routine provides the line

$$\hat{t}_{(i)} = 5.94 * E_{i, 100}, \qquad i = 1, \ldots, 100$$

Accordingly, the slope of a straight line fitted to the points provides an estimate of the true mean and standard deviation, $\beta = 5$. An estimate of the median is

$$\hat{t}(0.693) = 0.693 \times 5.9413 = 4.117$$

The true median is $M_e = 3.465$.

Figure 14.5 shows a probability plot of $n = 100$ values generated from a Weibull distribution with parameters $\nu = 2$ and $\beta = 2.5$.

Least-squares fitting of a straight line to these points yields the line

$$\hat{y}_{(i)} = 0.856 + 0.479 W_{i, 100}, \qquad i = 1, \ldots, 100$$

Accordingly, we obtain the following estimates:

$$\hat{\nu} = 1/0.479 = 2.087$$
$$\hat{\beta} = \exp(0.856) = 2.354$$

F I G U R E   14.4
QQ plot of a sample of 100 values from $E(5)$

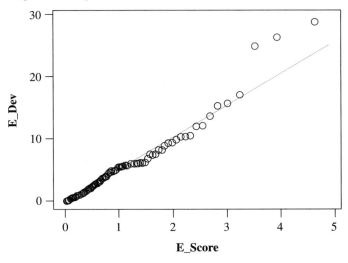

F I G U R E   14.5
QQ plot of a sample of 100 values from $W(2, 2.5)$

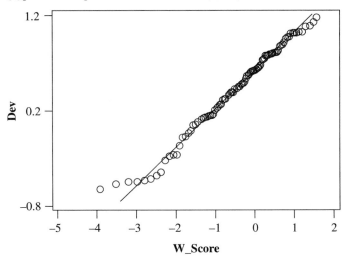

$$\text{Median} = \exp(0.856 - (0.3665)(0.479))$$
$$= 1.975$$

The true median is equal to $\beta(\ln 2)^{1/2} = 2.081$. The estimate of the mean is

$$\hat{\mu} = \hat{\beta}\Gamma(1 + 0.479)$$
$$= \hat{\beta} \times 0.479 \times \Gamma(0.479) = 2.080$$

The true mean is $\mu = \beta\Gamma(1.5) = 2.216$. Finally, an estimate of the standard deviation is

$$\hat{\sigma} = \hat{\beta}(\Gamma(1.958) - \Gamma^2(1.479))^{1/2}$$
$$= \hat{\beta}[0.958 \times \Gamma(0.958) - (0.479 \times \Gamma(0.479))^2]^{1/2} = 1.054$$

The true value is $\sigma = \beta(\Gamma(2) - \Gamma^2(1.5))^{1/2} = 1.158$. ∎

If observations are censored from the left or from the right, we plot the quantiles only from the uncensored part of the sample. The plotting positions take into consideration the number of censored values from the left and from the right. For example, if $n = 20$ and the 2 smallest observations are censored, the plotting positions are as follows:

i	$t_{(i)}$	$\frac{i}{n+1}$
1	—	—
2	—	—
3	$t_{(3)}$	$\frac{3}{21}$
⋮		
20	$t_{(20)}$	$\frac{20}{21}$

14.6
Nonparametric Estimation of Reliability

A nonparametric method called the *Kaplan-Meier* method yields an estimate, called the **product limit (PL) estimate** of the reliability function, without an explicit reference to the life distribution. The estimator of the reliability function at time t will be denoted by $\hat{R}_n(t)$ when n is the number of units put on test at time $t = 0$. If all the failure times $0 < t_1 < t_2 < \cdots < t_n < \infty$ are known, then the PL estimator is equivalent to

$$\hat{R}_n(t) = 1 - F_n(t) \tag{14.6.1}$$

where $F_n(t)$ is the empirical c.d.f. defined earlier.

In some cases, either random or nonrandom censoring or withdrawals occur, and we do not have complete information on the exact failure times. Suppose that $0 < t_1 < t_2 < \cdots < t_k < \infty$, $k \leq n$, are the failure times and $w = n - k$ is the total number of withdrawals. Let $I_j = (t_{j-1}, t_j)$, $j = 1, \ldots, k+1$, with $t_0 = 0$, $t_{k+1} = \infty$, be the time intervals between recorded failures. Let W_j be the number of withdrawals during the time interval I_j. The PL estimator of the reliability function is then

$$\hat{R}_n(t) = I\{t < t_1\} + \sum_{i=2}^{k+1} I\{t_{i-1} \leq t \leq t_i\} \prod_{j=1}^{i-1} \left(1 - \frac{1}{n_{j-1} - w_j/2}\right) \tag{14.6.2}$$

where $n_0 = n$, and n_l is the number of operating units just prior to failure time t_l.

Usually, when units are tested in the laboratory under controlled conditions, there may be no withdrawals. This is not the case, however, if tests are conducted in field conditions, and units on test may be lost, withdrawn, or destroyed for reasons other than the failure phenomenon under study.

Suppose now that systems are installed in the field as they are purchased (random times). We decide to make a follow-up study of the systems for a period of two years. The time till failure of systems participating in the study is recorded. We assume that each system operates continuously from the time of installment until its failure. If a system has not failed by the end of the study period, the only information available is the length of time it has been operating. This is a case of multiple censoring. At the end of the study period, we have the following observations $\{(T_i, \delta_i), \, i = 1, \dots, n\}$, where n is the number of systems participating in the study, T_i is the length of operation of the ith system (TTF or time till censoring), and $\delta_i = 1$ if the ith observation is not censored and $\delta_i = 0$ otherwise.

Let $T_{(1)} \le T_{(2)} \le \cdots \le T_{(n)}$ be the order statistic of the operation times and let $\delta_{j_1}, \delta_{j_2}, \dots, \delta_{j_n}$ be the δ values corresponding to the ordered T values, where j_i is the index of the ith order statistic $T_{(i)}$—that is, $T_{(i)} = T_j, \, (i = 1, \dots, n)$.

The PL estimator of $R(t)$ is given by

$$\hat{R}_n(t) = I\{t < T_{(1)}\}$$

$$+ \sum_{i=1}^{n} I\{T_{(i)} \le t < T_{(i+1)}\} \prod_{j=1}^{i} \left(1 - \frac{\delta_j}{n - j + 1}\right) \tag{14.6.3}$$

Another situation prevails in the laboratory or in field studies when the exact failure times cannot be recorded. Let $0 < t_1 < t_2 < \cdots < t_k < \infty$ be fixed inspection times. Let w_i be the number of withdrawals and f_i the number of failures in the time interval $I_i \, (i = 1, \dots, k + 1)$. In this case, the formula is modified to be

$$\hat{R}_n(t) = I\{t < t_1\}$$

$$+ \sum_{i=2}^{k+1} I\{t_{i-1} \le t < t_i\} \prod_{j=1}^{i-1} \left(1 - \frac{f_j}{n_{j-1} - \frac{w_j}{2}}\right) \tag{14.6.4}$$

This version of the estimator of $R(t)$, when the inspection times are fixed (not random failure times), is called the **actuarial estimator**.

In the following example, we illustrate these estimators of the reliability function.

EXAMPLE **14.12**   Before being shipped to a customer, a machine is tested for a 5-day period (120 hr) or till its first failure, whichever comes first. Twenty such machines were tested consecutively. Table 14.1 gives the ordered time till failure or time till censor (TTF/TTC) of the 20 machines, the factors $(1 - \delta_i/(n - i + 1))$, and the PL estimator $\hat{R}(t_i)$, $i = 1, \dots, 20$. ∎

TABLE **14.1**

Failure times (hr) and PL estimates

i	$T_{(i)}$	$\left(1 - \dfrac{\delta_i}{n-i+1}\right)$	$\hat{R}(T_i)$
1	4.78	.950	.95
2	8.37	.947	.9
3	8.76	.944	.85
4	13.77	.941	.8
5	29.20	.937	.75
6	30.53	.933	.7
7	47.96	.928	.65
8	59.22	.923	.6
9	60.66	.916	.55
10	62.12	.909	.5
11	67.06	.9	.45
12	92.15	.888	.4
13	98.09	.875	.35
14	107.60	.857	.3
15	120	1	.3
16	120	1	.3
17	120	1	.3
18	120	1	.3
19	120	1	.3
20	120	1	.3

14.7 Estimation of Life Characteristics

In Chapter 8, we studied the estimation of parameters of distributions and of functions of the parameters. We discussed point estimators and confidence intervals. In particular, we discussed unbiased estimators, least-squares estimators, maximum likelihood estimators, and Bayes' estimators. All these methods of estimation can be applied in reliability studies. In this section, we discuss the maximum likelihood estimation of the parameters of some common life distributions, such as the exponential and the Weibull, and some nonparametric techniques for censored and uncensored data.

14.7.1 Maximum Likelihood Estimators for Exponential TTF Distributions

We start with the case of uncensored observations. Thus, let $T_1, T_2, \ldots, T_n$ be i.i.d. random variables distributed like $E(\beta)$. Let $t_1, \ldots, t_n$ be their sample realization (random sample). The likelihood function of β, $0 < \beta < \infty$, is

$$L(\beta; \mathbf{t}) = \frac{1}{\beta^n} \exp\left\{-\frac{1}{\beta}\sum_{i=1}^{n} t_i\right\}$$ (14.7.1)

It is easy to check that the maximum likelihood estimator (MLE) of β is the sample mean

$$\hat{\beta}_n = \overline{T}_n = \frac{1}{n}\sum_{i=1}^{n} T_i$$ (14.7.2)

$\overline{T}_n$ is distributed like $G\left(n, \frac{\beta}{n}\right)$. Thus, $E\{\hat{\beta}_n\} = \beta$ and $V\{\hat{\beta}_n\} = \beta^2/n$. From the relationship between the gamma and the χ^2 distributions, we have that $\hat{\beta}_n$ is distributed like $(\beta/2n)\chi^2[2n]$. Thus, a $1 - \alpha$ level confidence interval for β, based on the MLE $\hat{\beta}_n$, is

$$\left(\frac{2n\hat{\beta}_n}{\chi^2_{1-\alpha/2}[2n]}, \frac{2n\hat{\beta}_n}{\chi^2_{\alpha/2}[2n]}\right)$$ (14.7.3)

For large samples, we can use the normal approximation $\hat{\beta}_n \pm z_{1-\alpha/2}(\hat{\beta}_n/\sqrt{n})$.

EXAMPLE **14.13**   The failure times of 20 electric generators (in hr) are:

121.5	1425.5	2951.2	5637.9
1657.2	592.1	10609.7	9068.5
848.2	5296.6	7.5	2311.1
279.8	7201.9	6853.7	6054.3
1883.6	6303.9	1051.7	711.5

Exponential probability plotting of these data yields a scatter around the line with slope 3866.17 and $R^2 = .95$. The exponential model fits the failure times quite well. The MLE estimator of the MTTF, β, yields $\hat{\beta}_{20} = 3543.4$ (hr). Notice that the MLE is different, but not significantly, from the graphical estimate of β. Indeed, the standard error of $\hat{\beta}_{20}$ is SE $= \hat{\beta}_{20}/\sqrt{20} = 792.328$.

The confidence interval, at level .95, for β is given by (2388.5, 5801.0). The normal approximation to the confidence interval is (1990.4, 5096.4). The sample size is not sufficiently large for the normal approximation to be effective.   ∎

When observations are time censored by a fixed constant, t^*, let K_n denote the number of uncensored observations. K_n is a random variable having the binomial distribution $B(n, 1 - \exp\{-t^*/\beta\})$. Let $\hat{p}_n = K_n/n$. $\hat{p}_n$ is a consistent estimator of $1 - \exp\{-t^*/\beta\}$. Hence, a consistent estimator of β is

$$\tilde{\beta}_n = -t^*/\log(1 - \hat{p}_n)$$ (14.7.4)

This estimator is not an efficient one because it is not based on observed failures. Moreover, if $K_n = 0$, $\tilde{\beta}_n$ is undefined. Using the expansion method shown in Section 13.2, we obtain that the asymptotic variance of $\tilde{\beta}_n$ is

$$AV\{\tilde{\beta}_n\} \cong \frac{\beta^4}{nt^{*2}} \cdot \frac{1 - e^{-t^*/\beta}}{e^{-t^*/\beta}} \tag{14.7.5}$$

The likelihood function of β in this time censoring case is

$$L(\beta; K_n, \mathbf{T}_n) = \frac{1}{\beta^{K_n}} \exp\left\{ -\frac{1}{\beta} \left(\sum_{i=1}^{K_n} T_{(i)} + t^*(n - K_n) \right) \right\} \tag{14.7.6}$$

Also here, if $K_n = 0$, the MLE of β is undefined. If $K_n \geq 1$, the MLE is

$$\hat{\beta}_n = \frac{S_{n,K_n}}{K_n} \tag{14.7.7}$$

where $S_{n,K_n} = \sum_{i=1}^{K_n} T_i + (n - K_n)t^*$ is the **total time on test** *(TTT)* of the n units.

Theoretical evaluation of the properties of the MLE $\hat{\beta}_n$ is complicated. We can, however, get information on its behavior by simulation. Program CENEXP.EXE in subdirectory ISTAT\CHA14 performs such a simulation.

EXAMPLE **14.14**   For a sample of size $n = 50$, with $\beta = 1000$ and $t^* = 2000$, $\Pr\{K_{50} = 0\} = \exp\{-100\} \cong 0$. Thus, we expect that $\hat{\beta}_n$ will exist in all the simulation runs. File CENEXP.DAT contains 100 MLEs of β obtained by this simulation. For these 100 estimates, we have mean = 1025.5, median = 996.7, and standard deviation = 156.9. The standard deviation of $\hat{\beta}_n$, according to the previous formula, with the preceding values of n, β, and t^*, is 178.7.   ∎

For further reading on the properties of MLE under time censoring, see Zacks (1992, p. 125).

Under frequency censoring, the situation is simpler. Suppose that the censoring is at the rth failure. The total time on test is $S_{n,r} = \sum_{i=1}^{r} T_{(i)} + (n - r)T_{(r)}$. In this case,

$$S_{n,r} \text{ is distributed like } \frac{\beta}{2}\chi^2[2r] \tag{14.7.8}$$

and the MLE $\hat{\beta}_{n,r} = S_{n,r}/r$ is an unbiased estimator of β, with variance

$$V\{\hat{\beta}_{n,r}\} = \frac{\beta^2}{r}$$

Or, $$SE\{\hat{\beta}_{n,r}\} = \frac{\beta}{\sqrt{r}} \tag{14.7.9}$$

If we wish to have a certain precision, so that $SE\{\hat{\beta}_{n,r}\} = \gamma\beta$, then $r = 1/\gamma^2$. Obviously, $n \geq r$.

Suppose we pay for the test c_2 \$ per unit and c_1 \$ per time unit for the duration of the test. Then, the total cost of the test is

$$C_{n,r} = c_1 T_{(r)} + c_2 n \qquad (14.7.10)$$

For a given r, we choose n to minimize the expected total cost. The resulting formula is

$$n^0 \cong \frac{r}{2}\left(1 + \left(1 + \frac{4c_1}{rc_2}\beta\right)^{1/2}\right) \qquad (14.7.11)$$

The problem is that optimal sample size n^0 depends on the unknown β. If we have some prior estimate of β, it could be used to determine a good starting value for n.

EXAMPLE **14.15**   Consider a design of a life testing experiment with frequency censoring and exponential distribution of the TTF. We require that $\mathrm{SE}\{\hat{\beta}_{n,r}\} = 0.2\beta$. Accordingly, $r = (1/0.2)^2 = 25$. Suppose we want to minimize the total expected cost, at $\beta = 100$ (hr), where $c_1 = c_2 = 2$ \$. Then,

$$n^0 \cong \frac{25}{2}\left(1 + \left(1 + \frac{4}{25}100\right)^{1/2}\right) = 64$$

The expected duration of this test is

$$E\{T_{(25)}\} = 100 \sum_{i=1}^{25} \frac{1}{65-i} = 49.0 \text{ (hr)} \quad \blacksquare$$

14.7.2 Maximum Likelihood Estimation of the Weibull Parameters

Let $t_1, \ldots, t_n$ be uncensored failure times of n random variables having a Weibull distribution $W(\nu, \beta)$. The likelihood function of (ν, β) is

$$L(\nu, \beta; \mathbf{t}) = \frac{\nu^n}{\beta^{n\nu}}\left(\prod_{i=1}^{n} t_i\right)^{\nu-1} \exp\left\{-\sum_{i=1}^{n}\left(\frac{t_i}{\beta}\right)^{\nu}\right\} \qquad (14.7.12)$$

$0 < \beta, \nu < \infty$. The MLE of ν and β are the solutions $\hat{\beta}_n, \hat{\nu}_n$ of the equations

$$\hat{\beta}_n = \left(\frac{1}{n}\sum_{i=1}^{n} t_i^{\hat{\nu}_n}\right)^{1/\hat{\nu}_n} \qquad (14.7.13)$$

and

$$\hat{\nu}_n = \left[\frac{\sum\limits_{i=1}^{n} t_i^{\hat{\nu}_n}\log t_i}{\sum\limits_{i=1}^{n} t_i^{\hat{\nu}_n}} - \frac{1}{n}\sum_{i=1}^{n}\log(t_i)\right]^{-1} \qquad (14.7.14)$$

All logarithms are on base e (ln).

The equation for $\hat{v}_n$ is solved iteratively by the recursive equation

$$\hat{v}^{(j+1)} = \left[\frac{\sum_{i=1}^{n} t_i^{\hat{v}^{(j)}} \log(t_i)}{\sum_{i=1}^{n} t_i^{\hat{v}^{(j)}}} - \frac{1}{n} \sum_{i=1}^{n} \log(t_i) \right]^{-1} \qquad j = 0, 1, \ldots \qquad \text{(14.7.15)}$$

where $\hat{v}^{(0)} = 1$.

To illustrate, we simulated a sample of $n = 50$ failure times from $W(2.5, 10)$. To obtain the MLE, we have to continue the iterative process until the results converge. We show here the obtained values as functions of the number of iterations:

Number of iterations	$\hat{\beta}$	$\hat{v}$
10	11.437	2.314
20	9.959	2.367
30	9.926	2.368
40	9.925	2.368

It seems that 40 iterations yield sufficiently accurate solutions.

Confidence intervals for $\hat{v}_n$, $\hat{\beta}_n$ can be determined for large samples by using large-sample approximation formulas for the standard errors of the MLE, which are (see Zacks, 1992, p. 147)

$$SE\{\hat{\beta}_n\} \cong \frac{\hat{\beta}_n}{\sqrt{n}\,\hat{v}_n} \cdot 1.053 \qquad \text{(14.7.16)}$$

and

$$SE\{\hat{v}_n\} \cong .780 \frac{\hat{v}_n}{\sqrt{n}} \qquad \text{(14.7.17)}$$

The large-sample confidence limits are

$$\hat{\beta}_n \pm z_{1-\alpha/2} SE\{\hat{\beta}_n\} \qquad \text{(14.7.18)}$$

and

$$\hat{v}_n \pm z_{1-\alpha/2} SE\{\hat{v}_n\} \qquad \text{(14.7.19)}$$

In this numerical example, we obtained the MLE $\hat{\beta}_{50} = 9.925$ and $\hat{v}_{50} = 2.368$. Using these values, we obtain the large-sample approximate confidence intervals, with level of confidence $1 - \alpha = .95$, to be $(8.702, 11.148)$ for β and $(1.856, 2.880)$ for v.

We can obtain bootstrapping confidence intervals by using program SM-LEWEIB.EXE, which provides parametric bootstrap MLE. The resulting data are in file SMLEWEIB.DAT. The bootstrap confidence limits are the $\alpha/2$-th and $(1 - \alpha/2)$th quantile of the simulated values. Applying this program, we obtain the confidence intervals $(8.378, 11.201)$ for β and $(1.914, 3.046)$ for v. The difference between these confidence intervals and those using the large-sample approximation is not significant.

Maximum likelihood estimation in censored cases is more complicated and will not be discussed here. Estimates of ν and β in the censored case can be obtained from the intercept and slope of the regression line in the QQ plot.

Reliability Demonstration

Reliability demonstration is a procedure for testing whether the reliability of a given device (system) at a certain age is sufficiently high. More precisely, a time point t_0 and a desired reliability R_0 are specified, and we wish to test whether the reliability of the device at age t_0, $R(t_0)$, satisfies the requirement that $R(t_0) \geq R_0$. If the life distribution of the device is completely known, including all parameters, there is no problem of reliability demonstration—we compute $R(t_0)$ exactly and determine whether $R(t_0) \geq R_0$. If, as is generally the case, however, either the life distribution or its parameters are unknown, then the problem of reliability demonstration is that of obtaining suitable data and using them to test the statistical hypothesis that $R(t_0) \geq R_0$ versus the alternative that $R(t_0) < R_0$. Thus, the theory of testing statistical hypotheses provides the tools for reliability demonstration. In this section, we review some of the basic notions of hypothesis testing as they pertain to reliability demonstration.

In the following subsections, we develop several tests of interest in reliability demonstration. Procedures for obtaining confidence intervals for $R(t_0)$ that were discussed in the previous sections can be used to test hypotheses. Specifically, the procedure involves computing the upper confidence limit of a $(1 - 2\alpha)$-level confidence interval for $R(t_0)$ and comparing it with the value R_0. If the upper confidence limit exceeds R_0, then the null hypothesis $H_0 : R(t_0) \geq R_0$ is accepted; otherwise it is rejected. This test will have a significance level of α.

For example, if the specification of the reliability at age $t = t_0$ is $R = .75$ and the confidence interval for $R(t_0)$, at level of confidence $\gamma = .90$, is $(.80, .85)$, the hypothesis H_0 can be immediately accepted at a level of significance of $\alpha = (1 - \gamma)/2 = .05$. There is a duality between procedures for testing hypotheses and confidence intervals.

14.8.1 Binomial Testing

A random sample of n devices is put on life test simultaneously. Let J_n be the number of failures in the time interval $[0, t_0)$, and $K_n = n - J_n$. We have seen that K_n is distributed like $B(n, R(t_0))$. Thus, if H_0 is true—that is, $R(t_0) \geq R_0$—the values of K_n will tend to be larger, in a probabilistic sense. Thus, we test H_0 by specifying a critical value C_α and rejecting H_0 whenever $K_n \leq C_\alpha$. The critical value C_α is chosen as the largest value satisfying

$$F_B(C_\alpha; n, R_0) \leq \alpha$$

The OC function of this test, as a function of the true reliability R, is

$$\text{OC}(R) = \Pr\{K_n > C_\alpha \mid R(t_0) = R\}$$
$$= 1 - F_B(C_\alpha; n, R)$$

(14.8.1)

If n is large, then we can apply the normal approximation to the binomial c.d.f. In these cases, we can determine C_α to be the integer most closely satisfying

$$\Phi\left(\frac{C_\alpha + 1/2 - nR_0}{(nR_0(1 - R_0))^{1/2}}\right) = \alpha \qquad (14.8.2)$$

Generally, this will be given by

$$C_\alpha = \text{Integer closest to } \{nR_0 - 1/2 - z_{1-\alpha}(nR_0(1 - R_0))^{1/2}\} \qquad (14.8.3)$$

where $z_{1-\alpha} = \Phi^{-1}(1 - \alpha)$. The OC function of this test in the large-sample case is approximated by

$$OC(R) \cong \Phi\left(\frac{nR - C_\alpha - 1/2}{(nR(1 - R))^{1/2}}\right) \qquad (14.8.4)$$

The normal approximation is quite accurate whenever $n > 9/(R(1 - R))$.

If, in addition to specifying α, we specify that the test have type II error probability β when $R(t_0) = R_1$, then the normal approximation provides us with a formula for the necessary sample size:

$$n \cong \frac{(z_{1-\alpha}\sigma_0 + z_{1-\beta}\sigma_1)^2}{(R_1 - R_0)^2} \qquad (14.8.5)$$

where $\sigma_i^2 = R_i(1 - R_i)$, $i = 0, 1$.

EXAMPLE 14.16   Suppose we wish to test at significance level $\alpha = .05$ the null hypothesis that the reliability at age 1000 (hr) of a particular system is at least 85%. If the reliability is 80% or less, we want to limit the probability of accepting the null hypothesis to $\beta = .10$. Our test is to be based on K_n, the number of systems of a random sample of n surviving at least 1000 hours of operation. Setting $R_0 = .85$ and $R_1 = .80$, we have $\sigma_0 = .357$, $\sigma_1 = .4$, $z_{.95} = 1.645$, and $z_{.90} = 1.282$. Substituting into equation 14.8.5, we obtain that the necessary sample size is $n = 484$. The critical value is $C_{.05} = 397$.

In binomial testing, we may need very large samples to satisfy the specifications of the test. If, in this problem, we reduce the sample size to $n = 100$, then $C_{.05} = 79$. However, now the probability of accepting the null hypothesis when $R = .80$ is $OC(.8) = \Phi(.125) = .55$, which is considerably higher than the corresponding probability of .10 under $n = 484$.   ∎

14.8.2 Exponential Distributions

Suppose we know that the life distribution is exponential $E(\beta)$, but β is unknown. The hypotheses

$$H_0 : R(t_0) \geq R_0$$

versus

$$H_1 : R(t_0) < R_0$$

can be rephrased in terms of the unknown parameter β as

$$H_0 : \beta \geq \beta_0$$

versus

$$H_1 : \beta < \beta_0$$

where $\beta_0 = -t_0/\ln R_0$. Let $t_1, \ldots, t_n$ be the values of a (complete) random sample of size n. Let $\bar{t}_n = 1/n \sum_{i=1}^{n} t_i$. The hypothesis H_0 is rejected if $\bar{t}_n < C_\alpha$, where

$$C_\alpha = \frac{\beta_0}{2n} \chi^2_\alpha[2n] \tag{14.8.6}$$

The OC function of this test, as a function of β, is

$$\begin{aligned} OC(\beta) &= \Pr\{\bar{t}_n > C_\alpha \mid \beta\} \\ &= \Pr\{\chi^2[2n] > \frac{\beta_0}{\beta} \chi^2_\alpha[2n]\} \end{aligned} \tag{14.8.7}$$

If we require that at $\beta = \beta_1$ the OC function of the test will assume the value γ, then the sample size n should satisfy

$$\frac{\beta_0}{\beta_1} \chi^2_\alpha[2n] \geq \chi^2_{1-\gamma}[2n]$$

The quantiles of $\chi^2[2n]$, for $n \geq 15$, can be approximated by the formula

$$\chi^2_p[2n] \cong \frac{1}{2}(\sqrt{4n} + z_p)^2 \tag{14.8.8}$$

Substituting this approximation and solving for n, we obtain the approximation

$$n \cong \frac{1}{4} \frac{(z_{1-\gamma} + z_{1-\alpha}\sqrt{\zeta})^2}{(\sqrt{\zeta} - 1)^2} \tag{14.8.9}$$

where $\zeta = \beta_0/\beta_1$.

EXAMPLE **14.17**   Suppose that in Example 14.16, we know that the system lifetimes are exponentially distributed. If our decision were now based on $\bar{t}_n$, it is interesting to examine how many systems would have to be tested in order to achieve the same error probabilities as before.

Because $\beta = -t/\ln R(t)$, the value of the parameter β under $R(t_0) = R(1000) = .85$ is $\beta_0 = -1000/\ln(.85) = 6153$ (hr), whereas its value under $R(t_0) = .80$ is $\beta_1 = -1000/\ln(.80) = 4481$ (hr). Substituting these values into equation 14.8.9, along with $\alpha = .05$ and $\gamma = .10$, we obtain the necessary sample size $n \cong 87$. If we use equation 9.3.4, we obtain $n \cong 484$.

Thus, we see that the additional knowledge that the lifetime distribution is exponential, along with the use of complete lifetime data on the sample, allows us to achieve a greater than fivefold increase in efficiency in terms of the sample size necessary to achieve the desired error probabilities. ∎

Note that if the sample is censored at the rth failure, then all the formulas just developed apply after replacing n by r, and $\bar{t}_n$ by $\hat{\beta}_{n,r} = T_{n,r}/r$.

EXAMPLE **14.18**   Suppose the reliability at age $t = 250$ (hr) should be at least $R_0 = .85$. Let $R_1 = .75$. The corresponding values of β_0 and β_1 are 1538 (hr) and 869 (hr), respectively. Suppose that the sample is censored at the $r = 25$th failure. Let $\hat{\beta}_{n,r} = S_{n,r}/25$ be the MLE of β. H_0 is rejected, with level of significance $\alpha = .05$, if

$$\hat{\beta}_{n,r} \leq \frac{1538}{50} \chi^2_{.05}[50] = 1069 \text{ (hr)}$$

The type II error probability of this test, at $\beta = 869$, is

$$\text{OC}(869) = \Pr\{\chi^2[50] > \frac{1538}{869} \chi^2_{.05}[50]\}$$

$$= \Pr\{\chi^2[50] > 61.5\}$$

$$\cong 1 - \Phi\left(\frac{61.5 - 50}{\sqrt{100}}\right)$$

$$= .125 \quad \blacksquare$$

14.8.3   Sequential Reliability Testing

Sometimes in reliability demonstration an overriding concern is keeping the number of items tested to a minimum, subject to whatever accuracy requirements are imposed. This could be the case, for example, when testing very complex and expensive systems. In such cases, it may be worthwhile applying a sequential testing procedure, where items are tested one at a time in sequence until the procedure indicates that testing can stop and a decision be made. Such an approach would also be appropriate when testing prototypes of some new design, which are being produced one at a time at a relatively slow rate.

In Chapter 9 we introduced the Wald SPRT for testing hypotheses with binomial data. Here we reformulate this test for reliability testing.

14.8.3.1   The SPRT for Binomial Data

Without any assumptions about the lifetime distribution of a device, we can test hypotheses concerning $R(t_0)$ by simply observing whether or not a device survives to age t_0. Letting K_n represent the number of devices among n randomly selected ones surviving to age t_0, we have K_n is distributed like $B(n, R(t_0))$. The likelihood ratio is given by

$$\lambda_n = \left(\frac{1 - R_1}{1 - R_0}\right)^n \left(\frac{R_1(1 - R_0)}{R_0(1 - R_1)}\right)^{K_n} \tag{14.8.10}$$

Thus,

$$\ln \lambda_n = n \ln\left(\frac{1 - R_1}{1 - R_0}\right) - K_n \ln\left(\frac{R_0(1 - R_1)}{R_1(1 - R_0)}\right)$$

It follows that the SPRT can be expressed in terms of K_n as follows:

Continue sampling if $-h_1 + sn < K_n < h_2 + sn$

Accept H_0 if $K_n \geq h_2 + sn$

Reject H_0 if $K_n \leq -h_1 + sn$

where

$$\begin{cases} s = \ln\left(\frac{1 - R_1}{1 - R_0}\right) \Big/ \ln\left(\frac{R_0(1 - R_1)}{R_1(1 - R_0)}\right) \\[2mm] h_1 = \ln\left(\frac{1 - \gamma}{\alpha}\right) \Big/ \ln\left(\frac{R_0(1 - R_1)}{R_1(1 - R_0)}\right) \\[2mm] h_2 = \ln\left(\frac{1 - \alpha}{\gamma}\right) \Big/ \ln\left(\frac{R_0(1 - R_1)}{R_1(1 - R_0)}\right) \end{cases} \tag{14.8.11}$$

α and γ are the prescribed probabilities of type I and type II errors. Note that if we plot K_n vs. n, the accept and reject boundaries are parallel straight lines with common slope s and intercepts h_2 and $-h_1$, respectively.

The OC function of this test is expressible (approximately) in terms of an implicit parameter ψ. Letting

$$R^{(\psi)} = \begin{cases} \dfrac{1 - \left(\dfrac{1 - R_1}{1 - R_0}\right)^\psi}{\left(\dfrac{R_1}{R_0}\right)^\psi - \left(\dfrac{1 - R_1}{1 - R_0}\right)^\psi} & \psi \neq 0 \\[6mm] s & \psi = 0 \end{cases} \tag{14.8.12}$$

we have that the OC function at $R(t_0) = R^{(\psi)}$ is given by

$$\text{OC}(R^{(\psi)}) \cong \begin{cases} \dfrac{\left(\dfrac{1 - \gamma}{\alpha}\right)^\psi - 1}{\left(\dfrac{1 - \gamma}{\alpha}\right)^\psi - \left(\dfrac{\gamma}{1 - \alpha}\right)^\psi} & \psi \neq 0 \\[8mm] \dfrac{\ln\left(\dfrac{1 - \gamma}{\alpha}\right)}{\ln\left(\dfrac{(1 - \alpha)(1 - \gamma)}{\alpha\gamma}\right)} & \psi = 0 \end{cases} \tag{14.8.13}$$

It is easily verified that for $\psi = 1$, $R^{(\psi)}$ equals R_0 and $\mathrm{OC}(R^{(\psi)})$ equals $1 - \alpha$, whereas for $\psi = -1$, $R^{(\psi)} = R_1$ and $\mathrm{OC}(R^{(\psi)}) = \gamma$.

The expected sample size, or **average sample number (ASN)**, as a function of $R^{(\psi)}$, is given by

$$
\mathrm{ASN}(R^{(\psi)}) \cong
\begin{cases}
\dfrac{\ln \dfrac{1 - \gamma}{\alpha} - \mathrm{OC}(R^{(\psi)}) \ln \left(\dfrac{(1 - \alpha)(1 - \gamma)}{\alpha \gamma} \right)}{\ln \dfrac{1 - R_1}{1 - R_0} - R^{(\psi)} \ln \left(\dfrac{R_0(1 - R_1)}{R_1(1 - R_0)} \right)} & \psi \neq 0 \\[6mm]
\dfrac{h_1 h_2}{s(1 - s)}, & \psi = 0
\end{cases}
\tag{14.8.14}
$$

The ASN function will typically have a maximum at some value of R between R_0 and R_1, and decrease as R moves away from the point of maximum in either direction.

EXAMPLE 14.19   Consider Example 14.17, where we had $t = 1000$ (hr), $R_0 = .85$, $R_1 = .80$, $\alpha = .05$, $\gamma = .10$. Suppose now that a system is tested sequentially, and we apply the binomial SPRT. Each time the system survives 1000 hr of operation without failure, K_n is increased by 1. We continue testing the system until K_n crosses one of the boundaries. The parameters of the boundary lines are $s = .826$, $h_1 = 8.30$, and $h_2 = 6.46$.

The OC and ASN functions of the test are given in Table 14.2 for selected values of ψ.

Compare the values in the ASN column to the sample size required for the corresponding fixed-sample test, $n = 483$. It is clear that the SPRT effects a considerable saving in sample size, particularly when $R(t_0)$ is less than R_1 or greater than R_0. Note also that the maximum ASN value occurs when $R(t_0)$ is near s.   ∎

14.8.3.2   The SPRT for Exponential Lifetimes

When the lifetime distribution is known to be exponential, we have seen the increase in efficiency gained by measuring the actual failure times of the parts being tested. By using a sequential procedure based on these failure times, further gains in efficiency can be achieved.

Expressing the hypotheses in terms of the parameter β of the lifetime distribution $E(\beta)$, we want to test $H_0 : \beta \geq \beta_0$ vs. $H_1 : \beta < \beta_0$, with significance level α and type II error probability γ when $\beta = \beta_1$, where $\beta_1 < \beta_0$. Letting $\mathbf{t}_n = (t_1, \ldots, t_n)$ be the times till failure of the first n parts tested, the likelihood ratio statistic is

$$
\lambda_n(\mathbf{t}_n) = \left(\frac{\beta_0}{\beta_1} \right)^n \exp \left(- \left(\frac{1}{\beta_1} - \frac{1}{\beta_0} \right) \sum_{i=1}^{n} t_i \right)
\tag{14.8.15}
$$

Thus,

$$
\ln \lambda_n(\mathbf{t}_n) = n \ln(\beta_0 / \beta_1) - \left(\frac{1}{\beta_1} - \frac{1}{\beta_0} \right) \sum_{i=1}^{n} t_i
$$

TABLE   14.2

OC and ASN values for the SPRT

ψ	$R^{(\psi)}$	$OC(R^{(\psi)})$	$ASN(R^{(\psi)})$
−2.0	.7724	.0110	152.0
−1.8	.7780	.0173	167.9
−1.6	.7836	.0270	186.7
−1.4	.7891	.0421	208.6
−1.2	.7946	.0651	234.1
−1.0	.8000	.1000	263.0
−0.8	.8053	.1512	294.2
−0.6	.8106	.2235	325.5
−0.4	.8158	.3193	352.7
−0.2	.8209	.4357	370.2
0.0	.8259	.5621	373.1
0.2	.8309	.6834	360.2
0.4	.8358	.7858	334.8
0.6	.8406	.8629	302.6
0.8	.8453	.9159	269.1
1.0	.8500	.9500	238.0
1.2	.8546	.9709	210.8
1.4	.8590	.9833	187.8
1.6	.8634	.9905	168.6
1.8	.8678	.9946	152.7
2.0	.8720	.9969	139.4

The SPRT rules are accordingly,

$$\text{Continue sampling if } -h_1 + sn < \sum_{i=1}^{n} t_i < h_2 + sn$$

$$\text{Accept } H_0 \text{ if } \sum_{i=1}^{n} t_i \geq h_2 + sn$$

$$\text{Reject } H_0 \text{ if } \sum_{i=1}^{n} t_i \leq -h_1 + sn$$

where

$$s = \ln(\beta_0/\beta_1) \left/ \left(\frac{1}{\beta_1} - \frac{1}{\beta_0} \right) \right.$$

$$h_1 = \ln((1-\gamma)/\alpha) \left/ \left(\frac{1}{\beta_1} - \frac{1}{\beta_0} \right) \right.$$

$$h_2 = \ln((1-\alpha)/\gamma) \left/ \left(\frac{1}{\beta_1} - \frac{1}{\beta_0} \right) \right.$$

(14.8.16)

Thus, if we plot $\sum_{i=1}^{n} t_i$ vs. n, the accept and reject boundaries are again parallel straight lines.

As before, let ψ be an implicit parameter, and define

$$
\beta^{(\psi)} = \begin{cases} \dfrac{(\beta_0/\beta_1)^{\psi} - 1}{\psi\left(\dfrac{1}{\beta_1} - \dfrac{1}{\beta_0}\right)} & \psi \neq 0 \\[2em] \dfrac{\ln(\beta_0/\beta_1)}{\dfrac{1}{\beta_1} - \dfrac{1}{\beta_0}} & \psi = 0 \end{cases}
$$

(14.8.17)

Then the OC and ASN functions are approximately given by

$$
OC(\beta^{(\psi)}) \cong \begin{cases} \dfrac{((1-\gamma)/\alpha)^{\psi} - 1}{((1-\gamma)/\alpha)^{\psi} - (\gamma/(1-\alpha))^{\psi}} & \psi \neq 0 \\[2em] \dfrac{\ln((1-\gamma)/\alpha)}{\ln((1-\alpha)(1-\gamma)/\alpha\gamma)} & \psi = 0 \end{cases}
$$

(14.8.18)

and

$$
ASN(\beta^{(\psi)}) \cong \begin{cases} \dfrac{\ln((1-\gamma)/\alpha) - OC(\beta^{(\psi)})\ln((1-\alpha)(1-\gamma)/\alpha\gamma)}{\ln(\beta_0/\beta_1) - \beta^{(\psi)}\left(\dfrac{1}{\beta_1} - \dfrac{1}{\beta_0}\right)} & \psi \neq 0 \\[2em] \dfrac{h_1 h_2}{s^2} & \psi = 0 \end{cases}
$$

(14.8.19)

Note that when $\psi = 1$, $\beta^{(\psi)} = \beta_0$, whereas when $\psi = -1$, $\beta^{(\psi)} = \beta_1$.

EXAMPLE 14.20   Continuing Example 14.17, recall we had $\alpha = .05$, $\gamma = .10$, $\beta_0 = 6153$, $\beta_1 = 4481$. The parameters of the boundaries of the SPRT are $s = 5229$, $h_1 = 47{,}663$, $h_2 = 37{,}124$. The OC and ASN functions for selected values of ψ are given in Table 14.3.

In Example 14.17, we saw that the fixed-sample test with the same α and γ requires a sample size of $n = 87$. Thus, in the exponential case as well, we see that the SPRT can result in substantial savings in sample size.   ∎

It is obviously impractical to perform a sequential test like the one described in the preceding example by running one system, waiting till it fails, renewing it, and running it again and again until a decision can be made. In the example, if the MTTF of the system is close to the value of $\beta_0 = 6153$ (hr), it takes, on average, 256 days between failures and on average, 36 failures till a decision is reached. This trial may take over 25 years. There are three ways to overcome this problem. The first is to put several systems on test simultaneously. Thus, if in the trial described in the preceding example 25 systems are tested simultaneously, the expected duration of the test will be reduced to 1 year. Another way is to consider a test based on a continuous time process, not on discrete samples of failure times. The third possibility of reducing the expected test duration is to perform accelerated life testing. In the following sections we discuss these alternatives.

TABLE   14.3

OC and ASN values for the SPRT

ψ	$\beta^{(\psi)}$	$\mathbf{OC}(\beta^{(\psi)})$	$\mathbf{ASN}(\beta^{(\psi)})$
−2.0	3872	.0110	34.3
−1.8	3984	.0173	37.1
−1.6	4101	.0270	40.2
−1.4	4223	.0421	43.8
−1.2	4349	.0651	47.9
−1.0	4481	.1000	52.4
−0.8	4618	.1512	57.1
−0.6	4762	.2235	61.4
−0.4	4911	.3193	64.7
−0.2	5067	.4357	66.1
0.0	5229	.5621	64.7
0.2	5398	.6834	60.7
0.4	5575	.7858	54.8
0.6	5759	.8629	48.1
0.8	5952	.9159	41.5
1.0	6153	.9500	35.6
1.2	6363	.9709	30.6
1.4	6582	.9833	26.4
1.6	6811	.9905	22.9
1.8	7051	.9946	20.1
2.0	7301	.9969	17.8

14.8.3.3   The SPRT for Poisson Processes

Suppose we put n systems on test starting at $t = 0$. Suppose also that any system that fails is instantaneously renewed, and at the renewal time it is as good as new. In addition, we assume that the life characteristics of the systems are identical, the TTF of each system is exponential (with the same β) and failures of different systems are independent of each other.

Under these assumptions, the number of failures in each system, in the time interval $(0, t]$ is a Poisson random variable with mean λt, where $\lambda = 1/\beta$.

Let $X_n(t) =$ Total number of failures among all the n systems during the time interval $(0, t]$. $X_n(t)$ is distributed like a Poisson r.v. with mean $n\lambda t$, and the collection $\{X_n(t); \ 0 < t < \infty\}$ is called a *Poisson process*. We add the initial condition that $X_n(0) = 0$.

The random function $X_n(t)$, $0 < t < \infty$, is a nondecreasing step function that jumps one unit at each random failure time of the system. The random functions $X_n(t)$ satisfy in addition:

i   For any $t_1 < t_2$, $X_n(t_2) - X_n(t_1)$ is independent of $X_n(t_1)$.

ii   For any $t_1, t_2, 0 < t_1 < t_2 < \infty$, $X_n(t_2) - X_n(t_1)$ is distributed like a Poisson r.v. with mean $n\lambda(t_2 - t_1)$.

We now develop the SPRT based on the random functions $X_n(t)$.

The hypotheses $H_0 : \beta \geq \beta_0$ versus $H_1 : \beta \leq \beta_1$, for $0 < \beta_1 < \beta_0 < \infty$, are translated into the hypotheses $H_0 : \lambda \leq \lambda_0$ versus $H_1 : \lambda \geq \lambda_1$, where $\lambda = 1/\beta$. The likelihood ratio at time t is

$$\Lambda(t; X_n(t)) = \left(\frac{\lambda_1}{\lambda_0}\right)^{X_n(t)} \exp\{-nt(\lambda_1 - \lambda_0)\} \qquad \text{(14.8.20)}$$

The test continues as long as the random graph of $(T_n(t), X_n(t))$ is between the two linear boundaries

$$b_U(t) = h_2 + sT_n(t) \qquad 0 \leq t < \infty$$

and

$$b_L(t) = -h_1 + sT_n(t), \qquad 0 \leq t < \infty$$

where $T_n(t) = nt$ is the total time on test at t,

$$h_1 = \frac{\ln\left(\frac{1-\alpha}{\gamma}\right)}{\ln\left(\frac{\lambda_1}{\lambda_0}\right)}$$

$$h_2 = \frac{\ln\left(\frac{1-\gamma}{\alpha}\right)}{\ln\left(\frac{\lambda_1}{\lambda_0}\right)} \qquad \text{(14.8.21)}$$

and

$$s = \frac{\lambda_1 - \lambda_0}{\ln\left(\frac{\lambda_1}{\lambda_0}\right)}$$

The instant $X_n(t)$ jumps above $b_U(t)$ the test terminates and H_0 is rejected; the instant $X_n(t) = b_L(t)$ the test terminates and H_0 is accepted. Acceptance of H_0 requires that the reliability meets the specified requirement. Rejection of H_0 may lead to additional engineering modification to improve system reliability.

The OC function of this sequential test is the same as that in the exponential case. Let τ denote the random time of termination. It can be shown that $\Pr_\lambda\{\tau < \infty\} = 1$ for all $0 < \lambda < \infty$. The expected deviation of the test is given approximately by

$$E_\lambda\{\tau\} = \frac{1}{\lambda n} E_\lambda\{X_n(\tau)\} \qquad \text{(14.8.22)}$$

where

$$E_\lambda\{X_n(\tau)\} \cong \begin{cases} \dfrac{h_2 - OC(\lambda)(h_1 + h_2)}{1 - s/\lambda} & \text{if } \lambda \neq s \\[2mm] h_1 h_2 & \text{if } \lambda = s \end{cases} \qquad \text{(14.8.23)}$$

Notice that the last formula yields the same values as the formula in the exponential case for $\lambda = 1/\beta^{(\psi)}$. The SPRT of the previous section can terminate only after a failure, whereas the SPRT based on $X_n(t)$ may terminate while crossing the lower boundary $b_L(t)$ before a failure occurs.

The minimal time required to accept H_0 is $\tau_0 = h_1/ns$. In the case of Example 14.20, with $n = 20$, $\lambda_0 = \frac{1}{\beta_0}$ and $\lambda_1 = \frac{1}{\beta_1}$, we obtain $\tau_0 = 9.11536/(20 \times 0.0001912) = 2383.2$ (hr)—that is, over 99 days of testing without any failure. The SPRT may, in addition, be frequency censored by fixing a value x^* so that as soon as $X_n(t) \geq x^*$ the test terminates and H_0 is rejected. In Example 14.19, we see that the expected number of failures at termination may be as large as 66. We can censor the test at $x^* = 50$. This will reduce the expected duration of the test but will increase the probability of a type I error, α. Special programs are available for computing the operating characteristics of such censored tests, but they are beyond the scope of this text.

14.9
Accelerated Life Testing

It is often impractical to test highly reliable systems or components under normal operating conditions because no failures may be observed for long periods of time. Recall that in accelerated life testing, systems are subjected to higher than normal stress conditions in order to generate failures. The question is how to relate failure distributions under higher than normal stress conditions to those under normal conditions?

Accelerated life testing is used by engineers to test materials like food and drugs, lubricants, concrete and cement, building materials, and nuclear reactor materials. The stress conditions are, generally, mechanical load, vibrations, high temperatures, high humidity, high contamination, and so on. Accelerated testing is used for semiconductors, including transistors, electronic devices such as diodes, random access memories, and so on. Refer to Nelson (1990) for a survey of methods and applications. The statistical methodology of accelerated life testing is similar to the methods described earlier in this chapter, including graphical analysis and maximum likelihood estimation. The reader is referred to Nelson (1990) and Mann, Schafer, and Singpurwalla (1974) for details. Following are some of the models used to relate failures under various stress conditions.

14.9.1   The Arrhenius Temperature Model

The *Arrhenius temperature model* is widely used when product failure time is sensitive to high temperature. Applications include electrical insulations and dielectric (see Goba, 1969); solid state and semiconductors (see Peck and Trapp 1978); battery cells; lubricants and greases; plastics; incandescent light filaments.

The Arrhenius law states that the rate of simple chemical reaction depends on temperature as follows:

$$\lambda = A \exp\{-E/(kT)\} \tag{14.9.1}$$

where E is the activation energy (in electron volts); k is the Boltzmann constant, 8.6171×10^{-5} electron volts per $^\circ$C; T is the absolute Kelvin temperature ($273.16 + {}^\circ$C); and A is a product parameter that depends on the test conditions and failure characteristics. In applying the Arrhenius model to failure times distribution, we find the Weibull-Arrhenius life distribution, in which the scale parameter β of the Weibull distribution is related to temperature, T, according to the function

$$\beta(T) = \lambda \exp \left\{ A + \frac{B}{T} \right\} \qquad B > 0 \qquad\qquad \text{(14.9.2)}$$

where λ, A, and B are fitted empirically to the data.

14.9.2 Other Models

Another model is called the *log-normal Arrhenius model*, in which the log failure time is normally distributed with mean $A + B/T$ and variance σ^2. According to this model, the expected failure time is $\exp\{A + B/T + \sigma^2/2\}$. Another model prevalent in the literature relates expected failure time to a stress level V, according to the *inverse power model*

$$\text{MTTF}(V) = \frac{C}{V^p} \qquad C > 0 \qquad\qquad \text{(14.9.3)}$$

The statistical data analysis methodology is to fit an appropriate model to the data, usually by maximum likelihood estimation, and then predict the MTTF of the system under normal conditions, or some reliability or availability function. Tolerance intervals for the predicted value should be determined.

Burn-In Procedures

Many products show a relatively high frequency of early failures. For example, if a product has an exponential distribution of the TTF with MTTF of $\beta = 10{,}000$ (hr), we do not expect more than 2% of the product to fail within the first 200 hr. Nevertheless, many products designed for a high value of MTTF show a higher than expected number of early failures. This phenomenon led to the theory that the hazard rate function of products is typically a U-shaped function. In its early life, the product is within a phase with monotone decreasing hazard rate. This phase is called the "infant mortality" phase. After this phase, the product enters a phase of "maturity," in which the hazard rate function is almost constant. **Burn-in procedures** are designed to screen (burn) weak products within the plant by setting the product in operation for several days, in order to give it a chance to fail in the plant and not in the field, where loss due to failure is high.

How long should a burn-in procedure last? Jensen and Petersen (1982) discuss this and other issues in designing burn-in procedures. We refer the reader to this book for more details. Here we present only some basic ideas.

Burn-in procedures discussed by Jensen and Petersen are based on a model of a mixed life distribution. For example, suppose that experience shows that the life distribution of a product is Weibull, $W(v, \beta_1)$. A small proportion of units manufactured may have a generally short life, for various reasons, which is given by another Weibull distribution—say, $W(v, \beta_0)$—with $\beta_0 < \beta_1$. Thus, the life distribution of a randomly chosen product has a distribution that is a mixture of $W(v, \beta_0)$ and $W(v, \beta_1)$—that is,

$$F(t) = 1 - \left[p \exp\left\{ -\left(\frac{t}{\beta_0}\right)^v \right\} + (1-p) \exp\left\{ -\left(\frac{t}{\beta_1}\right)^v \right\} \right] \qquad \textbf{(14.10.1)}$$

for $t > 0$ p is the probability that the product's life distribution is $W(v_1, \beta_0)$. The objective of the burn-in is to give units having the $W(v, \beta_0)$ distribution an opportunity to fail in the plant. The units that do not fail during the burn-in have, for their remaining life, a life distribution closer to the desired $W(v, \beta_1)$.

Suppose that a burn-in continues for t^* time units. The conditional distribution of the time till failure T, given that $\{T > t^*\}$, is

$$F^*(t) = \frac{\int_{t^*}^t f(u)\,du}{1 - F(t^*)} \qquad t \geq t^* \qquad \textbf{(14.10.2)}$$

The c.d.f. $F^*(t)$ of units surviving the burn-in starts at t^*—that is, $F^*(t^*) = 0$ and has MTTF

$$\beta^* = t^* + \int_{t^*}^{\infty} (1 - F^*(t))\,dt \qquad \textbf{(14.10.3)}$$

We illustrate this in the following example on mixtures of exponential life times.

EXAMPLE **14.21**   Suppose that a product is designed to have an exponential life distribution, with mean of $\beta = 10,000$ (hr). A proportion $p = .05$ of the products come out of the production process with a short MTTF of $\gamma = 100$ (hr). Suppose that all products go through a burn-in for $t^* = 200$ (hr).

The c.d.f. of the TTF of units that did not fail during the burn-in is

$$F^*(t) = 1 - \frac{.05 \exp\{-\frac{t}{100}\} + .95 \exp\{-\frac{t}{10,000}\}}{.05 \exp\{-\frac{200}{100}\} + .95 \exp\{-\frac{200}{10,000}\}}$$

$$= 1 - \frac{1}{.93796} \left[.05 \exp\left\{ -\frac{t}{100} \right\} + .95 \exp\left\{ -\frac{t}{10,000} \right\} \right]$$

for $t \geq 200$. The mean time till failure for units surviving the burn-in is thus

$$\beta^* = 200 + \frac{1}{0.93796} \int_{200}^{\infty} \left(.05 \exp\left\{ -\frac{t}{100} \right\} + .95 \exp\left\{ -\frac{t}{10,000} \right\} \right) dt$$

$$= 200 + \frac{5}{.93796} \exp\left\{ -\frac{200}{100} \right\} + \frac{9500}{.93796} \exp\left\{ -\frac{200}{10,000} \right\}$$

$$= 10,128.53 \text{ (hr)}$$

A unit surviving 200 hours of burn-in is expected to operate an additional 9928.53 hours in the field. The expected life of these units without the burn-in is $0.05 \times 200 + 0.95 \times 10,000 = 9510$ (hr). The burn-in of 200 hours in the plant is expected to increase the mean life of the product in the field by 418 hours. Whether this increase in the MTTF justifies the burn-in depends on the relative cost of burn-in in the plant to the cost of failures in the field. The proportion p of "short-life" units also plays an important role. If this proportion is $p = .1$ rather than .05, the burn-in increases the MTTF in the field from 9020 hours to 9848.95 hours. One can easily verify that for $p = .2$, if the income for an hour of operation of one unit in the field is $C_p = 5\$$ and the cost of the burn-in per unit is 0.15\$ per hour, then the length of burn-in that maximizes the expected profit is about 700 hours. ∎

14.11
Chapter Highlights

Chapter 13 examined design decisions of product and process developers that are aimed at optimizing the quality and robustness of products and processes. This chapter looks at performance over time and discusses basic notions of repairable and nonrepairable systems. Graphical and nonparametric techniques are presented, together with classical parametric techniques for estimating life distributions. Special sections cover reliability demonstrational procedures, sequential reliability testing, burn-in procedures, and accelerated life testing. Design and testing of reliability is a crucial activity for organizations at the top of the quality ladder.

The main concepts and definitions introduced in this chapter include:

- Availability
- Life distributions
- Accelerated life testing
- Time categories

- Up time
- Down time
- Intrinsic availability
- Operational readiness
- Reliability function
- Mean time to failure
- Hazard function
- Failure rate function
- Structure function
- Time till failure
- Time till repair
- Cycle time
- Renewal function
- Censored data
- Product limit estimate
- Average sample number
- Burn-in procedure

14.12
Exercises

14.1.1 During 600 hours of manufacturing time, a machine was up 510 hours. It had 100 failures, which required a total of 11 hours of repair time. What is the MTTF of this machine? What is its mean time till repair (MTTR)? What is the intrinsic availability?

14.1.2 The frequency distribution of the lifetime in a random sample of $n = 2000$ solar cells under accelerated life testing is the following:

t (thousands of hr)	0–1	1–2	2–3	3–4	4–5	5–
Proportional frequency		0.15	0.25	0.25	0.10	0.10	0.15

The relationship of the scale parameters of the life distributions between normal and accelerated conditions is 10:1.

a Estimate the reliability of the solar cells at age $t = 4.0$ (yr).

b What proportion of solar cells are expected to survive 40,000 hr among those that survive 20,000 hr?

14.1.3 The c.d.f. of the lifetime (months) of a piece of equipment is

$$F(t) = \begin{cases} t^4/20736 & 0 \le t < 12 \\ 1 & 12 \le t \end{cases}$$

a What is the failure rate function of this equipment?

b What is the MTTF?

c What is the reliability of the equipment at 4 months?

14.1.4 The reliability of a system is

$$R(t) = \exp\{-2t - 3t^2\} \qquad 0 \le t < \infty$$

a What is the failure rate of this system at age $t = 3$?

b Given that the system reached the age of $t = 3$, what is its reliability for two additional time units?

14.2.1 An aircraft has four engines but can land using only two engines.

a Assuming that the reliability of each engine for the duration of a mission is $R = .95$ and that engine failures are independent, compute the mission reliability of the aircraft.

b What is the mission reliability of the aircraft if at least one functioning engine must be on each wing?

14.2.2

a Draw a block diagram of a system having the structure function

$$R_{sys} = \psi_s(\psi_p(\psi_{M_1}, \psi_{M_2}), R_6),$$
$$\psi_{M_1} = \psi_p(R_1, R_2, R_3),$$
$$\psi_{M_2} = \psi_2(R_4, R_5)$$

b Determine R_{sys} if all the components act independently and have the same reliability, $R = .8$.

14.2.3 Consider a system of n components in a series structure. Let $R_1, \ldots, R_n$ be the reliabilities of the components. Show that

$$R_{sys} \ge 1 - \sum_{i=1}^{n}(1 - R_i)$$

14.2.4 A 4-out-of-8 system has identical components whose life lengths T (weeks) are independent and identically distributed like a Weibull $W\left(\frac{1}{2}, 100\right)$. What is the reliability of the system at $t_0 = 5$ weeks?

14.2.5 A system consists of a main unit and two standby units. The lifetimes of these units are exponential with mean $\beta = 100$ (hr). The standby units undergo no failure while idle. Switching will take place when required. What is the MTTF of the system? What is the reliability function of this system?

14.3.1 Suppose that the TTF in a renewal cycle has a $W(\alpha, \beta)$ distribution and that the TTR has a log-normal distribution $LN(\mu, \sigma)$. Assume further that TTF and TTR are independent. What are the mean and standard deviation of a renewal cycle?

14.3.2 Suppose that a renewal cycle has the normal distribution $N(100, 10)$. Determine the p.d.f. of $N_R(200)$.

14.3.3 Let the renewal cycle C be distributed like $N(100, 10)$. Approximate $V(1000)$.

14.3.4 Derive the renewal density $v(t)$ for a renewal process with C is distributed like $N(100, 10)$.

14.3.5 Two identical components are connected in parallel. The system is not repaired until both components fail. Assuming that the TTF of each component is exponentially distributed, $E(\beta)$, and the total repair time is $G(2, \gamma)$, derive the Laplace transform of the availability function $A(t)$ of the system.

14.3.6 Simulate a sample of 100 TTF of a system comprising two components connected in parallel, where the life distribution of each component (in hours) is $E(100)$. [You can use MINITAB for this purpose.] Save the simulated sample under \ISTAT\DATA\SAMPLE1.DAT. Similarly, simulate a sample of 100 repair times (in hours) having a $G(2, 1)$ distribution. Save this sample in \ISTAT\DATA\SAMPLE2.DAT. Use program RENEWDIS.EXE to estimate the expected value and variance of the number of renewals in 2000 (hr).

14.4.1 In a given life test, $n = 15$ units are placed to operate independently. The time till failure of each unit has an exponential distribution with mean 2000 (hr). The life test terminates immediately after the 10th failure. How long is the test expected to last?

14.4.2 If n units are put on test and their TTF are exponentially distributed with mean β, the time elapsed between the rth and $(r+1)$st failure—that is, $\Delta_{n,r} = T_{n,r+1} - T_{n,r}$—is exponentially distributed with mean $\beta/(n-r)$, $r = 0, 1, \ldots, n-1$. Also, $\Delta_{n,0}, \Delta_{n,2}, \ldots, \Delta_{n,n-1}$ are independent. What is the variance of $T_{n,r}$? Use this result to compute the variance of the test length in Exercise 14.4.1.

14.4.3 Consider Exercise 14.4.2. How would you estimate unbiasedly the scale parameter β if the r failure times $T_{n,1}, T_{n,2}, \ldots, T_{n,r}$ are given? What is the variance of this unbiased estimator?

14.5.1 Simulate a random sample of 100 failure times following the Weibull distribution $W(2.5, 10)$. Save the sample in file C:\ISTAT\DATA \RELSIM.DAT. Draw a Weibull probability plot of the data. Estimate the parameters of the distribution from the parameters of the linear regression fitted to the QQ plot.

14.5.2 The following is a random sample of measures of the compressive strength of 20 concrete cubes (kg/cm^2).

94.9, 106.9, 229.7, 275.7, 144.5, 112.8, 159.3,

153.1, 270.6, 322.0, 216.4, 544.6, 266.2, 263.6,

138.5, 79.0, 114.6, 66.1, 131.2, 91.1

Make a log-normal QQ plot of these data and estimate the mean and standard deviation of this distribution.

14.5.3 The following data represent the time till first failure (days) of electrical equipment. The data were censored after 400 days.

13, 157, 172, 176, 249, 303, 350, 400$^+$, 400$^+$

(Censored values appear as x^+.) Make a Weibull QQ plot of these data and estimate the median of the distribution.

14.6.1 Make a PL (Kaplan-Meier) estimate of the reliability function of an electronic device, based on 50 failure times in file ELECFAIL.DAT.

14.7.1 Assuming that the failure times in file ELECFAIL.DAT come from an exponential distribution $E(\beta)$, compute the MLE of β and of $R(50; \beta) = \exp\{-50/\beta\}$. [The MLE of a function of a parameter is obtained by substituting the MLE of the parameter in the function.] Determine confidence intervals for β and for $R(50; \beta)$ at confidence level .95.

14.7.2 The following are values of 20 random variables having an exponential distribution $E(\beta)$. The values are censored at $t^* = 200$.

96.88, 154.24, 67.44, 191.72, 173.36, 200, 140.81,

200, 154.71, 120.73, 24.29, 10.95, 2.36, 186.93,

57.61, 99.13, 32.74, 200, 39.77, 39.52

Determine the MLE of β. Use program CENEXP.EXE, with β equal to the MLE, to estimate the standard deviation of the MLE and to obtain a confidence interval for β at level $1 - \alpha = .95$. [This simulation is called an empirical bootstrap.]

14.7.3 Determine n^0 and r for a frequency censoring test for an exponential distribution where the cost of a unit is 10 times bigger than the cost per time unit of testing. We want SE$\{\hat{\beta}_n\} = 0.1\beta$, and the expected cost should be minimized at $\beta = 100$ (hr). What is the expected cost of this test at $\beta = 100$ when $c_1 = \$1$ (hr)?

14.7.4 File WEIBUL.DAT contains the values of a random sample of size $n = 50$ from a Weibull distribution.

a Obtain the MLEs of the scale and shape parameters β and ν.

b Use program SMLEWEIB.EXE, with the MLEs of β and ν, to obtain the parametric EBD of $\hat{\beta}$, $\hat{\nu}$, with $M = 500$ runs. Estimate from this distribution the standard deviations of $\hat{\beta}$ and $\hat{\nu}$. Compare these estimates to the large-sample approximations.

14.8.1 In binomial life testing by a fixed-size sample, how large should the sample be in order to discriminate between $R_0 = .99$ and $R_1 = .90$, with $\alpha = \beta = .01$? [α and β denote the probabilities of error of type I and II.]

14.8.2 Design the Wald SPRT for binomial life testing in order to discriminate between $R_0 = .99$ and $R_1 = .90$, with $\alpha = \beta = .01$. What is the expected sample size, ASN, if $R = .9$?

14.8.3 Design a Wald SPRT for exponential life distribution in order to discriminate between $R_0 = .99$ and $R_1 = .90$, with $\alpha = \beta = .01$. What is the expected sample size, ASN, when $R = .90$?

14.8.4 Assume $n = 20$ computer monitors are put on accelerated life testing. The test is a SPRT for Poisson processes, based on the assumption that the TTF of a monitor in those conditions is exponentially distributed. The monitors are considered to be satisfactory if their MTBF is $\beta \geq 2000$ (hr) and considered to be unsatisfactory if $\beta \leq 1500$ (hr). What is the expected length of the test if $\beta = 2000$ (hr).

14.10.1 A product has an exponential lifetime with MTTF $\beta = 1000$ (hr). 1% of the products come out of production with MTTFs of $\gamma = 500$ (hr). A burn-in of $t^* = 300$ (hr) takes place. What is the expected life of units surviving the burn-in? Is such a long burn-in justified?

References

Aoki, M. (1989). *Optimization of stochastic systems: Topics in discrete-time dynamics* (2nd ed.). New York: Academic Press.

Barnard, G. A. (1959). Control charts and stochastic processes. *J. Roy. Statist. Soc.*, B, *21*, 239–267.

Box, G. E. P., Hunter, W. G., and Hunter, S. J. (1978). *Statistics for experimenters*. New York: John Wiley.

Box, G. E. P., and Kramer, T. (1992). Statistical process monitoring and feedback adjustment—a discussion. *Technometrics, 34*, 251–285.

Bratley, P., Fox, B. L., and Schrage, L. E. (1983). *A guide to simulation*. New York: Springer-Verlag.

Cochran, W. G. (1977) *Sampling techniques*. New York: John Wiley.

Conover, W. J. (1980). *Practical nonparametric statistics* (2nd ed.). New York: John Wiley.

Cusumano, M. (1991) *Japan's software factories: A challenge to U.S. management*. Oxford: Oxford University Press.

Daniel, C., and Wood. F. S. (1971). *Fitting equations to data: Computer analysis of multifactor data for scientists and engineers*. New York: John Wiley.

Dehand, K. (1989) *Quality control, robust design, and the Taguchi method*. Pacific Grove, CA: Wadsworth and Brooks/Cole.

Deming, W. E. (1982). *Quality, productivity and the competitive position*. Cambridge, MA: Center for Advanced Engineering Studies.

Deming, W. E. (1991). *Out of the crisis*. Boston: MIT Press.

Devlin, S. J., Gnanadesikan, R., and Ketterning, J. R. (1975). Robust estimation and outlier detection with correlation coefficients. *Biometrika, 62*, 531–545.

Dodge, H. F., and Romig, H. G. (1929). A method of sampling inspection. *Bell System Technical Journal, 8*, 613–631.

Dodge, H. F., and Romig, H. G. (1959). *Sampling inspection tables* (2nd ed.). New York: John Wiley.

Draper, N. R., and Smith, H. (1981). *Applied regression analysis* (2nd ed.). New York: John Wiley.

Duncan, A. J. (1956). The economic design of $\overline{X}$ charts used to maintain current control of a process. *Jour. Amer. Statist. Assoc., 51*, 228–242.

Duncan, A. J. (1971). The economic design of $\overline{X}$ charts when there is a multiplicity of assignable causes. *Jour. Amer. Statist. Assoc., 66*, 107–121.

Duncan, A. J. (1978). The economic design of p-charts to maintain current control of a process: Some numerical results. *Technometrics, 20*, 235–243.

Duncan, A. J. (1986). *Quality control and industrial statistics* (5th ed.). Homewood, IL: Irwin.

Efron, B, and Tibshirani, R. J. (1993). *An introduction to the bootstrap*. New York: Chapman and Hall.

Feigenbaum, A. V. (1983). *Total quality control* (3rd ed.). New York: McGraw-Hill.

Fuchs, C., and Kenett, R. S. (1997). *Multivariate quality control*. New York: Marcel Dekker. (In preparation.)

Gertsbakh, I. B. (1989). *Statistical reliability theory*. New York: Marcel Dekker.

Ghosh, S. (1990). *Statistical design and analysis of industrial experiments*. New York: Marcel Dekker.

Gibra, I. N. (1971). Economically optimal determination of the parameters of X-bar control chart. *Management Science, 17*, 635–646.

Gnanadesikan, R. (1977). *Methods of statistical data analysis of multivariate observations*. New York: John Wiley.

Goba, F. A. (1969). Biography on thermal aging of electrical insulation. *IEEE Trans. on Electrical Insulation, EI-4*, 31–58.

Godfrey, A. B. (1986). The history and evolution of quality in AT&T. *AT&T Technical Journal, 65*, 9–20.

Goldsmith, P. L. and Whitfield, H. (1961). Average run lengths in cumulative chart quality control schemes. *Technometrics, 3*, 11–20.

Grant, E. L. and Leavenworth, R. S. (1980). *Statistical quality control* (5th ed.). New York: McGraw-Hill.

Graybill, F. A. (1969). *Introduction to matrices with applications in statistics*. Belmont, CA: Wadsworth.

H-107, Single-level continuous sampling procedures and tables for inspection by attribute. (1959). Washington, DC: Government Printing Office.

H-108, Sampling procedures and tables for life and reliability testing (based on the exponential distribution). (1959) U.S. Department of Defense, *Quality Control and Reliability Handbook*, Washington, DC: Government Printing Office.

Hettmansperger, T. P. (1984). *Statistical inference based on ranks*. New York: John Wiley.

Hoadley, B. (1981). The quality measurement plan. *Bell System Technical Journal, 60*, 215–271.

Hoadley, B. (1986). Quality measurement plan. *Encyclopedia of Statistical Sciences*, Vol. 7. New York: John Wiley.

Hoadley, B. (1988). QMP/USP—A modern approach to statistical quality auditing. *Handbook of Statistics*, Vol. 7, pp. 353–373. P. R. Krishnaiah, Ed. New York: Elsevier Science Publishers.

Hoaglin, D. C., Mosteller, F., and Tukey, J. W., (Eds.) (1983). *Understanding robust and exploratory data analysis*. New York: John Wiley.

Hotelling, H. (1931). The generalization of student's ratio. *Annals of Mathematical Statistics, 2*, 360–378.

Ishikawa, K. (1986). *Guide to quality control* (2nd ed.). New York: Asian Productivity Organization, UNIPAB Kraus International Publications.

James, M., and Stein, C. (1961). Estimation with quadratic loss. *Proc. Fourth Berkeley Symp. Statist. and Prob., 2*, 361–379.

Jensen, F., and Petersen, N. E. (1982). *Burn-in: An engineering approach to the design and analysis of burn-in procedures*. New York: John Wiley.

John, P. W. M. (1990). *Statistical methods in engineering and quality assurance*. New York: John Wiley.

Johnson, N. L. (1961). A simple theoretical approach to cumulative sum control charts. *Jour. Amer. Statist. Assoc., 56*, 835–840.

Johnson, N. L., and Leone, F. C. (1962). Cumulative sum control charts, Part I. *Quality Control, 18*, 15–21.

Johnson, N. L. and Leone, F. C. (1963). Cumulative sum control charts, Part II. *Quality Control, 19*, 29–36.

Johnson, N. L., and Leone, F. C. (1964). Cumulative sum control charts, Part III. *Quality Control*, *19*, 22–28.

Johnson, N. L., and Leone, F. C. (1964). *Statistics and experimental design in engineering and physical sciences*. New York: John Wiley.

Johnson, R. A., and Wichern, D. W. (1982). *Applied multivariate statistical analysis*. Englewood Cliffs, NJ: Prentice Hall.

Juran, J. M. (Ed.) (1979). *Quality control handbook* (3rd ed.). New York: McGraw-Hill.

Juran, J. M. (Ed.) (1995). *A history of managing for quality*. Milwaukee, WI: ASQLC Quality Press.

Juran, J. M., and Gryna, F. M. (1988). *Juran's quality control handbook* (4th ed.). New York: McGraw-Hill.

Kacker, R. N. (1985). Off-line quality control, parameter design and the Taguchi method. *Jour. of Quality Tech.*, *17*, 176–209.

Kacker, R. N. (1986). Taguchi's quality philosophy analysis and commentary. *Quality Progress*, *18*, 21–29.

Kelly, T., Kenett, R. S., Newton, E., Roodman, G., and Wowk, A. (1991). Total quality management also applies to a school of management. *Proceedings of the 9th IMPRO Conference*. Atlanta, GA.

Kenett, R. S. (1983). On an exploratory analysis of contingency tables. *The Statistician*, *32*, 395–403.

Kenett, R. S. (1991). Two methods for comparing Pareto charts. *Journal of Quality Technology*, *23*, 27–31.

Kenett, R. S., and Pollak, M. (1996). Data analytic aspects of the Shyryaev-Roberts control chart. *Journal of Applied Statistics*, *23*, 125–137.

Kenett, R. S., and Zacks, S. (1992). Process tracking under random changes in the mean. Technical report. School of Management, Binghamton University, Binghamton, NY.

Kotz, S., and Johnson, N. L. (1985). *Encyclopedia of Statistical Sciences*. New York: John Wiley.

Kotz, S., and Johnson, N. L. (1994). *Process capability indices*. New York: Chapman and Hall.

Lehmann, E. L. (1975). *Nonparametrics: Statistical methods based on ranks*. San Francisco: Holden-Day.

Liebesman, B. S., and Saperstein, B. (1983). A proposed attribute skip-lot sampling program. *Journal of Quality Technology*, *15*, 130–140.

Lin, K. M., and Kacker, R. N. (1989). Optimizing the wave soldering process. In Khorsrow Dehand (Ed.), *Quality control robust design and the Taguchi method* (143–157). Pacific Grove, CA: Wadsworth and Brooks/Cole.

Lucas, J. M. (1982). Combined Shewhart—CUSUM quality control schemes. *Journal of Quality Technology*. *14*, 51–59.

Lucas, J. M., and Crosier, R. B. (1982). Fast initial response for CUSUM quality control scheme: Give your CUSUM a headstart. *Technometrics*, *24*, 199–205.

Mann, N. R., Schafer, R. E., and Singpurwalla, N. D. (1974). *Methods for statistical analysis of reliability and life data*. New York: John Wiley.

MIL-STD-105E, Sampling procedures and tables for inspection by attributes. (1989). Washington, DC: Government Printing Office.

MIL-STD-414, Sampling procedures and tables for inspection by variables for percent defectives. (1957). Washington, DC: Government Printing Office.

Mosteller, F., and Tukey, J., (1977). *Data analysis and regression*. Boston: Addison-Wesley.

Myers, R. H. (1976). Response surface methodology. Blacksburg, VA: Author.

Myers, R. H., and Montgomery, D. C. (1995). *Response surface methodology: Process and product optimization using designed experiments*. New York: John Wiley.

Nelson, W. (1990). *Accelerated testing: Statistical models, test plans and data analysis*. New York: John Wiley.

Oikawa, T., and Oka, T. (1987). New techniques for approximating the stress in pad-type nozzles attached to a spherical shell. *Transactions of the Amer. Soc. of Mechan. Eng.*, May, 188–192.

Page, E. S. (1954). Continuous inspection schemes. *Biometrika, 41*, 100–114.

Page, E. S. (1962). Cumulative sum schemes using gauging. *Technometrics, 4*, 97–109.

Page, E. S. (1962). A modified control chart with warning limits. *Biometrika, 49*, 171–176.

Pao, T. W., Phadke, M. S., and Sherrerd, C. S. (1985). Computer response time optimization using orthogonal array experiments. IEEE International Communication Conference, Chicago. *Conference Record, 2*, 890–895.

Peck, D. S., and Trapp, O. D. (1978). *Accelerated testing handbook*. Portola Valley, CA: Technology Associations.

Phadke, M. S. (1989). *Quality engineering using robust design*. Englewood Cliffs, NJ: Prentice Hall.

Phadke, M. S., Kackar, R. N., Speeney, D. V., and Grieco, M. J. (1983). Quality control in integrated circuits using experimental design. *Bell System Technical Journal, 62*, 1275–1309.

Powers, J. D. (1987). *The Japanese competition: Phase 2*. Ann Arbor, MI: Center for Japanese Studies, University of Michigan.

Quinlan, J. (1985). Product improvement by application of Taguchi methods. Third Supplier Symposium on Taguchi Methods, American Supplier Institute, Inc., Dearborn, MI.

Roberts, S. W. (1966). A comparison of some control chart procedures. *Technometrics, 8*, 411–430.

Rodriguez, R. N. (1992). Recent developments in process capability analysis. *Journal of Quality Technology, 24*, 176–187.

Rosseeuw, P. J., and Leroy, A. M. (1987). Robust regression and outlier detection. New York: John Wiley.

Ryan, B. F., and Joiner, B. L. (1994). *Minitab Handbook* (3rd ed.). Belmont, CA: Duxbury Press.

Ryan, T. P. (1988). *Statistical methods for quality improvement*. New York: John Wiley.

Scheffe, H. (1959). *The analysis of variance*, New York: John Wiley.

Searle, S. R. (1966). *Matrix algebra for biological sciences*. New York: John Wiley.

Sen, A., and Srivastava, M. (1990). *Regression analysis: Theory, methods, and applications*. New York: Springer-Verlag.

Shewhart, W. A. (1926). Quality control charts, *Bell System Technical Journal, 5*, 593–603.

Shiryayev, A. N. (1963). On optimum methods in quickest detection problem. *Theory of Prob. and Appl., 8*, 22–46.

Shiryayev, A.N. (1978). *Optimal stopping rules*. New York: Springer-Verlag.

Taguchi, G. (1987). *Systems of experimental design*, Vol 1–2. D. Clausing (Ed.). New York: UNIPUB/Kraus International Publications.

Taguchi, G, and Taguchi, S. (1987). *Taguchi methods: Orthogonal arrays and linear graphs*. Dearborn, MI: American Supplier Institute.

Vardeman, S. B. (1994). *Statistics for engineering problem solving*. Boston: PWS.

Velleman, P., and Hoaglin, D. (1981). *Applications, basics, and computing of exploratory data analysis*. Boston: Duxbury Press.

Wadsworth, H. M., Stephens, K. S., and Godfrey, A. B. (1986). *Modern methods for quality control and improvement*. New York: John Wiley.

Walton, M. (1986). *The Deming management method*. New York: Perigee Books, Putnam.

Weindling, J. I. (1967). Statistical properties of a general class of control charts treated as a Markov process. Ph.D. dissertation, Columbia University, New York.

Weindling, J. I., Littauer, S. B., and de Oliviera, J. (1970). Mean action time of the X-bar control chart with warning limits. *Journal of Quality Technology, 2*, 79–85.

Welch, W. J., Tat-Kwan Yu, Sung Mo Kang, and Sacks, J. (1990). Computer experiments for quality control by parameter design. *Journal of Quality Technology, 22*, 15–22.

Womack, J. P., Jones, D. T., and Roos, D. (1990). *The machine that changed the world*. New York: Rawson Associates.

Yashchin, E. (1985a). On the design and analysis of Shewhart-CUSUM control schemes. *IBM Jour. of Res. Develop., 29*, 377–391.

Yashchin, E. (1985b). On a unified approach to the analysis of two-sided cumulative sum control schemes with headstarts. *Advances in Applied Probability, 17*, 562–593.

Yashchin, E. (1991a). Package CONTRAD for design, analysis and running of CUSUM - Shewhart control schemes. Research Report, RC11880. New York: IBM Research Division.

Yashchin, E. (1991b). Some aspects of the theory of statistical control schemes. *IBM Journal of Research Development*, *31*, 199–205.

Zacks, S. (1991). Detection and change-point problems. *Handbook of Sequential Analysis*, Ch 23. In B. K. Ghosh and P. K. Sen (Eds.), New York: Marcel Dekker.

Zacks, S. (1992). *Introduction to reliability analysis: Probability models and statistical methods*. New York: Springer-Verlag.

Review of Matrix Algebra for Statistics with MINITAB Computations

In this appendix, we summarize some useful properties and techniques of matrix operations that are useful in statistical analysis. Basic results are presented but not proven. The interested student can study the subject from one of the many textbooks on matrix algebra. Matrix algebra for statistics is discussed in Graybill (1969). We will use MINITAB to perform matrix calculations.

An array of n rows and 1 column is called an n-dimensional **vector**. We denote a vector with lower-case bold type letter $\mathbf{a} = (a_1, a_2, \ldots, a_n)$, where a_i is the ith component of vector $\mathbf{a}$. The transpose of an n-dimensional vector $\mathbf{a}$ is an array of 1 row and n columns. We denote the transpose of $\mathbf{a}$ by $\mathbf{a}'$. We can add (subtract) only vectors of the same dimension. Thus $\mathbf{a} + \mathbf{b}$ is a vector $\mathbf{c}$ whose ith component is $c_i = a_i + b_i$ $(i = 1, \ldots, n)$. The vector $\mathbf{0}$ is one whose components are all equal to zero. Obviously, $\mathbf{a} + (-\mathbf{a}) = \mathbf{0}$. If $\mathbf{a}$ and $\mathbf{b}$ are vectors of the same dimension, n, then the **inner product** of $\mathbf{a}$ and $\mathbf{b}$ is $\mathbf{a}'\mathbf{b} = \sum_{i=1}^{n} a_i b_i$. The Euclidean norm (length) of a vector $\mathbf{a}$ is $||\mathbf{a}|| = (\mathbf{a}'\mathbf{a})^{1/2}$. If α is a scalar (real number) and $\mathbf{a}$ a vector, then $\alpha\mathbf{a}$ is a vector whose ith component is αa_i $(i = 1, \ldots, n)$. If $\mathbf{a}_1, \mathbf{a}_2, \ldots, \mathbf{a}_k$ are n-dimensional vectors and $\alpha_1, \ldots, \alpha_k$ are scalars, then $\sum_{i=1}^{k} \alpha_i \mathbf{a}_i$ is an n-dimensional vector.

A vector $\mathbf{b}$ is *linearly independent* of vectors $\mathbf{a}_1, \ldots, \mathbf{a}_k$ if there exist no scalars $\alpha_1, \ldots, \alpha_k$ such that $\mathbf{b} = \sum_{i=1}^{k} \alpha_i \mathbf{a}_i$. k vectors $\mathbf{a}_1, \ldots, \mathbf{a}_k$ are *mutually independent* if no one is a linear combination of the other $k - 1$ vectors.

k mutually independent n-dimensional vectors $(1 \leq k \leq n)$ span a k-dimensional linear subspace $\mathcal{S}_k = \{\mathbf{b} : \mathbf{b} = \sum_{j=1}^{k} \alpha_j \mathbf{a}_j, \text{ all finite } \alpha_1, \ldots, \alpha_k\}$. Obviously $\mathcal{S}_1 \subset \mathcal{S}_2 \subset \cdots \subset \mathcal{S}_n$.

547

A matrix of order $n \times k$ is an array of n rows and k columns—that is, $\mathbf{A} = (\mathbf{a}_1, \mathbf{a}_2, \ldots, \mathbf{a}_k)$. An $n \times n$ matrix is called a *squared matrix*. An $n \times k$ matrix $\mathbf{A}$ will be denoted also as

$$\mathbf{A} = (a_{ij}, i = 1, \ldots, n, j = 1, \ldots, k)$$

If $\mathbf{A}$ and $\mathbf{B}$ are matrices of the same order, then $\mathbf{A} \pm \mathbf{B} = \mathbf{C}$, where

$$c_{ij} = a_{ij} \pm b_{ij} \qquad i = 1, \ldots, n; \quad j = 1, \ldots, k$$

An $n \times n$ matrix $\mathbf{A}$, such that

$$a_{ij} = \begin{cases} a_i & \text{if } i = j \\ 0 & \text{if } i \neq j \end{cases}$$

is called a *diagonal matrix*. The diagonal matrix $\mathbf{I}_n$, such that its diagonal elements are all equal to 1, is called the *identity matrix* of order n. The vector $\mathbf{1}_n$ will denote an n-dimensional vector of 1s. The matrix $\mathbf{J}_n$ will denote an $n \times n$ matrix whose elements are all equal to 1. Notice that $\mathbf{1}'_n \mathbf{1}_n = n$.

The product of $\mathbf{A}$ and $\mathbf{B}$ is defined if the number of columns of $\mathbf{A}$ is equal to the number of rows of $\mathbf{B}$. The product of an $n \times k$ matrix $\mathbf{A}$ by an $k \times l$ matrix $\mathbf{B}$ is an $n \times l$ matrix C such that

$$c_{ij} = \sum_{m=1}^{k} a_{im} b_{mj} \qquad i = 1, \ldots, n \quad j = 1, \ldots, l$$

Notice that two squared matrices of the same order can always be multiplied. Also

$$\mathbf{J}_n = \mathbf{1}_n \mathbf{1}'_n$$

We illustrate these operations by using MINITAB.

In MINITAB, matrices are denoted by $M1$, $M2$, and so on. Calculations with matrices can be found in the menu, under "Calc." Matrices can be read from a text file or from the keyboard. For example, to read in the matrix

$$\mathbf{A} = \begin{bmatrix} 13 & 7 & 25 \\ -4 & 15 & 5 \\ 6 & -2 & 20 \end{bmatrix}$$

we use the command

```
MTB> Read 3 3 M1.
DATA> 13 7 25
DATA> −4 15 5
DATA> 6 −2 20
```

The data are entered one row at a time. We can also copy several columns of equal size into a matrix—for example,

```
MTB> copy C1-C3 M5.
```

To obtain the transpose of **A** we write

```
MTB> transpose M1 M2
```

$\mathbf{A}'$ resides in $M2$. We can now compute $\mathbf{B} = \mathbf{A}'\mathbf{A}$ by

```
MTB> multiply M2 M1 M3.
```

The matrix B is

$$\mathbf{B} = \begin{bmatrix} 221 & 19 & 425 \\ 19 & 278 & 210 \\ 425 & 210 & 1050 \end{bmatrix}$$

Notice that the matrix **B** is symmetric—that is, $b_{ij} = b_{ji}$ for all $i \neq j$.

Suppose that **A** is an $n \times k$ matrix, and $k \leq n$. The rank of **A** is the maximal number of mutually independent column vectors of **A**.

An $n \times n$ matrix of rank n is called **nonsingular**. Every nonsingular matrix **A** has an inverse $\mathbf{A}^{-1}$ satisfying

$$\mathbf{A}\mathbf{A}^{-1} = \mathbf{A}^{-1}\mathbf{A} = \mathbf{I}$$

To obtain the inverse of a matrix using MINITAB, we write

```
MTB> inverse M3 M4
```

The inverse of the preceding matrix **B** is

$$\mathbf{B}^{-1} = \begin{bmatrix} 0.03277 & 0.00916 & -0.01510 \\ 0.00916 & 0.00680 & -0.00507 \\ -0.01510 & -0.00507 & 0.00808 \end{bmatrix}$$

A squared matrix **H** is called *orthogonal* if $\mathbf{H}^{-1} = \mathbf{H}'$. An example of orthogonal matrix is

$$\mathbf{H} = \begin{bmatrix} \dfrac{1}{\sqrt{3}} & -\dfrac{1}{\sqrt{2}} & \dfrac{1}{\sqrt{6}} \\[2ex] \dfrac{1}{\sqrt{3}} & 0 & -\dfrac{2}{\sqrt{6}} \\[2ex] \dfrac{1}{\sqrt{3}} & \dfrac{1}{\sqrt{2}} & \dfrac{1}{\sqrt{6}} \end{bmatrix}$$

Use MINITAB to verify that indeed $\mathbf{H}^{-1} = \mathbf{H}'$.

For every real symmetric matrix $\mathbf{A}$, there exists an orthogonal matrix $\mathbf{H}$ such that

$$\mathbf{HAH}' = \boldsymbol{\Lambda}$$

where $\boldsymbol{\Lambda}$ is a diagonal matrix. The elements λ_{ii} $(i = 1, \dots, n)$ of $\boldsymbol{\Lambda}$ are called the *eigenvalues* of $\mathbf{A}$. Corresponding to the eigenvalue λ_{ii}, there is an *eigenvector* $\mathbf{x}_i$ satisfying the linear equations

$$\mathbf{Ax}_i = \lambda_{ii}\mathbf{x}_i \qquad (i = 1, \dots, n)$$

The eigenvalues and vectors of a symmetric matrix are obtained by the command

```
MTB> eigen M3 C1 M8
```

Here, the eigenvalues of $M3$ are stored in $C1$ and the eigenvectors are put in the matrix $M8$. The eigenvalues of $\mathbf{B}$ are thus

1269.98,   255.73,   23.29

and the matrix of eigenvectors is

$$\mathbf{P} = \begin{bmatrix} 0.371082 & 0.334716 & -0.866177 \\ 0.199177 & -0.939762 & -0.277847 \\ 0.907001 & 0.069367 & 0.415377 \end{bmatrix}$$

Notice that $\mathbf{P}$ is an orthogonal matrix.

The rank of a *symmetric* matrix $\mathbf{A}$ is equal to the number of eigenvalues different from zero. This is one way of checking whether a symmetric matrix is nonsingular. A symmetric matrix $\mathbf{Q}$ is called *positive definite* if all its eigenvalues are positive. If all the eigenvalues are nonnegative, the symmetric matrix is said to be *nonnegative definite*.

A symmetric matrix $\mathbf{A}$ is called *idempotent* if $\mathbf{A}^2 = \mathbf{A}$. If $\mathbf{A}$ is idempotent, then all its eigenvalues are either 1 or 0. Indeed, because $\mathbf{A}$ is symmetric, there exists an orthogonal matrix $\mathbf{P}$ such that

$$\mathbf{PAP}' = \boldsymbol{\Lambda} = \mathbf{PA}^2\mathbf{P}' = \mathbf{PAP}'\mathbf{PAP}' = \boldsymbol{\Lambda}^2$$

Thus, $\lambda_{ii} = \lambda_{ii}^2$ for every $i = 1, \dots, n$. This implies that $\lambda_{ii} = 0, 1$ for all $i = 1, \dots, n$. The matrix $\mathbf{A} = \mathbf{I}_n - (1/n)\mathbf{J}_n$ is an idempotent symmetric matrix, where n is the order of $\mathbf{A}$. $(n-1)$ eigenvalues of $\mathbf{A}$ are equal to 1 and one eigenvalue is equal to 0.

Let $\mathbf{A}$ be an $n \times n$ symmetric matrix, and let $\mathbf{x}$ be an $n \times 1$ vector. The scalar

$$Q = \mathbf{x}'\mathbf{A}\mathbf{x} = \sum_{i=1}^{n}\sum_{j=1}^{n} a_{ij}x_i x_j$$

is called a *symmetric quadratic form*. If $\mathbf{A}$ is also nonnegative definite, then $Q = \mathbf{x}'\mathbf{A}\mathbf{x} \geq 0$. Indeed, let $\mathbf{H}$ be an orthogonal matrix such that $\mathbf{HAH'} = \mathbf{\Lambda}$. Let $\mathbf{y} = \mathbf{Hx}$. Then $\mathbf{x} = \mathbf{H'y}$ and

$$Q = \mathbf{x}'\mathbf{A}\mathbf{x} = \mathbf{y}'\mathbf{HAH'y}$$
$$= \mathbf{y}'\mathbf{\Lambda}\mathbf{y} = \sum_{i=1}^{n} \lambda_{ii}y_i^2 \geq 0$$

because $\lambda_{ii} \geq 0$ for all $i = 1, \ldots, n$. Consider, for example, a sample of n values $x_1, \ldots, x_n$. Let $\mathbf{x}$ be an $n \times 1$ vector consisting of these values. Let $\overline{X}_n$ be the sample mean: $\overline{X}_n = (1/n)\mathbf{1}'_n\mathbf{x}$. The sum of squares of deviations from $\overline{X}_n$ can be written as

$$Q = \sum_{i=1}^{n}(X_i - \overline{X})^2 = \mathbf{x}'\mathbf{A}\mathbf{x}$$

where $\mathbf{A}$ is the idempotent matrix $\mathbf{I}_n - (1/n)\mathbf{J}_n$. This is a nonnegative definite quadratic form. Because rank $(\mathbf{A}) = n - 1$, we say that Q has $n - 1$ degrees of freedom.

Another important result is: *If* $X_1, \ldots, X_n$ *are i.i.d. with a* $N(0, \sigma)$ *distribution, then* $\mathbf{x}'\mathbf{A}\mathbf{x}$ *is distributed like* $\sigma^2 \chi^2[\nu]$ *if and only if* $\mathbf{A}$ *is idempotent, and* $rank(\mathbf{A}) = \nu$. Thus, in random samples from a normal distribution, $Q = \sum_{i=1}^{n}(X_i - \overline{X}_n)^2$ is distributed like $\sigma^2 \chi^2[n - 1]$.

Simulators

$$\parallel$$

The Piston Simulator The computer programs

TURB1.EXE	TURB4.EXE	TURB21.EXE	TURB31.EXE
TURB2.EXE	TURB5.EXE	TURB22.EXE	TURB32.EXE
TURB3.EXE	TURBEXP.EXE	TURB23.EXE	TURB33.EXE

are all designed around a core simulator of a piston moving within a cylinder (see Figure 10.1). The piston's linear motion is transformed into circular motion by connecting a linear rod to a disk. The faster the piston moves inside the cylinder, the quicker the disk rotation and therefore the faster the engine will run. The piston's performance is measured by the time it takes to complete one cycle, in seconds. Several factors can affect the piston's performance. They are—together with their symbols and measurement units—as follows:

M = The impact pressure determined by the piston weight (kg)

S = The piston surface area (m^2)

V_0 = The initial volume of gas inside the piston (m^3)

P = The gas pressure inside the piston (N/m^3)

k = The spring coefficient (N/m)

V = The volume of gas inside the piston after it moves x from its initial position x_0 (m^3)

P_0 = The atmospheric pressure (N/m^2)

T = The surrounding ambient temperature (K)

T_0 = The filling gas temperature (K)

We measure the time to complete one cycle as

Y = The piston cycle time (sec)

The forces on the piston cause the piston to move. Initially, the piston is positioned at X_0 such that

$$Mg = kX_0 \qquad g \sim 9.81$$

After moving X, the gas volume becomes V

$$S(X_0 - X) = V_0 - V$$

so that

$$X = X_0 + \frac{V - V_0}{S}$$

Now

$$PS - kX - Mg = P_0 S$$

$$PS - k\left(X_0 + \frac{V - V_0}{S}\right) - Mg = P_0 S$$

Hence,

$$\frac{P_0 V_0}{z} - kz = A$$

where $z = V/S$ and $A = P_0 S + 2Mg - (kV_0/S)$, or

$$P_0 V_0 - kz^2 = Az$$

Thus, the volume of gas at equilibrium, V, satisfies the equation

$$\frac{V}{S} = \frac{-A + \sqrt{A^2 + 4kP_0 V_0}}{2k}$$

The cycle time, y, can be computed as

$$y = 2\pi \sqrt{\frac{M}{K}}$$

where $K = k + S^2 (P_0 V_0 / T_0)(T/V^2)$ so that

$$V = \frac{S}{2k}\left(\sqrt{A^2 + 4k\frac{P_0 V_0}{T_0}T} - A\right)$$

$$A = P_0 S + 2Mg - \frac{kV_0}{S}$$

and

$$y = 2\pi \sqrt{\frac{M}{k + S^2 \dfrac{P_0 V_0}{T_0}\dfrac{T}{V^2}}}$$

which can be expressed as a function of seven parameters:

$$y = f(P_0, S, M, k, V_0, T, T_0)$$

The maximum and minimum values of these parameters are listed in the following table. These seven parameters can be measured with varying degrees of precision.

The measurement precision can be assessed in terms of standard deviations. The actual numbers used in the simulator are as follows:

Parameter	Precision
Piston weight, M	0.1 kg
Piston surface area, S	0.01 m^2
Initial gas volume, V_0	0.0005 m^3
Internal gas pressure, P	not measured
Spring coefficient, k	50 N/m
Internal gas volume, V	not measured
Atmospheric pressure, P_0	0.01 N/m^2
Ambient temperature, T	0.3 K
Filling gas temperature, T_0	0.3 K

Randomness in cycle times is induced by measurement errors in the seven control parameters.

The Power Circuit Simulator The computer programs

>   ELECT1.EXE
>
>   ELECT2.EXE
>
>   ELECT3.EXE

are all designed around a core simulator of a voltage conversion power circuit. The simulator was motivated by an experiment performed in 1974 at the Skin Nippon Denki Company as it was reported and analyzed by Taguchi in his book on systems of experimental design (Taguchi, 1987).

The target output voltage of the power circuit y is 220 volts DC. The input voltage is 100 volts AC. The circuit consists of 10 resistances, labeled A to J, and 3 transistors, labeled K to M. These components can be purchased at low grades, with 10% tolerances, or at more expensive high grades, with 5% tolerances. Quoting Taguchi, "The problem is how to go about efficiently calculating how much y changes as a result of changes in $A, B, \ldots$" (1987, p. 381).

The output voltage of the circuit is

$$y = \frac{136.67\left(a + \dfrac{b}{Z(10)}\right) + d(c+e)\dfrac{g}{f} - h}{1 + d\dfrac{e}{f} + b\left[\dfrac{1}{Z(10)} + .006\left(1 + \dfrac{13.67}{Z(10)}\right)\right] + .08202a}$$

where

$$a = \frac{Z(2)}{Z(1) + Z(2)}$$

$$b = \frac{1}{Z(12)Z(13)}\left(Z(3) + \frac{Z(1)Z(2)}{Z(1) + Z(2)}\right) + Z(9)$$

$$c = Z(5) + Z(7)/2$$

$$d = Z(11)\frac{Z(1)Z(2)}{Z(1) + Z(2)}$$

$$e = Z(6) + Z(7)/2$$

$$f = (c + e)(1 + Z(11))Z(8) + ce$$

$$g = 0.6 + Z(8)$$

$$h = 1.2$$

$Z(1), \ldots, Z(10)$ are resistances (in ohms) of the 10 transistors, and $Z(11)$, $Z(12)$, $Z(13)$ are the h_{FE} values of the three transistors. These parameters are given in data file PARELCT.DAT.

Information on
Data Files and Software

1   The software provided with this book should be installed to your hard disk. We recommend you keep the subdirectory names and structures as provided on the CD-ROM; however, only C:\ISTAT\DATA is absolutely required in that format.

2   Throughout the book, ASCII data files, BASIC executable files, and MINITAB macros are invoked. The BASIC programs have been compiled with QuickBASIC Version 4.5, and in order to run, they need files with both the .EXE and .OBJ extensions. MINITAB macros have an .MTB extension. Appendix VI contains S-PLUS procedures that are equivalent to the BASIC programs and MINITAB macros used in this book. These procedures are stored as text files in subdirectory C:\ISTAT\SPLUS. In order to use these procedures from within S-PLUS, you have to assign them with the source command. For example, boot.mean ← source (file = "C:\\ISTAT\\SPLUS\\ BOOTMEAN.TXT").

3   The BASIC programs are executable files with an .EXE extension. BASIC programs get inputs either from the keyboard or from files stored in subdirectory C:\ISTAT\DATA. The BASIC programs produce outputs that are directed to the monitor and to output files stored in subdirectory C:\ISTAT\DATA.

4   The data files are stored as ASCII files in C:\ISTAT\DATA with a .DAT extension. ASCII files cannot be read directly by MINITAB and have to be imported.

5   The data files are also stored in a format directly accessible by MINITAB. These copies carry the same name as the ASCII data file except for their extensions. Subdirectory C:\ISTAT\MTW contains MINITAB retrievable files with the .MTW extension.

6 MINITAB cannot read files with labels on the first line. When reading an ASCII file, MINITAB requires you to know ahead of time the number of columns in the file.

7 Executable BASIC programs and MINITAB macros are stored in subdirectories with names corresponding to the chapters where they are used.

8 ACCEPT.EXE, in subdirectory CHA9, is a program that provides three types of acceptance sampling designs. These are: single-stage sampling for attributes, double-stage sampling for attributes, and sequential sampling for attributes.

IV

Detailed Information on Data Files

Alphabetical list of 70 data files:

C:\ISTAT\DATA

ALMPIN.DAT
ALMPIN2.DAT
ANOVTEST.DAT
AVAILDIS.DAT
BOOT1SMP.DAT
BOOT2SMP.DAT
BOOTCORR.DAT
CAR.DAT
CENEXP.DAT
COAL.DAT
CONTACT.DAT
CUSARLN1.DAT
CUSARLN2.DAT
CUSARLN3.DAT
CUSARLN4.DAT
CUSARLN.DAT
CYCLT.DAT
DESCOEF.DAT
DESMAT.DAT
DESMATQ.DAT
DISTRUNS.DAT
DOJO1935.DAT
DOW1941.DAT

EFFECTS.DAT
ELECFAIL.DAT
ELECINDEX.DAT
FILMSP.DAT
FRACF1.DAT
FRACF2.DAT
FRACF3.DAT
FRACFAC.DAT
FULFAC.DAT
GASOL.DAT
GASOL2.DAT
HADPAS.DAT
HYBRID.DAT
HYBRID1.DAT
HYBRID2.DAT
HYPERG.DAT
MPG.DAT
OELECT.DAT
OELECT1.DAT
OELECT2.DAT
OELECT3.DAT
OTURB.DAT
OTURB1.DAT
OTURB2.DAT
OTURB3.DAT
OTURB31.DAT

OTURB32.DAT

OTURB33.DAT

OTURB4.DAT

OTURB5.DAT

OTURBEXP.DAT

PARELCT.DAT

PLACE.DAT

PRED.DAT

RENEWDIS.DAT

SAMPLE1.DAT

SAMPLE2.DAT

SHROPERN.DAT

SMLEWEIB.DAT

SOCELL.DAT

SOLDEF.DAT

STEELROD.DAT

STRESS.DAT

STUDENT.DAT

TSQ.DAT

TURBEXP.DAT

VARTEST.DAT

VENDOR.DAT

WEIBULL.DAT

YARNSTRG.DAT

C:\ISTAT\MTW

ALMPIN.MTW

ALMPIN2.MTW

CAR.MTW

COAL.MTW

CYCLT.MTW

DOJO1935.MTW

ELECINDX.MTW

FILMSP.MTW

GASOL.MTW

GASOL2.MTW

HADPAS.MTW

HYBRID.MTW

HYBRID1.MTW

HYBRID2.MTW

OELECT.MTW

OELECT1.MTW

OELECT2.MTW

OELECT3.MTW

OTURB.MTW

OTURB1.MTW

OTURB2.MTW

OTURB3.MTW

OTURB31.MTW

OTURB32.MTW

OTURB33.MTW

OTURB4.MTW

OTURB5.MTW

OTURBEXP.MTW

PLACE.MTW

PRED.MTW

SOCELL.MTW

SOLDEF.MTW

STEELROD.MTW

STRESS.MTW

TURBEXP.MTW

VENDOR.MTW

YARNSTRG.MTW

List of files, by chapter:

Chapter 2

CYCLT.DAT

STEELROD.DAT

YARNSTRG.DAT

OELECT.DAT

Chapter 3

PLACE.DAT

ALMPIN.DAT

HADPAS.DAT

SOCELL.DAT

CAR.DAT

Chapter 4

HYPERG.DAT

Short description of data files:

ALMPIN.DAT   Records of 6 dimension variables measured in mm on 70 aluminium pins used in airplanes, in order of production.

CAR.DAT   Records on 109 different car models, including number of cylinders, origin, turn diameter, horsepower, and number of miles per gallon in city driving

COAL.DAT   Number of coal mine disasters in England, per year, from 1850 to 1961 (111 years)

CYCLT.DAT   50 cycle times (in sec) of a piston operating at fixed operating conditions set at the minimal levels of seven control factors

DOJOI935.DAT   The Dow-Jones financial index for the 300 business days of 1935

ELECTINDEX.DAT   A binary transformation of OELEC.DAT

FILMSP.DAT   Film speed of 217 rolls of film

GASOL.DAT   32 measurements of distillation properties of crude oils. Five variables are measured: crude oil gravity, crude oil vapor pressure, crude oil ASTM (American Society for Testing and Materials) 10% point, gasoline ASTM endpoint, and yield.

GASOL2.DAT   A subset of GASOL.DAT

HADPAS.DAT   Records in ohms of 5 resistances on 192 hybrid microcircuits. The hybrids are manufactured on ceramic substrates containing 6 hybrids each.

HYBRID.DAT   A subset of HADPAS.DAT

HYBRID1.DAT A subset of HYBRID2.DAT

HYBRID2.DAT A subset of HYBRID.DAT

OELECT.DAT 99 measurements (in volts) of a rectifying circuit

OTURB.DAT 100 cycle times of a piston

OTURB1.DAT 50 cycle times of a piston operating at the maximal level of 7 control factors. This file will be changed by running TURB1.EXE at different levels of the control factors.

OTURB2.DAT 20 sample means and standard deviations of piston cycle times (in sec) from subgroups (samples) of size 5.

OTURB5.DAT Averages over 5 replicas from a two-factorial experiment with the piston simulator.

PLACE.DAT Position errors of 416 electronic components on 26 printed circuit boards. Sixteen components are placed on each board and, in addition to the board number, each observation consists of position errors on the x-axis and y-axis and the orientation errors.

PRED.DAT Records on the number of soldering points and number of defective soldering points on 1000 printed circuit boards

RENEWDIS.DAT 100 renewal times per cycle. Output of program RENEWDIS.EXE

SAMPLE1.DAT 50 failure times from $G(2, 100)$

SAMPLE2.DAT 50 repair times

SMLEWEIB.DAT Results of parametric bootstrap of Weibull MLE

SOCELL.DAT Short circuit current (ISC) of 16 solar cells measured at three time epochs, one month apart

SOLDEF.DAT Solder defects on 380 printed circuit boards of varying size

STEELROD.DAT 50 measurements (in cm) of the length of steel rods produced by a process adjusted to produce rods of length 20 cm

STRESS.DAT Results from a 33-factorial experiment on stress levels of a membrane.

TURBEXP.DAT Results from a 33-factorial experiment with 5 replicas with the piston simulator.

VENDOR.DAT Number of cycles required until latch failure in 30 floppy disk drives from three different disk vendors.

YARNSTRG.DAT 100 values of $y = \ln(x)$, where x is the yarn-strength (in lb 22 yarns) of woolen fibers

File Name	Number of rows	Number of columns	Labels in first row	File Name	Number of rows	Number of columns	Labels in first row
ALMPIN.DAT	70	7	Y	HYBRID1.DAT	32	1	N
ALMPIN2.DAT	70	2	N	HYBRID2.DAT	32	2	N
ANOVTEST.DAT	ns	1	Y	HYPERG.DAT	n	3	N
AVAILDIS.DAT	ns	1	N	MPG	58	3	N
BOOT1SMP.DAT	ns	2	N	OELECT.DAT	99	1	N
BOOT2SMP.DAT	ns	2	Y	OELECT1.DAT	n+1	1	Y
BOOTCORR.DAT	ns	1	N	OELECT2.DAT	ns+1	3	Y
CAR.DAT	109	5	N	OELECT3.DAT	33	2	Y
CENEXP.DAT	ns	1	N	OTURB.DAT	100	1	N
COAL.DAT	111	1	N	OTURB1.DAT	n	1	Y
CONTACT.DAT	100	1	N	OTURB2.DAT	n	3	Y
CUSARLN.DAT	ns	1	N	OTURB3.DAT	128	9	N
CUSARLN1.DAT	ns	1	N	OTURB31.DAT	17	9	Y
CUSARLN2.DAT	ns	1	N	OTURB32.DAT	33	9	Y
CUSARLN3.DAT	ns	1	N	OTURB33.DAT	65	9	Y
CUSARLN4.DAT	ns	1	N	OTURB4.DAT	28	6	N
CYCLT.DAT	50	1	N	OTURB5.DAT	32	2	N
DESCOEF.DAT	32	5	N	OTURBEXP.DAT	45	3	N
DESMAT.DAT	32	13	N	PARELECT.DAT	13	2 (,)	N
DESMATQ.DAT	28	4	N	PLACE.DAT	416	4	N
DISTRUNS.DAT	ns	3	N	PRED.DAT	1000	2	N
DOJO1935.DAT	300	1	N	RENEWDIS.DAT	100	1	N
DOW1941.DAT	302	1	N	SAMPLE1.DAT	50	1	N
EFFECTS.DAT	14	2	Y	SAMPLE2.DAT	50	1	N
ELECFAIL.DAT	50	1	N	SHROPREN.DAT	10	1	N
ELECTINDEX.DAT	99	1	N	SMLEWEIB.DAT	100	2	N
FILMSP.DAT	217	1	N	SOCELL.DAT	16	3	N
FRACF1.DAT	16	7	N	SOLDEF.DAT	380	1	N
FRACF2.DAT	32	7	N	STEELROD.DAT	50	1	N
FRACF3.DAT	64	7	N	STRESS.DAT	27	4	N
FRACFAC.DAT	8	4	N	STUDENT.DAT	n	1	Y
FULFAC.DAT	128	7(,)	N	TSQ.DAT	368	1	N
GASOL.DAT	32	5	N	TURBEXP.DAT	45	3	N
GASOL2.DAT	32	2	N	VARTEST.DAT	n	1	N
HADPAS.DAT	193	8	Y	VENDOR.DAT	10	3	N
HYBRID.DAT	32	3	N	YARNSTRG.DAT	100	1	N

V

Detailed Information on BASIC Executable Files and MINITAB Macros

C:\ISTAT\CHA4

   HYPERG.EXE

C:\ISTAT\CHA7

   ANOVTEST.EXE
   BINOPRED.MTB
   BOOT1SMP.EXE
   BOOT1SMP.MTB
   BOOT2SMP.EXE
   BOOTCORR.EXE
   BOOTPERC.MTB
   BOOTREGR.MTB
   BOOTREGR.MTB
   CONFINT.MTB
   CONTPRED.MTB
   RANDTEST.EXE
   STUDTEST.EXE
   VARTEST.EXE

C:\ISTAT\CHA9

   ACCEPT.EXE

C:\ISTAT\CHA10

   ELECT1.EXE

ELECT2.EXE
TURB1.EXE
TURB2.EXE
TURB21.EXE
TURB22.EXE
TURB23.EXE

C:\ISTAT\CHA11

   CUSARLB.EXE
   CUSARLN.EXE
   CUSARLP.EXE
   CUSPERB.EXE
   CUSPERN.EXE
   CUSPERP.EXE
   DISTRUNS.EXE
   SHROPERP.EXE
   SHROPERN.EXE
   SHROARLP.EXE
   SHROARLN.EXE

C:\ISTAT\CHA12

   ANOVTEST.EXE

GENEDSN.EXE
RANDTES2.MTB
RANDTES3.MTB
RPCOMP.MTB
TURB3.EXE
TURB31.EXE
TURB32.EXE
TURB33.EXE
TURB4.EXE
TURB5.EXE
TURBEXP.EXE

C:\ISTAT\CHA13

   ELECT3.EXE

C:\ISTAT\CHA14

   ARRSAMP.EXE
   AVAILDIS.EXE
   CENEXP.EXE
   REL.EXE
   RENEWDIS.EXE
   SMLEWEIB.EXE

Program Name	Inputs	Outputs
ACCEPT.EXE	Keyboard	Screen + Printer
ANOVTEST.EXE	$n \times k$ data file, ns	ANOVTEST.DAT
BINOPRED.MTB	$C1, k7 = n, k1 =$ future, $k3 \ k4 =$ tollev. $k8 = LTT, k9 = UTI$	
BOOT1SMP.EXE	$n1$ data file, ns	BOOT1SMP.DAT
BOOT1SMP.MTB	$C1 =$ data	$k1 =$ mean, $k2 =$ std
BOOT2SMP.EXE	$n1 \times n2$ data file, ns	BOOT2SMP.DAT
BOOTCORR.EXE	$n1 \times 2$ data file, ns	BOOTCORR.DAT
BOOTPERC.MTB	$C1 =$ data	$k1 = Q1, k2 = Q2, k3 = Q3$
BOOTREGR.MTB	$C1 + X, C4 =$ res., $k1 = a, k2 = b$	$k7 = b, k8 = a$
CONFINT.MTB	$C1 =$ data, $k1 = n$	$k2 = LCI, k3 = UCI$
CONTPRED.MTB	$C1 =$ data	$k1 = LTI, k2 = UTI$
CUSARLB.EXE	ns, n, teta, $K+, h+, K-, h-$	Screen
CUSARLN.EXE	ns, amu, std, $K+, h+, K-, h-$	Screen
CUSARLP.EXE	ns, lamda, $K+, h+, K-, h-$	Screen
CUSPERB.EXE	ns, n, teta, teta2, $K+, h+, K-, h-$, tau	Screen
CUSPERN.EXE	ns, amu, amu2, $K+, h+, K-, h-$, tau	Screen
CUSPERP.EXE	ns, lamda, lamda2, $K+, h+, K-, h-$, tau	Screen
ELECT1.EXE	PARELCT.DAT, n	OELECT1.DAT
ELECT2.EXE	PARELCT.DAT, n, ns	OELECT2.DAT
ELECT3.EXE	PARELCT.DAT, DESMAT.DAT, n	OELECT3 + EFFECTS
GENEDSN.EXE	n, generators and block	FRACFAC.DAT
HYPERG.EXE	N, M, n	HYPERG.DAT
RANDTEST.MTB	3×10 data file	$k2 = R$
RPCOMP.MTB	$1 \times n$ data file	$k2 = D$
STUDTEST.EXE	n1, mu0, ns	STUDTEST.DAT
TURB1.EXE	7 parameters, n	OTURB1.DAT
TURB2.EXE	7 parameters, n, ns	OTURB2.DAT
TURB21.EXE	7 parameters, n, ns	OTURB21.DAT
TURB22.EXE	7 parameters, n, ns	OTURB22.DAT
TURB23.EXE	7 parameters, n, ns	OTURB23.DAT
TURB3.EXE	n, FULFAC.DAT	OTURB3.DAT
TURB31.EXE	n, FRAC1.DAT	OTURB31.DAT
TURB32.EXE	n, FRAC2.DAT	OTURB32.DAT
TURB33.EXE	n, FRAC3.DAT	OTURB33.DAT
TURB4.EXE	n, DESMATQ.DAT	OTURB4.DAT
TURB5.EXE	n, DESCOEF.DAT	OTURB5.DAT
TURBEXP.EXE	5 parameters	OTURBEXP.DAT
VARTEST.EXE	$k \times ni$ data files, ns	VARTEST.DAT

VI

S-PLUS Procedures

1 The S-PLUS functions that follow are stored in C:\ISTAT\SPLUS with a .txt extension. In order to incorporate them into the S-PLUS environment, use the source function. For example, trimcorr ← source (file = "C:\\ISTAT\\SPLUS\\trimcorr.text").

2 Open the graphic window using:

win.graph( )

Minimize the graphics window. Resize the windows so that both the commands window and the graphics icon are visible.

3 To list the procedures in your S-PLUS environment, type:

objects.summary( )

Chapter 2

Section 2.3.1

To draw a random sample of size n, with or without replacement, from a data object x:

$y \leftarrow$ sample(x, n, replace $= F/T$)

Example: $n = 10$ numbers are drawn at random from the set $\{0, \dots, 99)$ without replacement,

$y \leftarrow$ sample($c(0 : 99)$, 10, replace $= F$)

To obtain the frequency distribution of the sample values use table(y)

Section 2.4

To import (load) a data set in ASCII, use

$y \leftarrow$ scan("c:\\istat\\data\\filename.dat")

Section 2.4.2

To plot the histogram of a data set y, use

hist(y, k)

k is the number of bins. The default is $[\log(n)]$.

Section 2.4.3

To describe a data set y, use

summary(y)

The size of the data set is obtained by

$n \leftarrow$ length(y)

Section 2.4.4

The mean and variance of a sample in y are obtained by the S-PLUS functions

xbar $\leftarrow$ mean(y)

svar $\leftarrow$ var(y)

Because the standard deviation is an important sample statistic, we construct the function

```
std ←
function(x)
{
    sqrt(var(x))
}
```

To compute the standard deviation of y, we can use std(y)

The skewness, kurtosis, and interquartile range of a sample can be computed by the functions:

```
skewness ←
function(x)
{
    m3 <- mean((x - mean(x))^3)
    skw <- m3/(std(x)^3)
    skw
}
```

```
kurtosis ←
function(x)
{
    m4 <- mean((x - mean(x))^4)
    kurt <- m4/(var(x)^2)
    kurt
}
```

```
iqr ←
function(x)
{
    n <- length(x)
    q1 <- (n + 1)/4
    q3 <- 3 * q1
    xs <- sort(x)
    d <- xs[floor(q3)] - xs[floor(q1)]
    d
}
```

Section 2.6.1

The box and whiskers plot of data in x is obtained by the S-PLUS function

boxplot(x)

The outlier is marked with a line. The width of the box is proportional to sqrt(n).

Section 2.6.2

The quantile plot has the S-PLUS function

qqplot(x, y)

This will give a plotting of the quantiles of a y sample against those of an x sample. In the quantile plot discussed in Section 2.6.2, the quantiles of x are plotted against an array y that consists of the proportions $(1/(n + 1), \ldots, n/(n + 1))$. We can obtain this by the commands

$y \leftarrow c(1 : n)/(n + 1)$

qqplot(x, y)

For such quantile plotting, we have the function

```
sqplot ←
function(x)
{
    n <- length(x)
    y <- c(1:n)/(n + 1)
    qqplot(x,y)
}
```

The following function yields four plots on one screen. These are the histogram, boxplot, density estimated function, and qqnormal plotting.

```
eda.shape ←
function(x)
{
    oldpars <- par(mfrow = c(2, 2))
    hist(x,25)
    boxplot(x)
    iqd <- summary(x)[5] - summary(x)[2]
    plot(density(x, width = 2 * iqd),
```

```
        xlab = "x", ylab = " ", type
          = "1")
   qqnorm(x)
   qqline(x)
}
```

Section 2.6.3

The stem-and-leaf diagram of sample data in x can be obtained by

stem(x)

We can control the display by using the available options [see help (stem)]. For example, use

stem(oelect)

and

stem(oelect, scale $= -1$,

head $= T$, depth $= T$)

Section 2.6.4

We give here three functions for robust statistics. These are the α-trimmed mean, α-trimmed std, and an estimate of the standard deviation based on the IQR.

```
trimmean ←
function(x,alfa)
{
     n <- length(x)
     i1 <- floor((n * alfa)/2) + 1
     i2 <- floor(n * (1 - alfa/2))
     xs <- sort(x)
     xst <- xs[i1:i2]
     mean(xst)
}
```

```
trimstd ←
function(x, alfa)
{
     n <- length(x)
     i1 <- floor((n * alfa)/2) + 1
```

```
     i2 <- floor(n * (1 - alfa/2))
     xs <- sort(x)
     xst <- xs[i1:i2]
     std(xst)
}
```

```
staniqr ←
function(x)
{
     iqr(x)/1.349
}
```

Section 2.8

Following are a few clues for solving the exercises with S-PLUS.

2.1.2 One can generate 50 numbers from a uniform distribution on (-10,10) by the command

runif(50, -10, 10)

After the command plot(x, y) write

abline(5, 2.5)

This will draw on the scatterplot the line with intercept 5 and slope 2.5.

2.1.3 Use the command

rbinom(50, 1, .9)

2.4.4 To read the file CAR.DAT, use the command

auto ← as.matrix(read.table(

"c:\\istat\\data\\car.dat", header $= F$))

Object auto is now a matrix of 109 rows and 5 columns. To check the dimensionality of the matrix, use the command dim(auto). You can now define the objects

auto.nucycl ← auto[,1]

auto.origin ← auto[,2]

etc.

Chapter 3

Section 3.1

Side-by-side boxplots are obtained by the command

boxplot(x, y)

Here y is the vector of group labels or numbers—for example,

boxplot(x.dev,boardn)

plots the box and whiskers of x.dev in data file PLACE.DAT by their board number. boardn is the first column of PLACE.DAT, x.dev is its second column, and so on.

The scatterplot of y vs x can be conditioned on a third variable, z, by the command

coplot(y $x|z$)

Multiple scatterplots are obtained by the command

pairs(x)

where x is a data matrix.

Section 3.2.2

Three-dimensional spinning and brushing can be done by

brush(x)

Section 3.3.1

To obtain a bivariate frequency distribution of two continuous variables in data vectors x, y, we have to code the data into k categories. k does not have to be the same for x and y. This coding can be obtained by the function code(x,catg,xc), which follows. The argument catg is a vector consisting of the right-hand limits of the intervals of x. For example, if we wish to code x into 4 intervals

$$\{(x < 100), (100 \leq x < 200),$$
$$(200 \leq x < 300), (300 \leq x)\}$$

we use

catg $\leftarrow c(100, 200, 300)$

The vector xc can be intialized as $xc \leftarrow c(1:n)$, where n is the length of x. The code function is:

```
code ←
function(x, y, xc)
{
    jm <- length(y)
    for(i in 1:length(x)) {
        if(x[i] < y[1])
        xc[i]< − 1
    }
    for(i in 1:length(x)) {
    for(j in 2:jm) {
        if(x[i] >= y[j - 1] &&
            x[i] < y[j]
            )
            c[i] <- j
        }
        if(x[i] >= y[jm])
        xc[i] <- jm + 1
    }
    xc
}
```

In a similar manner, y is coded and the bivariate frequency distribution over the specified intervals can then be obtained by the function table(xc, xy).

Example: To obtain the bivariate frequency distribution of x.dev and y.dev of data set PLACE.DAT, we first multiply these vectors by 10^5 and then we use the following commands:

cx.dev $\leftarrow$ as.matrix(x.dev $* 10^5$, nrow $= 416$)

cy.dev $\leftarrow$ as.matrix(y.dev $* 10^5$, nrow $= 416$)

catg $\leftarrow c(-200, -100, 0, 100, 200, 300, 400)$

$cxs \leftarrow$ cx.dev

$cys \leftarrow$ cy.dev

$a \leftarrow$ code(cx.dev,catg, cxs)

$b \leftarrow$ code(cy.dev,catg, cys)

table(a, b)

Section 3.4.1

The covariance matrix of a data matrix x is obtained by var(x) and the correlation matrix by cor(x). We can use the function

cor(x, trim $= \alpha$)

to obtain the α trimming of the data vectors and to avoid the influence of extremes.

Section 3.4.2

The linear model $y = \alpha + \beta * x + \epsilon$ is written in S-PLUS by the formula $y \sim x$. To obtain the least-squares estimates of the regression line, use:

fm$< - lm(y \sim x)$

summary(*fm*)

fmres or resid(fm) will give the vector of residuals around the regression. fmfit or fitted(fm) will give the vector of fitted values.

To plot the scatter of x, y and the regression line, use:

plot(x, y)

abline(*fm*)

simple regression of y on x can be performed by the method of least-squares estimation also by using the function

my.lse(x, y)

x and y should be $n \times 1$ vectors. The resulting object is a list of the LSE coefficients and the standard deviation around the regression line SE.

```
my.lse ←
function(x,y)
{
    n <- length(x)
    x <- cbind(jay(n,1),x)
    s <- t(x) %*% x
    b <- inv(s) %*% t(x) %*% y
    yhat <- x %*% b
    vyx <- sum((y - yhat)^2)/(n-2)
    ser <- sqrt(vyx)
    res <- list(b,ser)
    names(res) <- c("LSE","SE")
    res
}
```

Section 3.4.2.2

Prediction limits for a regression line can be plotted by using the function

```
predint ←
function(x, y, alfa)
{
    n <- length(x)
    cf <- qt(1 - alfa/2, n - 2)
    lse <- my.lse(x, y)
    a <- lse$LSE[1]
    b <- lse$LSE[2]
    xc <- as.matrix(x[, 2], nrow =
        length(y))
    yh <- a + b * xc
    ste <- lse$SE
    upper <- yh + cf * ste * sqrt(1 + 1/
        n + ((xc - mean(xc))^2)/sum((
        xc - mean(xc))^2))
    lower <- yh - cf * ste * sqrt(1 + 1/
        n + ((xc - mean(xc))^2)/sum((
        x - mean(xc))^2))
    plot(xc, y, main =
        "Linear Regression With
        Prediction Lines"
        )
    abline(a, b)
    lines(xc, upper)
    lines(xc, lower)
}
```

x should be an $n \times 1$ vector and y should be an $n \times 1$ vector.

Section 3.4.2.3

The robust correlation $r*$ can be computed by the function

```
trimcorr ←
function(x, y, alfa)
{
    z1 <- (x - trim.mean(x,alfa))/
        trim.std(x,alfa)
    z1 <- z1 + (y - trim.mean(y,alfa))/
        trim.std(y,alfa)
```

```
        z2 <- (x - trim.mean(x,alfa))/
        trim.std(x,alfa)
        z2 <- z2 - (y - trim.mean(y,alfa))/
        trim.std(y, alfa)
        v1 <- (trim.std(z1,alfa))^2
        v2 <- (trim.std(z2,alfa))^2
        rs <- (v1 - v2)/(v1 + v2)
        rs
   }
```

The S-PLUS function $\text{cor}(x,\text{trim}=\alpha)$ yields different results from the preceding.

To compute the Spearman rank-order correlation, use

$$\text{cor}(\text{rank}(x), \text{rank}(y))$$

Section 3.5.2.1

For computing the chi-square statistic of a contingency table, use

$$\text{chisq.test}(x)$$

The object x has the contingency table as a matrix.

Chapter 4

Section 4.2

A step function can be plotted by the command

$$\text{plot}(\text{stepfun}(x, y), \text{type} = \text{``}l\text{''})$$

Section 4.3.1

The c.d.f. of the binomial distribution is given by

$$\text{pbinom}(j, n, p)$$

Thus, to compute all values of $B(j; n, p)$, use

$$\text{pbinom}(c(0:n), n, p)$$

The quantiles are given by

$$\text{qbinom}(p, n, p)$$

To obtain the .1, .25, .5, .75 and .9th quantiles of the binomial, use

$$\text{qbinom}(c(.1, .25, .5, .75, .9), n, p)$$

Section 4.3.2

The hypergeometric distribution $H(N,M,n)$ is computed by

$$\text{dhyper}(j, M, N - M, n)$$
$$\text{phyper}(j, M, N - M, n)$$
$$\text{qhyper}(j, M, N - M, n)$$

dhyper is the p.d.f.

Section 4.3.3

The Poisson distribution function:

$$\text{dpois}(j, \text{lambda})$$
$$\text{ppois}(j, \text{lambda})$$
$$\text{qpois}(p, \text{lambda})$$

Section 4.3.4

The negative binomial distribution:

$$\text{dnbinom}(j, k, p)$$
$$\text{pnbinom}(j, k, p)$$
$$\text{qnbinom}(p, k, p)$$

Section 4.4.1

The uniform distribution:

$$\text{dunif}(x, a, b)$$
$$\text{punif}(x, a, b)$$
$$\text{qunif}(p, a, b)$$

for $a \le x \le b, 0 < p < 1$.

Section 4.4.2

The normal distribution:

$$\text{dnorm}(x, \text{mu}, \text{sigma})$$
$$\text{pnorm}(x, \text{mu}, \text{sigma})$$
$$\text{qnorm}(p, \text{mu}, \text{sigma})$$

Mu and sigma are the mean and standard deviation of the distribution.

The log-normal distribution:

$$\text{dlnorm}(x, \text{mu}, \text{sigma})$$

Section 4.4.3

The exponential distribution:

$\text{dexp}(x, \lambda)$

$\text{pexp}(x, \lambda)$

$\text{qexp}(p, \lambda)$

$\lambda = 1/\text{mean}$.

Section 4.4.4

The gamma distribution with scale parameter $\beta = 1$ and shape parameter vu:

$\text{dgamma}(x, vu)$

$\text{pgamma}(x, vu)$

$\text{qgamma}(p, vu)$

The Weibull distribution with scale parameter $\beta = 1$ and shape parameter vu:

$\text{dweibull}(x, vu)$

$\text{pweibull}(x, vu)$

$\text{qweibull}(p, vu)$

The gamma function is $\Gamma(vu)$.

Section 4.4.5

The beta distribution:

$\text{dbeta}(x, v1, v2)$

$\text{pbeta}(x, v1, v2)$

$\text{qbeta}(p, v1, v2)$

Chapter 5

The following functions resample from a data vector with or without replacement. Here x is a data vector, n is the size of x. We distinguish between:

1. $\text{sample}(n)$ selects a random permutation of $\{1, 2, \ldots, n\}$
2. $\text{sample}(x)$ randomly permutes x
3. $\text{sample}(x, \text{replace} = T)$ yields a sample from x with replacement (a bootstrap sample)
4. $\text{sample}(x, n)$ samples n elements from x without replacement

5. $\text{sample}(x, n, \text{replace} = T)$ samples n from x with replacement

Section 5.1.2

The function for obtaining m means of RSWR of size n from x is

```
smean ←
function(x, n, m)
{
amean ← c(1 : m)
for(i in 1 : m)
{
amean[i] ← mean(sample(x, n,replace= T)
}
amean
}
```

Section 5.7

For Exercise 5.2.4, use the function $\text{sample.corr}(x, y, n, m)$ that follows: x, y should be vectors.

```
sample.corr ←
function(x,y,n,m)
{
    ns<- length(x)
    acorr<- c(1:m)
    for(i in 1:m){
        a<- sample(c(1:ns),n,replace = T)
        ax <- x[a, ]
        ay <- y[a, ]
        acorr[i] <- cor(ax,ay)
    }
    acorr
}
```

Chapter 6

Section 6.3

1. Confidence interval for μ, normal distribution, std unknown

```
contint.normal ←
function(x,al)
{
        n <- length(x)
        cf <- qt(1 - al/2, n - 1)
        upper <- mean(x) + (cf * std(x))/sqrt(n)
        lower <- mean(x) - (cf * std(x))/sqrt(n)
        conint <- c(lower, upper)
        conint
}
```

2 Confidence interval for variance, normal distribution

```
conint.sigmsq ←
function(x,alfa)
{
        n <- length(x)
        cf1 <- qchisq(alfa/2, n - 1)
        cf2 <- qchisq(1 - alfa/2, n - 1)
        lower <- ((n - 1) * var(x))/cf2
        upper <- ((n - 1) * var(x))/cf1
        conint.sigmsq <- c(lower, upper)
        conint.sigmsq
}
```

3 Confidence interval for the parameter p of the binomial distribution

```
conint.bino ←
function(x, n,alfa)
{
        df1 <- 2 * (x + 1)
        df2 <- 2 * (n - x)
        cf1 <- qf(1 - alfa/2, df2 + 2, df1 -
            2)
        cf2 <- qf(1 - alfa/2, df1, df2)
        lower <- x/(x + cf1 * (n - x + 1))
        upper <- (cf2 * (x + 1))/(n - x +
            cf2 * (x + 1))
        conint.bino <- c(lower, upper)
        conint.bino
}
```

4 Confidence interval for the mean of a Poisson distribution; x is a sample of n observations

```
conint.pois ←
function(x, alfa)
{
        n <- length(x)
        ft <- 2 * sum(x) + 2
        upper <- qchisq(1 - alfa/2, ft)/(2 *
            n)
        lower <- qchisq(alfa/2, ft)/(2 * n)
        conint.pois <- c(lower, upper)
        conint.pois
}
```

Section 6.4

Tolerance intervals $(1 - \alpha, 1 - \beta)$ for normal samples

```
tolint.norm ←
function(x,alfa,beta)
{
        n <- length(x)
        za <- qnorm(1 - alfa/2, 0, 1)
        zb <- qnorm(1 - beta/2, 0, 1)
        tab <- zb/(1 - za^2/(2 * n))
        tab <- tab + (za * sqrt(1 + zb^2/2 -
            za^2/(2 * n)))/(sqrt(n) * (1 -
            za^2/(2 * n)))
        lower <- mean(x) - tab * std(x)
        upper <- mean(x) + tab * std(x)
        tolint.norm <- c(lower, upper)
        tolint.norm
}
```

Section 6.5

The *t*-test is done by the S-PLUS function t.test(x, options). The main features are:

1 t.test(x) Two-sided, one-sample test. The null hypothesis is that the mean of x is 0.

2 t.test(x, y,alternative="less",paired= T)
One-sided, paired *t*-test. The null hypothesis is $\delta = 0$.

3 t.test(x, y, δ) Two-sided standard two-sample *t*-test, comparing x and y. The null hypothesis is that the mean difference = delta.

4 t.test(x, y,var.equal= F,con.level= $1 - \alpha$) Two-sided Welch-modified two-sample test. The null hypothesis is that the means of x and y are the same.

Section 6.6

Normal quantile plotting:

qqnorm(x)

qqline(x)

Section 6.7.1

The chi-square test for goodness of fit for the normal distribution:

```
chi.nfit ←
function(x, catg)
{
    n <- length(x)
    k <- length(catg)
    mu <- mean(x)
    sig <- std(x)
    ax <- x
    ax <- code(x, catg, ax)
    frey <- as.matrix(table(ax), nrow =
        length(catg))
    exp <- c(1:k)
    exp[1] <- n * pnorm(catg[1], mu, sig
        )
    for(i in 2:k) {
        exp[i] <- n * (pnorm(catg[i],
        mu, sig) - pnorm(
        catg[i - 1], mu, sig
        ))
    }
    exp <- as.matrix(exp, nrow = k)
    chi <- sum(frey^2/exp) - n
    pval <- 1 - pchisq(chi, k)
```

```
    res <- list(chi, pval)
    names(res) <- c("CHISQ", "P-value")
    res
}
```

x is the data vector, catg is a vector of upper limits of class intervals. To apply this function, you have to check first that table( ) gives nonzero frequencies to all categories of catg. If this is not the case, change the number of categories and their limits in catg.

Chapter 7

Section 7.2

We give here several functions to create EBD of sample means, standard deviations, medians, first and third quartiles, correlations, and regression coefficients.

1

```
boot.mean ←
function(x, m)
{
    n <- length(x)
    amean <- c(1:m)
    for(i in 1:m) {
        amean[i] <- mean(sample(x, n, replace
            = T))
    }
    amean
}
```

2

```
boot.median ←
function(x, m)
{
    n <- length(x)
    amedian <- c(1:m)
    for(i in 1:m) {
```

```
            amedian[i] <- median(sample(
            x, n, replace = T))
        }
        amedian
}
```

3

```
boot.std <-
function(x, m)
        n <- length(x)
        astd <- c(1:m)
        for(i in 1:m) {
                astd[i] <- std(sample(x, n,
                replace = T))
        }
        astd
}
```

4

```
boot.quartiles <-
function(x, m)
{
        n <- length(x)
        aq1 <- c(1:m)
        aq3 <- c(1:m)
        for(i in 1:m) {
                sx <- sample(x, n, replace
                = T)
        aq1[i] <- quantile(sx, 0.25)
        aq3[i] <- quantile(sx, 0.75)
        }
        res <- cbind(aq1, aq3)
        res
}
```

5

```
boot.corr <-
function(x, y, m)
{
        n <- length(x)
```

```
        acorr <- c(1:m)
        ns <- length(x)
        for(i in 1:m) {
                a <- sample(c(1:ns), n,
                replace = T)
                ax <- x[a, ]
                ay <- y[a, ]
                acorr[i] <- cor(ax, ay)
        }
        acorr
}
```

6

```
boot.regr <-
function(x, y, m)
{
        n <- length(x)
        b0 <- c(1:m)
        b1 <- c(1:m)
        ser <- c(1:m)
        for(i in 1:m) {
                a <- sample(c(1:n), n,
                replace = T)
                sx <- x[a, ]
                sy <- y[a, ]
                b1[i] <- (cor(sx, sy) * std(
                sy))/std(sx)
                b0[i] <- mean(sy) - b1[i] *
                mean(sx)
                ser[i] <- std(sy) * sqrt(1 -
                cor(sx, sy)^2)
        }
        res <- cbind(b0, b1, ser)
        res
}
```

Section 7.3

Bootstrap Test Procedures:
 1 Studentized test for the mean. The test is
    two-sided . The null hypothesis is that the
    mean is equal to the argument μ. m is the
    number of replicas. Use

```
boot.student ←
function(x,mu, m)
{
    n <- length(x)
    avg <- mean(x)
    sig <- std(x)
    tn <- ((avg - mu) * sqrt(n))/sig
    tsx <- c(1:m)
    for(i in 1:m) {
        xs <- sample(x, n, replace
            = T)
        avgs <- mean(xs)
        sigs <- std(xs)
        tsx[i] <- ((avgs - avg) *
            sqrt(n))/sigs
    }
    prop <- 0 * jay(m, 1)
    for(i in 1:m) {
        if(abs(tsx[i]) >= abs(tn))
            prop[i] <- 1
    }
    pvalue <- sum(prop)/m
    res <- list(avg, sig, tn, pvalue)
    names(res) <- c("S.Mean", "S.STD",
        "t-value", "P-Value")
    res
}
```

```
    da <- a1 - a2
    tsts <- c(1:m)
    for(i in 1:m) {
        xs <- sample(x, n1, replace
            = T)
        ys <- sample(y, n2, replace
            = T)
        as1 <- mean(xs)
        as2 <- mean(ys)
        vs1 <- var(xs)
        vs2 <- var(ys)
        tsts[i] <- (as1 - as2 - da)/
            sqrt(vs1/n1 + vs2/n2
            )
    }
    prop <- 0 * jay(m, 1)
    for(i in 1:m) {
        if(abs(tsts[i]) >= abs(tst)
            )
            prop[i] <- 1
    }
    pvalue <- sum(prop)/m
    res <- list(da, tst, pvalue)
    names(res) <- c("Diff. Means",
        "t-Diff.", "P-Value")
    res
}
```

2 Studentized test of the difference between two sample means. x and y are the data vectors of two samples. m is the number of replicas. d stands for the specified difference between the population means, under the null hypothesis.

3 A test of significance of the difference between k sample variances. The data are in a matrix of $n \times k$, where k is the number of samples. All samples are of size n. The test is based on the EBD of max $S^2 / \min S^2$.

```
boot.2smp ←
function(x, y, d, m)
{
    n1 <- length(x)
    n2 <- length(y)
    a1 <- mean(x)
    a2 <- mean(y)
    sq1 <- var(x)
    sq2 <- var(y)
    tst <- (a1 - a2 - d)/sqrt(sq1/n1 +
        sq2/n2)
```

```
boot.vartes ←
function(x, m)
{
    ft <- max(diag(var(x)))/min(diag(var(
        x)))
    n <- nrow(x)
    k <- ncol(x)
    fts <- c(1:m)
    for(i in 1:m) {
        xs <- jay(n, k)
        for(j in 1:k) {
```

```
        xs[, j] <- sample(x[,
           j], n,
              replace = T)
      }
      fts[i] <- max(diag(var(xs)))/
           min(diag(var(xs)))
    }
    prop <- 0 * jay(m, 1)
    for(i in 1:m) {
      if(fts[i] >= ft)
          prop[i] <- 1
    }
    pvalue <- sum(prop)/m
    res <- list(ft, pvalue)
    names(res) <- c("FMAX", "P-Value")
    res
}
```

4 Bootstrap ANOVA. This is to test the significance of the difference between k sample means. x is an $n \times k$ data matrix consisting of the k samples. m is the number of replicas.

```
boot.anova ←
function(x, m)
{
    n <- nrow(x)
    k <- ncol(x)
    nk <- n * k
    av <- c(1:k)
    for(i in 1:k) {
        av[i] <- mean(x[, i])
    }
    avg <- mean(av)
    fst <- (n * var(av))/mean(diag(var(x
        )))
    fsts ← c(1:m)
    for(i in 1:m) {
        xs <- jay(n, k)
        avs <- c(1:k)
        for(j in 1:k) {
            xs[, j] <- sample(x[,
               j], n,
                  replace = T)
            avs[j] <- mean(xs[,
               j])
        }
    }
```

```
        avsg <- mean(avs)
        devs <- c(1:k)
        for(j in 1:k) {
            devs[j] <- n * (avs[
               j] - av[j])^2
        }
        fsts[i] <- (sum(devs) - nk * (
    avg - avsg)^2)/((k -
              1) * mean(diag(var(
              xs))))
    }
    prop <- 0 * jay(m, 1)
    for(i in 1:m) {
        if(fsts[i] >= fst)
            prop[i] <- 1
    }
    pvalue <- mean(prop)
    res <- list(fst, pvalue)
    names(res) <- c("F for ANOVA",
        "P-Value")
    res
}
```

Section 7.4

1 Binomial tolerance intervals:

```
boot.binopred ←
function(x,alfa,beta, m)
{
    n <- length(x)
    pred <- jay(m, 1)
    for(i in 1:m) {
        xs <- sample(x, n, replace
            = T)
        pred[i] <- mean(xs)
    }
    pred1 <- quantile(pred,alfa/2)
    pred2 <- quantile(pred, 1 - alfa/2)
    toler1 <- qbinom(beta/2, n, pred1)
    toler2 <- qbinom(1 - beta/2, n,
        pred2)
    res <- list(toler1, toler2)
    names(res) <- c("Lower", "Upper")
    res
}
```

2 Tolerance limits for continuous distributions:

```
boot.contpred ←
function(x,alfa,beta,m)
{
    n <- length(x)
    cint <- jay(m, 2)
    for(i in 1:m) {
        xs <- sample(x, n, replace
            = T)
        cint[i, 1] <- quantile(xs,
            beta/2)
        cint[i, 2] <- quantile(xs, 1 -
            beta/2)
    }
    lower <- quantile(cint[, 1],alfa/2)
    upper <- quantile(cint[, 2], 1 -
    alfa/2)
    res <- list(lower, upper)
    names(res) <- c("Lower", "Upper")
    res
}
```

Section 7.5.2

The randomization test for comparing the means of two random samples *x*, *y*:

```
rand.test ←
function(x, y, m)
{
    n1 <- length(x)
    n2 <- length(y)
    nc <- n1 + n2
    a1 <- c(1:n1)
    a2 <- c((n1 + 1):nc)
    diff <- mean(x) - mean(y)
    ax <- c(1:nc)
    indx <- c(1:nc)
    for(i in 1:n1) {
        ax[i] <- x[i]
    }
    for(i in (n1 + 1):nc) {
        ax[i] <- y[i - n1]
    }
    ds <- c(1:m)
```

```
    for(i in 1:m) {
        ac <- sample(indx, nc,
            replace = F)
        ac1 <- ac[1:n1]
        ac2 <- ac[(n1 + 1):nc]
        xs <- ax[ac1]
        ys <- ax[ac2]
        ds[i] <- mean(xs) - mean(ys)
    }
    prop <- 0 * jay(m, 1)
    for(i in 1:m) {
        if(abs(ds[i]) >= abs(diff))
            prop[i] <- 1
    }
    pvalue <- mean(prop)
    res <- list(diff, pvalue)
    names(res) <- c("DIFF", "P-Value")
    res
}
```

Section 7.5.3

The Wilcoxon test can be done by the S-PLUS function wilcox.test

Chapter 8

The multiple regression procedure is performed by using the S-PLUS functions lm( ) (linear models) or glm( ) (general linear model). The linear model is specified by the formula

$$y \sim x1 + x2 + \cdots + xk$$

If you wish to fit a polynomial model, the formula to use is

$$y \sim \text{poly}(x1, 2) + \text{poly}(x2, 2) + \cdots$$

It is desirable to save the list of output in an object—for example, in

regr< − lm();

Then writing

summary(regr)

yields all the important statistics of the regression.

The residuals of the regression can be obtained by regr$res, the fitted values by regr$fit.

Section 8.2

The partial correlation of y and $x1$, given $x2$, can be computed by the function:

```
part.cor ←
function(y, x1, x2)
{
      n <- length(y)
      xs2 <- cbind(jay(n, 1), x2)
      b1 <- inv(t(xs2) %*% xs2)
            %*% t(xs2) %*% y
      e1 <- y - xs2 %*% b1
      b2 <- inv(t(xs2) %*% xs2) %*%
            t(xs2) %*% x1
      e2 <- x1 - xs2 %*% b2
      r <- cor(e1, e2)
      r
}
```

Section 8.4

The function xleverage computes the x-leverage of each point:

```
xleverage ←
function(x)
{
      n <- nrow(x)
      sx <- cbind(jay(n, 1), x)
      hx <- sx %*% inv(t(sx) %*% sx) %*% t(sx)
      lev <- diag(hx)
      lev
}
```

The argument x is a matrix of k regressors.

The standardized residuals $e*$ are given by:

```
stdres ←
function(y, x)
{
      n <- length(y)
      xs <- cbind(jay(n, 1), x)
      b <- inv(t(xs) %*% xs) %*% t(xs) %*% y
```

```
      yhat <- xs %*% b
      err <- y - yhat
      ht <- xleverage(x)
      sg <- sqrt(sum((y - yhat)^2)/(n - ncol(xs)))
      sag <- err/(sg * sqrt(1 - ht))
      sag <- matrix(sag, ncol = 1)
      sag
}
```

The Cook distance function is:

```
cookdist ←
function(y, x)
{
      n <- length(y)
      k <- ncol(x)
      xs <- cbind(jay(n, 1), x)
      b <- inv(t(xs) %*% xs) %*% t(xs) %*% y
      yhat <- xs %*% b
      syx <- sqrt(sum((y - yhat)^2)/(n - k - 1))
      ds <- t(xs) %*% xs
      cookd <- c(1:n)
      a <- c(2:n)
      xs1 <- xs[a, ]
      y1 <- y[a, ] b1 <- inv(t(xs1) %*% xs1)
            %*% t(xs1) %*% y1
      cookd[1] <- t(b1 - b) %*% ds %*%
            (b1 - b)/(syx * (k + 1))
      for(i in 2:(n - 1)) {
            a <- c(1:i - 1, i + 1, n)
            xs1 <- xs[a, ]
            y1 <- y[a, ]
            b1 <- inv(t(xs1) %*% xs1)
                  %*% t(xs1) %*% y1
            cookd[i] <- t(b1 - b) %*% ds
                  %*% (b1 - b)/(syx * (k + 1))
      }
      a <- c(1:n - 1)
      xs1 <- xs[a, ]
      y1 <- y[a, ]
      b1 <- inv(t(xs1) %*% xs1) %*% t(xs1)
            %*% y1
      cookd[n] <- t(b1 - b) %*% ds %*%
            (b1 - b)/(syx * (k + 1))
      cookd
}
```

VII

Solutions

We present here short solutions to selected exercises.

Chapter 2

2.1.1 The following are the commands and the MINITAB output:

```
MTB> random 50 c1;
SUBC> integer 1 6.
MTB> table c1
Tabulated Statistics
    x      Count
    1       11
    2       10
    3        9
    4       10
    5        4
    6        6
   ALL      50
MTB>
```

The expected frequency in each cell, under randomness, is $50/6 = 8.3$. You will get different numerical results, due to randomness.

2.2.1 As shown in Figure A.2.2.1, the measurements on instrument 1, $\odot$, seem to be accurate but less precise than those on instrument 2, $*$. Instrument 2 seems to have an upward bias (inaccurate).

2.3.3

a $26^5 = 11,881,376$

b $7,893,600$

c $26^3 = 17,576$

d $2^{10} = 1,024$

e $\binom{10}{5} = 252$

2.4.3 The following are MINITAB commands and output. Column $C1$ contains the yearly number of disasters.

```
MTB> Read 'C:\ISTAT\DATA\COAL.DAT' c1.
Entering data from file:
C:\ISTAT\DATA\COAL.DAT
  111 rows read.
MTB> table c1
Tabulated Statistics
    x      Count
    0       33
    1       28
    2       18
    3       17
    4        6
    5        5
    6        3
    7        1
   ALL     111
```

2.4.5 $X_{(1)} = 66$; $Q_1 = 102$; $M_e = X_{(109)} = 105$; $Q_3 = 109$; $X_{(217)} = 118$; $x_{.8} = 110$; $x_{.9} = 111$; $x_{.99} = 115.64$.

2.4.7

Origin	Mean	Std
US (1)	20.931	3.598
Europe (2)	19.500	2.624
Asia (3)	23.108	4.280

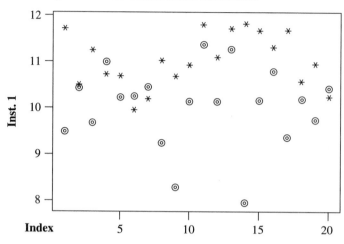

FIGURE A.2.2.1

Scatter diagram of measurements on two instruments

2.4.8   Coefficient of variation = 0.084

2.5.1 $\overline{X} = 104.59$; $S = 6.55$

Interval	Actual frequency	Pred. frequency
$\overline{X} \pm S$	173	147.56
$\overline{X} \pm 2S$	205	206.15
$\overline{X} \pm 3S$	213	216.35

The discrepancies between the actual frequencies and the predicted frequencies are due to the fact that the distribution of film speed is neither symmetric nor bell-shaped.

2.6.3 $\overline{T}_\alpha = 0.4056$; $S_\alpha = 0.0988$; $\alpha = .10$

2.6.6   Lower whisker starts at $\max(1587, 1511.7) = 1587 = X_{(1)}$; upper whisker ends at $\min(2427, 2440.5) = 2427 = X_{(n)}$. There are no outliers.

Chapter 3

3.1.3   Figures A.3.1.3a and A.3.1.3b present the multiple boxplots of Res3 by hybrid and a matrix plot of all Res variables. From Figure A.3.1.3a, we can learn that the conditional distributions of Res at different hybrids are different. Figure A.3.1.3b reveals that Res3 and Res7 are positively correlated. Res20 is generally larger than the corresponding Res14. Res18 and Res20 seem to be negatively associated.

3.3.2   The joint frequency distribution of Res3 and Res14 is given in the following table:

codres3	1	2	3	4	5	6	All
			codres14				
1	11	11	5	2	0	0	29
2	3	33	16	11	9	0	72
3	3	28	24	12	8	1	76
4	0	2	6	5	1	0	14
5	0	0	0	1	0	0	1
ALL	17	74	51	31	18	1	192

The intervals for Res3 start at 1580 and end at 2580 with a length of 200. The intervals of Res14 start at 900 and end at 2700 with a length of 300.

3.4.1

Data Set	Intercept (a)	Slope (b)	R^2
1	3.00	0.500	.667
2	2.97	0.509	.658
3	2.98	0.501	.667
4	3.02	0.499	.667

Notice the influence of the point (19, 12.5) on the regression in data set 4. Without this point, the correlation between x and y is zero.

3.4.3

a   The robust correlations r_{XY}^*, with trimming of 10% (5%

from each tail), are as follows:

	Turn D.	Horse P.	MPG
Turn D.	1	.560	−.599
Horse P.	.560	1	−.797
MPG	−.599	−.797	1

b The Spearman rank-order correlations are as follows:

	Turn D.	Horse P.	MPG
Turn D.	1	.551	−.549
Horse P.	.551	1	−.797
MPG	−.549	−.797	1

F I G U R E   A.3.1.3a
Multiple boxplot of Res3 by hybrid

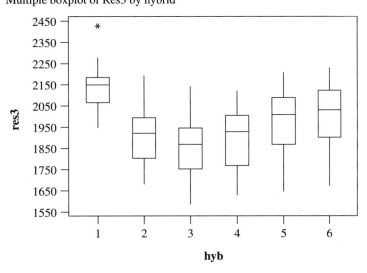

F I G U R E   A.3.1.3b
Matrix plot of res variables

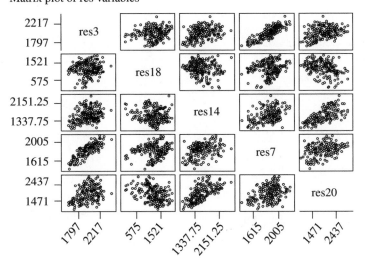

3.5.1   Because the frequencies in rows 1, 2 and columns 1, 2 of both tables are very small, we collapse the m into 3-×-3 tables.

(Question 3 × Question 1)

Q_3	Q_1			
	3	4	5	
3	12	6	1	19
4	13	23	13	49
5	2	15	100	117
	27	44	114	185

For this table, we obtain

$X^2 = 96.624$ (chi-square)

$\Phi^2 = 0.5223$ (mean-squared contingency)

$T = 0.3616$ (Tschuprow's index)

$C = 0.5110$ (Cramers index)

(Question 3 × Question 2)

$Q3$	$Q1$			
	3	4	5	
3	10	6	1	17
4	12	7	5	24
5	1	30	134	165
	23	43	140	206

For this table,

$\chi^2 = 108.282$ (chi-square)

$\Phi^2 = 0.5256$ (mean-squared contingency)

$T = 0.3625$ (Tschuprow's index)

$C = 0.5126$ (Cramér's index)

Chapter 4

4.1.1

a $S = \{(w_1, \ldots, w_{20}); w_j = G, D, j = 1, \ldots, 20\}$

b $2^{20} = 1,048,576$

c $A_n = \{(w_1, \ldots, w_{20}) : \sum_{j=1}^{20} I\{w_j = G\} = n\}, \quad n = 0, \ldots,$ 20, where $I\{A\} = 1$ if A is true and $I\{A\} = 0$

otherwise. The number of elementary events in A_n is $\binom{20}{n} = [20!/n!(20 - n)!]$.

4.1.3

a $S = \{(i_1, \ldots, i_{30}) : i_j = 0, 1, j = 1, \ldots, 30\}$

b $A_{10} = \{(1, 1, \ldots, 1, i_{11}, i_{12}, \ldots, i_{30}) : i_j = 0, 1, j = 11, \ldots, 30\}$. $|A_{10}| = 2^{20} = 1,048,576$. [$|A_{10}|$ is the number of elements in A_{10}.]

c $B_{10} = \{(i_1, \ldots, i_{30}) : i_j = 0, 1 \text{ and } \sum_{j=1}^{30} i_j = 10\}$ $|B_{10}| = \binom{30}{10} = 30,045,015$. No, A_{10} and B_{10} have only one common element.

4.1.6 $\bigcup_{i=1}^{n} A_i = S$ and $A_i \cap A_j = \emptyset$ for all $i \neq j$.

$$B = B \cap S = B \cap \left(\bigcup_{i=1}^{n} A_i \right)$$

$$= \bigcup_{i=1}^{n} A_i B$$

4.1.8   We showed in Exercise 4.1.6 that $B = \bigcup_{i=1}^{n} A_i B$. Moreover, because $\{A_1, \ldots, A_n\}$ is a partition, $A_i B \cap A_j B = (A_i \cap A_j) \cap B = \emptyset \cap B = \emptyset$ for all $i \neq j$. Hence, from axiom 3:

$$\Pr\{B\} = \Pr\left\{ \bigcup_{i=1}^{n} A_i B \right\} = \sum_{i=1}^{n} \Pr\{A_i B\}$$

4.1.10 $\Pr\{150 < T < 280\} = 0.2258$

4.1.11 $\dfrac{\binom{10}{2}\binom{10}{2}\binom{15}{2}\binom{5}{2}}{\binom{40}{8}} = 0.02765$

4.1.13 $N = 1000, M = 900, n = 10$

i $\Pr\{X \geq 8\} = \sum_{j=8}^{10} \binom{10}{j}(.9)^j(.1)^{10-j}$

$= .9298$

ii $\Pr\{X \geq 8\} = \sum_{j=8}^{10} \dfrac{\binom{900}{j}\binom{100}{10-j}}{\binom{1000}{10}} = .9308$

4.1.15 $\Pr\{T > 300 \mid T > 200\} = .6065$

4.1.17   Because A and B are independent

$$\Pr\{A \cap B\} = \Pr\{A\}\Pr\{B\}$$
$$= (1 - \Pr\{A^c\})(1 - \Pr\{B^c\})$$
$$= 1 - \Pr\{A^c\} - \Pr\{B^c\} + \Pr\{A^c\}\Pr\{B^c\}$$

By DeMorgan's law,

$$(A \cap B)^c = A^c \cup B^c$$
$$\Pr\{(A \cap B)^c\} = \Pr\{A^c \cup B^c\}$$
$$= \Pr\{A^c\} + \Pr\{B^c\} - \Pr\{A^c \cap B^c\}$$

However,

$$Pr\{(A \cap B)^c\} = 1 - Pr\{A \cap B\}$$
$$= 1 - (1 - Pr\{A^c\} - Pr\{B^c\} + Pr\{A^c \cap B^c\})$$
$$= Pr\{A^c\} + Pr\{B^c\} - Pr\{A^c \cap B^c\}$$

From this and the preceding result, we get $Pr\{A^c \cap B^c\} = Pr\{A^c\}Pr\{B^c\}$. Hence, A^c and B^c are independent.

4.1.19
$$Pr\{A \cup B\} = Pr\{A\} + Pr\{B\} - Pr\{A \cap B\}$$
$$= Pr\{A\} + Pr\{B\} - Pr\{A\}Pr\{B\}$$
$$= Pr\{A\}(1 - Pr\{B\}) + Pr\{B\}$$
$$= Pr\{B\}(1 - Pr\{A\}) + Pr\{A\}$$

4.1.21 Let n be the number of people in the party. The probability that all their birthdays fall on different days is

$$Pr\{D_n\} = \prod_{j=1}^{n}\left(\frac{365 - j + 1}{365}\right)$$

i If $n = 10$, $Pr\{D_{10}\} = .8831$

ii If $n = 23$, $Pr\{D_{23}\} = .4927$ Thus, the probability of at least 2 persons with the same birthday when $n = 23$ is $1 - Pr\{D_{23}\} = .5073 > \frac{1}{2}$.

4.1.23 $\prod_{j=1}^{10}\left(1 - \frac{1}{24 - j + 1}\right) = .5833$

4.1.25 .0889

4.1.27 Without loss of generality, assume that the stick is of length 1. Let x, y, and $(1 - x - y)$, $0 < x, y < 1$, be the length of the three pieces. Obviously, $0 < x + y < 1$. All points in $S = \{(x, y) : x, y > 0, x + y < 1\}$ are uniformly distributed. For the three pieces to form a triangle, the following three conditions should be satisfied:

i $x + y > (1 - x - y)$

ii $x + (1 - x - y) > y$

iii $y + (1 - x - y) > x$

The set of points (x, y) satisfying (i), (ii), and (iii) is bounded by a triangle of area $\frac{1}{8}$. S is bounded by a triangle of area $\frac{1}{2}$. Hence, the required probability is $\frac{1}{4}$.

4.1.28 $Pr\{Hit\} = \frac{1}{\pi} \tan^{-1}\left(\frac{h}{a}\right)$

4.1.29 $1 - (.999)^{100} - 100 \times (.001) \times (.999)^{99} - \binom{100}{2} \times (.001)^2(.999)^{98} = 0.0001505$

4.2.2

a $\sum_{x=0}^{\infty}\frac{5^x}{x!} = e^5$. Hence, $\sum_{x=0}^{\infty}p(x) = 1$.

b $Pr\{X \le 1\} = e^{-5}(1 + 5) = .0404$

c $Pr\{X \le 7\} = .8666$

4.2.4
$$E\{X\} = \int_0^{\infty}(1 - F(x))dx$$
$$= \int_0^{\infty}e^{-x^2/2\sigma^2}dx$$
$$= \sigma\sqrt{\frac{\pi}{2}}$$

4.2.6 $\frac{15}{16}$

4.2.8
$$\mu_l^* = E\{(X - \mu_1)^l\}$$
$$= \sum_{j=0}^{l}(-1)^j\binom{l}{j}\mu_1^j\mu_{l-j}$$

When $j = l$, the term is $(-1)^l\mu_1^l$. When $j = l - 1$, the term is $(-1)^{l-1}l\mu_1^{l-1}\mu_1 = (-1)^{l-1}l\mu_1^l$. Thus, the sum of the last two terms is $(-1)^{l-1}(l - 1)\mu_1^l$. Hence,

$$\mu_l^* = \sum_{j=0}^{l-2}(-1)^j\binom{l}{j}\mu_1^j\mu_{l-j} + (-1)^{l-1}(l - 1)\mu_1^l$$

4.2.10 $M_X(t) = \frac{1}{t(b - a)}(e^{tb} - e^{ta})$, $-\infty < t < \infty$, $a < b$

$$E\{X\} = \frac{a + b}{2}$$
$$V\{X\} = \frac{(b - a)^2}{12}$$

4.3.1 Use the commands:

```
MTB> Set C1
DATA> 1(0:20/1)1
DATA> end.
MTB> pdf C1 C2;
SUBC> binomial 20 .17.
MTB> cdf C1 C3;
SUBC> binomial 20 .17.
```

The p.d.f. and c.d.f. values will be in columns $C2$ and $C3$.

4.3.3 $E\{X\} = 15.75$; $\sigma = 3.1996$

4.3.5 Notice first that

$$\lim_{\substack{n \to \infty \\ np \to \lambda}} b(0; n, p) = \lim_{n \to \infty}\left(1 - \frac{\lambda}{n}\right)^n = e^{-\lambda}$$

Second, for all $j = 0, 1, \ldots$,

$$\frac{b(j + 1; n, p)}{b(j; n, p)} = \frac{n - j}{j + 1} \cdot \frac{p}{1 - p}$$

Thus, by induction on j, for $j > 0$

$$\lim_{\substack{n \to \infty \\ np \to \lambda}} b(j; n, p) = \lim_{\substack{n \to \infty \\ np \to \lambda}} b(j - 1; n, p) \frac{n - j + 1}{j} \cdot \frac{p}{1 - p}$$

$$= e^{-\lambda} \frac{\lambda^{j-1}}{(j-1)!} \lim_{\substack{n \to \infty \\ np \to \lambda}} \frac{(n - j + 1)p}{j\left(1 - \frac{\lambda}{n}\right)}$$

$$= e^{-\lambda} \frac{\lambda^{j-1}}{(j-1)!} \cdot \frac{\lambda}{j} = e^{-\lambda} \frac{\lambda^j}{j!}$$

4.3.7 $\quad E\{X\} = 20 \cdot \dfrac{350}{500} = 14$

$V\{X\} = 20 \cdot \dfrac{350}{500} \cdot \dfrac{150}{500} \left(1 - \dfrac{19}{499}\right) = 4.0401$

4.3.9 $\quad \Pr\{R\} = 1 - H(3; 100, 10, 20) +$

$\displaystyle\sum_{i=1}^{3} h(i; 100, 10, 20)[1 - H(3 - i; 80, 10 - i, 40)] = .87395$

4.3.11 $\quad$ .1912

4.3.13 $\quad$ Geometric with parameter $p = .00038$.
$E\{N\} = 2631.6; \sigma_N = 2631.08$

4.3.15 $\quad$ The m.g.f. of NB (p, k) is

$$M(t) = \sum_{i=0}^{\infty} \binom{k + i - 1}{k - 1} p^k ((1 - p)e^t)^i$$

for $t < -\log(1 - p)$. Thus

$$M(t) = \frac{p^k}{(1 - (1 - p)e^t)^k}$$

$$\sum_{i=0}^{\infty} \binom{k + i - 1}{k - 1} (1 - (1 - p)e^t)^k ((1 - p)e^t)^i$$

Hence,

$$M(t) = \left[\frac{p}{1 - (1 - p)e^t}\right]^k \qquad t < -\log(1 - p)$$

4.3.17 $\quad$ If there are n chips, $n > 50$, the probability of at least 50 good ones is $1 - B(49; n, .998)$. Thus, n is the smallest integer > 50 for which $B(49; n, .998) < .05$. It is sufficient to order 51 chips.

4.4.1 $\quad$ For $0 < y < 1$,

$$\Pr\{F(X) \le y\} = \Pr\{X \le F^{-1}(y)\} = F(F^{-1}(y)) = y$$

Hence, the distribution of $F(X)$ is uniform on $(0, 1)$. Conversely, if U has a uniform distribution on $(0, 1)$, then

$$\Pr\{F^{-1}(U) \le x\} = \Pr\{U \le F(x)\} = F(x)$$

4.4.3 $\quad \Pr\{X \le x\} = \Pr\{-\log(U) \le x\} = \Pr\{U \ge e^{-x}\} = 1 - e^{-x}$

4.4.5 $\quad \mu + Z_{.9}\sigma = 15$
$\mu + Z_{.99}\sigma = 20$
$(Z_{.99} - Z_{.9})\sigma = 5$

Thus,

$\sigma = 4.785622$
$\mu = 8.866979$

4.4.7 $\quad \mu_d = 10.0658$ (mm)

4.4.9 $\quad \Pr\{X > 300\} = \Pr\{\log X > 5.7038\} = 1 - \Phi(.7038) = .24078$

4.4.11 $\quad$ The quantiles of $E(\beta)$ are $x_p = -\beta \log(1 - p)$. Thus, $Q_1 = 0.2877\beta$, $M_e = 0.6931\beta$, $Q_3 = 1.3863\beta$.

4.4.13 $\quad M(t) = \dfrac{1}{\beta} \displaystyle\int_0^{\infty} e^{tx - x/\beta} dx$

$$= \frac{1}{\beta} \int_0^{\infty} e^{-\frac{(1 - t\beta)}{\beta}x} dx$$

$$= (1 - t\beta)^{-1} \qquad \text{for } t < \frac{1}{\beta}$$

4.4.15 $\quad G(t; k, \lambda) = \dfrac{\lambda^k}{(k+1)!} \displaystyle\int_0^t x^{k-1} e^{-\lambda x} dx$

$$= \frac{\lambda^k}{k!} t^k e^{-\lambda t} + \frac{\lambda^{k+1}}{k!} \int_0^t x^k e^{-\lambda x} dx$$

$$= \frac{\lambda^k}{k!} t^k e^{-\lambda t} + \frac{\lambda^{k+1}}{(k+1)!} t^{k+1} e^{-\lambda t} +$$

$$\frac{\lambda^{k+2}}{(k+1)!} \int_0^{\infty} x^{k+1} e^{-\lambda x} dx = \cdots =$$

$$= e^{-\lambda t} \sum_{j=k}^{\infty} \frac{(\lambda t)^j}{j!}$$

$$= 1 - e^{-\lambda t} \sum_{j=0}^{k-1} \frac{(\lambda t)^j}{j!}$$

4.4.17 $\quad$ The moment-generating function of the sum of independent random variables is the product of their respective m.g.f.'s. Thus, if $X_1, \ldots, X_k$ are i.i.d. $E(\beta)$, using the result of Exercise 4.4.13, show that the moment-generating function of S is

$$M_S(t) = \prod_{i=1}^{k} (1 - \beta t)^{-1} = (1 - \beta t)^{-k} \qquad t < \frac{1}{\beta}$$

where $S = \displaystyle\sum_{i=1}^{k} X_i$. However, $(1 - \beta t)^{-k}$ is the m.g.f. of $G(k, \beta)$.

4.4.19 $\quad \Pr\{W(1.5, 500) \ge 600\} = \Pr\{W(1.5, 1) \ge \frac{6}{5}\} = .2686$.

4.4.21 $\quad X$ is distributed like beta(v, v). The first four moments are

$$\mu_1 = v/2v = \frac{1}{2}$$

$$\mu_2 = \frac{B(v + 2, v)}{B(v, v)} = \frac{v + 1}{2(2v + 1)}$$

$$\mu_3 = \frac{B(v+3, v)}{B(v, v)} = \frac{(v+1)(v+2)}{2(2v+1)(2v+2)}$$

$$\mu_4 = \frac{B(v+4, v)}{B(v, v)} = \frac{(v+1)(v+2)(v+3)}{2(2v+1)(2v+2)(2v+3)}$$

The variance is

$$\sigma^2 = 1/4(2v+1)$$

and the fourth central moment is

$$\mu_4^* = \mu_4 - 4\mu_3 \cdot \mu_1 + 6\mu_2 \cdot \mu_1^2 - 3\mu_1^4$$

$$= \frac{3}{16(3 + 8v + 4v^2)}$$

Finally, the index of kurtosis is

$$\beta_2 = \frac{\mu_4^*}{\sigma^4} = \frac{3(1+2v)}{3+2v}$$

4.5.2 The marginal p.d.f. of Y is $f(y) = e^{-y}$, $y > 0$—that is, Y is distributed like $E(1)$. The conditional p.d.f. of X, given $Y = y$, is $f(x \mid y) = \frac{1}{y}e^{-x/y}$. Thus, $E\{X \mid Y = y\} = y$, $E\{X\} = E\{Y\} = 1$

$$E\{XY\} = E\{YE\{X \mid Y\}\}$$
$$= E\{Y^2\} = 2$$

Hence, $\text{cov}(X, Y) = E\{XY\} - E\{X\}E\{Y\} = 1$. The variance of Y is $\sigma_Y^2 = 1$. The variance of X is

$$\sigma_X^2 = E\{V\{X \mid Y\}\} + V\{E\{X \mid Y\}\}$$
$$= E\{Y^2\} + V\{Y\} = 2 + 1 = 3$$

The correlation between X and Y is $\rho_{XY} = \frac{1}{\sqrt{3}}$.

4.5.4 $J \mid N$ is distributed like $B(N, p)$; N is distributed like $P(\lambda)$. $E\{N\} = \lambda$, $V\{N\} = \lambda$, $E\{J\} = \lambda p$.

$$V\{J\} = E\{V\{J \mid N\}\} + V\{E\{J \mid N\}\}$$
$$= E\{Np(1-p)\} + V\{Np\}$$
$$= \lambda p(1-p) + p^2 \lambda = \lambda p.$$

$$E\{JN\} = E\{NE\{J \mid N\}\} = pE\{N^2\}$$
$$= p(\lambda + \lambda^2)$$

$$\text{cov}(J, N) = p\lambda(1+\lambda) - p\lambda^2 = p\lambda$$

$$\rho_{JN} = \frac{p\lambda}{\lambda\sqrt{p}} = \sqrt{p}$$

4.6.1

a $B(20; 3500, .005) = .7699$

b Binomial $B\left(3485, \frac{.004}{.999}\right)$

c $\lambda = 3485 \times \frac{0.004}{0.999} = 13.954$, $\Pr\{J_2 \leq 15 \mid J_3 = 15\} \cong P(15; 13.954) = .6739$

4.6.3 $V\{Y \mid X\} = 150$, $\sigma_Y^2 = 200$, $V\{Y \mid X\} = \sigma_Y^2(1 - \rho_{XY}^2)$; $|\rho_{XY}| = 0.5$. The sign of ρ_{XY} cannot be determined.

4.7.2 J is distributed like $B(10, .95)$. If $\{J = j\}, j > 1, X_{(1)}$ is the minimum of a sample of j i.i.d. $E(10)$. Thus, $X_{(1)} \mid J = j$

is distributed like $E\left(\dfrac{10}{j}\right)$.

a $\Pr\{J = k, X_{(1)} \leq x\} = b(k; 10, .95)(1 - e^{-kx/10})$, $k = 1, 2, \ldots, 10$.

b $\Pr\{J \geq 1\} = 1 - (.05)^{10} = 1$

$$\Pr\{X_{(1)} \leq x \mid J \geq 1\}$$

$$= \sum_{k=1}^{10} b(k; 10, .95)(1 - e^{-kx/10})$$

$$= 1 - \sum_{k=1}^{10} \binom{10}{k}(.95e^{-x/10})^k(.05)^{10-k}$$

$$= 1 - [.05 + .95e^{-x/10}]^{10} + (.05)^{10}$$

$$= 1 - (.05 + .95e^{-x/10})^{10}$$

4.8.1 $T + \dfrac{1}{2}W = X + Y + \dfrac{1}{2}X - \dfrac{1}{2}Y$

$$= \frac{3}{2}X + \frac{1}{2}Y$$

$$V\left\{T + \frac{1}{2}W\right\} = \beta^2 \left(\left(\frac{3}{2}\right)^2 + \left(\frac{1}{2}\right)^2\right)$$

$$= 2.5\beta^2$$

4.8.3 $V\{\alpha X + \beta Y\} = \alpha^2 \sigma_X^2 + \beta^2 \sigma_Y^2 + 2\alpha\beta \, \text{cov}(X, Y)$
$$= \alpha^2 \sigma_X^2 + \beta^2 \sigma_Y^2 + 2\alpha\beta\rho_{XY}\sigma_X\sigma_Y$$

4.8.5 Let U_1, U_2, X be independent random variables; U_1, U_2 i.i.d. $N(0, 1)$. Then

$$\Phi^2(X) = \Pr\{U_1 \leq X, U_2 \leq X \mid X\}$$

Hence,

$$E\{\Phi^2(X)\} = \Pr\{U_1 \leq X, U_2 \leq X\}$$
$$= \Pr\{U_1 - X \leq 0, U_2 - X \leq 0\}$$

$(U_1 - X, U_2 - X)$ have a bivariate normal distribution with means $(-\mu, -\mu)$ and variance–covariance matrix

$$V = \begin{bmatrix} 1 + \sigma^2 & \sigma^2 \\ \sigma^2 & 1 + \sigma^2 \end{bmatrix}$$

Accordingly,

$$E\{\Phi^2(X)\} = \Phi_2\left(\frac{\mu}{\sqrt{1+\sigma^2}}, \frac{\mu}{\sqrt{1+\sigma^2}}; \frac{\sigma^2}{1+\sigma^2}\right)$$

4.8.7 Let $F_2(x) = \int_{-\infty}^{x} f_2(z)dz$ be the c.d.f. of X_2. Because X_1 and X_2 are independent,

$$\Pr\{Y \leq y\} = \int_{-\infty}^{\infty} f_1(x)\Pr\{X_2 \leq y - x\}dx$$

$$= \int_{-\infty}^{\infty} f_1(x)F_2(y - x)dx$$

Therefore, the p.d.f. of Y is

$$g(y) = \frac{d}{dy} \Pr\{Y \le y\}$$

$$= \int_{-\infty}^{\infty} f_1(x) f_2(y - x) dx$$

4.8.9 X_1, X_2 are i.i.d. $E(1)$ and $Y = X_1 - X_2$.

$$\Pr\{Y \le y\} = \int_0^{\infty} e^{-x} \Pr\{X_1 \le y + x\} dx$$

Notice that $-\infty < y < \infty$ and $\Pr\{X_1 \le y + x\} = 0$ if $x + y < 0$. Let $a^+ = \max(a, 0)$. Then

$$\Pr\{Y \le y\} = \int_0^{\infty} e^{-x}(1 - e^{-(y+x)^+}) dx$$

$$= 1 - \int_0^{\infty} e^{-x-(y+x)^+} dx$$

$$= \begin{cases} 1 - \frac{1}{2} e^{-y} & \text{if } y \ge 0 \\ \frac{1}{2} e^{-|y|} & \text{if } y < 0 \end{cases}$$

Thus, the p.d.f. of Y is

$$g(y) = \frac{1}{2} e^{-|y|} \quad -\infty < y < \infty$$

4.9.2 X is distributed like $B(200, .15)$; $\mu = np = 30$, $\sigma = \sqrt{np(1 - p)} = 5.0497$

$$\Pr\{25 < X < 35\} \doteq \Phi\left(\frac{34.5 - 30}{5.0497}\right) - \Phi\left(\frac{25.5 - 30}{5.0497}\right)$$

$$= .6271$$

4.9.4 X is distributed like beta$(3, 5)$; $\mu = E\{X\} = \frac{3}{8} = 0.375$

$$\sigma = \sqrt{V\{X\}} = \left(\frac{15}{576}\right)^{1/2} = 0.161374$$

$$\Pr\{|\overline{X}_{200} - .375| < .2282\} \cong 2\Phi\left(\frac{0.2282}{0.161374}\right) - 1$$

$$= .8427$$

4.10.1 $t_{.95}[10] = 1.81246$; $t_{.95}[15] = 1.7530$; $t_{.95}[20] = 1.7247$

4.10.3   By definition, $t[n]$ is distributed like $[N(0, 1)/\sqrt{(\chi^2[n])/n}]$, where $N(0, 1)$ and $\chi^2[n]$ are independent.

$t^2[n]$ is distributed like $\dfrac{(N(0, 1))^2}{\frac{\chi^2[n]}{n}}$ is distributed like $F[1, n]$

Thus, because

$$\Pr\{F[1, n] \le F_{1-\alpha}[1, n]\} = 1 - \alpha$$

$$\Pr\{-\sqrt{F_{1-\alpha}[1, n]} \le t[n] \le \sqrt{F_{1-\alpha}[1, n]}\} = 1 - \alpha$$

it follows that $\sqrt{F_{1-\alpha}[1, n]} = t_{1-\alpha/2}[n]$, or $F_{1-\alpha}[1, n] = t_{1-\alpha/2}^2[n]$.

4.10.5

$$V\{t[v]\} = V\left\{\frac{N(0, 1)}{\sqrt{\chi^2[v]/v}}\right\}$$

$$= E\left\{V\left\{\frac{N(0, 1)}{\sqrt{\frac{\chi^2[v]}{v}}} \,\middle|\, \chi^2[v]\right\}\right\} +$$

$$V\left\{E\left\{\frac{N(0, 1)}{\sqrt{\frac{\chi^2[v]}{v}}} \,\middle|\, \chi^2[v]\right\}\right\}$$

Because $N(0, 1)$ and $\chi^2[v]$ are independent,

$$V\left\{\frac{N(0, 1)}{\sqrt{\frac{\chi^2[v]}{v}}} \,\middle|\, \chi^2[v]\right\} = \frac{v}{\chi^2[v]}$$

and

$$E\left\{\frac{N(0, 1)}{\sqrt{\frac{\chi^2[v]}{v}}} \,\middle|\, \chi^2[v]\right\} = 0$$

Thus,

$$V\{t[v]\} = vE\left\{\frac{1}{\chi^2[v]}\right\}$$

Because $\chi^2[v]$ is distributed like $G\left(\frac{v}{2}, 2\right)$ is distributed like $2G\left(\frac{v}{2}, 1\right)$,

$$E\left\{\frac{1}{\chi^2[v]}\right\} = \frac{1}{2} \cdot \frac{1}{\Gamma\left(\frac{v}{2}\right)} \int_0^{\infty} x^{v-2} e^{-x} dx$$

$$= \frac{1}{2} \cdot \frac{\Gamma\left(\frac{v}{2} - 1\right)}{\Gamma\left(\frac{v}{2}\right)} = \frac{1}{2} \cdot \frac{1}{\frac{v}{2} - 1} = \frac{1}{v - 2}$$

Finally, $V\{t[v]\} = \dfrac{v}{v - 2}$, $v > 2$.

Chapter 5

5.2.1   Define the binary random variables: (r.v.'s)

$$I_{ij} = \begin{cases} 1 & \text{if } x_j \text{ is selected at the } i\text{th sampling} \\ 0 & \text{otherwise} \end{cases}$$

The r.v.'s $X_1, \ldots, X_n$ are given by

$$X_i = \sum_{j=1}^{N} x_j I_{ij} \quad i = 1, \ldots, n$$

Because sampling is RSWR,

$$\Pr\{X_i = x_j\} = \frac{1}{N} \quad \text{for all } i = 1, \ldots, n \; j = 1, \ldots, N$$

Hence, $\Pr\{X_i \le x\} = \hat{F}_N(x) \; \forall \; x$, and all $i = 1, \ldots, n$. Moreover, by definition of RSWR, the vectors $\mathbf{I}_i = (I_{i1}, \ldots, I_{iN})$, $i = 1, \ldots, n$ are mutually independent. Hence, $X_1, \ldots, X_n$ are i.i.d., having a common c.d.f. $\hat{F}_N(x)$.

5.2.3 By the CLT $(0 < \sigma_N^2 < \infty)$,

$$\Pr\{\sqrt{n}|\bar{X}_n - \mu_N| < \delta\} \cong 2\Phi\left(\frac{\delta}{\sigma_N}\right) - 1$$

as $n \to \infty$.

5.2.5 Use the following macro:

```
sample 50 C3 C9
let k1 = median(C9)
stack C6 k1 C6
sample 50 C4 C9
let k1 = median(C9)
stack C7 k1 C7
sample 50 C5 C9
let k1 = median(C9)
stack C8 k1 C8
end
```

Import the data into columns $C1$–$C5$. Then, fill $C6(1)$, $C7(1)$, $C8(1)$ by the medians of $C3$, $C4$, $C5$.

5.2.7 The required sample size is a solution of the equation

$$0.002 = 2 \cdot 1.96 \cdot \sqrt{\frac{P(1-P)}{n}\left(1 - \frac{n-1}{N}\right)}$$

The solution is $n = 1611$.

5.4.1 $L(n_1, \ldots, n_k; \lambda) = \sum_{i=1}^{k} W_i^2[\tilde{\sigma}_{N_i}^2/n_i] - \lambda\left(n - \sum_{i=1}^{k} n_i\right)$.

Partial differentiation of L with respect to $n_1, \ldots, n_k$ and λ yields the following equations:

$$\frac{W_i^2\tilde{\sigma}_{N_i}^2}{n_i^2} = \lambda \quad i = 1, \ldots, k$$

$$\sum_{i=1}^{k} n_i = n$$

Equivalently,

$$n_i = \frac{1}{\sqrt{\lambda}} W_i \tilde{\sigma}_{N_i} \quad i = 1, \ldots, k$$

$$n = \frac{1}{\sqrt{\lambda}} \sum_{i=1}^{k} W_i \tilde{\sigma}_{N_i}$$

Thus

$$n_i^0 = n \frac{W_i \tilde{\sigma}_{N_i}}{\sum_{j=1}^{k} W_j \tilde{\sigma}_{N_j}} \quad i = 1, \ldots, k$$

5.5.2 The model is

$$y_i = \beta_0 + \beta_1 x_i + e_i \quad i = 1, \ldots, N$$

$E\{e_i\} = 0$, $V\{e_i\} = \sigma^2 x_i$, $i = 1, \ldots, N$. Given a sample $\{(X_1, Y_1), \ldots, (X_n, Y_n)\}$, we estimate β_0 and β_1 by the weighted LSE, because the variances of y_i depend on x_i,

$i = 1, \ldots, N$. These weighted LSE are values $\hat{\beta}_0$ and $\hat{\beta}_1$, minimizing

$$Q = \sum_{i=1}^{n} \frac{1}{X_i}(Y_i - \beta_0 - \beta_1 X_i)^2$$

which are given by

$$\hat{\beta}_1 = \frac{\bar{Y}_n \cdot \frac{1}{n}\sum_{i=1}^{n}\frac{1}{X_i} - \frac{1}{n}\sum_{i=1}^{n}\frac{Y_i}{X_i}}{\bar{X}_n \cdot \frac{1}{n}\sum_{i=1}^{n}\frac{1}{X_i} - 1}$$

and

$$\hat{\beta}_0 = \frac{1}{\sum_{i=1}^{n}\frac{1}{x_i}}\left(\sum_{i=1}^{n}\frac{Y_i}{X_i} - n\hat{\beta}_i\right)$$

It is easily shown that $E\{\hat{\beta}_1\} = \beta_1$ and $E\{\hat{\beta}_0\} = \beta_0$. Thus, an unbiased predictor of μ_N is

$$\hat{\mu}_N = \hat{\beta}_0 + \hat{\beta}_1 \bar{x}_N$$

Chapter 6

6.1.2 424

6.2.1 $\hat{\xi}_p = \bar{X}_n + Z_p\hat{\sigma}_n$, where $\hat{\sigma}_n^2 = (1/n)\sum(X_i - \bar{X}_n)^2$

6.2.3 Let $M_1 = (1/n)\sum_{i=1}^{n}X_i$ and $M_2 = \frac{1}{n}\sum_{i=1}^{n}X_i^2$.

$\hat{\sigma}_n^2 = M_2 - M_1^2$. Then

$$\hat{v}_1 = M_1(M_1 - M_2)/\hat{\sigma}_n^2$$

$$\hat{v}_2 = (M_1 - M_2)(1 - M_1)/\hat{\sigma}_n^2$$

6.2.5 Because Y_i are uncorrelated,

$$V\{\hat{\beta}_1\} = \sum_{i=1}^{n} w_i^2 V\{Y_i\}$$

$$= \sigma^2 \sum_{i=1}^{n} \frac{(x_i - \bar{x}_n)^2}{SS_x^2}$$

$$= \frac{\sigma^2}{SS_x}$$

where $SS_x = \sum_{i=1}^{n}(x_i - \bar{x}_n)^2$.

6.2.6 $$V\{\hat{\beta}_0\} = \sigma^2\left(\frac{1}{n} + \frac{\bar{x}^2}{SS_x}\right)$$

$$\text{cov}(\hat{\beta}_0, \hat{\beta}_1) = -\sigma^2 \frac{\bar{x}_n}{SS_x}$$

6.2.7 $$\rho_{\hat{\beta}_0, \hat{\beta}_1} = -\frac{\sigma^2\bar{x}_n}{\sigma^2 SS_x\left[\left(\frac{1}{n} + \frac{\bar{x}_n^2}{SS_x}\right)\frac{1}{SS_x}\right]^{1/2}}$$

$$= -\frac{\bar{x}_n}{\left(\frac{1}{n}\sum_{i=1}^{n}x_i^2\right)^{1/2}}$$

6.2.9   Because v is known, the likelihood of β is

$$L(\beta) = C_n \frac{1}{\beta^{nv}} e^{-\sum_{i=1}^{n} X_i / \beta} \qquad 0 < \beta < \infty$$

where C_n does not depend on β. The log-likelihood function is

$$l(\beta) = \log C_n - nv \log \beta - \frac{1}{\beta} \sum_{i=1}^{n} X_i$$

The score function is

$$l'(\beta) = -\frac{nv}{\beta} + \frac{\sum_{i=1}^{n} X_i}{\beta^2}$$

Equating the score to 0 and solving for β, we obtain the MLE $\hat{\beta} = (1/nv) \sum_{i=1}^{n} X_i = (1/v) \overline{X}_n$. The variance of $\hat{\beta}$ is $\frac{\beta^2}{nv}$.

6.2.11

a   The likelihood function of μ and β is

$$L(\mu, \beta) = I\{X_{(1)} \geq \mu\} \frac{1}{\beta^n} \exp \left\{ -\frac{1}{\beta} \sum_{i=1}^{n} (X_{(i)} - X_{(1)}) \right.$$
$$\left. - \frac{n}{\beta} (X_{(1)} - \mu) \right\}$$

for $-\infty < \mu \leq X_{(1)}, 0 < \beta < \infty$. $L(\mu, \beta)$ is maximized by $\hat{\mu} = X_{(1)}$—that is,

$$L^*(\beta) = \sup_{\mu \leq X_{(1)}} L(\mu, \beta) = \frac{1}{\beta^n} \exp \left\{ -\frac{1}{\beta} \sum_{i=1}^{n} (X_{(i)} - X_{(1)}) \right\}$$

where $X_{(1)} < X_{(2)} < \cdots < X(n)$ are the ordered statistics.

b   Furthermore, $L^*(\beta)$ is maximized by

$$\hat{\beta}_n = \frac{1}{n} \sum_{i=2}^{n} (X_{(i)} - X_{(1)})$$

The MLE are $\hat{\mu} = X_{(1)}, \hat{\beta}_n = \frac{1}{n} \sum_{i=2}^{n} (X_{(i)} - X_{(1)})$.

c   The distribution of $X_{(1)}$ is like that of $\mu + E\left(\frac{\beta}{n}\right)$, with p.d.f.

$$f_{(1)}(x; \mu, \beta) = I\{x \geq \mu\} \frac{n}{\beta} e^{-\frac{n}{\beta}(x - \mu)}$$

Thus, the joint p.d.f. of $(X_1, \ldots, X_n)$ is factored to a product of the p.d.f. of $X_{(1)}$ and a function of $\hat{\beta}_n$, which does not depend on $X_{(1)}$ (neither on μ). This implies that $X_{(1)}$ and $\hat{\beta}_n$ are independent.

$$V\{\hat{\mu}\} = V\{X_{(1)}\} = \frac{\beta^2}{n^2}$$

It can be shown that $\hat{\beta}_n$ is distributed like $\frac{1}{n} G(n - 1, \beta)$. Accordingly,

$$V\{\hat{\beta}_n\} = \frac{n - 1}{n^2} \beta^2 = \frac{1}{n} \left(1 - \frac{1}{n} \right) \beta^2$$

6.3.2   The OC function is

$$OC(p) = B(2; 30, p) \qquad 0 < p < 1$$

6.3.4

a $Pr\left\{ \overline{X}_n < \mu_0 - \frac{3\sigma}{\sqrt{n}} \right\} + Pr\left\{ \overline{X}_n > \mu_0 + \frac{3\sigma}{\sqrt{n}} \right\}$
$$= \Phi(-3) + 1 - \Phi(3) = 2\Phi(-3)$$
$$= .0027$$

b   .9474   **c**   .1587   **d**   .1777

6.3.6   The MINITAB command and output are:

```
MTB> TTest 4.0 'ISC/t²';
SUBC> Alternative 1.
T-Test of the Mean
Test of mu = 4.000 vs mu > 4.000

Variable  N  Mean  StDev  SE Mean    T  P-Value
ISC/t²   16  4.209  0.418   0.104   2.01   0.032
MTB>
```

H_0 that $\mu \leq 4$ is rejected.

6.3.8 $OC(\sigma^2) = \Pr \left\{ S^2 \leq \frac{\sigma_0^2}{n - 1} \chi_{.9}^2[n - 1] \right\}$
$$= \Pr \left\{ \chi^2[30] \leq \frac{\sigma_0^2}{\sigma^2} \chi_{.9}^2[30] \right\}$$

6.3.10   The power function is

$$\psi(\sigma^2) = \Pr \left\{ S^2 \geq \frac{\sigma_0^2}{n - 1} \chi_{1-\alpha}^2[n - 1] \right\}$$
$$= \Pr \left\{ \chi^2[n - 1] \geq \frac{\sigma_0^2}{\sigma^2} \chi_{1-\alpha}^2[n - 1] \right\}$$

6.4.1

a

```
MTB> TInterval 99.0 'Sample'.
Confidence Intervals

Variable  N   Mean   StDev  SE Mean    99.0% C.I.
Sample   20  20.760  0.975   0.218   (20.137,21.384)
```

b   99% confidence interval for σ^2 is $(0.468, 2.638)$

c   99% confidence interval for σ is $(0.684, 1.624)$

6.4.3 $X = 17$, $n = 20$, $X \sim B(20, p)$. 95% confidence interval for p is $(0.6211, 0.9679)$

6.5.1 $n = 20$, $\sigma = 5$, $\overline{Y}_{20} = 13.75$. $\beta = .10$, $\alpha = .05$; $Z_{.95} = 1.645$; $Z_{.975} = 1.96$; the tolerance limits are $(3.334, 24.166)$.

6.5.3 $Y_{(1)} = 1.151$, $Y_{(100)} = 5.790$. For $n = 100$ and

$\beta = .10$, we have $1 - \alpha = .988$. For $\beta = .05$, $1 - \alpha = .847$, the tolerance interval is $(1.151, 5.790)$.

6.6.2

a See Figure A.6.6.2a.

b In both cases the hypotheses of normality are not rejected. See Figure A.6.6.2b.

FIGURE A.6.6.2a
Normal probability plot for turn diameter

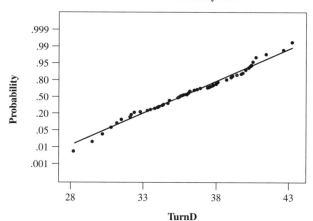

FIGURE A.6.6.2b
Normal probability plot of log-horsepower

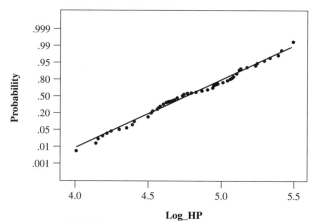

6.7.1   Frequency distribution for turn diameter

Interval	Observed	Expected	$(0 - E)^2/E$
–31	11	8.1972	0.9583
31–32	8	6.3185	0.4475
32–33	9	8.6687	0.0127
33–34	6	10.8695	2.1815
34–35	18	12.4559	2.4677
35–36	8	13.0454	1.9513
36–37	13	12.4868	0.0211
37–38	6	10.9234	2.2191
38–39	9	8.7333	0.0081
39–40	8	6.3814	0.4106
40–	13	9.0529	1.7213
Total	109	—–	12.399

The expected frequencies were computed for $N(35.5138, 3.3208)$. $\chi^2 = 12.4$, d.f. $= 8$; P-value is .135. The differences from normal are not significant!

6.7.2 $D^*_{109} = 0.0702$. For $\alpha = .05$, $k^*_\alpha = .895/(\sqrt{109} - .01 + (.85/\sqrt{109})) = .0851$. The deviations from the normal distribution are not significant!

6.8.2   beta(9, 11)

6.8.4

a $G\left(84, \dfrac{50}{51}\right)$

b $G_{.025}\left(84, \dfrac{50}{51}\right) = 65.6879$

$G_{.975}\left(84, \dfrac{50}{51}\right) = 100.873$

6.8.6   (43.235, 60.765). This is also an HPD.

Chapter 7

7.1.1   The macro for this problem is:

```
sample 64 c5 c6;
replace.
let k1 = mean(c6)
stack c7 k1 c7
end
```

The SE $\{\bar{X}\} = S/8 = .48965$. The standard deviation of column $C7$ is 0.4951. This is a resampling estimate of SE $\{\bar{X}\}$. The normal probability plot of the means in $C7$ is in Figure A.7.1.1.

FIGURE   A.7.1.1
Normal probability plot of resampling means, $n = 64$. The resampling distribution is approximately normal.

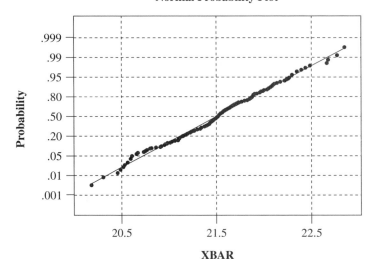

Normal Probability Plot

Average: 21.5015
Std Dev: 0.495095
N of data: 200

Kolmogorov-Smirnov Normality Test
D+: 0.034   D–: 0.059   D: 0.059
Approximate p value > 0.090

7.1.3 $\hat{P} \cong .25$. The mean $\overline{X} = 37.203$ is not significantly larger than 37.

7.3.1

a 95% bootstrap confidence interval for the mean is $(0.557, 0.762)$. The one for the STD is $(0.338, 0.396)$.

b See figures Figure A.7.3.1a and Figure A.7.3.1b.

7.3.3 Running program BOOTCORR.EXE 1000 times yields an EBD whose histogram is given in Figure A.7.3.3. The bootstrap confidence interval is $(0.931, 0.994)$.

7.3.5 The mean of the sample is $\overline{X}_{50} = 0.652$. The studentized difference from $\mu_0 = 0.55$ is $t = 1.943$.

a Bootstrap P-level is $P^* = .057$.

b μ is very close to the lower bootstrap confidence limit (0.557). The null hypothesis $H_0 : \mu = 0.55$ is rejected.

FIGURE   A.7.3.1a
Histogram of EBD of 1000 sample means

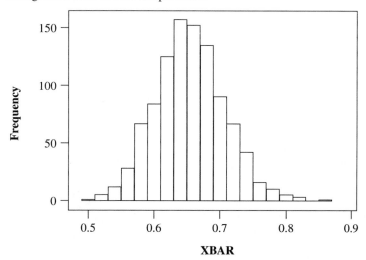

FIGURE   A.7.3.1b
Histogram of EBD of 1000 sample STDs

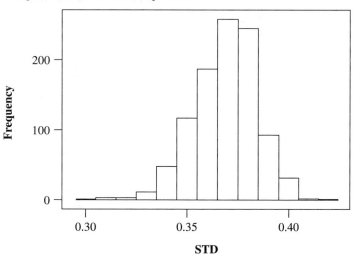

FIGURE   A.7.3.3
Histogram of the resample correlations between ISC1 and ISC2, $M = 1000$

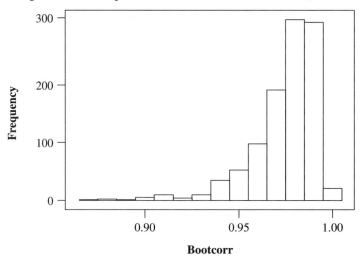

Bootcorr

7.3.7   The mean of sample 1 (Diam1) is $\overline{X}_1 = 9.993$; the mean of sample 2 (Diam2) is $\overline{X}_2 = 9.987$. The studentized difference is $t = 1.912$. The P^*-level is .063 ($M = 1000$). The variance of sample 1 is $S_1^2 = 0.00027$. The variance of Sample 2 is $S_2^2 = 0.000324$. The variance ratio is $F = S_2^2/S_1^2 = 1.2016$. The bootstrap level for variance ratios is $P^* = .756$.

7.4.1   The tolerance intervals of the number of defective items in future batches of size $N = 50$, with $\alpha = .05$ and $\beta = .05$, are:

	Limits	
p	Lower	Upper
.20	2.5	18
.10	0	15
.05	0	10

[Remember to initiate correctly the constants $k1, k3, k4, k7$.]

7.4.3   (.95, .95) tolerance interval for OTURB.DAT is (0.238, 0.683).

7.5.2   The sample from file OELECT.DAT is put into column $C1$.

```
MTB> WTest 220.0 C1;
SUBC> Alternative 0.
Wilcoxon Signed Rank Test
```

```
TEST OF MEDIAN = 220.0 VERSUS MEDIAN
N.E. 220.0

        N FOR  WILCOXON          ESTIMATED
    N   TEST   STATISTIC P-VALUE  MEDIAN
C1  99    99      1916.0   0.051    219.2
```

The null hypothesis is rejected with P-value equal to 0.051.

Chapter 8

8.1.2   The following is MINITAB output for the command

```
MTB> regr C5 on 2 C3 C4
Regression Analysis
The regression equation is
MPG = 38.3 − 0.251 TurnD − 0.0631 HorseP
```

Predictor	Coef	Stdev	t-ratio	p
Constant	38.264	2.654	14.42	0.000
TurnD	−0.25101	0.8377	−3.00	0.003
HorseP	−0.063076	0.006926	−9.11	0.000

$s = 2.491$ R-sq $= 60.3\%$ R-sq(adj) $= 59.6\%$

We see that only 60% of the variability in MPG is explained by the linear relationship with TurnDiam, and Horsepower. Both variables contribute significantly to the regression.

8.2.1 -0.279

8.2.2   The partial regression equation is $\hat{e}_1 = -0.251\hat{e}_2$.

8.3.2

```
MTB> regr c5 4 c1 − c4
```
Regression Analysis
The regression equation is

$$y = -6.8 + 0.227x_1 + 0.554x_2 - 0.150x_3 + 0.155x4$$

Predictor	Coef	Stdev	t-ratio	p
Constant	−6.82	10.12	−0.67	0.506
x1	0.22725	0.09994	2.27	0.031
x2	0.5537	0.3698	1.50	0.146
x3	−0.14954	0.02923	−5.12	0.000
x4	0.154650	0.006446	23.99	0.000

s = 2.234   R-sq = 96.2%   R-sq(adj)= 95.7%

a   The regression equation is

$$\hat{y} = -6.8 + 0.227x_1 + 0.554x_2 - 0.150x_3 + 0.155x_4$$

b $R^2 = .962$

c   The regression coefficient of x_2 is not significant.

d   Running the multiple regression again, without x_2, we obtain the equation

$$\hat{y} = 4.03 + 0.222x_1 - 0.187x_3 + 0.157x_4$$

with $R^2 = .959$. Variables x_1, x_3, and x_4 are important.

e   Normal probability plotting of the residuals $\hat{e}$ from the equation of part d shows that they are normally distributed.

8.3.3

a
$$(H) = (X)(B)$$
$$= (X)[(X)'(X)]^{-1}(X)'$$
$$H^2 = (X)[(X)'(X)]^{-1}(X)'(X)[(X)'(X)]^{-1}(X)'$$
$$= H$$

b
$$(Q) = I - (H)$$
$$(Q)^2 = (I - (H))(I - (H))$$
$$= I - (H) - (H) + (H)^2$$
$$= I - (H) = Q$$
$$s_e^2 = \mathbf{y}'(Q)(Q)\mathbf{y}/(n - k - 1)$$
$$= \mathbf{y}'(Q)\mathbf{y}/(n - k - 1)$$

8.3.4 $\hat{\mathbf{y}} = (X)\hat{\boldsymbol{\beta}} = (X)(B)\mathbf{y} = (H)\mathbf{y}$
$$\hat{\mathbf{e}} = Q\mathbf{y} = (I - (H))\mathbf{y}$$
$$\hat{\mathbf{y}}'\hat{\mathbf{e}} = \mathbf{y}'(H)(I - (H))\mathbf{y}$$
$$= \mathbf{y}'(H)\mathbf{y} - \mathbf{y}'(H)^2\mathbf{y}$$
$$= 0$$

8.3.6   From the basic properties of the cov(X, Y) operator,

$$\text{cov}\left(\sum_{i=1}^{n} \beta_i X_i, \sum_{j=1}^{n} \gamma_j X_j\right) = \sum_{i=1}^{n}\sum_{j=1}^{n} \beta_i \gamma_j \text{cov}(X_i, X_j)$$

$$= \sum_{i=1}^{n}\sum_{j=1}^{n} \beta_i \gamma_j \ \Sigma_{ij}$$

$$= \boldsymbol{\beta}'(\Sigma)\boldsymbol{\gamma}$$

8.3.8 $\Sigma(\mathbf{Y}) = \sigma^2 I; \mathbf{b} = (B)\mathbf{Y}$,

$$\Sigma(\mathbf{b}) = (B)\Sigma(\mathbf{Y})(B)' = \sigma^2(B)(B)'$$
$$= \sigma^2[(\mathbf{X})'(\mathbf{X})]^{-1} \cdot \mathbf{X}'\mathbf{X}[(\mathbf{X})'(\mathbf{X})]^{-1}$$
$$= \sigma^2[(\mathbf{X})'(\mathbf{X})]^{-1}$$

8.4.1   The regression of Y on X_1 is:

```
MTB> regr c5 1 c1
```
Regression Analysis
The regression equation is

$$Y = 81.0 + 1.86X_1$$

Predictor	Coef	Stdev	t-ratio	p
Constant	81.042	4.955	16.36	0.000 s
X$_1$	1.8603	0.5294	3.51	0.005

= 10.79   R-sq = 52.9%   R-sq(adj) = 48.6%.
Analysis of Variance

SOURCE	DF	SS	MS	F	p
Regression	1	1437.0	1437.0	12.35	0.005
Error	11	1279.9	116.4		
Total	12	2716.9			

$F = 12.35$ (In the 1st stage F is equal to the partial−F.)
$SSE_1 = 1279.9$, $R^2_{Y|(X_1)} = 0.529$.

The regression of Y on x_1 and x_2 is:

```
MTB> regr c5 2 c1 c2
```
Regression Analysis
The regression equation is

$$Y = 51.3 + 1.45X_1 + 0.678X_2$$

Predictor	Coef	Stdev	t-ratio	p
Constant	51.278	2.576	19.90	0.000
X$_1$	1.4526	0.1338	10.85	0.000
X$_2$	0.67804	0.05171	13.11	0.000

s = 2.652   R-sq = 97.4%   R-sq(adj) = 96.9%
Analysis of Variance

SOURCE	DF	SS	MS	F	p
Regression	2	2646.6	1323.3	188.08	0.000
Error	10	70.4	7.0		
Total	12	2716.9			

SOURCE	DF	SEQ SS
X_1	1	1437.0
X_2	1	1209.5

Here, $R^2_{y|(X_1,X_2)} = .974$, $SSE_2 = 70.4$, $s^2_{e_2} = 7.04$, Partial

$$-F = \frac{2716.9(.974 - .529)}{7.04} = 171.735.$$ Notice that SEQ SS

for $X_2 = 2716.9(.974 - .529) = 1209.3$.

The regression of Y on X_1, X_2, X_3 is:

```
MTB> regr c5 3 c1 c2 c3
```
Regression Analysis
The regression equation is

$$Y = 45.7 + 1.74X_1 + 0.673X_2 + 0.316X_3$$

Predictor	Coef	Stdev	t-ratio	p
Constant	45.658	4.237	10.78	0.000
X_1	1.7389	0.2172	8.00	0.000
X_2	0.67321	0.04814	13.98	0.000
X_3	0.3158	0.9165	1.61	0.142

$s = 2.465$ R-sq $= 98.0\%$ R-sq(adj) $= 97.3\%$
Analysis of Variance

SOURCE	DF	SS	MS	F	p
Regression	3	2662.26	887.42	146.10	0.000
Error	9	54.67	6.07		
Total	12	2716.92			

SOURCE	DF	SEQ SS
X_1	1	1437.02
X_2	1	1209.54
X_3	1	15.69

Here $R^2_{y|(X_1,X_2,X_3)} = .980$. Partial $-F = \dfrac{2716.92(.980 - .974)}{6.07}$

$= 2.686$. The SEQ SS of X_3 is 15.69. The .95-quantile of $F[1, 9]$ is 5.117. Thus, the contribution of X_3 is not significant.

The regression of Y on X_1, X_2, X_3, X_4 is:

```
MTB> regr C5 4 C1 − C4
```
Regression Analysis
NOTE X_2 is highly correlated with other predictor variables
NOTE X_4 is highly correlated with other predictor variables
The regression equation is

$$Y = 84.8 + 1.32X_1 + 0.275X_2 - 0.123X_3 - 0.384X_4$$

Predictor	Coef	Stdev	t-ratio	p
Constant	84.78	53.19	1.69	0.150
X_1	1.3166	0.6142	2.14	0.064
X_2	0.2747	0.5424	0.51	0.626
X_3	−0.1230	0.6280	−0.20	0.850
X_4	−0.3840	0.5204	−0.74	0.482

$s = 2.529$ R-sq $= 98.1\%$ R-sq(adj) $= 97.2\%$
Analysis of Variance

SOURCE	DF	SS	MS	F	p
Regression	4	2665.74	666.43	104.16	0.000
Error	8	51.18	6.40		
Total	12	2716.92			

SOURCE	DF	SEQ SS
X_1	1	1437.02
X_2	1	1209.54
X_3	1	15.69
X_4	1	3.48

Partial $-F = 3.48/6.40 = 0.544$. The effect of X_4 is not significant.

8.8.2

a

```
MTB> AOVOneway 'Exp1' 'Exp2' 'Exp3'
```
One-Way Analysis of Variance
Analysis of Variance

Source	DF	SS	MS	F	p
Factor	2	3.3363	1.6682	120.92	0.000
Error	27	0.3725	0.0138		
Total	29	3.7088			

Individual 95% CIs for Mean
Based on Pooled StDev

Level	N	Mean	StDev	-------+---------+---------+---------
Exp1	10	2.5170	0.0392	(--*-)
Exp2	10	2.7740	0.1071	(-*--)
Exp3	10	1.9740	0.1685	(--*-)

Pooled StDev = 0.1175   2.10   2.40   2.70

b The bootstrap program ANOVTEST.EXE gave, for $M = 1000$, $F^* = 120.917$, with P-level $P^* = 0$. This is identical to the result in part a.

Chapter 9

9.2.1

i $n = 1878, c = 13$

ii $n = 702, c = 12$

iii $n = 305, c = 7$

9.2.3 $OC(p) = H(7; 2500, M_p, 305)$. For $p = .025$, $M_p = 62$, $OC(.025) = 0.5091$

9.3.1 The large-sample approximation yields $n^* = 292$, $c^* = 11$. The exact plan is $n = 311, c = 12$. Notice that the actual risks of the plan (n^*, c^*) are $\alpha^* = .0443$ and $\beta^* = .0543$. The actual risks of the exact plan are $\alpha = .037$ and $\beta = .049$.

9.4.1

a $\alpha = .021, \beta = .0532$

b 228.7

c $n = 253, c = 7, \alpha^* = .0283, \beta^* = .0486$

9.5.1 $OC(.02) = .95 = 1 - \alpha$; $\qquad OC(.06) = .05 = \beta$. $ASN(0.02) = 140$, $ASN(0.06) = 99$, $ASN(0.035) = 191$.

9.6.1 $n = 70, k = 1.986$

9.6.2 $\xi = 14.9$, $\quad \bar{x} = 14.9846$, $\quad S = 0.019011$. $\quad k = 2.742375$, $\bar{X} - kS = 14.9325 > \xi$. The lot is accepted. $OC(0.03) \cong 0.4812$.

9.7.1 $(139, 3)$; $N = 500$; $OC(p) = H(3; 500, M_p, 139)$; $R^* = H(2; 499, M_p - 1, 138)/H(3; 500, M_p, 139)$

p	AOQ	ATI
.01	0.0072	147.1
.02	0.0112	244.2
.03	0.0091	369.3
.05	0.0022	482.0

9.9.1 The total sample size from 10 consecutive lots is 1000. Thus, from Table 9.16, S_{10} should be less than 5. The last two samples should each have fewer than two defective items. Hence, the probability for qualification is

$$QP = b^2(1; 100, .01)B(2; 800, .01)$$
$$+ 2b(1; 100, .01)b(0; 100, .01)B(3; 800, .01)$$
$$+ b^2(0; 100, .01)B(4; 800, .01) = .0263$$

Chapter 10

10.2.3

a 15.87%   **b** 29.1%   **c** 50%

10.3.1 Perform MINITAB capability analysis. Although $Cp = .93$, there is room for improvement by designing the process with reduced variability; see Chapter 13.

10.4.1 $n = 99$, $\bar{X} = 219.25$, $S = 4.004$, $\xi_U = 230$, $\xi_L = 210$; $\hat{C}_{pl} = .77$, $\hat{C}_{pu} = .895$, $\hat{C}_{pk} = .77$; $F_{.975}[1, 98] = 5.1818$. Confidence intervals for C_{pl} and C_{pu} at .975 confidence level are:

	Lower Limit	**Upper Limit**
C_{pl}	.6434	.9378
C_{pu}	.7536	1.0847

Finally, the .95 confidence interval for C_{p_k} is (.6434, .9378).

10.4.4

a 1.8247

b control limits for $\overline{X}$: (1.8063, 1.8430)

c $C_{pk} = .44$

10.7.2 *Part I*: See control charts in Figure A.10.7.2a The change after the 50th cycle time is so big that the means of the subgroup are immediately out of the control limits. The shift has no influence on the ranges within subgroups.

Part II: See control charts in Figure A.10.7.2b. The effect of adding the random number is to increase the mean of all cycle times by 0.5. The variability is increased too.

FIGURE A.10.7.2a
$\overline{X}$ and R control charts

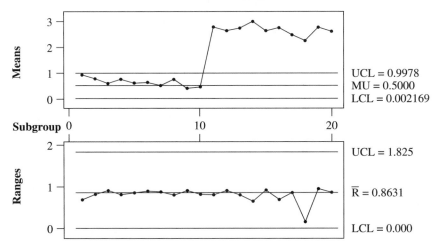

FIGURE A.10.7.2b
$\overline{X}$ and R control charts

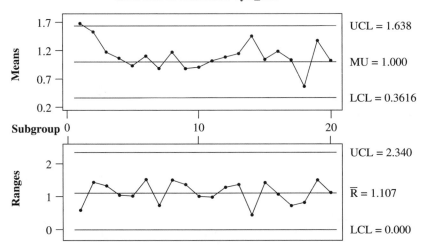

Chapter 11

11.1.1

a $Q_1 = 10; M_e = 12; Q_3 = 14$

b $\mu_R = 13; \sigma_R = 2.3452$

c .8657

d .8644.

11.1.3

a $E\{R^*\} = 33$

b $R^* = 34$ (See Figure A.11.1.3.)

c $\sigma^* = 2.9269$

$\alpha_u^* = 1 - \Phi\left(\dfrac{1}{2.9269}\right) = 0.3663$

The deviation is not significant.

d .102

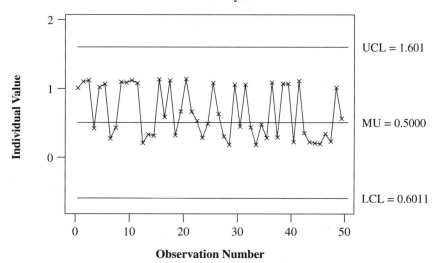

FIGURE A.11.1.3
Runs up and down in CYCLT.DAT

I Chart for cyclt

UCL = 1.601

MU = 0.5000

LCL = 0.6011

Individual Value

Observation Number

11.1.5 The following is MINITAB output. K for XBAR is the grand mean; K for STD is the mean of the 50 STDs.

```
MTB> Runs 'XBAR'.
Runs Test
XBAR
K = 0.4295
The observed no. of runs = 26
The expected no. of runs = 25.9600
26 Observations above K 24 below
        The test is significant at 0.9909
        Cannot reject at alpha = 0.05
MTB> Runs 'STD'.
Runs Test
STD
K = 0.1261
The observed no. of runs = 26
```

```
The expected no. of runs = 25.6400
28 Observations above K 22 below
        The test is significant at 0.9169
        Cannot reject at alpha = 0.05
MTB>
```

11.1.9 Let θ be the probability of not detecting the shift in a given day. Solving $\theta^5 = 0.2$, we get $\theta = 0.72478$. We approximate n by solving

$$\theta \geq \Phi\left(\frac{0.01 + 3\sqrt{\frac{0.01\times0.99}{n}} - 0.05}{\sqrt{\frac{0.05\times0.95}{n}}}\right)$$

$$- \Phi\left(\frac{0.01 - 3\sqrt{\frac{0.01\times0.99}{n}} - 0.05}{\sqrt{\frac{0.05\times0.95}{n}}}\right)$$

The solution is $n = 16$ (using equation (11.3.7), $n \cong 18$).

11.2.1 370.3; 1.75; 1.0

11.2.2 33.4

11.3.1 $h^0 = 34$ (min) $= 0.57$ (hr)

11.3.4

a 4.5 (hr)   **b** 7

c For a shift of size $\delta = 1$, option 1 detects it, on average, after 4.5 hr. Option 2 detects it, on average, after $2 \times 1.77 = 3.5$ hr. Option 2 is preferred.

11.5.2

a The CUSUM has the parameters $K^+ = 221$ and $h^+ = -\log(0.001)\frac{\sigma^2}{2} = 88.19$. The values of S^+ are 0.39, 0, 0, 10.88, 26.16, 52.72, 71.39, 74.06, 82.19, 92.33,... Thus, a shift above 222 is detected after the 9th observation. If $\alpha = .01$, then detection is right after the 5th observation. Try also the MINITAB CUSUM control chart, with the V-mask option (see Figure A.11.5.2). Note that the V-mask is an alternative graphical proedure for detecting points of change by CUSUM.

b ARL(0) = 133.19 ± 9.56, for $K^+ = 221$, $h^+ = 88.19$, $K^- = 219$, $h^- = -88.19$, $\theta_0 = 220$, $\sigma = 9.5$

c

$\delta = \dfrac{\theta_1 - \theta_0}{\sigma}$	PFA	CED
0.21	.09	47.86 ± 7.02
0.52	.07	17.33 ± 4.21
1.57	.09	5.37 ± 3.28

11.5.4 PFA = .006, CED = 3.81

11.5.6 The modified Shewhart chart with $a = 3$, $w = 1.5$, $r = 2.2$, and $n = 7$ yields ARL(0) = 136. This is close to the ARL(0) of the CUSUM of Exercise 11.5.2. It is interesting that for this control scheme, ARL(0.52) = 6.96 and ARL(1.57) = 1.13. These are significantly smaller than those of the CUSUM.

11.6.2 Execution of the macro yields that the first $W_m > 99$ is at $m = 31$.

11.6.3 Running the program $M = 100$ times yields the estimate ARL(0) = 221.8 ± 45.3. The boxplot is given in Figure A.11.6.3.

Chapter 12

12.3.1 Main effects: τ_i^A, τ_j^B, τ_k^C, each at 3 levels; first-order interactions: τ_{ij}^{AB}, τ_{ik}^{AC}, τ_{jk}^{BC}, each at 9 levels; second-order interaction: τ_{ijk}^{ABC}, at 27 levels.

12.4.1 The test can be performed by determining donfidence interval for the difference.

MTB> Read 'C:\ISTAT\DATA\SOCELL.DAT' c1-c3.
Entering data from file:
C:\ISTAT\DATA\SOCELL.DAT
       16 rows read.
MTB> let c4 = c2 − c1
MTB> TInterval 95.0 'Diff'.

FIGURE A.11.5.2
CUSUM V-mask control chart

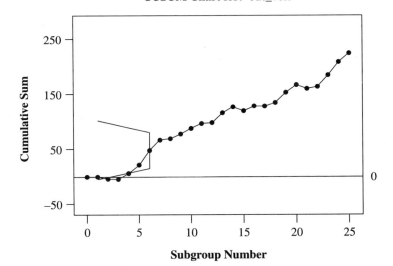

CUSUM Chart for: out_volt

F I G U R E   A.11.6.3
Boxplot of RL for the Shiryayev-Roberts procedure

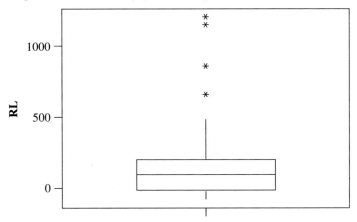

The difference between the two means is significantly greater than zero.

12.4.2   The randomization paired comparison with $M = 100$ yielded a P-value estimate $\hat{P} = .11$. The t-test yields P-value $P = .12$. The two methods give the same result.

12.4.3

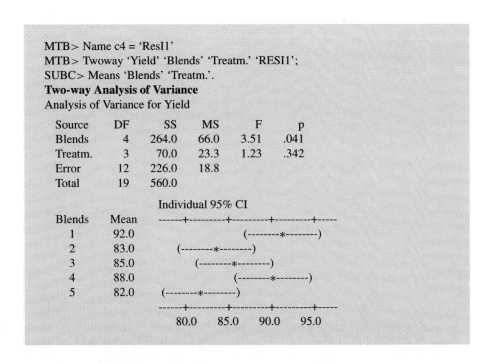

```
MTB> Name c4 = 'ResI1'
MTB> Twoway 'Yield' 'Blends' 'Treatm.' 'RESI1';
SUBC> Means 'Blends' 'Treatm.'.
```
Two-way Analysis of Variance
Analysis of Variance for Yield

Source	DF	SS	MS	F	p
Blends	4	264.0	66.0	3.51	.041
Treatm.	3	70.0	23.3	1.23	.342
Error	12	226.0	18.8		
Total	19	560.0			

```
                          Individual 95% CI
Blends   Mean    ------+---------+---------+---------+-----
  1      92.0                           (--------*-------)
  2      83.0           (--------*-------)
  3      85.0              (--------*-------)
  4      88.0                     (--------*-------)
  5      82.0          (--------*-------)
                 ------+---------+---------+---------+-----
                     80.0      85.0      90.0      95.0
```

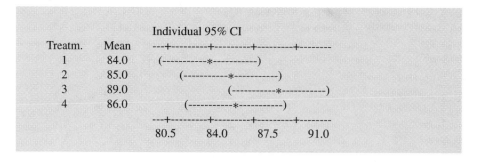

```
                                          Individual 95% CI
       Treatm.      Mean      ---+---------+---------+---------+--------
          1         84.0         (-----------*-----------)
          2         85.0           (-----------*-----------)
          3         89.0                       (-----------*-----------)
          4         86.0             (-----------*-----------)
                              ---+---------+---------+---------+--------
                                80.5      84.0      87.5      91.0
```

There are no significant differences between the treatments.
There are significant differences between blends.

12.6.1   An ANOVA of this Latin square is:

Source	DF	SS	MS	F	p
Treatments	3	388.75	129.58	1.155	.40
Rows	3	16.25	5.42	—	—
Columns	3	165.76	55.25	—	—
Error	6	673.25	112.21	—	—
Total	15	1244.00	—	—	—

The differences between the treatments are not significant.

12.7.1

```
MTB> Name c4 = 'RESI1'
MTB> Twoway 'Cyclet' 'Pist_Wg' 'Spring_C'
'RESI1';
SUBC> Means 'Pist_Wg' 'Spring_C'.
Two-way Analysis of Variance
Analysis of Variance for Cyclet
```

Source	DF	SS	MS	F	p
Pist_Wg	2	0.4749	0.2374	2.929	.07
Spring_C	2	0.0430	0.0215	0.266	.78
Interaction	4	0.2858	0.0715	0.895	.48
Error	36	2.8760	0.0799		
Total	44	3.6798			

The results are similar to those of Example 12.7.

12.8.2 $\mu, ACE, ABEF, ABCD, CBF, BDE, CDEF, AFD$.

12.9.1

a   The response function is $\hat{Y} = 62.0 + 2.39X_1 + 4.16X_2 - 0.363X_1X_2$, with $R^2 = .99$. The contour plot is given in Figure A.12.9.1.

b $\hat{\sigma}^2 = 0.13667$, 3 DF; the variance around the regression is $s^2_{y|(x)} = 0.2357$, 8 DF. The ratio $F = s^2_{y|(x)}/\hat{\sigma}^2 = 1.7246$, $P = .36$. There is no significant difference between the variances. Therefore, the test of goodness of fit does not reject the model.

Chapter 13

13.1.2   The regression analysis of SN on the seven factors (only main effects) is

```
MTB> regr c10 7 c1 − c7
Regression Analysis
The regression equation is
```

$$SN = 4.06 + 0.004x_1 - 0.407x_2 + 2.32x_3$$
$$+ 0.675x_4 - 0.176x_5$$
$$+ 0.125x_6 + 0.058x_7$$

Predictor	Coef	Stdev	t-ratio	p
Constant	4.0567	0.1172	34.60	.000
x_1	0.0037	0.1172	0.03	.975
x_2	−0.4066	0.1172	−3.47	.001
x_3	2.3176	0.1172	19.77	.000
x_4	0.6754	0.1172	5.76	.000
x_5	−0.1757	0.1172	−1.50	.137
x_6	0.1257	0.1172	1.07	.286
x_7	0.0576	0.1172	0.49	.624

$s = 1.326$ R-sq $= 78.6\%$ R-sq(adj) $= 77.3\%$

We see that only factors 2, 3, and 4 have a significant effect. The R^2 is only 78.6%. After adding to the regression equation the first-order interactions among X_2, X_3, and X_4, the regression analysis now is

FIGURE **A.12.9.1**

Contour plot of equal response

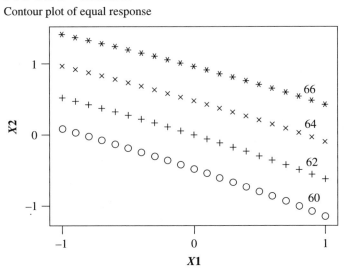

The regression equation is

$$SN = 4.06 - 0.407x_2 + 2.32x_3 + 0.675x_4$$
$$- 0.677x_2 * x_3 - 0.152x_2 * x_4 + 0.630x_3 * x_4$$

Predictor	Coef	Stdev	t-ratio	p
Constant	4.05670	0.08236	49.26	.000
x_2	−0.40663	0.08236	−4.94	.000
x_3	2.31758	0.08236	28.14	.000
x_4	0.67535	0.08236	8.20	.000
$x_2 * x_3$	−0.67662	0.08236	−8.22	.000
$x_2 * x_4$	−0.15230	0.08236	−1.85	.067
$x_3 * x_4$	0.63045	0.08236	7.66	.000

$s = 0.9318$ R-sq $= 89.3\%$ R-sq(adj) $= 88.8\%$

We see that the interactions $X_2 * X_3$ and $X_3 * X_4$ are very significant. The interaction $X_2 * X_4$ is significant at the 6.7% level. The regression equation with the interaction terms predicts the SN better; $R^2 = 89.3\%$.

13.2.1

i $$E\left\{\frac{X_1}{X_2}\right\} \cong \frac{\xi_1}{\xi_2} - \sigma_{12}\frac{1}{\xi_2^2} + \sigma_2^2\frac{\xi_1}{\xi_2^3}$$

$$V\left\{\frac{X_1}{X_2}\right\} \cong \sigma_1^2/\xi_2^2 + \sigma_2^2\xi_1^2/\xi_2^4 - 2\sigma_{12}\xi_1/\xi_2^3$$

ii $$E\left\{\log\frac{X_1^2}{X_2^2}\right\} \cong \log\left(\frac{\xi_1}{\xi_2}\right)^2 - \frac{\sigma_1^2}{\xi_1^2} + \frac{\sigma_2^2}{\xi_2^2}$$

$$V\left\{\log\left(\frac{X_1}{X_2}\right)^2\right\} \cong \frac{4\sigma_1^2}{\xi_1^2} + \frac{4\sigma_2^2}{\xi_2^2} - 8\frac{\sigma_{12}}{\xi_1\xi_2}$$

iii $$E\{(X_1^2 + X_2^2)^{1/2}\} \cong (\xi_1^2 + \xi_2^2)^{1/2}$$

$$+ \frac{\sigma_1^2}{2}(\xi_1^2 + \xi_2^2)^{-1/2}(1 - \xi_1^2(\xi_1^2 + \xi_2^2)^{-1})$$

$$+ \frac{\sigma_2^2}{2}(\xi_1^2 + \xi_2^2)^{-1/2}(1 - \xi_2^2(\xi_1^2 + \xi_2^2)^{-1})$$

$$- \sigma_{12}\xi_1\xi_2(\xi_1^2 + \xi_2^2)^{-3/2}$$

$$V\{(X_1^2 + X_2^2)^{1/2}\} \cong \frac{\sigma_1^2\xi_1^2}{(\xi_1^2 + \xi_2^2)}$$

$$+ \frac{\sigma_2^2\xi_2^2}{(\xi_1^2 + \xi_2^2)} + 2\frac{\sigma_{12}\xi_1\xi_2}{(\xi_1^2 + \xi_2^2)}$$

13.2.2   Approximation formulas yield: $E\{Y\} \cong 0.5159$ and $V\{Y\} \cong 0.06762$. Simulation with 5000 runs yields the estimates $E\{Y\} \cong 0.6089$, $V\{Y\} \cong 0.05159$. The first approximation of $E\{Y\}$ is significantly lower than the simulation estimate.

13.4.1   The coefficients in the regression analysis yield estimates of the main effects. Regression analysis of the full factorial yields:

The regression equation is

$$SN = 2.03 + 0.0019PW - 0.203PSA + 1.159IGV$$
$$+ 0.338SC - 0.0878AP$$
$$+ 0.0629AT + 0.0288FGT$$

Predictor	Coef	Stdev	t-ratio	p
Constant	2.02835	0.05862	34.60	.000
PW	0.00187	0.05862	0.03	.975
PSA	−0.20331	0.05862	−3.47	.001
IGV	1.15879	0.05862	19.77	.000
SC	0.33768	0.05862	5.76	.000
AP	−0.08785	0.05862	−1.50	.137
AT	0.06287	0.05862	1.07	.286
FGT	0.02880	0.05862	0.49	.624

$s = 0.6632$ R-sq = 78.6%   R-sq(adj) = 77.3%

Analysis of Variance

SOURCE	DF	SS	MS	F	p
Regression	7	193.365	27.624	62.81	.000
Error	120	52.773	0.440		
Total	127	246.138			

SOURCE	DF	SEQ SS
PW	1	0.000
PSA	1	5.291
IGV	1	171.878
SC	1	14.595
AP	1	0.988
AT	1	0.506
FGT	1	0.106

The three significant main effects are those of piston surface area (PSA), initial gas volume (IGV), and spring coefficients (SC). Regression analysis of the half factorial is:

The regression equation is

$$SN = 1.94 + 0.0127PW − 0.203PSA$$
$$+ 1.183IGV + 0.337SC − 0.102AP$$
$$+ 0.0529AT + 0.0329FGT$$

Predictor	Coef	Stdev	t-ratio	p
Constant	1.93648	0.06951	27.86	.000
PW	0.01266	0.06951	0.18	.856
PSA	−0.20334	0.06951	−2.93	.005
IGV	1.18261	0.06951	17.01	.000
SC	0.33718	0.06951	4.85	.000
AP	−0.10199	0.06951	−1.47	.148
AT	0.05294	0.06951	0.76	.450
FGT	0.03294	0.06951	0.47	.637

$s = 0.5561$ R-sq = 85.3%   R-sq(adj) = 83.4%

Analysis of Variance

SOURCE	DF	SS	MS	F	p
Regression	7	100.356	14.337	46.36	.000
Error	56	17.319	0.309		
Total	63	117.674			

SOURCE	DF	SEQ SS
PW	1	0.010
PSA	1	2.646
IGV	1	89.508
SC	1	7.276
AP	1	0.666
AT	1	0.179
FGT	1	0.069

The results are similar to those of the full factorial. Regression analysis of the quarter factorial gives:

The regression equation is

$$SN = 1.955 + 0.026PW − 0.244PSA + 1.033IGV$$
$$+ 0.374SC − 0.210AP$$
$$− 0.007AT + 0.014FGT$$

Predictor	Coef	Stdev	t-ratio	p
Constant	1.9547	0.1413	13.84	.000
PW	0.0256	0.1413	0.18	.858
PSA	−0.2443	0.1413	−1.73	.097
IGV	1.0329	0.1413	7.31	.000 $s =$
SC	0.3741	0.1413	2.65	.014
AP	−0.2104	0.1413	−1.49	.149
AT	−0.0070	0.1413	−0.05	.961
FGT	0.0142	0.1413	0.10	.921

0.7991 R-sq = 73.3%   R-sq(adj) = 65.5%

Analysis of Variance

SOURCE	DF	SS	MS	F	p
Regression	7	41.9745	5.9964	9.39	.000
Error	24	15.3264	0.6386		
Total	31	57.3009			

SOURCE	DF	SEQ SS
PW	1	0.0209
PSA	1	1.9092
IGV	1	34.1408
SC	1	4.4795
AP	1	1.4160
AT	1	0.0016
FGT	1	0.0065

Here the effect of PSA is significant only at a 10% *P*-level. Finally, the regression analysis of the one-eighth factorial is:

The regression equation is

$$SN = 2.06 − 0.079PW − 0.183PSA + 1.078IGV$$
$$+ 0.454SC + 0.374AP$$
$$− 0.014AT + 0.021FGT$$

Predictor	Coef	Stdev	t-ratio	p
Constant	2.0623	0.1659	12.43	.000
PW	−0.0789	0.1659	−0.48	.647
PSA	−0.1829	0.1659	−1.10	.302
IGV	1.0783	0.1659	6.50	.000
SC	0.4537	0.1659	2.74	.026
AP	0.374	0.1659	0.23	.827
AT	−0.0136	0.1659	−0.08	.937
FGT	0.0213	0.1659	0.13	.901

$s = 0.6635$ R-sq $= 86.5\%$ R-sq(adj) $= 74.7\%$

Analysis of Variance

SOURCE	DF	SS	MS	F	p
Regression	7	22.5646	3.2235	7.32	.006
Error	8	3.5217	0.4402		
Total	15	26.0862			

SOURCE	DF	SEQ SS
PW	1	0.0995
PSA	1	0.5351
IGV	1	18.6036
SC	1	3.2938
AP	1	0.0223
AT	1	0.0030
FGT	1	0.0073

Now the effect of PSA is not significant.

Chapter 14

14.1.1   MTTF = 5.1 hr; MTTR = 6.6 min; intrinsic availability = 0.979

14.1.3

a
$$h(t) = \frac{4t^3}{20736 \cdot \left(1 - \frac{t^4}{20736}\right)} \qquad 0 \le t < 12$$
$$= \infty \quad t > 12$$

b
$$\text{MTTF} = \int_0^{12} \left(1 - \frac{t^4}{20736}\right) dt$$
$$= 9.6 \text{ months}$$

c
$$R(4) = 1 - F(4)$$
$$= .9877$$

14.2.1

a $1 - B(1; 4, .95) = .9995$

b $(1 - B(0; 2, .95))^2 = (1 - .05^2)^2 = .9950$

14.2.4 $\theta(5) = \Pr\left\{W\left(\frac{1}{2}, 100\right) > 5\right\}$
$$= .79963$$
$$R^5 = 1 - B(3; 8, .7996) = .9895$$

14.3.1   Let C be the length of the renewal cycle; $E\{C\} = \beta\Gamma(1 + \frac{1}{\alpha}) + e^{\mu + \sigma^2/2};$ STD$(C) = \left[\beta^2\left(\Gamma\left(1 + \frac{2}{\alpha}\right) - \Gamma^2\left(1 + \frac{1}{\alpha}\right)\right) + e^{2\mu + \sigma^2}(e^{\sigma^2} - 1)\right]^{1/2}$

14.3.3 $V(1000) = \sum_{n=1}^{\infty} \Phi\left(\frac{100 - 10n}{\sqrt{n}}\right) \cong 9.501$

14.3.4 $v(t) = \frac{1}{10\sqrt{n}}\sum_{n=1}^{\infty} \phi\left(\frac{t - 100n}{10\sqrt{n}}\right)$; $\phi(z)$ is the p.d.f. of $N(0, 1)$.

14.3.5   TTF is distributed like $\max(E_1(\beta), E_2(\beta))$:
$$F(t) = (1 - e^{-t/\beta})^2$$
$$R(t) = 1 - (1 - e^{-t/\beta})^2$$
$$R^*(s) = \frac{\beta(3 + s\beta)}{(1 + s\beta)(2 + s\beta)}$$
$$f^*(s) = \frac{2}{(1 + \beta s)(2 + \beta s)}$$
$$g^*(s) = \frac{1}{(1 + s\gamma)^2}$$
$$A^*(s) = \beta(3 + s\beta)(1 + \gamma s)^2/s[3\beta + 4\gamma + (\beta^2 + 6\beta\gamma + 2\gamma^2)s + (2\beta^2\gamma + 3\beta\gamma^2)s^2 + \beta^2\gamma^2 s^3]$$

14.3.6   Estimate of $V(2000) = 14.27$, based on $M = 500$ runs

14.4.1   2069.8 hr

14.4.3   Let $S_{n,r} = \sum_{i=1}^{r} T_{n,i} + (n - r)T_{n,r}$. An unbiased estimator of β is $\hat{\beta} = \frac{S_{n,r}}{r}$; $V\{\hat{\beta}\} = \frac{\beta^2}{r}$

14.6.1   See Figure A.14.6.1.

14.7.2 $n = 20$, $K_n = 17$, $\hat{\beta} = 129.011$. Running program CENEXP.EXE, with $M = 500$ runs, the estimated STD of $\hat{\beta}$ is 34.912, with confidence interval (78.268, 211.312).

14.7.4

a   The MLEs are $\hat{\beta} = 27.0789$, $\hat{v} = 1.374$

b   The EBD estimates are SE$\{\hat{\beta}_{50}\} = 3.074$, SE$\{\hat{v}_{50}\} = 0.1568$. The asymptotic results are: SE$\{\hat{\beta}_{50}\} = 2.934$; SE$\{\hat{v}_{50}\} = 0.1516$.

14.8.2 $R_0 = .99$, $R_1 = .90$, $\alpha = \beta = .01$ $s = .9603$, $h_1 = 1.9163$, $h_2 = 1.9163$; ASN(.9) = 31.17.

FIGURE A.14.6.1

Kaplan-Meier estimator of the reliability of an electric device

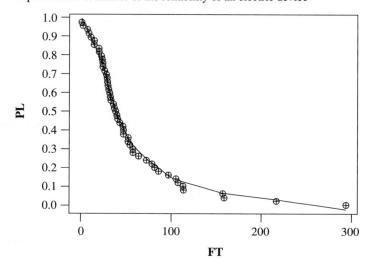

VIII

MINITAB Operations and Commands

MINITAB has a powerful command language that can be used for statistical analysis and other mathematical manipulations. It also contains excellent graphics devices and other options. Groups of commands can be put together into a macro for repetitive execution. In this appendix, we present some basic commands and principles for working with MINITAB. In recent versions for the PC (versions 10 and higher), the user can use Window capabilities to activate commands and macros by simple aim and click operations. MINITAB will show you automatically, in the session window, the commands that will be executed. A particularly strong feature of MINITAB is its choice of default options in graphical and numerical commands. In most cases, the user should not change the default options.

1 Columns, constants, and matrices

Data are stored in MINITAB either in columns, constants, or matrices. Columns are labeled C1, C2,...; constants are labeled K1, K2,...; and matrices are labeled M1, M2,.... We can insert sample data into columns manually from the keyboard or by importing data from a stored ASCII (text) file. For manual insertion of data from the keyboard, use the command:

MTB > Set C1

MINITAB will respond:

DATA>

Insert the data one number followed by another, with one blank separation. You do not have to worry about format. Continue on new lines as needed.

When you finish, write "end" on a new line. If you wish to insert a data file consisting of *n* records of *k* variables in this way, insert each variable in a separate column. At the end you will have *k* columns with *n* rows in each.

Many data files are stored in C:\ISTAT\DATA. Such data files can be imported by using the command **READ** or by choosing from the menu under File\Other Files. For example, file SAMPLE1.DAT contains data on one variable. It can be read into MINITAB by the command

MTB> read 'c:\istat\data\sample1.dat' C1

The data in the file will be automatically read into column C1. If you wish to read data from a file consisting of several variables in several columns, as, for example, the data in CAR.DAT, specify the columns to read and their addresses, as:

MTB> read 'c:\istat\data\car.dat' C1-C5

If the columns have titles in the data file, the first row of the columns in MINITAB will have * in each cell. These *'s can be deleted by using the EDIT option in the menu.

New columns can be created by manipulating existing columns. For example, if you wish column C2 to consist of the square roots of the numbers in column C1, write

MTB> Let C2=sqrt(C1)

Constants are inserted with the **LET** command. For example, suppose you want the number 100 to be in K1; write the command

MTB> Let K1=100

Constants can be defined also as a result of MINITAB computations. For example, if you wish the constant K2 to be the mean of the data in column C1, write

MTB> Let K2=mean(C1)

Many other examples are given in the text.

Matrices can be defined by inserting the entries manually or by copying columns of equal length. For example, suppose that columns C1–C5 consist of 20 rows. By writing

MTB> copy c1-c5 M1

you will get a matrix M1 of 20 rows and 5 columns. For additional examples of commands for operating on matrices, see Appendix I.

2 Saving data

For saving all the data created in a session, use the **SAVE** command in the menu, under File. If you wish, however, to write the data in columns C1–C2 into a file, use the command

MTB> write 'c:\istat\data\NAME.DAT' C1 C2

3 Writing a macro

We can write a macro outside MINITAB using an editor. For example,

```
sample k7 c1 c2;
replace.
let k2=mean(c2)
InvCDV k3 k5;
Binomial k1 k2.
InvCDF k4 k6;
Binomial k1 k2.
stack k5 c3 c3
stack k6 c4 c4
end
```

is the macro stored in file C:\ISTAT\CHA7\BINOPRED.MTB. This group of commands can be prepared ahead. The file should be saved as an MTB file. The macro can be written during a MINITAB session by writing

> MTB> store 'c:\istat\cha7\binopred.mtb'

In this way, the file name is specified and the macro will be saved under this name. Each line in the session will start now with STOR>. To end the macro, write END. MINITAB will respond with the prompt MTB>. Many examples of macros are given in the text. To execute a macro several times, write:

> MTB> exec 'c:\istat\cha7\binopred.mtb' 100

The group of commands in the macro will now be executed repeatedly 100 times.

4   Use of HELP
If you forget how to write a certain command, use the on-line HELP. For example,

> MTB> help sort

will open the Help window in which you can read all you need about the **SORT** command.

5   Commands and subcommands
Some commands have optional subcommands. If you wish to write a subcommand, write a ; (semicolon) after the main command and end the subcommand with . (period). For example,

> MTB> sort C1 C2 C3 C4;
> SUBC> by C1.

will sort the data in columns C1 and C2 simultaneously by C2, in ascending order, and put the sorted data in columns C3 and C4.

6   Some helpful commands
The **CODE** command replaces all values in a specified range in a given column, with a specified value. For example, suppose that column C1 contains 100 values that range between 100 and 200. If we wish to find the frequency distribution of these values in subintervals of length 10, we can write

```
MTB> code (100:109.9)1 (110:119.9)2 (120:129.9)3 (130:139.9)4
(140:149.9)5 (150:159.9)6 (160:169.9)7 (170:179.9)8 (180:189.9)9
(190:200)10 C1 C2
```

Column C2 now contains 100 rows of numbers between 1 and 10, according to the subintervals to which the numbers in the corresponding rows of C1 belong. Follow this by the command

```
MTB> table C2
```

and you will get the frequency distribution of the data in C1 according to the specified subintervals.

The **COPY** command allows you to copy columns into other columns, constants into constants, and so on. An important feature is the ability to copy several rows of a column. For example, if we wish to copy into C2 only the first 20 rows of C1, we write:

```
MTB> copy C1 C2;
SUBC> use 1:20.
```

Other important commands that were used frequently in the text are: **SAMPLE**, **RANDOM**, **DESCRIBE**, **PRINT**, **PLOT**, and so on.

List of Abbreviations

ANOVA	Analysis of variance
AOQ	Average outgoing quality
AOQL	Average outgoing quality limit
AQL	Acceptable quality level
ARL	Average run length
ASN	Average sample number
ATI	Average total inspection
BECM	Bayes' estimation of the current mean
BIBD	Balanced incomplete block design
BP	Bootstrap population
c.d.f.	Cumulative distribution function
CAD	Computer-aided design
CADD	Computer-aided drawing and drafting
CAM	Computer-aided manufacturing
CED	Conditional expected delay
CIM	Computer-integrated manufacturing
CLT	Central limit theorem
CNC	Computerized numerically controlled
CPA	Circuit pack assembly
CUSUM	Cumulative sum
DLM	Dynamic linear model
EBD	Empirical bootstrap distribution
EWMA	Exponentially weighted moving average

FIT	Failures per 10^9 hours
FPM	Failures per million
HPD	Highest posterior density
i.i.d.	Independent and identically distributed
IQR	Interquartile range
ISC	Short-circuit current of solar cells (in amperes)
KS	Kolmogorov-Smirnov
LCL	Lower control limit
LLN	Law of large numbers
LQL	Limiting quality level
LSE	Least-squares estimator
LSL	Lower specification limit
LTPD	Lot tolerance percent defective
LWL	Lower warning limit
m.g.f.	Moment-generating function
MLE	Maximum likelihood estimator
MSD	Mean-squared deviation
MSE	Mean-squared error
MTBF	Mean time between failures
MTTF	Mean time to failure
NB	Negative binomial
OC	Operating characteristic

p.d.f.	Probability density function	SPC	Statistical process control
PERT	Project evaluation and review technique	SPRT	Sequential probability ratio test
		SR	Shiryayev-Roberts statistic
PFA	Probability of false alarm	SSE	Sum of squares of errors
PL	Product limit estimate	SSR	Sum of squares around the regression model
PPM	Parts per million		
QMP	Quality measurement plan	SST	Total sum of squares
QQ plot	Quantile vs. quantile plot	STD	Standard deviation
RCBD	Randomized complete block design	TTC	Time till censoring
RL	Run length	TTF	Time till failure
RMSE	Root mean-squared error	TTR	Time till repair
RSWOR	Random sample without replacement	TTT	Total time on test
RSWR	Random sample with replacement	UCL	Upper control limit
r.v.	Random variable	USL	Upper specification limit
SE	Standard error	UWL	Upper warning limit
SL	Skip lot	WLLN	Weak law of large numbers
SLOC	Source lines of code	WSP	Wave-soldering process
SLSP	Skip lot sampling plans	WSR	Wilcoxon signed-rank test
SN	Signal-to-noise ratio		

X

The Greek Alphabet

A	α	alpha	N	ν	nu
B	β	beta	Ξ	ξ	xi
Γ	γ	gamma	O	o	omicron
Δ	δ	delta	Π	π	pi
E	ϵ	epsilon	P	ρ	rho
Z	ζ	zeta	Σ	σ	sigma
H	η	eta	T	τ	tau
Θ	θ	theta	Y	υ	upsilon
I	ι	iota	Φ	ϕ	phi
K	κ	kappa	X	χ	chi
Λ	λ	lambda	Ψ	ψ	psi
M	μ	mu	Ω	ω	omega

Author Index

Subject Index